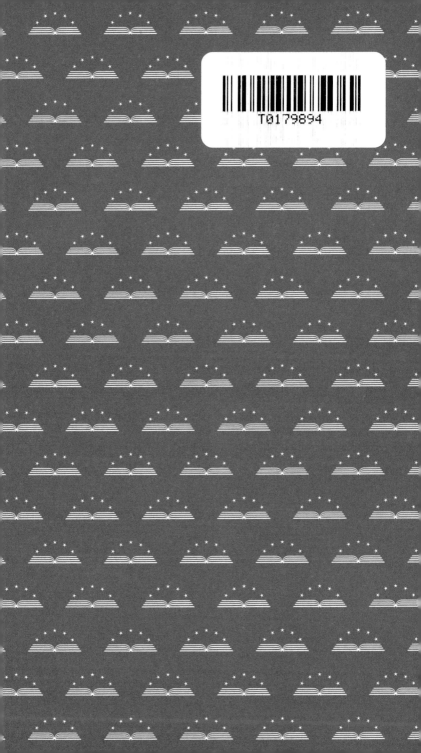

Library of America, a nonprofit organization,
champions our nation's cultural heritage
by publishing America's greatest writing in
authoritative new editions and providing resources
for readers to explore this rich, living legacy.

WENDELL BERRY

WENDELL BERRY

ESSAYS 1993–2017

INCLUDING
Life Is a Miracle

AND SELECTIONS FROM
Sex, Economy, Freedom & Community
Another Turn of the Crank
Citizenship Papers
The Way of Ignorance
What Matters?
Imagination in Place
It All Turns on Affection
Our Only World
The Art of Loading Brush

Jack Shoemaker, *editor*

THE LIBRARY OF AMERICA

Published in the United States by Library of America.
Visit our website at www.loa.org.

All texts reprinted by arrangement with Counterpoint Press.

This paper exceeds the requirements of
ANSI/NISO Z39.48–1992 (Permanence of Paper).

Distributed to the trade in the United States
by Penguin Random House Inc.
and in Canada by Penguin Random House Canada Ltd.

Library of Congress Control Number: 2018952819
ISBN 978–1–59853–608–9

Third Printing
The Library of America—317

Manufactured in the United States of America

Wendell Berry: Essays 1993–2017
is published with support from

THE GOULD FAMILY FOUNDATION

WALTER E. ROBB

Contents

SEX, ECONOMY,
FREEDOM & COMMUNITY
(1993)

Conservation and Local Economy

IN OUR RELATION to the land, we are ruled by a number of terms and limits set not by anyone's preference but by nature and by human nature:

I. Land that is used will be ruined unless it is properly cared for.

II. Land cannot be properly cared for by people who do not know it intimately, who do not know how to care for it, who are not strongly motivated to care for it, and who cannot afford to care for it.

III. People cannot be adequately motivated to care for land by general principles or by incentives that are merely economic—that is, they won't care for it merely because they think they should or merely because somebody pays them.

IV. People are motivated to care for land to the extent that their interest in it is direct, dependable, and permanent.

V. They will be motivated to care for the land if they can reasonably expect to live on it as long as they live. They will be more strongly motivated if they can reasonably expect that their children and grandchildren will live on it as long as they live. In other words, there must be a mutuality of belonging: they must feel that the land belongs to them, that they belong to it, and that this belonging is a settled and unthreatened fact.

VI. But such belonging must be appropriately limited. This is the indispensable qualification of the idea of land ownership. It is well understood that ownership is an incentive to care. But there is a limit to how much land can be owned before an owner is unable to take proper care of it. The need for attention increases with the intensity of use. But the *quality* of attention decreases as acreage increases.

VII. A nation will destroy its land and therefore itself if it
does not foster in every possible way the sort of thrifty,
prosperous, permanent rural households and communities that have the desire, the skills, and the means to care
properly for the land they are using.

In an age notoriously impatient of restraints, such a list of
rules will hardly be welcome, but that these *are* the rules of
land use I have no doubt. I am convinced of their authenticity both by common wisdom and by my own experience and
observation. The rules exist; the penalties for breaking them
are obvious and severe; the failure of land stewardship in this
country is the result of a general disregard for all of them.

As proof of this failure, there is no need to recite again the
statistics of land ruination. The gullies and other damages are
there to be seen. Very little of our land that is being used—for
logging, mining, or farming—is being well used. Much of our
land has never been well used. Those of us who know what we
are looking at know that this is true. And after observing the
worsening condition of our land, we have only to raise our eyes
a little to see the worsening condition of those who are using
the land and who are entrusted with its care. We must accept
as a fact that by now, our country (as opposed to our nation)
is characteristically in decline. War, depression, inflation, usury,
the attitudes of the industrial economy, social and educational
fashions—all have taken their toll. For a long time, the news
from everywhere in rural America has been almost unrelievedly
bad: bankruptcy, foreclosure, depression, suicide, the departure of the young, the loneliness of the old, soil loss, soil degradation, chemical pollution, the loss of genetic and specific
diversity, the extinction or threatened extinction of species, the
depletion of aquifers, stream degradation, the loss of wilderness, strip mining, clear-cutting, population loss, the loss of
supporting economies, the deaths of towns. Rural American
communities, economies, and ways of life that in 1945 were
thriving and, though imperfect, full of promise for an authentic human settlement of our land are now as effectively destroyed as the Jewish communities of Poland; the means of
destruction were not so blatantly evil, but they have proved
just as thorough. The news of rural decline and devastation

has been accompanied, to be sure, by a chorus of professional, institutional, and governmental optimists, who continue to insist that all is well, that we are making things worse only as a way of making things better, that farmers who failed are merely "inefficient producers" for whose failure the country is better off, that money and technology will fill the gaps, that government will fill the gaps, that science will soon free us from our regrettable dependence on the soil. We have heard that it is good business and good labor economics to destroy the last remnants of American wilderness. We have heard that the rural population is actually growing because city people are moving to the country and commuters are replacing farmers. We have heard that the rural economy can be repaired by moving the urban economy out into the country and by replacing rural work with work in factories and offices. And all the while the real conditions of the rural land and the rural people have been getting worse.

Of the general condition of the American countryside, my own community will serve well enough as an example. The town of Port Royal, Kentucky, has a population of about one hundred people. The town came into existence as a trading center, serving the farms in a few square miles of hilly country on the west side of the Kentucky River. It has never been much bigger than it is now. But whereas now it is held together by habit or convenience, once it was held together by a complex local economy. In my mother's childhood, in the years before World War I, there were sixteen business and professional enterprises in the town, all serving the town and the surrounding farms. By the time of my own childhood, in the years before World War II, the number had been reduced to twelve, but the town and its tributary landscape were still alive as a community and as an economy. Now, counting the post office, the town has five enterprises, one of which does not serve the local community. There is now no market for farm produce in the town or within forty miles. We no longer have a garage or repair shop of any kind. We have had no doctor for forty years and no school for thirty. Now, as a local economy and therefore as a community, Port Royal is dying.

What does the death of a community, a local economy, cost its members? And what does it cost the country? So far as I know, we have no economists who are interested in such costs. Nevertheless, when you must drive ten or twenty or more miles to reach a doctor or a school or a mechanic or to find parts for farm machinery, the costs exist, and they are increasing. As they increase, they make the economy of every farm and household less tenable.

As people leave the community or, remaining in the place, drop out of the local economy, as the urban-industrial economy more and more usurps the local economy, as the scale and speed of work increase, care declines. As care declines, the natural supports of the human economy and community also decline, for whatever is used, is used destructively.

We in Port Royal are part of an agricultural region surrounded by cities that import much of their food from distant places. Though we urgently need crops that can be substituted for tobacco, we produce practically no vegetables or other foods for consumption in our region. Having no local food economy, we produce a less and less diverse food supply for the general market. This condition implies and virtually requires the abuse of our land and our people, and they are abused. We are also part of a region that is abundantly and diversely forested, and we have no forest economy. We have no local wood products industry. This makes it almost certain that our woodlands and their owners will be abused, and they are abused.

We provide, moreover, a great deal of recreation for our urban neighbors—hunting, fishing, boating, and the like—and we have the capacity to provide more. But for this we receive little or nothing, and sometimes we suffer damage.

In our region, furthermore, there has been no public effort to preserve the least scrap of land in its pristine condition. And the last decade or so of agricultural depression has caused much logging of the few stands of mature forest in private hands. Now, if we want our descendants to know what the original forest was like—that is, to know the original nature of our land—we must start from scratch and grow the necessary examples over the next two or three hundred years.

My part of rural America is, in short, a colony, like every other part of rural America. Almost the whole landscape of this country—from the exhausted cotton fields of the plantation South to the eroding wheatlands of the Palouse, from the strip mines of Appalachia to the clear-cuts of the Pacific slope—is in the power of an absentee economy, once national and now increasingly international, that is without limit in its greed and without mercy in its exploitation of land and people. Between the prosperity of this vast centralizing economy and the prosperity of any local economy or locality, there is now a radical disconnection. The accounting that measures the wealth of corporations, great banks, and national treasuries takes no measure of the civic or economic or natural health of places like Port Royal, Kentucky; Harpster, Ohio; Indianola, Iowa; Matfield Green, Kansas; Wolf Hole, Arizona; or Nevada City, California—and it does not intend to do so.

In 1912, according to William Allen White, "the county in the United States with the largest assessed valuation was Marion County, Kansas. . . . Marion County happened to have a larger per capita of bank deposits than any other American county. . . . Yet no man in Marion County was rated as a millionaire, but the jails and poorhouses were practically empty. The great per capita of wealth was actually distributed among the people who earned it." This, of course, is the realization of that dream that is sometimes called Jeffersonian but is really the dream of the economically oppressed throughout human history. And because this was a rural county, White was not talking just about bank accounts; he was talking about real capital—usable property. That era and that dream are now long past. Now the national economy, which is increasingly a global economy, no longer prospers by the prosperity of the land and people but by their exploitation.

The Civil War made America safe for the moguls of the railroads and of the mineral and timber industries who wanted to be free to exploit the countryside. The work of these industries and their successors is now almost complete. They have dispossessed, disinherited, and moved into the urban economy almost the entire citizenry; they have defaced and plundered the countryside. And now this great corporate enterprise, thoroughly uprooted and internationalized, is moving toward the exploitation of the whole world under the shibboleths of "globalization," "free trade," and "new world order." The proposed revisions in the General Agreement on Tariffs and Trade are intended solely to further this exploitation. The aim is simply and unabashedly to bring every scrap of productive land and every worker on the planet under corporate control.

The voices of the countryside, the voices appealing for respect for the land and for rural community, have simply not been heard in the centers of wealth, power, and knowledge. The centers have decreed that the voice of the countryside shall be that of Snuffy Smith or Li'l Abner, and only that voice have they been willing to hear.

"The business of America is business," a prophet of our era too correctly said. Two corollaries are clearly implied: that the business of the American government is to serve, protect, and defend business; and that the business of the American people is to serve the government, which means to serve business. The costs of this state of things are incalculable. To start with, people in great numbers—because of their perception that the government serves not the country or the people but the corporate economy—do not vote. Our leaders, therefore, are now in the curious—and hardly legitimate—position of asking a very substantial number of people to cheer for, pay for, and perhaps die for a government that they have not voted for.

But when the interests of local communities and economies are relentlessly subordinated to the interests of "business," then two further catastrophes inevitably result. First, the people are increasingly estranged from the native wealth, health, knowledge, and pleasure of their country. And, second, the country itself is destroyed.

It is not possible to look at the present condition of our land and people and find support for optimism. We must not fool ourselves. It is altogether conceivable that we may go right along with this business of "business," with our curious religious faith in technological progress, with our glorification of our own greed and violence always rationalized by our indignation at the greed and violence of others, until our land, our world, and ourselves are utterly destroyed. We know from history that massive human failure is possible. It is foolish to assume that we will save ourselves from any fate that we have made possible simply because we have the conceit to call ourselves *Homo sapiens.*

On the other hand, we want to be hopeful, and hope is one of our duties. A part of our obligation to our own being and to our descendants is to study our life and our condition, searching always for the authentic underpinnings of hope. And if we look, these underpinnings can still be found.

For one thing, though we have caused the earth to be seriously diseased, it is not yet without health. The earth we have before us now is still abounding and beautiful. We must learn again to see that present world for what it is. The health of nature is the primary ground of hope—if we can find the humility and wisdom to accept nature as our teacher. The pattern of land stewardship is set by nature. This is why we must have stable rural economies and communities; we must keep alive in every place the human knowledge of the nature of that place. Nature is the best farmer and forester, for she does not destroy the land in order to make it productive. And so in our wish to preserve our land, we are not without the necessary lessons, nor are we without instruction, in our cultural and religious tradition, necessary to learn those lessons.

But we have not only the example of nature; we have still, though few and widely scattered, sufficient examples of competent and loving human stewardship of the earth. We have, too, our own desire to be healthy in a healthy world. Surely, most of us still have, somewhere within us, the fundamental human wish to die in a world in which we have been glad to live. And we *are*, in spite of much evidence to the contrary, somewhat sapient. We *can* think—if we will. If we know carefully enough who, what, and where we are, and if we keep the scale of our work small enough, we can think responsibly.

These assets are not the gigantic, technical, and costly equipment that we tend to think we need, but they are enough. They are, in fact, God's plenty. Because we have these assets, which are the supports of our legitimate hope, we can start from where we are, with what we have, and imagine and work for the healings that are necessary.

But we must begin by giving up any idea that we can bring about these healings without fundamental changes in the way we think and live. We face a choice that is starkly simple: we must change or be changed. If we fail to change for the better, then we will be changed for the worse. We cannot blunder our way into health by the same sad and foolish hopes by which we have blundered into disease. We must see that the standardless aims of industrial communism and industrial capitalism equally have failed. The aims of productivity, profitability, efficiency, limitless growth, limitless wealth, limitless power, limitless mechanization and automation can enrich and empower the few (for a while), but they will sooner or later ruin us all. The gross national product and the corporate bottom line are utterly meaningless as measures of the prosperity or health of the country.

If we want to succeed in our dearest aims and hopes as a people, we must understand that we cannot proceed any further without standards, and we must see that ultimately the standards are not set by us but by nature. We must see that it is foolish, sinful, and suicidal to destroy the health of nature for the sake of an economy that is really not an economy at all but merely a financial system, one that is unnatural, undemocratic, sacrilegious, and ephemeral. We must see the error of our effort to live by fire, by burning the world in order to live in it. There is no plainer symptom of our insanity than our avowed intention to maintain by fire an unlimited economic growth. Fire destroys what nourishes it and so in fact imposes severe limits on any growth associated with it. The true source and analogue of our economic life is the economy of plants, which never exceeds natural limits, never grows beyond the power of its place to support it, produces no waste, and enriches and preserves itself by death and decay. We must learn to grow like a tree, not like a fire. We must repudiate what Edward Abbey called "the ideology of the cancer cell": the idiotic ideology of "unlimited economic growth" that pushes blindly toward the limitation of massive catastrophe.

We must give up also our superstitious conviction that we can contrive technological solutions to all our problems. Soil loss, for example, is a problem that embarrasses all of our technological pretensions. If soil were all being lost in a huge slab somewhere, that would appeal to the would-be heroes of "science and technology," who might conceivably engineer a glamorous, large, and speedy solution—however many new problems they might cause in doing so. But soil is not usually lost in slabs or heaps of magnificent tonnage. It is lost a little at a time over millions of acres by the careless acts of millions of people. It cannot be saved by heroic feats of gigantic technology but only by millions of small acts and restraints, conditioned by small fidelities, skills, and desires. Soil loss is ultimately a cultural problem; it will be corrected only by cultural solutions.

The aims of production, profit, efficiency, economic growth, and technological progress imply, as I have said, no social or ecological standards, and in practice they submit to none. But there is another set of aims that does imply a standard, and these aims are freedom (which is pretty much a synonym for personal and local self-sufficiency), pleasure (that is, our gladness to be alive), and longevity or sustainability (by which we signify our wish that human freedom and pleasure may last). The standard implied by all of these aims is health. They depend ultimately and inescapably on the health of nature; the idea that freedom and pleasure can last long in a diseased world is preposterous. But these good things depend also on the health of human culture, and human culture is to a considerable extent the knowledge of economic and other domestic procedures—that is, ways of work, pleasure, and education— that preserve the health of nature.

In talking about health, we have thus begun to talk about community. But we must take care to see how this standard of health enlarges and clarifies the idea of community. If we speak of a *healthy* community, we cannot be speaking of a community that is merely human. We are talking about a neighborhood of humans in a place, plus the place itself: its soil, its water, its air, and all the families and tribes of the nonhuman creatures that belong to it. If the place is well preserved, if its entire membership, natural and human, is present in it, and if the human economy is in practical harmony with the nature of the place, then the community is healthy. A diseased community will be suffering natural losses that become, in turn, human losses. A healthy community is sustainable; it is, within reasonable limits, self-sufficient and, within reasonable limits, self-determined —that is, free of tyranny.

Community, then, is an indispensable term in any discussion of the connection between people and land. A healthy community is a form that includes all the local things that are connected by the larger, ultimately mysterious form of the Creation. In speaking of community, then, we are speaking of a complex connection not only among human beings or between humans and their homeland but also between the human economy and nature, between forest or prairie and field or orchard, and between troublesome creatures and pleasant ones. *All* neighbors are included.

From the standpoint of such a community, any form of land abuse—a clear-cut, a strip mine, an overplowed or overgrazed field—is as alien and as threatening as it would be from the standpoint of an ecosystem. From such a standpoint, it would be plain that land abuse reduces the possibilities of local life, just as do chain stores, absentee owners, and consolidated schools.

One obvious advantage of such an idea of community is that it provides a common ground and a common goal between conservationists and small-scale land users. The long-standing division between conservationists and farmers, ranchers, and other private small-business people is distressing because it is to a considerable extent false. It is readily apparent that the economic forces that threaten the health of ecosystems and the survival of species are equally threatening to economic democracy and the survival of human neighborhoods.

I believe that the most necessary question now—for conservationists, for small-scale farmers, ranchers, and businesspeople, for politicians interested in the survival of democracy, and for consumers—is this: What must be the economy of a healthy community based in agriculture or forestry? It *cannot* be the present colonial economy in which only "raw materials" are exported and *all* necessities and pleasures are imported. To be healthy, land-based communities will need to add value to local products, they will need to supply local demand, and they will need to be reasonably self-sufficient in food, energy, pleasure, and other basic requirements.

Once a person understands the necessity of healthy local communities and community economies, it becomes easy to imagine a range of reforms that might bring them into being.

It is at least conceivable that useful changes might be started or helped along by consumer demand in the cities. There is, for example, already evidence of a growing concern among urban consumers about the quality and the purity of food. Once this demand grows extensive and competent enough, it will have the power to change agriculture—if there is enough left of agriculture, by then, to be changed.

It is even conceivable that our people in Washington might make decisions tending toward sustainability and self-sufficiency in local economies. The federal government could do much to help, if it would. Its mere acknowledgment that problems exist would be a promising start.

But let us admit that urban consumers are not going to be well informed about their economic sources very soon and that a federal administration enlightened about the needs and problems of the countryside is not an immediate prospect.

The real improvements then must come, to a considerable extent, from the local communities themselves. We need local revision of our methods of land use and production. We need to study and work together to reduce scale, reduce overhead, reduce industrial dependencies; we need to market and process local products locally; we need to bring local economies into harmony with local ecosystems so that we can live and work with pleasure in the same places indefinitely; we need to substitute ourselves, our neighborhoods, our local resources, for expensive imported goods and services; we need to increase cooperation among all local economic entities: households, farms, factories, banks, consumers, and suppliers. If we are serious about reducing government and the burdens of government, then we need to do so by returning economic self-determination to the people. And we must not do this by inviting destructive industries to provide "jobs" in the community; we must do it by fostering economic democracy. For example, as much as possible of the food that is consumed locally ought to be locally produced on small farms, and then processed in small, non-polluting plants that are locally owned. We must do everything possible to provide to ordinary citizens the opportunity to own a small, usable share of the country. In that way, we will put local capital to work locally, not to exploit and destroy the land but to use it well. This is not work just for the privileged, the well-positioned, the wealthy, and the powerful. It is work for everybody.

I acknowledge that to advocate such reforms is to advocate a kind of secession—not a secession of armed violence but a quiet secession by which people find the practical means and the strength of spirit to remove themselves from an economy that is exploiting them and destroying their homeland. The great, greedy, indifferent national and international economy is killing rural America, just as it is killing America's cities— it is killing our country. Experience has shown that there is no use in appealing to this economy for mercy toward the earth or toward any human community. All true patriots must find ways of opposing it.

Conservation Is Good Work

THERE ARE, as nearly as I can make out, three kinds of conservation currently operating. The first is the preservation of places that are grandly wild or "scenic" or in some other way spectacular. The second is what is called "conservation of natural resources"—that is, of the things of nature that we intend to use: soil, water, timber, and minerals. The third is what you might call industrial troubleshooting: the attempt to limit or stop or remedy the most flagrant abuses of the industrial system. All three kinds of conservation are inadequate, both separately and together.

Right at the heart of American conservation, from the beginning, has been the preservation of spectacular places. The typical American park is in a place that is "breathtakingly" beautiful or wonderful and of little apparent economic value. Mountains, canyons, deserts, spectacular landforms, geysers, waterfalls—these are the stuff of parks. There is, significantly, no prairie national park. Wilderness preserves, as Dave Foreman points out, tend to include much "rock and ice" and little marketable timber. Farmable land, in general, has tempted nobody to make a park. Wes Jackson has commented with some anxiety on the people who charge blindly across Kansas and eastern Colorado, headed for the mountains west of Denver. These are nature lovers and sightseers, but they are utterly oblivious of or bored by the rich natural and human history of the Plains. The point of Wes Jackson's anxiety is that the love of nature that limits itself to the love of places that are "scenic" is implicitly dangerous. It is dangerous because it tends to exclude unscenic places from nature and from the respect that we sometimes accord to nature. This is why so much of the landscape that is productively used is also abused; it is used solely according to standards dictated by the financial system and not at all according to standards dictated by the nature of the place. Moreover, as we are beginning to see, it is going to be extremely difficult to make enough parks to preserve vulnerable species and the health of ecosystems or large watersheds.

"Natural resources," the part of nature that we are going

to use, is the part outside the parks and preserves (which, of course, we also use). But "conservation of natural resources" is now in confusion because it is a concept that has received much lip service but not much thought or practice. Part of the confusion is caused by thinking of "natural resources" as belonging to one category when, in fact, they belong to two: surface resources, like soils and forests, which can be preserved in use; and underground resources, like coal or oil, which cannot be. The one way to conserve the minable fuels and materials that use inevitably exhausts is to limit use. At present, we have no intention of limiting such use, and so we cannot say that we are at all interested in the conservation of exhaustible resources. Surface or renewable resources, on the other hand, can be preserved in use so that their yield is indefinitely sustainable.

Sustainability is a hopeful concept not only because it is a present necessity but because it has a history. We know, for example, that some agricultural soils have been preserved in continuous use for several thousand years. We know, moreover, that it is possible to improve soil in use. And it is clear that a forest can be used in such a way that it remains a forest, with its biological communities intact and its soil undamaged, while producing a yield of timber. But the methods by which exhaustible resources are extracted and used have set the pattern also for the use of sustainable resources, with the result that now soils and forests are not merely being used but are being used up, exactly as coal seams are used up.

Since the sustainable use of renewable resources depends on the existence of settled, small local economies and communities capable of preserving the local knowledge necessary for good farming and forestry, it is obvious that there is no simple, easy, or quick answer to the problem of the exhaustion of sustainable resources. We probably are not going to be able to conserve natural resources so long as our extraction and use of the goods of nature are wasteful and improperly scaled, or so long as these resources are owned or controlled by absentees, or so long as the standard of extraction and use is profitability rather than the health of natural and human communities.

Because we are living in an era of ecological crisis, it is understandable that much of our attention, anxiety, and energy is focused on exceptional cases, the outrages and extreme abuses of the industrial economy: global warming, the global assault on the last remnants of wilderness, the extinction of species, oil spills, chemical spills, Love Canal, Bhopal, Chernobyl, the burning oil fields of Kuwait. But a conservation effort that concentrates only on the extremes of industrial abuse tends to suggest that the only abuses are the extreme ones when, in fact, the earth is probably suffering more from many small abuses than from a few large ones. By treating the spectacular abuses as exceptional, the powers that be would like to keep us from seeing that the industrial system (capitalist or communist or socialist) is in itself and by necessity of all of its assumptions extremely dangerous and damaging and that it exists to support an extremely dangerous and damaging way of life. The large abuses exist within and because of a pattern of smaller abuses.

Much of the Sacramento River is now dead because a carload of agricultural poison was spilled into it. The powers that be would like us to believe that this colossal "accident" was an exception in a general pattern of safe use. Diluted and used according to the instructions on the label, they will tell us, the product that was spilled is harmless. They neglect to acknowledge any of the implications that surround the accident: that if this product is to be used in dilution almost everywhere, it will have to be manufactured, stored, and transported in concentration somewhere; that even in "harmless" dilution, such chemicals are getting into the water, the air, the rain, and into the bodies of animals and people; that when such a product is distributed to the general public, it will inevitably be spilled in its concentrated form in large or small quantities and that such "accidents" are anticipated, discounted as "acceptable risk," and charged to nature and society by the powers that be; that such chemicals are needed, in the first place, because the scale, the methods, and the economy of American agriculture are all monstrously out of kilter; that such chemicals are used to replace the work and intelligence of people; and that such a deformed agriculture is made necessary, in the first place, by the public's demand for a diet that is at once cheap and luxurious—too cheap to support adequate agricultural communities or good agricultural methods or good maintenance of agricultural land and yet so goofily self-indulgent as to demand, in every season, out-of-season foods produced by earth-destroying machines and chemicals.

We tend to forget, too, in our just and necessary outrage at the government-led attack on public lands and the last large tracts of wilderness, that for the very same reasons and to the profit of the very same people, thousands of woodlots are being abusively and wastefully logged.

Here, then, are three kinds of conservation, all of them urgently necessary and all of them failing. Conservationists have won enough victories to give them heart and hope and a kind of accreditation, but they know better than anybody how immense and how baffling their task has become. For all their efforts, our soils and waters, forests and grasslands are being used up. Kinds of creatures, kinds of human life, good natural and human possibilities are being destroyed. Nothing now exists anywhere on earth that is not under threat of human destruction. Poisons are everywhere. Junk is everywhere.

These dangers are large and public, and they inevitably cause us to think of changing public policy. This is good, so far as it goes. There should be no relenting in our efforts to influence politics and politicians. But in the name of honesty and sanity we must recognize the limits of politics. It is, after all, much easier to improve a policy than it is to improve the community the policy attempts to affect. And it is also probable that some changes required by conservation cannot be politically made and that some necessary changes will have to be made by the governed without the help or approval of the government.

I must admit here that my experience over more than twenty years as part of an effort to influence agricultural policy has not been encouraging. Our arguments directed at the government and the universities by now remind me of the ant crawling up the buttocks of the elephant with love on his mind. We have not made much impression. My conclusion, I imagine, is the same as the ant's, for these great projects, once undertaken, are hard to abandon: we have got to get more radical.

However destructive may be the policies of the government and the methods and products of the corporations, the root of the problem is always to be found in private life. We must learn to see that every problem that concerns us as conservationists always leads straight to the question of how we live. The world is being destroyed, no doubt about it, by the greed of the rich and powerful. It is also being destroyed by popular demand. There are not enough rich and powerful people to consume the whole world; for that, the rich and powerful need the help of countless ordinary people. We acquiesce in the wastefulness and destructiveness of the national and global economics by acquiescing in the wastefulness and destructiveness of our own households and communities. If conservation is to have a hope of succeeding, then conservationists, while continuing their effort to change public life, are going to have to begin the effort to change private life as well.

The problems we are worried about are caused not just by other people but by ourselves. And this realization should lead directly to two more. The first is that solving these problems is not work merely for so-called environmental organizations and agencies but also for individuals, families, and local communities. We are used to hearing about turning off unused lights, putting a brick in the toilet tank, using water-saving shower heads, setting the thermostat low, sharing rides, and so forth—pretty dull stuff. But I'm talking about actual jobs of work that are interesting because they require intelligence and because they are accomplished in response to interesting questions: What are the principles of household economy, and how can they be applied under present circumstances? What are the principles of a neighborhood or a local economy, and how can they be applied under present circumstances? What do people already possess in their minds and bodies, in their families and neighborhoods, in their dwellings and in their local landscape, that can replace what is now being supplied by our consumptive and predatory so-called economy? What can we supply to ourselves cheaply or for nothing that we are now paying dearly for? To answer such questions requires more intelligence and involves more pleasure than all the technological breakthroughs of the last two hundred years.

Second, the realization that we ourselves, in our daily economic life, are causing the problems we are trying to solve ought to show us the inadequacy of the language we are using to talk about our connection to the world. The idea that we live in something called "the environment," for instance, is utterly preposterous. This word came into use because of the pretentiousness of learned experts who were embarrassed by the religious associations of "Creation" and who thought "world" too mundane. But "environment" means that which surrounds or encircles us; it means a world separate from ourselves, outside us. The real state of things, of course, is far more complex and intimate and interesting than that. The world that environs us, that is around us, is also within us. We are made of it; we eat, drink, and breathe it; it is bone of our bone and flesh of our flesh. It is also a Creation, a holy mystery, made for and to some extent by creatures, some but by no means all of whom are humans. This world, this Creation, belongs in a limited sense to us, for we may rightfully require certain things of it—the things necessary to keep us fully alive as the kind of creature we are—but we also belong to it, and it makes certain rightful claims on us: that we care properly for it, that we leave it undiminished not just to our children but to all the creatures who will live in it after us. None of this intimacy and responsibility is conveyed by the word *environment*.

That word is a typical product of the old dualism that is at the root of most of our ecological destructiveness. So, of course, is "biocentrism." If life is at the center, what is at the periphery? And for that matter, *where* is the periphery? "Deep ecology," another bifurcating term, implies that there is, a couple of layers up, a shallow ecology that is not so good—or that an ecosystem is a sort of layer cake with the icing on the bottom. Not only is this language incapable of giving a description of our relation to the world; it is also academic, artificial, and pretentious. It is the sort of language used by a visiting expert who does not want the local people to ask any questions. (I am myself an anthropobiotheointerpenetrist and a gastrointeroenvironmentalist, but I am careful to say so only in the company of other experts.)

No settled family or community has ever called its home place an "environment." None has ever called its feeling for its home place "biocentric" or "anthropocentric." None has ever thought of its connection to its home place as "ecological," deep or shallow. The concepts and insights of the ecologists are of great usefulness in our predicament, and we can hardly escape the need to speak of "ecology" and "ecosystems." But the terms themselves are culturally sterile. They come from the juiceless, abstract intellectuality of the universities which was invented to disconnect, displace, and disembody the mind. The real names of the environment are the names of rivers and river valleys; creeks, ridges, and mountains; towns and cities; lakes, woodlands, lanes, roads, creatures, and people.

And the real name of our connection to this everywhere different and differently named earth is "work." We are connected by work even to the places where we don't work, for all places are connected; it is clear by now that we cannot exempt one place from our ruin of another. The name of our *proper* connection to the earth is "good work," for good work involves much giving of honor. It honors the source of its materials; it honors the place where it is done; it honors the art by which it is done; it honors the thing that it makes and the user of the made thing. Good work is always modestly scaled, for it cannot ignore either the nature of individual places or the differences between places, and it always involves a sort of religious humility, for not everything is known. Good work can be defined only in particularity, for it must be defined a little differently for every one of the places and every one of the workers on the earth.

The name of our present society's connection to the earth is "bad work"—work that is only generally and crudely defined, that enacts a dependence that is ill understood, that enacts no affection and gives no honor. Every one of us is to some extent guilty of this bad work. This guilt does not mean that we must indulge in a lot of breast-beating and confession; it means only that there is much good work to be done by every one of us and that we must begin to do it. All of us are responsible for bad work, not so much because we do it ourselves (though we all do it) as because we have it done for us by other people.

Here we are bound to see our difficulty as almost overwhelming. What proxies have we issued, and to whom, to use the earth on our behalf? How, in this global economy, are we to render anything like an accurate geographical account of our personal economies? How do we take our lives from this earth that we are so anxious to protect and restore to health?

Most of us get almost all the things we need by buying them; most of us know only vaguely, if at all, where those things come from; and most of us know not at all what damage is involved in their production. We are almost entirely dependent on an economy of which we are almost entirely ignorant. The provenance, for example, not only of the food we buy at the store but of the chemicals, fuels, metals, and other materials necessary to grow, harvest, transport, process, and package that food is almost necessarily a mystery to us. To know the full economic history of a head of supermarket cauliflower would require an immense job of research. To be so completely and so ignorantly dependent on the present abusive food economy certainly defines us as earth abusers. It also defines us as potential victims.

Living as we now do in almost complete dependence on a global economy, we are put inevitably into a position of ignorance and irresponsibility. No one can know the whole globe. We can connect ourselves to the globe as a whole only by means of a global economy that, without knowing the earth, plunders it for us. The global economy (like the national economy before it) operates on the superstition that the deficiencies or needs or wishes of one place may safely be met by the ruination of another place. To build houses here, we clear-cut the forests there. To have air-conditioning here, we strip-mine the mountains there. To drive our cars here, we sink our oil wells there. It is an absentee economy. Most people aren't using or destroying what they can see. If we cannot see our garbage or the grave we have dug with our energy proxies, then we assume that all is well. The issues of carrying capacity and population remain abstract and not very threatening to most people for the same reason. If this nation or region cannot feed its population, then food can be imported from other nations or regions. All the critical questions affecting our use of the earth are left to be answered by "the market" or the law of supply and demand, which proposes no limit on either supply or demand. An economy without limits is an economy without discipline.

Conservationists of all kinds would agree, I think, that no discipline, public or private, is implied by the industrial economy and that none is practiced by it. The implicit wish of the industrial economy is that producers might be wasteful, shoddy, and irresponsible and that consumers might be gullible, extravagant, and irresponsible. To fulfill this wish, the industrial economy employs an immense corps of hireling politicians, publicists, lobbyists, admen, and adwomen. The consequent ruin is notorious: we have been talking about it for generations; it brought conservation into being. And conservationists have learned very well how to address this ruin as a public problem. There is now no end to the meetings and publications in which the horrifying statistics are recited, usually with the conclusion that pressure should be put on the government to do something. Often, this pressure has been applied, and the government has done something. But the government has not done enough and may never do enough. It is likely that the government *cannot* do enough.

The government's disinclination to do more than it does is explained, of course, by the government's bought-and-paid-for servitude to interests that do not want it to do more. But there may also be a limit of another kind: a government that could do enough, assuming it had the will, would almost certainly be a government radically and unpleasantly different from the one prescribed by our Constitution. A government undertaking to protect all of nature that is now abused or threatened would have to take total control of the country. Police and bureaucrats—and opportunities for malfeasance—would be everywhere. To wish only for a public or a political solution to the problem of conservation may be to wish for a solution as bad as the problem and still be unable to solve it.

The way out of this dilemma is to understand the ruin of nature as a problem that is both public and private. The failure of public discipline in matters of economy is only the other face of the failure of private discipline. If we have worked at the issues of public policy so long and so exclusively as to bring political limits into sight, then let us turn—not instead but also—to issues of private economy and see how far we can go in that direction. It is a direction that may take us further and produce more satisfactory and lasting results than the direction of policy.

The dilemma of private economic responsibility, as I have said, is that we have allowed our suppliers to enlarge our economic boundaries so far that we cannot be responsible for our effects on the world. The only remedy for this that I can see is to draw in our economic boundaries, shorten our supply lines, so as to permit us literally to know where we are economically. The closer we live to the ground that we live from, the more we will know about our economic life; the more we know about our economic life, the more able we will be to take responsibility for it. The way to bring discipline into one's personal or household or community economy is to limit one's economic geography.

This obviously opens up an agenda almost as daunting as the political agenda. The difference—a consoling one—is that when we try to influence policy, only large jobs must be done; whereas when we seek to reform private economies, the work is necessarily modest, and it can be started by anybody anywhere. What is required is the formation of local economic strategies and eventually of local economies—by which to resist abuses of natural and human communities by the larger economy. And, of course, in talking about the formation of local economies capable of using an earthly place without ruining it, we are talking about the reformation of people; we are talking about reviving good work as an economic force.

If we think of this task of rebuilding local economies as one large task that must be done in a hurry, then we will again be overwhelmed and will want the government to do it. If, on the other hand, we define the task as beginning the reformation of our private or household economies, then the way is plain. What we must do is use well the considerable power we have as consumers: the power of choice. We can choose to buy or not to buy, and we can choose what to buy. The standard by which we choose must be the health of the community—and by that we must mean the *whole* community: ourselves, the place where we live, and all the humans and other creatures who live there with us. In a healthy community, people will be richer in their neighbors, in neighborhood, in the health and pleasure of neighborhood, than in their bank accounts. It is better, therefore, even if the cost is greater, to buy near at hand than to buy at a distance. It is better to buy from a small, privately owned local store than from a chain store. It is better to buy a good product than a bad one. Do not buy anything you don't need. Do as much as you can for yourself. If you cannot do something for yourself, see if you have a neighbor who can do it for you. Do everything you can to see that your money stays as long as possible in the local community. If you have money to invest, try to invest it locally, both to help the local community and to keep from helping the larger economy that is destroying local communities. Begin to ask yourself how your money could be put at minimal interest into the hands of a young person who wants to start a farm, a store, a shop, or a small business that the community needs. This agenda can be followed by individuals and single families. If it is followed by people in groups—churches, conservation organizations, neighborhood associations, groups of small farmers, and the like—the possibilities multiply and the effects will be larger.

The economic system that most affects the health of the world and that may be most subject to consumer influence is that of food. And the issue of food provides an excellent example of what I am talking about. If you want to reform your own food economy, you can make a start without anybody's permission or help. If you have a place to do it, grow some food for yourself. Growing some of your own food gives you pleasure, exercise, knowledge, sales resistance, and standards. Your own food, if you grow it the right way, will taste good and so will cause you to wish to buy food that tastes good. So far as you can, buy food that is locally grown. Tell your grocer that you are interested in locally grown food. If you can't find locally grown food in stores, then see if you can deal directly with a local farmer. The value of this, for conservationists, is that when consumers are acquainted and friendly with their producers, they can influence production. They can know the land on which their food is produced. They can refuse to buy food that is produced with dangerous chemicals or by other destructive practices. As these connections develop, local agriculture will diversify, become more healthy and more stable, employ more people. As local demand increases and becomes more knowledgeable, small food-processing industries will enter the local economy. Everything that is done by the standard of community health will make new possibilities for good work —that is for the responsible use of the world.

The forest economy is not so obviously subject to consumer influence, but such influence is sorely needed. Both the forests themselves and their human communities suffer for the want of local forest economies—properly scaled wood-products industries that would be the basis of stable communities and would provide local incentives for the good use of the forest. People who see that they must depend on the forest for generations, in a complex local forest economy, will want the forest to last and be healthy; they will *not* want to see all the marketable timber ripped out of it as fast as possible. Both forest and farm communities would benefit from technologies that could be locally supplied and maintained. Draft horses, for example, are better than large machines, both for the woods and for the local economy.

The economy of recreation has hardly been touched as an issue of local economy and conservation, though conservationists and consumers alike have much to gain from making it such an issue. At present, there is an almost complete disconnection between the economic use of privately owned farm and forest land and its use for recreation. Such land is now much used by urban people for hunting and fishing, but mainly without benefit to the landowners, who therefore receive no incentive from this use to preserve wildlife habitat or to take the best care of their woodlands and stream margins. They need to receive such incentives. It is not beyond reason that public funds might be given to private landowners to preserve and enhance the recreational value—that is, the wildness—of their land. But since governments are unlikely to do this soon, the incentives need to be provided by consumer and conservation groups working in cooperation with farm groups. The rule of the food economy ought to apply to the recreation economy: find your pleasure and your rest as near home as possible. In Kentucky, for example, we have hundreds of miles of woodland stretching continuously along the sides of our creek and river valleys. Why should conservation and outdoor groups not pay an appropriate price to farmers to maintain hiking trails and campsites and preserve the forests in such places? The money that would carry a family to a vacation in a distant national park could thus be kept at home and used to help the local economy and protect the local countryside.

The point of all this is the use of local buying power, local gumption, and local affection to see that the best care is taken of the local land. This sort of effort would bridge the gap, now so destructive, between the conservationists and the small farmers and ranchers, and that would be one of its great political benefits. But the fundamental benefit would be to the world and ourselves. We would begin to protect the world not just by conserving it but also by living in it.

Christianity and the Survival of Creation

I CONFESS that I have not invariably been comfortable in front of a pulpit; I have never been comfortable behind one. To be behind a pulpit is always a forcible reminder to me that I am an essayist and, in many ways, a dissenter. An essayist is, literally, a writer who attempts to tell the truth. Preachers must resign themselves to being either right or wrong; an essayist, when proved wrong, may claim to have been "just practicing." An essayist is privileged to speak without institutional authorization. A dissenter, of course, must speak without privilege.

I want to begin with a problem: namely, that the culpability of Christianity in the destruction of the natural world and the uselessness of Christianity in any effort to correct that destruction are now established clichés of the conservation movement. This is a problem for two reasons.

First, the indictment of Christianity by the anti-Christian conservationists is, in many respects, just. For instance, the complicity of Christian priests, preachers, and missionaries in the cultural destruction and the economic exploitation of the primary peoples of the Western Hemisphere, as of traditional cultures around the world, is notorious. Throughout the five hundred years since Columbus's first landfall in the Bahamas, the evangelist has walked beside the conqueror and the merchant, too often blandly assuming that their causes were the same. Christian organizations, to this day, remain largely indifferent to the rape and plunder of the world and of its traditional cultures. It is hardly too much to say that most Christian organizations are as happily indifferent to the ecological, cultural, and religious implications of industrial economics as are most industrial organizations. The certified Christian seems just as likely as anyone else to join the military-industrial conspiracy to murder Creation.

The conservationist indictment of Christianity is a problem, second, because, however just it may be, it does not come from an adequate understanding of the Bible and the cultural traditions that descend from the Bible. The anti-Christian conservationists characteristically deal with the Bible by waving

28

it off. And this dismissal conceals, as such dismissals are apt to do, an ignorance that invalidates it. The Bible is an inspired book written by human hands; as such, it is certainly subject to criticism. But the anti-Christian environmentalists have not mastered the first rule of the criticism of books: you have to read them before you criticize them. Our predicament now, I believe, requires us to learn to read and understand the Bible in the light of the present fact of Creation. This would seem to be a requirement both for Christians and for everyone concerned, but it entails a long work of true criticism—that is, of careful and judicious study, not dismissal. It entails, furthermore, the making of very precise distinctions between biblical instruction and the behavior of those peoples supposed to have been biblically instructed.

I cannot pretend, obviously, to have made so meticulous a study; even if I were capable of it, I would not live long enough to do it. But I have attempted to read the Bible with these issues in mind, and I see some virtually catastrophic discrepancies between biblical instruction and Christian behavior. I don't mean disreputable Christian behavior, either. The discrepancies I see are between biblical instruction and allegedly respectable Christian behavior.

If because of these discrepancies Christianity were dismissible, there would, of course, be no problem. We could simply dismiss it, along with the twenty centuries of unsatisfactory history attached to it, and start setting things to rights. The problem emerges only when we ask, Where then would we turn for instruction? We might, let us suppose, turn to another religion—a recourse that is sometimes suggested by the anti-Christian conservationists. Buddhism, for example, is certainly a religion that could guide us toward a right respect for the natural world, our fellow humans, and our fellow creatures. I owe a considerable debt myself to Buddhism and Buddhists. But there are an enormous number of people—and I am one of them—whose native religion, for better or worse, is Christianity. We were born to it; we began to learn about it before we became conscious; it is, whatever we think of it, an intimate belonging of our being; it informs our consciousness, our language, and our dreams. We can turn away from it or against it, but that will only bind us tightly to a reduced version of it. A better possibility is that this, our native religion, should survive and renew itself so that it may become as largely and truly instructive as we need it to be. On such a survival and renewal of the Christian religion may depend the survival of the Creation that is its subject.

II

If we read the Bible, keeping in mind the desirability of those two survivals—of Christianity and the Creation—we are apt to discover several things about which modern Christian organizations have kept remarkably quiet or to which they have paid little attention.

We will discover that we humans do not own the world or any part of it: "The earth is the Lord's, and the fulness thereof: the world and they that dwell therein." There is in our human law, undeniably, the concept and right of "land ownership." But this, I think, is merely an expedient to safeguard the mutual belonging of people and places without which there can be no lasting and conserving human communities. This right of human ownership is limited by mortality and by natural constraints on human attention and responsibility; it quickly becomes abusive when used to justify large accumulations of "real estate," and perhaps for that reason such large accumulations are forbidden in the twenty-fifth chapter of Leviticus. In biblical terms, the "landowner" is the guest and steward of God: "The land is mine; for ye are strangers and sojourners with me."

We will discover that God made not only the parts of Creation that we humans understand and approve but all of it: "All things were made by him; and without him was not anything made that was made." And so we must credit God with the making of biting and stinging insects, poisonous serpents, weeds, poisonous weeds, dangerous beasts, and disease-causing microorganisms. That we may disapprove of these things does not mean that God is in error or that He ceded some of the work of Creation to Satan; it means that we are deficient in wholeness, harmony, and understanding—that is, we are "fallen."

We will discover that God found the world, as He made it, to be good, that He made it for His pleasure, and that He continues to love it and to find it worthy, despite its reduction and corruption by us. People who quote John 3:16 as an easy formula for getting to Heaven neglect to see the great difficulty implied in the statement that the advent of Christ was made possible by God's love for the world—not God's love for Heaven or for the world as it might be but for the world as it was and is. Belief in Christ is thus dependent on prior belief in the inherent goodness—the lovability—of the world.

We will discover that the Creation is not in any sense independent of the Creator, the result of a primal creative act long over and done with, but is the continuous, constant participation of all creatures in the being of God. Elihu said to Job that if God "gather unto himself his spirit and his breath; all flesh shall perish together." And Psalm 104 says, "Thou sendest forth thy spirit, they are created." Creation is thus God's presence in creatures. The Greek Orthodox theologian Philip Sherrard has written that "Creation is nothing less than the manifestation of God's hidden Being." This means that we and all other creatures live by a sanctity that is inexpressibly intimate, for to every creature, the gift of life is a portion of the breath and spirit of God. As the poet George Herbert put it:

> Thou art in small things great, not small in any . . .
> For thou art infinite in one and all.

We will discover that for these reasons our destruction of nature is not just bad stewardship, or stupid economics, or a betrayal of family responsibility; it is the most horrid blasphemy. It is flinging God's gifts into His face, as if they were of no worth beyond that assigned to them by our destruction of them. To Dante, "despising Nature and her goodness" was a violence against God. We have no entitlement from the Bible to exterminate or permanently destroy or hold in contempt anything on the earth or in the heavens above it or in the waters beneath it. We have the right to use the gifts of nature but not to ruin or waste them. We have the right to use what we need but no more, which is why the Bible forbids usury and great accumulations of property. The usurer, Dante said, "condemns Nature . . . for he puts his hope elsewhere."

William Blake was biblically correct, then, when he said that "everything that lives is holy." And Blake's great commentator Kathleen Raine was correct both biblically and historically when she said that "the sense of the holiness of life is the human norm."

The Bible leaves no doubt at all about the sanctity of the act of world-making, or of the world that was made, or of creaturely or bodily life in this world. We are holy creatures living among other holy creatures in a world that is holy. Some people know this, and some do not. Nobody, of course, knows it all the time. But what keeps it from being far better known than it is? Why is it apparently unknown to millions of professed students of the Bible? How can modern Christianity have so solemnly folded its hands while so much of the work of God was and is being destroyed?

III

Obviously, "the sense of the holiness of life" is not compatible with an exploitive economy. You cannot know that life is holy if you are content to live from economic practices that daily destroy life and diminish its possibility. And many if not most Christian organizations now appear to be perfectly at peace with the military-industrial economy and its "scientific" destruction of life. Surely, if we are to remain free and if we are to remain true to our religious inheritance, we must maintain a separation between church and state. But if we are to maintain any sense or coherence or meaning in our lives, we cannot tolerate the present utter disconnection between religion and economy. By "economy" I do not mean "economics," which is the study of money-making, but rather the ways of human housekeeping, the ways by which the human household is situated and maintained within the household of nature. To be uninterested in economy is to be uninterested in the practice of religion; it is to be uninterested in culture and in character. Probably the most urgent question now faced by people who would adhere to the Bible is this: What sort of economy would be responsible to the holiness of life? What, for Christians, would be the economy, the practices and the restraints, of "right livelihood"? I do not believe that organized Christianity now has any idea. I think its idea of a Christian economy is no more or less than the industrial economy—which is an economy firmly founded on the seven deadly sins and the breaking of all ten of the Ten Commandments. Obviously, if Christianity is going to survive as more than a respecter and comforter of profitable iniquities, then Christians, regardless of their organizations, are going to have to interest themselves in economy—which is to say, in nature and in work. They are going to have to give workable answers to those who say we cannot live without this economy that is destroying us and our world, who see the murder of Creation as the only way of life.

The holiness of life is obscured to modern Christians also by the idea that the only holy place is the built church. This idea may be more taken for granted than taught; nevertheless, Christians are encouraged from childhood to think of the church building as "God's house," and most of them could think of their houses or farms or shops or factories as holy places only with great effort and embarrassment. It is understandably difficult for modern Americans to think of their dwellings and workplaces as holy, because most of these are, in fact, places of desecration, deeply involved in the ruin of Creation.

The idea of the exclusive holiness of church buildings is, of course, wildly incompatible with the idea, which the churches also teach, that God is present in all places to hear prayers. It is incompatible with Scripture. The idea that a human artifact could contain or confine God was explicitly repudiated by Solomon in his prayer at the dedication of the Temple: "Behold, the heaven and the heaven of heavens cannot contain thee: how much less this house that I have builded?" And these words of Solomon were remembered a thousand years later by Saint Paul, preaching at Athens:

> God that made the world and all things therein, seeing that he is lord of heaven and earth, dwelleth not in temples made with hands . . .
>
> For in him we live, and move, and have our being; as certain also of your own poets have said.

Idolatry always reduces to the worship of something "made with hands," something confined within the terms of human work and human comprehension. Thus, Solomon and Saint Paul both insisted on the largeness and the at-largeness of God, setting Him free, so to speak, from *ideas* about Him. He is not to be fenced in, under human control, like some domestic creature; He is the wildest being in existence. The presence of His spirit in us is our wildness, our oneness with the wilderness of Creation. That is why subduing the things of nature to human purposes is so dangerous and why it so often results in evil, in separation and desecration. It is why the poets of our tradition so often have given nature the role not only of mother or grandmother but of the highest earthly teacher and judge, a figure of mystery and great power. Jesus' own specifications for his church have nothing at all to do with masonry and carpentry but only with people; his church is "where two or three are gathered together in my name."

The Bible gives exhaustive (and sometimes exhausting) attention to the organization of religion: the building and rebuilding of the Temple; its furnishings; the orders, duties, and paraphernalia of the priesthood; the orders of rituals and ceremonies. But that does not disguise the fact that the most significant religious events recounted in the Bible do not occur in "temples made with hands." The most important religion in that book is unorganized and is sometimes profoundly disruptive of organization. From Abraham to Jesus, the most important people are not priests but shepherds, soldiers, property owners, workers, housewives, queens and kings, manservants and maidservants, fishermen, prisoners, whores, even bureaucrats. The great visionary encounters did not take place in temples but in sheep pastures, in the desert, in the wilderness, on mountains, on the shores of rivers and the sea, in the middle of the sea, in prisons. And however strenuously the divine voice prescribed rites and observances, it just as strenuously repudiated them when they were taken to *be* religion:

> Your new moons and your appointed feasts my soul hateth: they are a trouble unto me; I am weary to bear them.
>
> And when you spread forth your hands, I will hide mine eyes from you: yea, when you make many prayers, I will not hear: your hands are full of blood.
>
> Wash you, make you clean; put away the evil of your doings from before mine eyes; cease to do evil;
>
> Learn to do well; seek judgment, relieve the oppressed, judge the fatherless, plead for the widow.

Religion, according to this view, is less to be celebrated in rituals than practiced in the world.

I don't think it is enough appreciated how much an outdoor book the Bible is. It is a "hypaethral book," such as Thoreau talked about—a book open to the sky. It is best read and understood outdoors, and the farther outdoors the better. Or that has been my experience of it. Passages that within walls seem improbable or incredible, outdoors seem merely natural. This is because outdoors we are confronted everywhere with wonders; we see that the miraculous is not extraordinary but the common mode of existence. It is our daily bread. Whoever really has considered the lilies of the field or the birds of the air and pondered the improbability of their existence in this warm world within the cold and empty stellar distances will hardly balk at the turning of water into wine—which was, after all, a very small miracle. We forget the greater and still continuing miracle by which water (with soil and sunlight) is turned into grapes.

It is clearly impossible to assign holiness exclusively to the built church without denying holiness to the rest of Creation, which is then said to be "secular." The world, which God looked at and found entirely good, we find none too good to pollute entirely and destroy piecemeal. The church, then, becomes a kind of preserve of "holiness," from which certified lovers of God assault and plunder the "secular" earth.

Not only does this repudiate God's approval of His work; it refuses also to honor the Bible's explicit instruction to regard the works of the Creation as God's revelation of Himself. The assignation of holiness exclusively to the built church is therefore logically accompanied by the assignation of revelation exclusively to the Bible. But Psalm 19 begins, "The heavens declare the glory of God; and the firmament sheweth his handiwork." The word of God has been revealed in facts from the moment of the third verse of the first chapter of Genesis: "Let there be light: and there was light." And Saint Paul states the rule: "The invisible things of him from the creation of the world are clearly seen, being understood by the things that are made." Yet from this free, generous, and sensible view of things, we come to the idolatry of the book: the idea that nothing is true that cannot be (and has not been already) written.

The misuse of the Bible thus logically accompanies the abuse of nature: if you are going to destroy creatures without respect, you will want to reduce them to "materiality"; you will want to deny that there is spirit or truth in them, just as you will want to believe that the only holy creatures, the only creatures with souls, are humans—or even only Christian humans.

By denying spirit and truth to the nonhuman Creation, modern proponents of religion have legitimized a form of blasphemy without which the nature- and culture-destroying machinery of the industrial economy could not have been built —that is, they have legitimized bad work. Good human work honors God's work. Good work uses no thing without respect, both for what it is in itself and for its origin. It uses neither tool nor material that it does not respect and that it does not love. It honors nature as a great mystery and power, as an in-dispensable teacher, and as the inescapable judge of all work of human hands. It does not dissociate life and work, or pleasure and work, or love and work, or usefulness and beauty. To work without pleasure or affection, to make a product that is not both useful and beautiful, is to dishonor God, nature, the thing that is made, and whomever it is made for. This is blasphemy: to make shoddy work of the work of God. But such is not possible when the entire Creation is understood as holy and when the works of God are understood as embodying and thus revealing His spirit.

In the Bible we find none of the industrialist's contempt or hatred for nature. We find, instead, a poetry of awe and rever-ence and profound cherishing, as in these verses from Moses' valedictory blessing of the twelve tribes:

And of Joseph he said, Blessed of the Lord be his land, for the precious things of heaven, for the dew, and for the deep that croucheth beneath,

And for the precious fruits brought forth by the sun, and for the precious things put forth by the moon,

And for the chief things of the ancient mountains, and for the precious things of the lasting hills,

And for the precious things of the earth and fullness thereof, and for the good will of him that dwelt in the bush.

I have been talking, of course, about a dualism that manifests itself in several ways: as a cleavage, a radical discontinuity, between Creator and creature, spirit and matter, religion and nature, religion and economy, worship and work, and so on. This dualism, I think, is the most destructive disease that afflicts us. In its best-known, its most dangerous, and perhaps its fundamental version, it is the dualism of body and soul. This is an issue as difficult as it is important, and so to deal with it we should start at the beginning.

The crucial test is probably Genesis 2:7, which gives the process by which Adam was created: "The Lord God formed man of the dust of the ground, and breathed into his nostrils the breath of life: and man became a living soul." My mind, like most people's, has been deeply influenced by dualism, and I can see how dualistic minds deal with this verse. They conclude that the formula for man-making is man = body + soul. But that conclusion cannot be derived, except by violence, from Genesis 2:7, which is not dualistic. The formula given in Genesis 2:7 is not man = body + soul; the formula there is soul = dust + breath. According to this verse, God did not make a body and put a soul into it, like a letter into an envelope. He formed man of dust; then, by breathing His breath into it, He made the dust live. The dust, formed as man and made to live, did not *embody* a soul; it *became* a soul. "Soul" here refers to the whole creature. Humanity is thus presented to us, in Adam, not as a creature of two discrete parts temporarily glued together but as a single mystery.

We can see how easy it is to fall into the dualism of body and soul when talking about the inescapable worldly dualities of good and evil or time and eternity. And we can see how easy it is, when Jesus asks, "For what is a man profited, if he shall gain the whole world, and lose his own soul?" to assume that he is condemning the world and appreciating the disembodied soul. But if we give to "soul" here the sense that it has in Genesis 2:7, we see that he is doing no such thing. He is warning that in pursuit of so-called material possessions, we can lose our understanding of ourselves as "living souls"—that is, as creatures of God, members of the holy community of Creation. We can lose the possibility of the atonement of that membership. For we are free, if we choose, to make a duality of our one living soul by disowning the breath of God that is our fundamental bond with one another and with other creatures.

But we can make the same duality by disowning the dust. The breath of God is only one of the divine gifts that make us living souls; the other is the dust. Most of our modern troubles come from our misunderstanding and misvaluation of this dust. Forgetting that the dust, too, is a creature of the Creator, made by the sending forth of His spirit, we have presumed to decide that the dust is "low." We have presumed to say that we are made of two parts: a body and a soul, the body being "low" because made of dust, and the soul "high." By thus valuing these two supposed-to-be parts, we inevitably throw them into competition with each other, like two corporations. The "spiritual" view, of course, has been that the body, in Yeats's phrase, must be "bruised to pleasure soul." And the "secular" version of the same dualism has been that the body, along with the rest of the "material" world, must give way before the advance of the human mind. The dominant religious view, for a long time, has been that the body is a kind of scrip issued by the Great Company Store in the Sky, which can be cashed in to redeem the soul but is otherwise worthless. And the predictable result has been a human creature able to appreciate or tolerate only the "spiritual" (or mental) part of Creation and full of semiconscious hatred of the "physical" or "natural" part, which it is ready and willing to destroy for "salvation," for profit, for "victory," or for fun. This madness constitutes the norm of modern humanity and of modern Christianity.

But to despise the body or mistreat it for the sake of the "soul" is not just to burn one's house for the insurance, nor is it just self-hatred of the most deep and dangerous sort. It is yet another blasphemy. It is to make nothing—and worse than nothing—of the great Something in which we live and move and have our being.

When we hate and abuse the body and its earthly life and joy for Heaven's sake, what do we expect? That out of this life that we have presumed to despise and this world that we have presumed to destroy, we would somehow salvage a soul capable of

eternal bliss? And what do we expect when with equal and opposite ingratitude, we try to make of the finite body an infinite
reservoir of dispirited and meaningless pleasures?

Times may come, of course, when the life of the body must
be denied or sacrificed, times when the whole world must literally be lost for the sake of one's life as a "living soul." But such
sacrifice, by people who truly respect and revere the life of the
earth and its Creator, does not denounce or degrade the body
but rather exalts it and acknowledges its holiness. Such sacrifice
is a refusal to allow the body to serve what is unworthy of it.

                                    V

If we credit the Bible's description of the relationship between
Creator and Creation, then we cannot deny the spiritual importance of our economic life. Then we must see how religious
issues lead to issues of economy and how issues of economy
lead to issues of art. By "art" I mean all the ways by which
humans make the things they need. If we understand that no
artist—no maker—can work except by reworking the works
of Creation, then we see that by our work we reveal what we
think of the works of God. How we take our lives from this
world, how we work, what work we do, how well we use the
materials we use, and what we do with them after we have used
them—all these are questions of the highest and gravest religious significance. In answering them, we practice, or do not
practice, our religion.

The significance—and ultimately the quality—of the work
we do is determined by our understanding of the story in
which we are taking part.

If we think of ourselves as merely biological creatures, whose
story is determined by genetics or environment or history or
economics or technology, then, however pleasant or painful
the part we play, it cannot matter much. Its significance is that
of mere self-concern. "It is a tale / Told by an idiot, full of
sound and fury, / Signifying nothing," as Macbeth says when
he has "supp'd full with horrors" and is "aweary of the sun."

If we think of ourselves as lofty souls trapped temporarily in
lowly bodies in a dispirited, desperate, unlovable world that we
must despise for Heaven's sake, then what have we done for

this question of significance? If we divide reality into two parts, spiritual and material, and hold (as the Bible does *not* hold) that only the spiritual is good or desirable, then our relation to the material Creation becomes arbitrary, having only the quantitative or mercenary value that we have, in fact and for this reason, assigned to it. Thus, we become the judges and inevitably the destroyers of a world we did not make and that we are bidden to understand as a divine gift. It is impossible to see how good work might be accomplished by people who think that our life in this world either signifies nothing or has only a negative significance.

If, on the other hand, we believe that we are living souls, God's dust and God's breath, acting our parts among other creatures all made of the same dust and breath as ourselves; and if we understand that we are free, within the obvious limits of mortal human life, to do evil or good to ourselves and to the other creatures—then all our acts have a supreme significance. If it is true that we are living souls and morally free, then all of us are artists. All of us are makers, within mortal terms and limits, of our lives, of one another's lives, of things we need and use.

This, Ananda Coomaraswamy wrote, is "the normal view," which "assumes . . . not that the artist is a special kind of man, but that every man who is not a mere idler or parasite is necessarily some special kind of artist." But since even mere idlers and parasites may be said to work inescapably, by proxy or influence, it might be better to say that everybody is an artist—either good or bad, responsible or irresponsible. Any life, by working or not working, by working well or poorly, inescapably changes other lives and so changes the world. This is why our division of the "fine arts" from "craftsmanship," and "craftsmanship" from "labor," is so arbitrary, meaningless, and destructive. As Walter Shewring rightly said, both "the plowman and the potter have a cosmic function." And bad art in any trade dishonors and damages Creation.

If we think of ourselves as living souls, immortal creatures, living in the midst of a Creation that is mostly mysterious, and if we see that everything we make or do cannot help but have an everlasting significance for ourselves, for others, and for the

world, then we see why some religious teachers have understood work as a form of prayer. We see why the old poets invoked the muse. And we know why George Herbert prayed, in
his poem "Mattens":

> Teach me thy love to know;
> That this new light, which now I see,
> May both the work and workman show.

Work connects us both to Creation and to eternity. This is the
reason also for Mother Ann Lee's famous instruction: "Do all
your work as though you had a thousand years to live on earth,
and as you would if you knew you must die tomorrow."

Explaining "the perfection, order, and illumination" of the
artistry of Shaker furniture makers, Coomaraswamy wrote,
"All tradition has seen in the Master Craftsman of the Universe
the exemplar of the human artist or 'maker by art,' and we are
told to be 'perfect, *even as* your Father in heaven is perfect.'"
Searching out the lesson, for us, of the Shakers' humble, impersonal, perfect artistry, which refused the modern divorce of
utility and beauty, he wrote, "Unfortunately, we do not desire
to be such as the Shaker was; we do not propose to 'work as
though we had a thousand years to live, and as though we
were to die tomorrow.' Just as we desire peace but not the
things that make for peace, so we desire art but not the things
that make for art. . . . we have the art that we deserve. If the
sight of it puts us to shame, it is with ourselves that the reformation must begin."

Any genuine effort to "re-form" our arts, our ways of making, must take thought of "the things that make art." We
must see that no art begins in itself; it begins in other arts,
in attitudes and ideas antecedent to any art, in nature, and
in inspiration. If we look at the great artistic traditions, as it
is necessary to do, we will see that they have never been divorced either from religion or from economy. The possibility
of an entirely secular art and of works of art that are spiritless
or ugly or useless is not a possibility that has been among us
for very long. Traditionally, the arts have been ways of making
that have placed a just value on their materials or subjects,
on the uses and the users of the things made by art, and on

the artists themselves. They have, that is, been ways of giving honor to the works of God. The great artistic traditions have had nothing to do with what we call "self-expression." They have not been destructive of privacy or exploitive of private life. Though they have certainly originated things and employed genius, they have no affinity with the modern cults of originality and genius. Coomaraswamy, a good guide as always, makes an indispensable distinction between genius in the modern sense and craftsmanship: "Genius inhabits a world of its own. The master craftsman lives in a world inhabited by other men; he has neighbors." The arts, traditionally, belong to the neighborhood. They are the means by which the neighborhood lives, works, remembers, worships, and enjoys itself.

But most important of all, now, is to see that the artistic traditions understood every art primarily as a skill or craft and ultimately as a service to fellow creatures and to God. An artist's first duty, according to this view, is technical. It is assumed that one will have talents, materials, subjects—perhaps even genius or inspiration or vision. But these are traditionally understood not as personal properties with which one may do as one chooses but as gifts of God or nature that must be honored in use. One does not dare to use these things without the skill to use them well. As Dante said of his own art, "far worse than in vain does he leave the shore . . . who fishes for the truth and has not the art." To use gifts less than well is to dishonor them and their Giver. There is no material or subject in Creation that in using, we are excused from using well; there is no work in which we are excused from being able and responsible artists.

VI

In denying the holiness of the body and of the so-called physical reality of the world—and in denying support to the good economy, the good work, by which alone the Creation can receive due honor—modern Christianity generally has cut itself off from both nature and culture. It has no serious or competent interest in biology or ecology. And it is equally uninterested in the arts by which humankind connects itself to nature. It manifests no awareness of the specifically Christian cultural lineages that connect us to our past. There is, for example, a splendid heritage of Christian poetry in English that most church members live and die without reading or hearing or hearing about. Most sermons are preached without any awareness at all that the making of sermons is an art that has at times been magnificent. Most modern churches look like they were built by robots without reference to the heritage of church architecture or respect for the place; they embody no awareness that work can be worship. Most religious music now attests to the general assumption that religion is no more than a vaguely pious (and vaguely romantic) emotion.

Modern Christianity, then, has become as specialized in its organizations as other modern organizations, wholly concentrated on the industrial shibboleths of "growth," counting its success in numbers, and on the very strange enterprise of "saving" the individual, isolated, and disembodied soul. Having witnessed and abetted the dismemberment of the households, both human and natural, by which we have our being as creatures of God, as living souls, and having made light of the great feast and festival of Creation to which we were bidden as living souls, the modern church presumes to be able to save the soul as an eternal piece of private property. It presumes moreover to save the souls of people in other countries and religious traditions, who are often saner and more religious than we are. And always the emphasis is on the individual soul. Some Christian spokespeople give the impression that the highest Christian bliss would be to get to Heaven and find that you are the only one there—that you were right and all the others wrong. Whatever its twentieth-century dress, modern Christianity as I know it is still at bottom the religion of Miss Watson, intent on a dull and superstitious rigmarole by which supposedly we can avoid going to "the bad place" and instead go to "the good place." One can hardly help sympathizing with Huck Finn when he says, "I made up my mind I wouldn't try for it."

Despite its protests to the contrary, modern Christianity has become willy-nilly the religion of the state and the economic status quo. Because it has been so exclusively dedicated to incanting anemic souls into Heaven, it has been made the tool of much earthly villainy. It has, for the most part, stood silently by while a predatory economy has ravaged the world, destroyed its natural beauty and health, divided and plundered its human communities and households. It has flown the flag and chanted the slogans of empire. It has assumed with the economists that "economic forces" automatically work for good and has assumed with the industrialists and militarists that technology determines history. It has assumed with almost everybody that "progress" is good, that it is good to be modern and up with the times. It has admired Caesar and comforted him in his depredations and defaults. But in its de facto alliance with Caesar, Christianity connives directly in the murder of Creation. For in these days, Caesar is no longer a mere destroyer of armies, cities, and nations. He is a contradicter of the fundamental miracle of life. A part of the normal practice of his power is his willingness to destroy the world. He prays, he says, and churches everywhere compliantly pray with him. But he is praying to a God whose works he is prepared at any moment to destroy. What could be more wicked than that, or more mad?

The religion of the Bible, on the contrary, is a religion of the state and the status quo only in brief moments. In practice, it is a religion for the correction equally of people and of kings. And Christ's life, from the manger to the cross, was an affront to the established powers of his time, just as it is to the established powers of our time. Much is made in churches of the "good news" of the Gospels. Less is said of the Gospels' bad news, which is that Jesus would have been horrified by just about every "Christian" government the world has ever seen. He would be horrified by our government and its works, and it would be horrified by him. Surely no sane and thoughtful person can imagine any government of our time sitting comfortably at the feet of Jesus while he is saying, "Love your enemies, bless them that curse you, do good to them that hate you, and pray for them that despitefully use you and persecute you." In fact, we know that one of the businesses of governments, "Christian" or not, has been to reenact the crucifixion. It has happened again and again and again. In *A Time for Trumpets*, his history of the Battle of the Bulge, Charles B. MacDonald tells how the SS Colonel Joachim Peiper was forced to withdraw from a bombarded château near the town of La Gleize, leaving behind a number of severely wounded soldiers of both armies. "Also left behind," MacDonald wrote, "on a whitewashed wall of one of the rooms in the basement was a charcoal drawing of Christ, thorns on his head, tears on his cheeks—whether drawn by a German or an American nobody would ever know." This is not an image that belongs to history but rather one that judges it.

Sex, Economy, Freedom, and Community

"It all turns on affection now," said Margaret. "Affection.
Don't you see?"
—E. M. Forster, *Howards End*

THE SEXUAL HARASSMENT phase of the Clarence Thomas
hearing was handled by the news media as if it were anomalous and surprising. In fact, it was only an unusually spectacular revelation of the destructiveness of a process that has been well established and well respected for at least two hundred years—the process, that is, of community disintegration. This process has been well established and well respected for so long, of course, because it has been immensely profitable to those in a position to profit. The surprise and dismay occasioned by the Thomas hearing were not caused by the gossip involved (for that, the media had prepared us very well) but by the inescapable message that this process of disintegration, so little acknowledged by politicians and commentators, can be severely and perhaps illimitably destructive.

In the government-sponsored quarrel between Clarence Thomas and Anita Hill, public life collided with private life in a way that could not have been resolved and that could only have been damaging. The event was depressing and fearful both because of its violations of due process and justice and because it was an attempt to deal publicly with a problem for which there is no public solution. It embroiled the United States Senate in the impossible task of adjudicating alleged offenses that had occurred in private, of which there were no witnesses and no evidence. If the hearing was a "lynching," as Clarence Thomas said it was, that was because it dealt a public punishment to an unconvicted and unindicted victim. But it was a peculiar lynching, all the same, for it dealt the punishment equally to the accuser. It was not a hearing, much less a trial; it was a storytelling contest that was not winnable by either participant.

Its only result was damage to all participants and to the nation. Public life obviously cannot be conducted in that way, and neither can private life. It was a public procedure that degenerated into a private quarrel. It was a private quarrel that became a public catastrophe.

Sexual harassment, like most sexual conduct, is extremely dangerous as a public issue. A public issue, properly speaking, can only be an issue about which the public can confidently know. Because most sexual conduct is private, occurring only between two people, there are typically no witnesses. Apart from the possibility of a confession, the public can know about it only as a probably unjudgeable contest of stories. (In those rare instances when a sexual offense occurs before reliable witnesses, then, of course, it is a legitimate public issue.)

Does this mean that sexual conduct is *only* private in its interest and meaning? It certainly does not. For if there is no satisfactory way to deal publicly with sexual issues, there is also no satisfactory way to deal with them in mere privacy. To make sense of sexual issues or of sex itself, a third term, a third entity, has to intervene between public and private. For sex is not and cannot be any individual's "own business," nor is it merely the private concern of any couple. Sex, like any other necessary, precious, and volatile power that is commonly held, is everybody's business. A way must be found to entitle everybody's legitimate interest in it without either violating its essential privacy or allowing its unrestrained energies to reduce necessary public procedures to the level of a private quarrel. For sexual problems and potentialities that have a more-than-private interest, what is needed are common or shared forms and solutions that are not, in the usual sense, public.

The indispensable form that can intervene between public and private interests is that of community. The concerns of public and private, republic and citizen, necessary as they are, are not adequate for the shaping of human life. Community alone, as principle and as fact, can raise the standards of local health (ecological, economic, social, and spiritual) without which the other two interests will destroy one another.

By community, I mean the commonwealth and common interests, commonly understood, of people living together in a place and wishing to continue to do so. To put it another way, community is a locally understood interdependence of local people, local culture, local economy, and local nature. (Community, of course, is an idea that can extend itself beyond the local, but it only does so metaphorically. The idea of a national or global community is meaningless apart from the realization of local communities.) Lacking the interest of or in such a community, private life becomes merely a sort of reserve in which individuals defend their "right" to act as they please and attempt to limit or destroy the "rights" of other individuals to act as they please.

A community identifies itself by an understood mutuality of interests. But it lives and acts by the common virtues of trust, goodwill, forbearance, self-restraint, compassion, and forgiveness. If it hopes to continue long as a community, it will wish to—and will have to—encourage respect for all its members, human and natural. It will encourage respect for all stations and occupations. Such a community has the power—not invariably but as a rule—to enforce decency without litigation. It has the power, that is, to influence behavior. And it exercises this power not by coercion or violence but by teaching the young and by preserving stories and songs that tell (among other things) what works and what does not work in a given place.

Such a community is (among other things) a set of arrangements between men and women. These arrangements include marriage, family structure, divisions of work and authority, and responsibility for the instruction of children and young people. These arrangements exist, in part, to reduce the volatility and the danger of sex—to preserve its energy, its beauty, and its pleasure; to preserve and clarify its power to join not just husband and wife to one another but parents to children, families to the community, the community to nature; to ensure, so far as possible, that the inheritors of sexuality, as they come of age, will be worthy of it.

But the life of a community is more vulnerable than public life. A community cannot be made or preserved apart from the loyalty and affection of its members and the respect and goodwill of the people outside it. And for a long time, these conditions have not been met. As the technological, economic, and political means of exploitation have expanded, communities have been more and more victimized by opportunists outside themselves. And as the salesmen, saleswomen, advertisers, and propagandists of the industrial economy have become more ubiquitous and more adept at seduction, communities have lost the loyalty and affection of their members. The community, wherever you look, is being destroyed by the desires and ambitions of both private and public life, which for want of the intervention of community interests are also destroying one another. Community life is by definition a life of cooperation and responsibility. Private life and public life, without the disciplines of community interest, necessarily gravitate toward competition and exploitation. As private life casts off all community restraints in the interest of economic exploitation or ambition or self-realization or whatever, the communal supports of public life also and by the same stroke are undercut, and public life becomes simply the arena of unrestrained private ambition and greed.

As our communities have disintegrated from external predation and internal disaffection, we have changed from a society whose ideal of justice was trust and fairness among people who knew each other into a society whose ideal of justice is public litigation, breeding distrust even among people who know each other.

Once it has shrugged off the interests and claims of the community, the public language of sexuality comes directly under the influence of private lust, ambition, and greed and becomes inadequate to deal with the real issues and problems of sexuality. The public dialogue degenerates into a stupefying and useless contest between so-called liberation and so-called morality. The real issues and problems, as they are experienced and suffered in people's lives, cannot be talked about. The public language can deal, however awkwardly and perhaps uselessly, with pornography, sexual hygiene, contraception, sexual harassment, rape, and so on. But it cannot talk about respect, responsibility, sexual discipline, fidelity, or the practice of love. "Sexual education," carried on in this public language, is and can only be a dispirited description of the working of a sort of anatomical machinery—and this is a sexuality that is neither erotic nor social nor sacramental but rather a cold-blooded, abstract procedure that is finally not even imaginable.

The conventional public opposition of "liberal" and "conservative" is, here as elsewhere, perfectly useless. The "conservatives" promote the family as a sort of public icon, but they will not promote the economic integrity of the household or the community, which are the mainstays of family life. Under the sponsorship of "conservative" presidencies, the economy of the modern household, which once required the father to work away from home—a development that was bad enough —now requires the mother to work away from home, as well. And this development has the wholehearted endorsement of "liberals," who see the mother thus forced to spend her days away from her home and children as "liberated"—though nobody has yet seen the fathers thus forced away as "liberated." Some feminists are thus in the curious position of opposing the mistreatment of women and yet advocating their participation in an economy in which everything is mistreated.

The "conservatives" more or less attack homosexuality, abortion, and pornography, and the "liberals" more or less defend them. Neither party will oppose sexual promiscuity. The "liberals" will not oppose promiscuity because they do not wish to appear intolerant of "individual liberty." The "conservatives" will not oppose promiscuity because sexual discipline would reduce the profits of corporations, which in their advertisements and entertainments encourage sexual self-indulgence as a way of selling merchandise.

The public discussion of sexual issues has thus degenerated into a poor attempt to equivocate between private lusts and public emergencies. Nowhere in public life (that is, in the public life that counts: the discussions of political and corporate leaders) is there an attempt to respond to community needs in the language of community interest.

And although we seem more and more inclined to look on education, even as it teaches less and is more overcome by violence, as the solution to all our problems (thus delaying the solution for a generation), there is really not much use in looking to education for the help we need. For education has become increasingly useless as it has become increasingly public. Real education is determined by community needs, not by public tests. Nor is community interest or community need going to receive much help from television and the other public media. Television is the greatest disrespecter and exploiter of sexuality that the world has ever seen; even if the network executives decide to promote "safe sex" and the use of condoms, they will not cease to pimp for the exceedingly profitable "sexual revolution." It is, in fact, the nature of the electronic media to blur and finally destroy all distinctions between public and community. Television has greatly accelerated the process, begun long ago, by which many communities have been atomized and congealed into one public. Nor is government a likely source of help. As political leaders have squirmed free of the claims and responsibilities of community life, public life has become their private preserve. The public political voice has become increasingly the voice of a conscious and self-serving duplicity: it is now, for instance, merely typical that a political leader can speak of "the preciousness of all life" while armed for the annihilation of all life. And the right of privacy, without the intervening claims and responsibilities of community life, has moved from the individual to the government and assumed the name of "official secrecy." Whose liberation is that?

In fact, there is no one to speak for the community interest except those people who wish to adhere to community principles. The community, in other words, must speak in its own interest. It must learn to defend itself. And in its self-defense, it may use the many powerful arguments provided for it by the failures of the private and public aims that have so nearly destroyed it.

The defenders of community should point out, for example, that for the joining of men and women there need to be many forms that only a community can provide. If you destroy the ideal of the "gentle man" and remove from men all expectations of courtesy and consideration toward women and children, you have prepared the way for an epidemic of rape and abuse. If you depreciate the sanctity and solemnity of marriage, not just as a bond between two people but as a bond between those two people and their forebears, their children, and their neighbors, then you have prepared the way for an epidemic of divorce, child neglect, community ruin, and loneliness. If you destroy the economies of household and community, then you destroy the bonds of mutual usefulness and practical dependence without which the other bonds will not hold.

If these and all other community-made arrangements between men and women are removed, if the only arrangements left between them are those of sex and sexual politics, instinct and polity without culture, then sex and politics are headed not only toward many kinds of private and public suffering but toward the destruction of justice, as in the confrontation of Clarence Thomas and Anita Hill.

II

But to deal with community disintegration as merely a matter of sexual disorder, however destructive that disorder may be, is misleading. The problem is far more complex than I have been able to suggest so far. There is much more to be said on the issue of sex, and I will return to it, but for the sake of both truth and clarity I must first examine other issues that are related and, in some respects, analogous to the issue of sex.

It is certain, as I have already said, that communities are destroyed both from within and from without: by internal disaffection and external exploitation. It is certain, too, that there have always been people who have become estranged from their communities for reasons of honest difference or disagreement. But it can be argued that community disintegration typically is begun by an aggression of some sort from the outside and that in modern times the typical aggression has been economic. The destruction of the community begins when its economy is made—not *dependent* (for no community has ever been entirely independent)—but *subject* to a larger external economy. As an example, we could probably do no better than the following account of the destruction of the local wool economy of the parish of Hawkshead in the Lake District of England:

> The . . . reason for the decline of the customary tenant must be sought in the introduction of machinery towards the end of the eighteenth century, which extinguished not only the local spinning and weaving, but was also the deathblow of the local market. Before this time, idleness at a fellside farm was unknown, for clothes and even linen were home-made, and all spare time was occupied by the youths in carding wool, while the girls spun the "garn" with distaff and wheel. . . . The sale of the yarn to the local weavers, and at the local market, brought important profits to the dalesman, so that it not only kept all hands busy, but put money

into his pocket. But the introduction of machinery for looms and for spinning, and consequent outside demand for fleeces instead of yarn and woven material, threw idle not only half of the family, but the local hand-weavers, who were no doubt younger sons of the same stock. Thus idleness took the place of thrift and industry among a naturally industrious class, for the sons and daughters of the 'statesmen, often too proud to go out to service, became useless encumbrances on the estates. Then came the improvement in agricultural methods [that is, technological innovations], which the 'statesman could not afford to keep abreast of . . . What else could take place but that which did? The estates became mortgaged and were sold, and the rich manufacturers, whose villas are on the margin of Windermere, have often enough among their servants the actual descendants of the old 'statesmen, whose manufactures they first usurped and whose estates they afterwards absorbed.

This paragraph sets forth the pattern of industrial exploitation of a locality and a local economy, a pattern that has prevailed for two hundred years. The industrialization of the eastern Kentucky coal fields early in the present century, though more violent, followed this pattern exactly. A decentralized, fairly independent local economy was absorbed and destroyed by an aggressive, monetarily powerful outside economy. And like the displaced farmers, spinners, and weavers of Hawkshead, the once-independent mountaineers of eastern Kentucky became the wage-earning servants of those who had dispossessed their parents, sometimes digging the very coal that their families had once owned and had sold for as little per acre as the pittance the companies paid per day. By now, there is hardly a rural neighborhood or town in the United States that has not suffered some version of this process.

The same process is destroying local economies and cultures all over the world. Of Ladakh, for example, Helena Norberg-Hodge writes:

In the traditional culture, villagers provided for their basic needs without money. They had developed skills that enabled them to grow barley at 12,000 feet and to manage yaks and other animals at even higher elevations.

> People knew how to build houses with their own hands from the materials of the immediate surroundings. The only thing they actually needed from outside the region was salt, for which they traded. They used money in only a limited way, mainly for luxuries.
>
> Now, suddenly, as part of the international money economy, Ladakhis find themselves ever more dependent —even for vital needs—on a system that is controlled by faraway forces. They are vulnerable to decisions made by people who do not even know that Ladakh exists. . . . For two thousand years in Ladakh, a kilo of barley has been a kilo of barley, but now you cannot be sure of its value.

This, I think, speaks for itself: if you are dependent on people who do not know you, who control the value of your necessities, you are not free, and you are not safe.

The industrial revolution has thus made universal the colonialist principle that has proved to be ruinous beyond measure: the assumption that it is permissible to ruin one place or culture for the sake of another. Thus justified or excused, the industrial economy grows in power and thrives on its damages to local economies, communities, and places. Meanwhile, politicians and bureaucrats measure the economic prosperity of their nations according to the burgeoning wealth of the industrial interests, not according to the success or failure of small local economies or the reduction and often hopeless servitude of local people. The self-congratulation of the industrialists and their political minions has continued unabated to this day. And yet it is a fact that the industrialists of Hawkshead, like all those elsewhere and since, have lived off the public, just as surely as do the despised clients of "welfare." They have lived off a public of industrially destroyed communities, and they have not compensated for this destruction by their ostentatious contributions to the art, culture, and education of the professional class, or by their "charities" to the poor. Nor is this state of things ameliorated by efforts to enable a local population to "participate" in the global economy, by "education" or any other means. It is true that local individuals, depending on their capital, intelligence, cunning, or influence, may be able to "participate" to their apparent advantage, but local communities and places can "participate" only as victims. The global economy does not exist to help the communities and localities of the globe. It exists to siphon the wealth of those communities and places into a few bank accounts. To this economy, democracy and the values of the religious traditions mean absolutely nothing. And those who wish to help communities to survive had better understand that a merely political freedom means little within a totalitarian economy. Ms. Norberg-Hodge says, of the relatively new influence of the global economy on Ladakh, that

> Increasingly, people are locked into an economic system that pumps resources out of the periphery into the center—from the nonindustrialized to the industrialized parts of the world, from the countryside to the city, from the poor to the rich. Often, these resources end up back where they came from as commercial products . . . at prices that the poor can no longer afford.

This is an apt description not just of what is happening in Ladakh but of what is happening in my own rural county and in every other rural county in the United States.

The situation in the wool economy of Hawkshead at the end of the eighteenth century was the same as that which, a little later, caused the brief uprising of those workers in England who were called Luddites. These were people who dared to assert that there were needs and values that justly took precedence over industrialization; they were people who rejected the determinism of technological innovation and economic exploitation. In them, the community attempted to speak for itself and defend itself. It happened that Lord Byron's maiden speech in the House of Lords, on February 27, 1812, dealt with the uprising of the Luddites, and this, in part, is what he said:

> By the adoption of one species of [weaving] frame in particular, one man performed the work of many, and the superfluous laborers were thrown out of employment. Yet it is to be observed, that the work thus executed was inferior in quality; not marketable at home, and merely hurried over with a view to exportation. . . .

> The rejected workmen . . . conceived themselves to
> be sacrificed to improvements in mechanism. In the
> foolishness of their hearts they imagined that the main-
> tenance and well-doing of the industrious poor were
> objects of greater importance than the enrichment of a
> few individuals by any improvement, in the implements
> of trade, which threw the workmen out of employment,
> and rendered the laborer unworthy of his hire.

The Luddites did, in fact, revolt not only against their own
economic oppression but also against the poor quality of the
machine work that had replaced them. And though they destroyed machinery, they "abstained from bloodshed or violence against living beings, until in 1812 a band of them was
shot down by soldiers." Their movement was suppressed by
"severe repressive legislation" and by "many hangings and
transportations."

The Luddites thus asserted the precedence of community
needs over technological innovation and monetary profit, and
they were dealt with in a way that seems merely inevitable in
the light of subsequent history. In the years since, the only
group that I know of that has successfully, so far, made the
community the standard of technological innovation has been
the Amish. The Amish have differed from the Luddites in that
they have not destroyed but merely declined to use the technologies that they perceive as threatening to their community.
And this has been possible because the Amish are an agrarian
people. The Luddites could not have refused the machinery
that they destroyed; the machinery had refused them.

The victory of industrialism over Luddism was thus overwhelming and unconditional; it was undoubtedly the most
complete, significant, and lasting victory of modern times. And
so one must wonder at the intensity with which any suggestion
of Luddism still is feared and hated. To this day, if you say
you would be willing to forbid, restrict, or reduce the use of
technological devices in order to protect the community—or
to protect the good health of nature on which the community
depends—you will be called a Luddite, and it will not be a
compliment. To say that the community is more important
than machines is certainly Christian and certainly democratic,
but it is also Luddism and therefore not to be tolerated.

Technological determinism, then, has triumphed. By now the rhetoric of technological determinism has become thoroughly mixed with the rhetoric of national mysticism so that a political leader may confidently say that it is "our destiny" to go to the moon or Mars or wherever we may go for the profit of those who will provide the transportation. Because this "destiny" has become the all-shaping superstition of our time and is therefore never debated in public, we have difficulty seeing that the triumph of technological determinism is the defeat of community. But it is a fact that in both private conversation and public dialogue, the community has neither status nor standing from which to plead in its own defense. There is no denying, of course, that "community" ranks with "family," "our land," and "our beloved country" as an icon of the public vocabulary; everybody is for it, and this means nothing. If individuals or groups have the temerity to oppose an actual item on the agenda of technological process because it will damage a community, the powers that be will think them guilty of Luddism, sedition, and perhaps insanity. Local and community organizations, of course, do at times prevail over the would-be "developers," but the status of these victories remains tentative. The powers that be, and the media as well, treat such victories as anomalies, never as the work of a legitimate "side" in the public dialogue.

Some time ago, for example, my friends in the Community Farm Alliance asked me to write a newspaper piece opposing the Bush administration's efforts to revise the General Agreement on Tariffs and Trade; the proposed revisions would destroy all barriers and restrictions to international trade in food, thus threatening the already precarious survival of farm communities in the industrial countries and also the local availability of food in Third World countries. I wrote the article and sent it to the editor of the op-ed page of the *New York Times*, who had invited me to make such submissions—and who rejected my article on the ground that the subject was "not sexy." I conclude that if the opponents of the GATT revisions should win, their victory will have an effect limited to the issue—that is, such a victory would not cause their point of view to be acknowledged, much less represented, among the powers that be, and the public standing of community interests would be no higher than before.

The triumph of the industrial economy is the fall of community. But the fall of community reveals how precious and how necessary community is. For when community falls, so must fall all the things that only community life can engender and protect: the care of the old, the care and education of children, family life, neighborly work, the handing down of memory, the care of the earth, respect for nature and the lives of wild creatures. All of these things have been damaged by the rule of industrialism, but of all the damaged things probably the most precious and the most damaged is sexual love. For sexual love is the heart of community life. Sexual love is the force that in our bodily life connects us most intimately to the Creation, to the fertility of the world, to farming and the care of animals. It brings us into the dance that holds the community together and joins it to its place.

In dealing with community, as in dealing with everything else, the industrial economy goes for the nucleus. It does this because it wants the power to cause fundamental change. To make sex the preferred bait of commerce may seem merely the obvious thing to do, once greed is granted its now conventional priority as a motive. But this could happen only after a probably instinctive sense of the sanctity and dignity of the body—the sense of its having been "fearfully and wonderfully made" had been destroyed. Once this ancient reticence had been broken down, then the come-on of the pimp could be instituted as the universal spiel of the marketplace; everything could be sold on the promise of instant, innocent sexual gratification, "no strings attached." Sexual energy cannot be made publicly available for commercial use—that is, prostituted—without destroying all of its communal or cultural forms: forms of courtship, marriage, family life, household economy, and so on. The devaluation of sexuality, like the devaluation of a monetary currency, destroys its correspondence to other values.

In the wake of the Thomas-Hill catastrophe in Washington, the *New York Times Sunday Magazine* contained a skin lotion advertisement that displayed a photograph of the naked torso of a woman. From a feminist point of view, this headless and footless body represents the male chauvinist's sexual ideal: a woman who cannot think and cannot escape. From a point of view somewhat more comprehensive—the point of view of community—it represents also the commercial ideal of the industrial economy: the completely seducible consumer, unable either to judge or to resist.

The headlessness of this lotionable lady suggests also another telling indication of the devaluation of sexual love in modern times—that is, the gravitation of attention from the countenance, especially the eyes, to the specifically sexual anatomy. The difference, of course, is that the countenance is both physical and spiritual. There is much testimony to this in the poetic tradition and elsewhere. Looking into one another's eyes, lovers recognize their encounter as a meeting not merely of two bodies but of two living souls. In one another's eyes, moreover, they see themselves reflected not narcissistically but as singular beings, separate and small, far inferior to the creature that they together make.

In this meeting of eyes, there is an acknowledgment that love is more than sex:

> 'This Ecstasy doth unperplex,'
>   We said, 'and tell us what we love;
> We see by this it was not sex;
>   We see we saw not what did move.'

These lines are from John Donne's poem "The Ecstasy," in which the lovers have been joined by the "double string" of their mutual gaze. This is not a disembodied love. Far from it. For love is finally seen in the poem as "that subtle knot, which makes us man" by joining body and soul together, just as it joins the two lovers. Sexual love is thus understood as both fact and mystery, physical motion and spiritual motive. That this complex love should be reduced simply to sex has always seemed a fearful thing to the poets. As late as our own century, for example, we have these lines from Wallace Stevens:

> If sex were all, then every trembling hand
> Could make us squeak, like dolls, the wished-for words.

The fear of the reduction of love to sex is obviously subject to distortion. At worst, this fear has caused a kind of joyless prudery. But at best, it has been a just fear, implying both an appropriate awe and a reluctance that humans should become the puppets of a single instinct. Such a puppetry is possible, and its sign, in modern terms, is the body as product, made delectably consumable by the application of lotions and other commercial liquids. But there is a higher, juster love of which the sign is the meeting of the eyes.

In the poetry that I know, the most graceful and the richest testimony to the power of the eyes is in act three, scene two, of *The Merchant of Venice*, where Portia says to Bassanio:

> Beshrew your eyes,
> They have o'erlook'd me and divided me,
> One half of me is yours, the other half yours,—
> Mine own I would say: but if mine then yours,
> And so all yours.

Portia is no headless woman. She is the brightest, most articulate character in the play. Witty and charming as these lines are, they are not frivolous. They attest to the sexual and the spiritual power of a look, which has just begun an endless conversation between two living souls. This speech of hers is as powerful as it is because she knows exactly what she is doing. What she is doing is giving herself away. She has entered into a situation in which she must find her life by losing it. She is glad, and she is frightened. She is speaking joyfully and fearfully of the self's suddenly irresistible wish to be given away. And this is an unconditional giving, on which, she knows, time and mortality will impose their inescapable conditions; she will have remembered the marriage ceremony with its warnings of difficulty, poverty, sickness, and death. There is nothing "safe" about this. This love has no place to happen except in this world, where it cannot be made safe. And this scene finally becomes a sort of wedding, in which Portia says to Bassanio, "Myself, and what is mine, to you and yours / Is now converted. . . . I give them with this ring."

There is no "sexism" or "double standard" in this exchange of looks and what it leads to. The look of which Portia speaks could have had such power only because it answered a look of hers; she has, in the same moment, "o'erlook'd" Bassanio. She is not speaking as a woman submitting to the power of a man, much less to seduction. She is speaking as one of a pair who are submitting to the redemptive power of love. Responding to Portia's pledge of herself to him, Bassanio cannot say much;

he is not so eloquent as she, and his mind has become "a wild of nothing, save of joy." He says only:

> when this ring
> Parts from this finger, then parts life from hence,—
> O then be bold to say Bassanio's dead!

But that is enough. He is saying, as his actual wedding will require him to say, that he will be true to Portia until death. But he is also saying, as she has said, that he is dead to his old life and can now live only in the new life of their love. He might have said, with Dante, "Of a surety I have now set my feet on that point of life, beyond the which he must not pass who would return."

This is a play about the precedence of affection and fidelity over profit, and so it is a play about the order of community. Lovers must not, like usurers, live for themselves alone. They must finally turn from their gaze at one another back toward the community. If they had only themselves to consider, lovers would not need to marry, but they must think of others and of other things. They say their vows to the community as much as to one another, and the community gathers around them to hear and to wish them well, on their behalf and on its own. It gathers around them because it understands how necessary, how joyful, and how fearful this joining is. These lovers, pledging themselves to one another "until death," are giving themselves away, and they are joined by this as no law or contract could ever join them. Lovers, then, "die" into their union with one another as a soul "dies" into its union with God. And so here, at the very heart of community life, we find not something to sell as in the public market but this momentous giving. If the community cannot protect this giving, it can protect nothing—and our time is proving that this is so.

We thus can see that there are two kinds of human economy. There is the kind of economy that exists to protect the "right" of profit, as does our present public economy; this sort of economy will inevitably gravitate toward protection of the "rights" of those who profit most. Our present public economy is really a political system that safeguards the private exploitation of the public wealth and health. The other kind of economy exists for the protection of gifts, beginning with the "giving in marriage," and this is the economy of community, which now has been nearly destroyed by the public economy.

There are two kinds of sexuality that correspond to the two kinds of economy. The sexuality of community life, whatever its inevitable vagaries, is centered on marriage, which joins two living souls as closely as, in this world, they can be joined. This joining of two who know, love, and trust one another brings them in the same breath into the freedom of sexual consent and into the fullest earthly realization of the image of God. From their joining, other living souls come into being, and with them great responsibilities that are unending, fearful, and joyful. The marriage of two lovers joins them to one another, to forebears, to descendants, to the community, to Heaven and earth. It is the fundamental connection without which nothing holds, and trust is its necessity.

Our present sexual conduct, on the other hand, having "liberated" itself from the several trusts of community life, is public, like our present economy. It has forsaken trust, for it rests on the easy giving and breaking of promises. And having forsaken trust, it has predictably become political. In private life, as in public, we are attempting to correct bad character and low motives by law and by litigation. "Losing kindness," as Lao-tzu said, "they turn to justness." The superstition of the anger of our current sexual politics, as of other kinds of anger, is that somewhere along the trajectory of any quarrel a tribunal will be reached that will hear all complaints and find for the plaintiff; the verdict will be that the defendant is entirely wrong, the plaintiff entirely right and entirely righteous. This, of course, is not going to happen. And because such "justice" cannot happen, litigation only prolongs itself. The difficulty is that marriage, family life, friendship, neighborhood, and other personal connections do not depend exclusively or even primarily on justice—though, of course, they all must try for it. They depend also on trust, patience, respect, mutual help, forgiveness—in other words, the *practice* of love, as opposed to the mere *feeling* of love.

As soon as the parties to a marriage or a friendship begin to require strict justice of each other, then that marriage or friendship begins to be destroyed, for there is no way to ad-judicate the competing claims of a personal quarrel. And so these relationships do not dissolve into litigation, really; they dissolve into a feud, an endless exchange of accusations and retributions. If the two parties have not the grace to forgive the inevitable offenses of close connection, the next best thing is separation and silence. But why should separation have come to be the virtually conventional outcome of close relationships in our society? The proper question, perhaps, is not why we have so much divorce, but why we are so unforgiving. The answer, perhaps, is that, though we still recognize the feeling of love, we have forgotten how to practice love when we don't feel it.

Because of our determination to separate sex from the practice of love in marriage and in family and community life, our public sexual morality is confused, sentimental, bitter, complexly destructive, and hypocritical. It begins with the idea of "sexual liberation": whatever people desire is "natural" and all right, men and women are not different but merely equal, and all desires are equal. If a man wants to sit down while a pregnant woman is standing or walk through a heavy door and let it slam in a woman's face, that is all right. Divorce on an epidemic scale is all right; child abandonment by one parent or another is all right; it is regrettable but still pretty much all right if a divorced parent neglects or refuses to pay child support; promiscuity is all right; adultery is all right. Promiscuity among teenagers is pretty much all right, for "that's the way it is"; abortion as birth control is all right; the prostitution of sex in advertisements and public entertainment is all right. But then, far down this road of freedom, we decide that a few lines ought to be drawn. Child molestation, we wish to say, is not all right, nor is sexual violence, nor is sexual harassment, nor is pregnancy among unmarried teenagers. We are also against venereal diseases, the diseases of promiscuity, though we tend to think that they are the government's responsibility, not ours.

In this cult of liberated sexuality, "free" of courtesy, ceremony, responsibility, and restraint, dependent on litigation and expert advice, there is much that is human, sad to say, but there is no sense or sanity. Trying to draw the line where we are trying to draw it, between carelessness and brutality, is like insisting that falling is flying—until you hit the ground—and then trying to outlaw hitting the ground. The pretentious, fantastical, and solemn idiocy of the public sexual code could not be better exemplified than by the now-ubiquitous phrase "sexual partner," which denies all that is implied by the names of "husband" or "wife" or even "lover." It denies anyone's responsibility for the consequences of sex. With one's "sexual partner," it is now understood, one must practice "safe sex" —that is, one must protect oneself, not one's partner or the children that may come of the "partnership."

But the worst hypocrisy of all is in the failure of the sexual libertarians to come to the defense of sexually liberated politicians. The public applies strenuously to public officials a sexual morality that it no longer applies to anyone privately and that it does not apply to other liberated public figures, such as movie stars, artists, athletes, and business tycoons. The prurient squeamishness with which the public and the public media poke into the lives of politicians is surely not an expectable result of liberation. But this paradox is not the only one. According to its claims, sexual liberation ought logically to have brought in a time of "naturalness," ease, and candor between men and women. It has, on the contrary, filled the country with sexual self-consciousness, uncertainty, and fear. Women, though they may dress as if the sexual millennium had arrived, hurry along our city streets and public corridors with their eyes averted, like hunted animals. "Eye contact," once the very signature of our humanity, has become a danger. The meeting ground between men and women, which ought to be safeguarded by trust, has become a place of suspicion, competition, and violence. One no longer goes there asking how instinct may be ramified in affection and loyalty; now one asks how instinct may be indulged with the least risk to personal safety.

Seeking to "free" sexual love from its old communal restraints, we have "freed" it also from its meaning, its responsibility, and its exaltation. And we have made it more dangerous. "Sexual liberation" is as much a fraud and as great a failure as the "peaceful atom." We are now living in a sexual atmosphere so polluted and embittered that women must look on virtually any man as a potential assailant, and a man must look on virtually any woman as a potential accuser. The idea that this situation can be corrected by the courts and the police only compounds the disorder and the danger. And in the midst of this acid rainfall of predation and recrimination, we presume to teach our young people that sex can be made "safe"—by the use, inevitably, of purchased drugs and devices. What a lie! Sex was never safe, and it is less safe now than it has ever been.

What we are actually teaching the young is an illusion of thoughtless freedom and purchasable safety, which encourages them to tamper prematurely, disrespectfully, and dangerously with a great power. Just as the public economy encourages people to spend money and waste the world, so the public sexual code encourages people to be spendthrifts and squanderers of sex. The basis of true community and household economy, on the other hand, is thrift. The basis of community sexuality is respect for everything that is involved—and respect, here as everywhere, implies discipline. By their common principles of extravagance and undisciplined freedom, our public economy and our public sexuality are exploiting and spending moral capital built up by centuries of community life—exactly as industrial agriculture has been exploiting and spending the natural capital built up over thousands of years in the soil.

In sex, as in other things, we have liberated fantasy but killed imagination, and so have sealed ourselves in selfishness and loneliness. Fantasy is of the solitary self, and it cannot lead us away from ourselves. It is by imagination that we cross over the differences between ourselves and other beings and thus learn compassion, forbearance, mercy, forgiveness, sympathy, and love—the virtues without which neither we nor the world can live.

Starting with economic brutality, we have arrived at sexual brutality. Those who affirm the one and deplore the other will have to explain how we might logically have arrived anywhere else. Sexual lovemaking between humans is not and cannot be the thoughtless, instinctual coupling of animals; it is not "recreation"; it is not "safe." It is the strongest prompting and the greatest joy that young people are likely to experience. Because it is so powerful, it is risky, not just because of the famous dangers of venereal disease and "unwanted pregnancy" but also because it involves and requires a giving away of the self that if not honored and reciprocated, inevitably reduces dignity and self-respect. The invitation to give oneself away is not, except for the extremely ignorant or the extremely foolish, an easy one to accept.

Perhaps the current revulsion against sexual harassment may be the beginning of a renewal of sexual responsibility and self-respect. It must, at any rate, be the beginning of a repudiation of the idea that sex among us is merely natural. If men and women are merely animals, it is hard to see how sexual harassment could have become an issue, for such harassment is no more than the instinctive procedure of male animals, who openly harass females, usually by unabashed physical display and contact; it is their way of asking who is and who is not in estrus. Women would not think such behavior offensive if we had not, for thousands of years, understood ourselves as specifically human beings—creatures who, if in some ways animal-like, are in other ways God-like. In asking men to feel shame and to restrain themselves—which one would not ask of an animal—women are implicitly asking to be treated as human beings in that full sense, as living souls made in the image of God. But any humans who wish to be treated and to treat others according to that definition must understand that this is not a kindness that can be conferred by a public economy or by a public government or by a public people. It can only be conferred on its members by a community.

III

Much of the modern assault on community life has been conducted within the justification and protection of the idea of freedom. Thus, it is necessary to try to see how the themes of freedom and community have intersected.

The idea of freedom, as Americans understand it, owes its existence to the inevitability that people will disagree. It is a way of guaranteeing to individuals and to political bodies the right to be different from one another. A specifically American freedom began with our wish to assert our differences from England, and its principles were then worked out in the effort to deal with differences among the states. The result is the Bill of Rights, of which the cornerstone is the freedom of speech. This freedom is not only the basic guarantee of political liberty but it also obligates public officials and private citizens alike to acknowledge the inherent dignity and worth of individual people. It exists only as an absolute; if it can be infringed at all, then probably it can be destroyed entirely.

But if it is an absolute, it is a peculiar and troubling one. It is not an absolute in the sense that a law of nature is. It is not absolute even as the moral law is. One person alone can uphold the moral law, but one person alone cannot uphold the freedom of speech. The freedom of speech is a public absolute, and it can remain absolute only so long as a sufficient segment of the public believes that it is and consents to uphold it. It is an absolute that can be destroyed by public opinion. This is where the danger lies. If this freedom is abused and if a sufficient segment of the public becomes sufficiently resentful of the abuses, then the freedom will be revoked. It is a freedom, therefore, that depends directly on responsibility. And so the First Amendment alone is not a sufficient guarantee of the freedom of speech.

As we now speak of it, freedom is almost always understood as a public idea having to do with the liberties of individuals. The public dialogue about freedom almost always has to do with the efforts of one group or another to wrest these individual liberties from the government or to protect them from another group. In this situation, it is inevitable that freedom will be understood as an issue of power. This is perhaps as necessary as it is unavoidable. But power is not the only issue related to freedom.

From another point of view, not necessarily incompatible, freedom has long been understood as the consequence of knowing the truth. When Jesus said to his followers, "Ye shall know the truth, and the truth shall make you free," he was not talking primarily about politics, but the political applicability of the statement has been obvious for a long time, especially to advocates of democracy. According to this line of thought, freedom of speech is necessary to political health and sanity because it permits speech—the public dialogue—to correct itself. Thomas Jefferson had this in mind when he said in his first inaugural address, "If there be any among us who would wish to dissolve this Union or to change its republican form, let them stand undisturbed as monuments of the safety with which error of opinion may be tolerated where reason is left free to combat it." The often-cited "freedom to be wrong" is thus a valid freedom, but it is a poor thing by itself; its validity comes from the recognition that error is real, identifiable as such, dangerous to freedom as to much else, and controvertible. The freedom to be wrong is valid, in other words, because it is the unexcisable other half of the freedom to be right. If freedom is understood as merely the privilege of the unconcerned and uncommitted to muddle about in error, then freedom will certainly destroy itself.

But to define freedom only as a public privilege of private citizens is finally inadequate to the job of protecting freedom. It leaves the issue too public and too private. It fails to provide a circumstance for those private satisfactions and responsibilities without which freedom is both pointless and fragile. Here as elsewhere, we need to interpose between the public and the private interests a third interest: that of the community. When there is no forcible assertion of the interest of community, public freedom becomes a sort of refuge for escapees from the moral law—those who hold that there is, in Mary McGrory's words, "no ethical transgression except an indictable one."

Public laws are meant for a public, and they vary, sometimes radically, according to forms of government. The moral law, which is remarkably consistent from one culture to another, has to do with community life. It tells us how we should treat relatives and neighbors and, by metaphorical extension, strangers. The aim of the moral law is the integrity and longevity of the community, just as the aim of public law is the integrity and longevity of a political body. Sometimes, the identities of community and political body are nearly the same, and in that case public laws are not necessary because there is, strictly speaking, no "public." As I understand the term, *public* means simply all the people, apart from any personal responsibility or belonging. A public building, for example, is a building which everyone may use but to which no one belongs, which belongs to everyone but not to anyone in particular, and for which no one is responsible except "public employees." A community, unlike a public, has to do first of all with belonging; it is a group of people who belong to one another and to their place. We would say, "We belong to our community," but never "We belong to our public."

I don't know when the concept of "the public," in our sense, emerged from the concept of "the people." But I am aware that there have been human situations in which the concept of "the public" was simply unnecessary. It is not quite possible, for example, to think of the Bushmen or the Eskimos as "publics" or of any parts of their homelands as "public places." And in the traditional rural villages of England there was no public place but rather a "common." A public, I suppose, becomes necessary when a political body grows so large as to include several divergent communities.

A public government, with public laws and a public system of justice, founded on democratic suffrage, is in principle a good thing. Ideally, it makes possible a just and peaceable settlement of contentions arising between communities. It also makes it possible for a mistreated member of a community to appeal for justice outside the community. But obviously such a government can fall short of its purpose. When a public government becomes identified with a public economy, a public culture, and public fashions of thought, it can become the tool of a public process of nationalism or "globalization" that is oblivious of local differences and therefore destructive of communities.

"Public" and "community," then, are different—perhaps radically different—concepts that under certain circumstances are compatible but that, in the present economic and technological monoculture, tend to be at odds. A community, when it is alive and well, is centered on the household—the family place and economy—and the household is centered on marriage. A public, when it is working in the best way—that is, as a political body intent on justice—is centered on the individual. Community and public alike, then, are founded on respect—the one on respect for the family, the other on respect for the individual. Both forms of respect are deeply traditional, and they are not fundamentally incompatible. But they are different, and that difference, once it is instituted in general assumptions, can be the source of much damage and much danger.

A household, according to its nature, will seek to protect and prolong its own life, and since it will readily perceive its inability to survive alone, it will seek to join its life to the life of a community. A young person, coming of age in a healthy household and community, will understand her or his life in terms of membership and service. But in a public increasingly disaffected and turned away from community, it is clear that individuals must be increasingly disinclined to identify themselves in such a way.

The individual, unlike the household and the community, always has two ways to turn: she or he may turn either toward the household and the community, to receive membership and to give service, or toward the relatively unconditional life of the public, in which one is free to pursue self-realization, self-aggrandizement, self-interest, self-fulfillment, self-enrichment, self-promotion, and so on. The problem is that—unlike a married couple, a household, or a community—one individual represents no fecundity, no continuity, and no harmony. The individual life implies no standard of behavior or responsibility.

I am indebted to Judith Weissman for the perception that there are two kinds of freedom: the freedom of the community and the freedom of the individual. The freedom of the community is the more fundamental and the more complex. A community confers on its members the freedoms implicit in familiarity, mutual respect, mutual affection, and mutual help; it gives freedom its proper aims; and it prescribes or shows the responsibilities without which no one can be legitimately free, or free for very long. But to confer freedom or any other benefits on its members, a community must also be free from outside pressure or coercion. It must, in other words, be so far as possible the cause of its own changes; it must change in response to its own changing needs and local circumstances, not in response to motives, powers, or fashions coming from elsewhere. The freedom of the individual, by contrast, has been construed customarily as a license to pursue any legal self-interest at large and at will in the domain of public liberties and opportunities.

These two kinds of freedom, so understood, are clearly at odds. In modern times, the dominant freedom has been that of the individual, and Judith Weissman believes—correctly, I think—that this self-centered freedom is still the aim of contemporary liberation movements:

> The liberation of the individual self for fulfillments, discoveries, pleasures, and joys, and the definition of oppression as mental and emotional constraints . . . this combination existing at the heart of Shelley's Romantic radicalism remains basically unchanged in later feminist writers. . . .

Freedom defined strictly as individual freedom tends to see itself as an escape from the constraints of community life

—constraints necessarily implied by consideration for the nature of a place; by consideration for the needs and feelings of neighbors; by kindness to strangers; by respect for the privacy, dignity, and propriety of individual lives; by affection for a place, its people, and its nonhuman creatures; and by the duty to teach the young.

But certain liberationist intellectuals are not the only ones who have demanded this sort of freedom. Almost everybody now demands it, as she or he has been taught to do by the schools, by the various forms of public entertainment, and by salespeople, advertisers, and other public representatives of the industrial economy. People are instructed to free themselves of all restrictions, restraints, and scruples in order to fulfill themselves as individuals to the utmost extent that the law allows. Moreover, we treat corporations as "persons"—an abuse of a metaphor if ever there was one!—and allow to them the same liberation from community obligations that we allow to individuals.

But there is a paradox in all this, and it is as cruel as it is obvious: as the emphasis on individual liberty has increased, the liberty and power of most individuals has declined. Most people are now finding that they are free to make very few significant choices. It is becoming steadily harder for ordinary people —the unrich, the unprivileged—to choose a kind of work for which they have a preference, a talent, or a vocation, to choose where they will live, to choose to work (or to live) at home, or even to choose to raise their own children. And most individuals ("liberated" or not) choose to conform not to local ways and conditions but to a rootless and placeless monoculture of commercial expectations and products. We try to be "emotionally self-sufficient" at the same time that we are entirely and helplessly dependent for our "happiness" on an economy that abuses us along with everything else. We want the liberty of divorce from spouses and independence from family and friends, yet we remain indissolubly married to a hundred corporations that regard us at best as captives and at worst as prey. The net result of our much-asserted individualism appears to be that we have become "free" for the sake of not much self-fulfillment at all.

However frustrated, disappointed, and unfulfilled it may be, the pursuit of self-liberation is still the strongest force now

operating in our society. It is the dominant purpose not only of
those feminists whose individualism troubles Judith Weissman
but also of virtually the entire population; it determines the
ethics of the professional class; it defines increasingly the ambitions of politicians and other public servants. This purpose
is publicly sanctioned and publicly supported, and it operates
invariably to the detriment of community life and community
values.

All the institutions that "serve the community" are publicly
oriented: the schools, governments and government agencies,
the professions, the corporations. Even the churches, though
they may have community memberships, do not concern
themselves with issues of local economy and local ecology on
which community health and integrity must depend. Nor do
the people in charge of these institutions think of themselves
as members of communities. They are itinerant, in fact or in
spirit, as their careers require them to be. These various public
servants all have tended to impose on the local place and the local people programs, purposes, procedures, technologies, and
values that originated elsewhere. Typically, these "services" involve a condescension to and a contempt for local life that are
implicit in all the assumptions—woven into the very fabric—of
the industrial economy.

A community, especially if it is a rural community, is understood by its public servants as provincial, backward and be-
nighted, unmodern, unprogressive, unlike "us," and therefore
in need of whatever changes are proposed for it by outside
interests (to the profit of the outside interests). Anyone who
thinks of herself or himself as a member of such a community
will sooner or later see that the community is under attack
morally as well as economically. And this attack masquerades
invariably as altruism: the community must be plundered, ex-
propriated, or morally offended for its own good—but its good
is invariably defined by the interest of the invader. The community is not asked whether or not it wishes to be changed, or
how it wishes to be changed, or what it wishes to be changed
into. The community is deemed to be backward and provincial, it is taught to believe and to regret that it is backward and
provincial, and it is thereby taught to welcome the purposes of
its invaders.

I have already discussed at some length the honored practice of community destruction by economic invasion. Now it will be useful to look at an example of the analogous practice of moral invasion. In 1989, Actors Theater of Louisville presented the premiere performance of Arthur Kopit's play *Bone-the-Fish*. The Louisville *Courier-Journal* welcomed Mr. Kopit and his play to town with the headline: "Arthur Kopit plans to offend almost everyone." In the accompanying article, Mr. Kopit is quoted as saying of his play:

> I am immodestly proud that it is written in consistently bad taste. It's about vile people who do vile things. They are totally loathsome, and I love them all. . . . I'm almost positive that it has something to offend everyone.

The writer of the article explains that "Kopit wrote 'Bone-the-Fish' out of a counterculture impulse, as a reaction against a complacency that he finds is corrupting American life."

My interest here is not in the quality or the point of Mr. Kopit's play, which I did not see (because I do not willingly subject myself to offense). I am interested in the article about him and his play merely as an example of the conventionality of the artistic intention to offend—and of the complacency of the public willingness not to be offended but passively to accept offense. Here we see the famous playwright coming from the center of culture to a provincial city, declaring his intention to "offend almost everyone," and here we see the local drama critic deferentially explaining the moral purpose of this intention. But the playwright makes three rather curious assumptions: (1) that the Louisville theater audience may be supposed (without proof) to be complacent and corrupt; (2) that they therefore deserve to be offended; and (3) that being offended will make them less complacent and less corrupt.

That Louisville theatergoers are more complacent and corrupt than theatergoers elsewhere (or than Mr. Kopit) is not an issue, for there is no evidence. That they deserve to be offended is not an issue for the same reason. That anyone's complacency and corruption can be corrected by being offended in a theater is merely a contradiction in terms, for people who are corrupted by complacency are by definition not likely to take offense. People who do take offense will be either fundamentally decent or aggressively corrupt. People who are fundamentally decent do not deserve to be offended and cannot be instructed by offense. People who are aggressively corrupt would perhaps see the offense but would not accept it. Mr. Kopit's preferred audience is therefore one that will applaud his audacity and pay no attention at all to his avowed didactic purpose—and this perhaps explains his love for "vile people."

If one thinks of Louisville merely as a public, there is not much of an issue here. Mr. Kopit's play is free speech, protected by the First Amendment, and that is that. If, however, one thinks of Louisville as a community or even as potentially a community, then the issue is sizable, and it is difficult. A public, as I have already suggested, is a rather odd thing; I can't think of anything else that is like it. A community is another matter, for it exists within a system of analogies or likenesses that clarify and amplify its meaning. A healthy community is like an ecosystem, and it includes—or it makes itself harmoniously a part of its local ecosystem. It is also like a household; it is the household of its place, and it includes the households of many families, human and nonhuman. And to extend Saint Paul's famous metaphor by only a little, a healthy community is like a body, for its members mutually support and serve one another.

If a community, then, is like a household, what are we to make of the artist whose intention is to offend? Would I welcome into my house any stranger who came, proud of his bad taste, professing his love for vile people and proposing to offend almost everyone? I would not, and I do not know anybody who would. To do so would contradict self-respect and respect for loved ones. By the same token, I cannot see that a community is under any obligation to welcome such a person. The public, so far as I can see, has no right to require a community to submit to or support statements that offend it.

I know that for a century or so many artists and writers have felt it was their duty—a mark of their honesty and courage—to offend their audience. But if the artist has a duty to offend, does not the audience therefore have a duty to be offended? If the public has a duty to protect speech that is offensive to the community, does not the community have the duty to respond, to be offended, and so defend itself against the offense? A community, as a part of a public, has no right to silence publicly protected speech, but it certainly has a right not to listen and to refuse its patronage to speech that it finds offensive. It is remarkable, however, that many writers and artists appear to be unable to accept this obvious and necessary limitation on their public freedom; they seem to think that freedom entitles them not only to be offensive but also to be approved and subsidized by the people whom they have offended.

These people believe, moreover, that any community attempt to remove a book from a reading list in a public school is censorship and a violation of the freedom of speech. The situation here involves what may be a hopeless conflict of freedoms. A teacher in a public school ought to be free to exercise his or her freedom of speech in choosing what books to teach and in deciding what to say about them. (This, to my mind, would certainly include the right to teach that the Bible is the word of God and the right to teach that it is not.) But the families of a community surely must be allowed an equal freedom to determine the education of their children. How free are parents who have no choice but to turn their children over to the influence of whatever the public will prescribe or tolerate? They obviously are not free at all. The only solution is trust between a community and its teachers, who will therefore teach as members of the community—a trust that in a time of community disintegration is perhaps not possible. And so the public presses its invasion deeper and deeper into community life under the justification of a freedom far too simply understood. It is now altogether possible for a teacher who is forbidden to teach the Bible to teach some other book that is not morally acceptable to the community, perhaps in order to improve the community by shocking or offending it. It is therefore possible that the future of community life in this country may depend on private schools and home schooling.

Does my objection to the intention to offend and the idea of improvement by offense mean that I believe it is invariably wrong to offend or that I think community and public life do not need improving? Obviously not. I do not mean at all to slight the issues of honesty and of artistic integrity that are involved. But I would distinguish between the intention to offend and the willingness to risk offending. Honesty and artistic integrity do not require anyone to intend to give offense, though they certainly may cause offense. The intention to offend, it seems to me, identifies the would-be offender as a public person. I cannot imagine anyone who is a member of a community who would purposely or gladly or proudly offend it, though I know very well that honesty might require one to do so.

Here we are verging on a distinction that had better be explicitly made. There is a significant difference between works of art made to be the vital possessions of a community (existing or not) and those made merely as offerings to the public. Some artists, and I am one of them, wish to live and work within a community, or within the hope of community, in a given place. Others wish to live and work outside the claims of any community, and these now appear to be an overwhelming majority. There is a difference between these two kinds of artists but not necessarily a division. The division comes when the public art begins to conventionalize an antipathy to community life and to the moral standards that enable and protect community life, as our public art has now done. Mr. Kopit's expressed eagerness to offend a local audience he does not know is representative of this antipathy. Our public art now communicates a conventional prejudice against old people, history, parental authority, religious faith, sexual discipline, manual work, rural people and rural life, anything local or small or inexpensive. At its worst, it glamorizes or glorifies drugs, promiscuity, pornography, violence, and blasphemy. Any threat to suppress or limit these public expressions will provoke much support for the freedom of speech. I concur in this. But as a community artist, I would like to go beyond my advocacy of the freedom of speech to deplore some of the uses that are made of it, and I wish that more of my fellow artists would do so as well.

I wish that artists and all advocates and beneficiaries of the First Amendment would begin to ask, for instance, how the individual can be liberated by disobeying the moral law, when the community obviously can be liberated only by obeying it. I wish that they would consider the probability that there is a direct relation between the public antipathy to community life and local ("provincial") places and the industrial destruction of communities and places. I wish, furthermore, that they could see that artists who make offensiveness an artistic or didactic procedure are drawing on a moral capital that they may be using up. A public is shockable or offendable only to the extent that it is already uncomplacent and uncorrupt—to the extent, in other words, that it is a community or remembers being one. What happens after the audience becomes used to being shocked and is therefore no longer shockable—as is apparently near to being the case with the television audience? What if offenses become stimulants—either to imitate the offenses or to avenge them? And what is the difference between the artist who wishes to offend the "provincials" and the industrialist or developer who wishes to dispossess them or convert them into a "labor force"?

The idea that people can be improved by being offended will finally have to meet the idea (espoused some of the time by some of the same people) that books, popular songs, movies, television shows, sex videos, and so on are "just fiction" or "just art" and therefore exist "for their own sake" and have no influence. To argue that works of art are "only" fictions or self-expressions and therefore cannot cause bad behavior is to argue also that they cannot cause good behavior. It is, moreover, to make an absolute division between art and life, experience and life, mind and body—a division that is intolerable to anyone who is at all serious about being a human or a member of a community or even a citizen.

Ananda Coomaraswamy, who had exhaustive knowledge of the traditional uses of art, wrote that "the purpose of any art . . . is to teach, to delight, *and above all to move*" (my emphasis). Of course art moves us! To assume otherwise not only contradicts the common assumption of teachers and writers from the earliest times almost until now; it contradicts everybody's experience. A cathedral, to mention only one of the most obvious examples, is a work of art made to cause a movement toward God, and this is in part a physical movement required by the building's structure and symbolism. But all works of any power move us, in both body and mind, from the most exalted music or poetry to the simplest dance tune. In fact, a dance tune is as good an example as a cathedral. An influence is cast over us, and we are moved. If we see that the influence is bad, we may be moved to reject it, but that is a second movement; it occurs only after we have felt the power of the influence.

People do not patronize the makers of pornographic films and sex videos because they are dispassionate appreciators of bad art; they do so because they wish to be moved. Perhaps the makers of pornographic films do not care what their products move their patrons to do. But if they do care, they are writing a check on moral capital to which they do not contribute. They trust that people who are moved by their work will not be moved to sexual harassment or child molestation or rape. They are banking heavily on the moral decency of their customers. And so are all of us who defend the freedom of speech. We are trusting—and not comfortably—that people who come under the influence of the sexual pandering, the greed, the commercial seductions, the moral oversimplification, the brutality, and the violence of our modern public arts will yet somehow remain under the influence of Moses and Jesus. I don't see how anyone can extend this trust without opposing in every way short of suppression the abuses and insults that are protected by it. The more a society comes to be divided in its assumptions and values, the more necessary public freedom becomes. But the more necessary public freedom becomes, the more necessary community responsibility becomes. This connection is unrelenting. And we should not forget that the finest works of art make a community of sorts of their audience. They do not divide people or justify or flatter their divisions; they define our commonwealth, and they enlarge it.

The health of a free public—especially that of a large nation under a representative government—depends on distrust. Thomas Jefferson thought so, and I believe he was right. In subscribing, generation after generation, to our Constitution, we extend to one another and to our government a trust that would be foolish if there were any better alternative. It is a breathtaking act of faith. And this trust is always so near to being misapplied that it cannot be maintained without distrust. People would fail it worse than they do if it were not for the constant vigilance and correction of distrust.

But a community makes itself up in more intimate circumstances than a public. And the health of a community depends absolutely on trust. A community knows itself and knows its place in a way that is impossible for a public (a nation, say, or a state). A community does not come together by a covenant, by a conscientious granting of trust. It exists by proximity, by neighborhood; it knows face to face, and it trusts as it knows. It learns, in the course of time and experience, what and who can be trusted. It knows that some of its members are untrustworthy, and it can be tolerant, because to know in this matter is to be safe. A community member can be trusted to be untrustworthy and so can be included. (A community can trust its liars to be liars, for example, and so enjoy them.) But if a community withholds trust, it withholds membership. If it cannot trust, it cannot exist.

One of the essential trusts of community life is that which holds marriages and families together. Another trust is that neighbors will help one another. Another is that privacy will be respected, especially the privacy of personal feeling and the privacy of relationships. All these trusts are absolutely essential, and all are somewhat fragile. But the most fragile, the most vulnerable to public invasion, is the trust that protects privacy. And in our time privacy has been the trust that has been most subjected to public invasion.

I am referring not just to the pryings and snoopings of our secret government, which contradict all that our public government claims to stand for, but also to those by now conventional publications of private grief, of violence to strangers, of the sexual coupling of strangers—all of which allow the indulgence of curiosity without sympathy. These all share in the evil of careless or malicious gossip; like careless or malicious gossip of any other kind, they destroy community by destroying respect for personal dignity and by destroying compassion. It is clear that no self-respecting human being or community would tolerate for a moment the representation of brutality or murder, on television or anywhere else, in such a way as to allow no compassion for the victim. But worst of all—and, I believe, involved in all—is the public prostitution of sex in guises of freedom ranging from the clinical to the commercial, from the artistic to the statistical.

One of the boasts of our century is that its artists—not to mention its psychologists, therapists, anthropologists, sociologists, statisticians, and pornographers—have pried open the bedroom door at last and shown us sexual love for what it "really" is. We have, we assume, cracked the shell of sexual privacy. The resulting implication that the shell is easily cracked disguises the probability that the shell is, in fact, not crackable at all and that what we have seen displayed is not private or intimate sex, not sexual love, but sex reduced, degraded, over-simplified, and misrepresented by the very intention to display it. Sex publicly displayed is public sex. Sex observed is not private or intimate and cannot be.

Could a voyeur conceivably crack the shell? No, for voyeurs are the most handicapped of all the sexual observers; they know only what they see. True intimacy, even assuming that it can be observed, cannot be known by an outsider and cannot be shown. An artist who undertakes to show the most intimate union of lovers—even assuming that the artist is one of the lovers—can only represent what she or he alone thinks it is. The intimacy, the union itself, remains unobserved. One cannot enter into this intimacy and watch it at the same time, any more than the mind can think about itself while it thinks about something else.

Is sexual love, then, not a legitimate subject of the imagination? It is. But the work of the imagination does not require that the shell be cracked. From Homer to Shakespeare, from the Bible to Jane Austen, we have many imaginings of the intimacy and power of sexual love that have respected absolutely its essential privacy and thus have preserved its intimacy and honored its dignity.

The essential and inscrutable privacy of sexual love is the sign both of its mystery and sanctity and of its humorousness. It is mysterious because the couple who are in it are lost in it. It is their profoundest experience of the being of the world and of their being in it and is at the same time an obliviousness to the world. This lostness of people in sexual love tends to be funny to people who are outside it. But having subscribed to the superstition that we have stripped away all privacy—and mystery and sanctity—from sex, we have become oddly humorless about it. Most people, for example, no longer seem to be aware of the absurdity of sexual vanity. Most people apparently see the sexual pretension and posturing of popular singers, athletes, and movie stars as some kind of high achievement, not the laughable inanity that it really is. Sexual arrogance, on the other hand, is not funny. It is dangerous, and there are some signs that our society has begun to recognize the danger. What it has not recognized is that the publication of sexual privacy is not only fraudulent but often also a kind of sexual arrogance, and a dangerous one.

Does this danger mean that any explicit representation of sexual lovemaking is inevitably wrong? It does not. But it means that such representations *can* be wrong and that when they are wrong, they are destructive.

The danger, I would suggest, is not in the representation but in the reductiveness that is the risk of representation and that is involved in most representations. What is so fearfully arrogant and destructive is the implication that what is represented, or representable, is all there is. In the best representations, I think, there would be a stylization or incompleteness that would convey the artist's honest acknowledgment that this is not all.

The best representations are surrounded and imbued with the light of imagination, so that they make one aware, with profound sympathy, of the two lives, not just the two bodies, that are involved; they make one aware also of the difficulty of full and open sexual consent between two people and of the history and the trust that are necessary to make possible that consent. Without such history and trust, sex is brutal, no matter what species is involved.

When sexual lovemaking is shown in art, one can respond intelligently to it by means of a handful of questions: Are the lovers represented as merely "physical" bodies or as two living souls? Does the representation make it possible to see why Eros has been understood not as an instinct or a "drive" but as a god? Are we asked to see this act as existing in and of and for itself or as joined to the great cycle of fertility and mortality? Does it belong to nature and to culture? Can we imagine this sweetness continuing on through the joys and difficulties of homemaking, the births and the upbringing of children, the deaths of parents and friends—through disagreements, hardships, quarrels, aging, and death? Does it encourage us to forget or to remember that "certainly it must some time come to pass that the very gentle Beatrice will die"?

And finally we must ask how the modern representations of lovemaking that we find in movies, books, paintings, sculptures, and on television measure up to the best love scenes that we know. The best love scene that I know is not explicitly sexual. It is the last scene of *The Winter's Tale*, in which Shakespeare brings onto the same stage, into the one light, young love in its astounding beauty, ardor, and hope, and old love with its mortal wrongs astoundingly graced and forgiven.

The relevance of such imagining is urgently practical; it is the propriety or justness that holds art and the world together. To represent sex without this fullness of imagination is to foreshadow the degradation and destruction of all that is not imagined. Just as the ruin of farmers, farming, and farmland may be predicted from a society's failure to imagine food in all its meanings and connections, so the failure to imagine sex in all its power and sanctity is to prepare the ruin of family and community life and of much else. In order to expose the privacy of sex, we have made of it another industrial special- ization, leaving it naked not only of clothes and of customary discretions and courtesies but also of all its cultural and natural connections.

There are, we must realize, kinds of nakedness that are sig- nificantly and sometimes ominously different from each other. To know this, we have only to study the examples that are be- fore us. There is—and who can ever forget it?—the nakedness of the photographs of prisoners in Hitler's death camps. This is the nakedness of absolute exposure to mechanical politics, politics gravitating toward the unimagining "efficiency" of machinery. I remember also a photograph of a naked small child running terrified down a dirt road in Vietnam, showing the body's absolute exposure to the indifference of air war, the appropriate technology of mechanical politics.

There is also the nakedness in advertising, in the worst kinds of fashionable or commercial art. This is the nakedness of free- market sexuality, the nakedness that is possible only in a society in which price is the only index of worth.

The nakedness of the death camps and of mechanical war denotes an absolute loss of dignity. In advertising, novels, and movies, the nakedness sometimes denotes a very significant and a very dangerous loss of dignity. Where the body has no dignity, where the sanctity of its own mystery and privacy is not recognized by a surrounding and protecting community, there can be no freedom. To destroy the dignity of the body —the dignity of any and every body—is to prepare the way for the enslaver, the rapist, the torturer, the user of cannon fodder. The nakedness or near-nakedness of some tribal peoples (I judge from the photographs that I have seen) is, in contrast, always dignified, and this dignity rests on a trust so complex and comprehensive as to be virtually unimaginable to us. The public nakedness of our own society involves no trust but only an exploitiveness that is inescapably economic and greedy. It is an abandonment of the self to self-exploitation and to exploitation by others.

There is also the nakedness of innocence, as, for example, in Degas's *Seated Bather Drying Herself*, in the Metropolitan Museum of Art, in which the body is shown in the unaware and unregarded coherence and mystery of its own being. This quality D. H. Lawrence saw and celebrated:

> People were bathing and posturing themselves on the beach
> and all was dreary, great robot limbs, robot breasts
> robot voices, robot even the gay umbrellas.
> But a woman, shy and alone, was washing herself under
>
>   a tap
> and the glimmer of the presence of the gods was like lilies
> and like water-lilies.

Finally, there is the nakedness of sexual candor. However easy or casual nakedness may have been made by public freedom, the nakedness of sexual candor is not possible except within the culturally delineated conditions that establish and maintain trust. And it is utterly private. It can be suggested in art but not represented. Any effort to represent it, I suspect, will inevitably be bogus. What must we do to earn the freedom of being unguardedly and innocently naked to someone? Our own and other cultures suggest that we must do a lot. We must make promises and keep them. We must assume many fearful responsibilities and do much work. We must build the household of trust.

It is the community, not the public, that is the protector of the possibility of this candor, just as it is the protector of other tender, vulnerable, and precious things—the childhood of children, for example, and the fertility of fields. These protections are left to the community, for they can be protected only by affection and by intimate knowledge, which are beyond the capacities of the public and beyond the power of the private citizen.

IV

If the word *community* is to mean or amount to anything, it must refer to a place (in its natural integrity) and its people. It must refer to a placed people. Since there obviously can be no cultural relationship that is uniform between a nation and a continent, "community" must mean a people locally placed and a people, moreover, not too numerous to have a common knowledge of themselves and of their place. Because places differ from one another and because people will differ somewhat according to the characters of their places, if we think of a nation as an assemblage of many communities, we are necessarily thinking of some sort of pluralism.

There is, in fact, a good deal of talk about pluralism these days, but most of it that I have heard is fashionable, superficial, and virtually worthless. It does not foresee or advocate a plurality of settled communities but is only a sort of indifferent charity toward a plurality of aggrieved groups and individuals. It attempts to deal liberally—that is, by the superficial courtesies of tolerance and egalitarianism—with a confusion of claims.

The social and cultural pluralism that some now see as a goal is a public of destroyed communities. Wherever it exists, it is the result of centuries of imperialism. The modern industrial urban centers are "pluralistic" because they are full of refugees from destroyed communities, destroyed community economies, disintegrated local cultures, and ruined local ecosystems. The pluralists who see this state of affairs as some sort of improvement or as the beginning of "global culture" are being historically perverse, as well as politically naive. They wish to regard liberally and tolerantly the diverse, sometimes competing claims and complaints of a rootless society, and yet they continue to tolerate also the ideals and goals of the industrialism that caused the uprooting. They affirm the pluralism of a society formed by the uprooting of cultures at the same time that they regard the fierce self-defense of still-rooted cultures as "fundamentalism," for which they have no tolerance at all. They look with wistful indulgence and envy at the ruined or damaged American Indian cultures so long as those cultures remain passively a part of our plurality, forgetting that these cultures, too, were once "fundamentalist" in their self-defense. And when these cultures again attempt self-defense—when they again assert the inseparability of culture and place—they are opposed by this pluralistic society as self-righteously as ever. The tolerance of this sort of pluralism extends always to the uprooted and passive, never to the rooted and active.

The trouble with the various movements of rights and liberties that have passed among us in the last thirty years is that they have all been too exclusive and so have degenerated too readily into special pleading. They have, separately, asked us to stop exploiting racial minorities or women or nature, and they have been, separately, right to do so. But they have not, separately or together, come to the realization that we live in a society that exploits, first, everything that is not ourselves and then, inevitably, ourselves. To ask, within this general onslaught, that we should honor the dignity of this or that group is to ask that we should swim up a waterfall.

Any group that takes itself, its culture, and its values seriously enough to try to separate, or to remain separate, from the industrial line of march will be, to say the least, unwelcome in the plurality. The tolerance of these doctrinaire pluralists always runs aground on religion. You may be fascinated by religion, you may study it, anthropologize and psychoanalyze about it, collect and catalogue its artifacts, but you had better not believe in it. You may put into "the canon" the holy books of any group, but you had better not think them holy. The shallowness and hypocrisy of this tolerance is exposed by its utter failure to extend itself to the suffering people of Iraq, who are, by the standards of this tolerance, fundamentalist, backward, unprogressive, and in general not like "us."

The problem with this form of pluralism is that it has no authentic standard; its standard simply is what one group or another may want at the moment. Its professed freedom is not that of community life but that of a political group acting on the pattern of individualism. To get farther toward a practicable freedom, the group must measure itself and its wants by standards external to itself. I assume that these standards must be both cultural and ecological. If people wish to be free, then they must preserve the culture that makes for political freedom, and they must preserve the health of the world.

There is an insistently practical question that any person and any group seriously interested in freedom must ask: Can land and people be preserved anywhere by means of a culture that is in the usual sense pluralistic? E. M. Forster, writing *Howards End* in the first decade of this century, doubted that they could. Nothing that has happened in the intervening eighty-odd years diminishes that doubt, and much that has happened confirms it.

A culture capable of preserving land and people can be made only within a relatively stable and enduring relationship between a local people and its place. Community cultures made in this way would necessarily differ, and sometimes radically so, from one place to another, because places differ. This is the true and necessary pluralism. There can, I think, be no national policy of pluralism or multiculturalism but only these pluralities of local cultures. And if these cultures are of any value and worthy of any respect, they will not be elective—not determined by mere wishes—but will be formed in response to local nature and local needs.

At present, the rhetoric of racial and cultural pluralism works against the possibility of a pluralism of settled communities, exactly as do the assumptions and the practices of national and global economies. So long as we try to think of ourselves as African Americans or European Americans or Asian Americans, we will never settle anywhere. For an authentic community is made less in reference to who we are than to where we are. I cannot farm my farm as a European American—or as an American, or as a Kentuckian—but only as a person belonging to the place itself. If I am to use it well and live on it authentically, I cannot do so by knowing where my ancestors came from (which, except for one great-grandfather, I do not know and probably can never know); I can do so only by knowing where I am, what the nature of the place permits me to do here, and who and what are here with me. To know these things, I must ask the place. A knowledge of foreign cultures is useful, perhaps indispensable, to me in my effort to settle here, but it cannot tell me where I am.

That there should be peace, commerce, and biological and cultural outcrosses among local cultures is obviously desirable and probably necessary as well. But such a state of things would be radically unlike what is now called pluralism. To start with, a plurality of settled communities could not be preserved by the present-day pluralists' easy assumption that all cultures are equal or of equal value and capable of surviving together by tolerance. The idea of equality is a good one, so long as it means "equality before the law." Beyond that, the idea becomes squishy and sentimental because of manifest inequalities of all kinds. It makes no sense, for example, to equate equality with freedom. The two concepts must be joined precisely and within strict limits if their association is to make any sense at all. Equality, in certain circumstances, is anything but free. If we have equality and nothing else—no compassion, no magnanimity, no courtesy, no sense of mutual obligation and dependence, no imagination—then power and wealth will have their way; brutality will rule. A general and indiscriminate egalitarianism is free-market culture, which, like free-market economics, tends toward a general and destructive uniformity. And tolerance, in association with such egalitarianism, is a way of ignoring the reality of significant differences. If I merely tolerate my neighbors on the assumption that all of us are equal, that means I can take no interest in the question of which ones of us are right and which ones are wrong; it means that I am denying the community the use of my intelligence and judgment; it means that I am not prepared to defer to those whose abilities are superior to mine, or to help those whose condition is worse; it means that I can be as self-centered as I please.

In order to survive, a plurality of true communities would require not egalitarianism and tolerance but knowledge, an understanding of the necessity of local differences, and respect. Respect, I think, always implies imagination—the ability to see one another, across our inevitable differences, as living souls.

FROM
ANOTHER TURN
OF THE CRANK
(1995)

Farming and the Global Economy

WE HAVE BEEN repeatedly warned that we cannot know where we wish to go if we do not know where we have been. And so let us start by remembering a little history.

As late as World War II, our farms were predominantly solar powered. That is, the work was accomplished principally by human beings and horses and mules. These creatures were empowered by solar energy, which was collected, for the most part, on the farms where they worked and so was pretty cheaply available to the farmer.

However, American farms had not become as self-sufficient in fertility as they should have been—or many of them had not. They were still drawing, without sufficient repayment, against an account of natural fertility accumulated over thousands of years beneath the native forest trees and prairie grasses.

The agriculture we had at the time of World War II was nevertheless often pretty good, and it was promising. In many parts of our country we had begun to have established agricultural communities, each with its own local knowledge, memory, and tradition. Some of our farming practices had become well adapted to local conditions. The best traditional practices of the Midwest, for example, are still used by the Amish with considerable success in terms of both economy and ecology.

Now that the issue of sustainability has arisen so urgently, and in fact so transformingly, we can see that the correct agricultural agenda following World War II would have been to continue and refine the already established connection between our farms and the sun and to correct, where necessary, the fertility deficit. There can be no question, now, that that is what we should have done.

It was, notoriously, not what we did. Instead, the adopted agenda called for a shift from the cheap, clean, and, for all practical purposes, limitless energy of the sun to the expensive, filthy, and limited energy of the fossil fuels. It called for the massive use of chemical fertilizers to offset the destruction of topsoil and the depletion of natural fertility. It called also for the displacement of nearly the entire farming population and the replacement of their labor and good farming practices by machines and toxic chemicals. This agenda has succeeded in its aims, but to the benefit of no one and nothing except the corporations that have supplied the necessary machines, fuels, and chemicals—and the corporations that have bought cheap and sold high the products that, as a result of this agenda, have been increasingly expensive for farmers to produce.

The farmers have not benefited—not, at least, as a class—for as a result of this agenda they have become one of the smallest and most threatened of all our minorities. Many farmers, sad to say, have subscribed to this agenda and its economic assumptions, believing that they would not be its victims. But millions, in fact, have been its victims—not farmers alone but also their supporters and dependents in our rural communities.

The people who benefit from this state of affairs have been at pains to convince us that the agricultural practices and policies that have almost annihilated the farming population have greatly benefited the population of food consumers. But more and more consumers are now becoming aware that our supposed abundance of cheap and healthful food is to a considerable extent illusory. They are beginning to see that the social, ecological, and even the economic costs of such "cheap food" are, in fact, great. They are beginning to see that a system of food production that is dependent on massive applications of drugs and chemicals cannot, by definition, produce "pure food." And they are beginning to see that a kind of agriculture that involves unprecedented erosion and depletion of soil, unprecedented waste of water, and unprecedented destruction of the farm population cannot by any accommodation of sense or fantasy be called "sustainable."

From the point of view, then, of the farmer, the ecologist, and the consumer, the need to reform our ways of farming is now both obvious and imperative. We need to adapt our farming much more sensitively to the nature of the places where the farming is done. We need to make our farming practices and our food economy subject to standards set not by the industrial system but by the health of ecosystems and of human communities.

The immediate difficulty in even thinking about agricultural reform is that we are rapidly running out of farmers. The tragedy of this decline is not just in its numbers; it is also in the fact that these farming people, assuming we will ever recognize our need to replace them, cannot be replaced anything like as quickly or easily as they have been dispensed with. Contrary to popular assumption, good farmers are not in any simple way part of the "labor force." Good farmers, like good musicians, must be raised to the trade.

The severe reduction of our farming population may signify nothing to our national government, but the members of country communities feel the significance of it—and the threat of it—every day. Eventually urban consumers will feel these things, too. Every day farmers feel the oppression of their long-standing problems: overproduction, low prices, and high costs. Farmers sell on a market that because of overproduction is characteristically depressed, and they buy their supplies on a market that is characteristically inflated—which is necessarily a recipe for failure, because farmers do not control either market. If they will not control production and if they will not reduce their dependence on purchased supplies, then they will keep on failing.

The survival of farmers, then, requires two complementary efforts. The first is entirely up to the farmers, who must learn —or learn again—to farm in ways that minimize their dependence on industrial supplies. They must diversify, using both plants and animals. They must produce, on their farms, as much of the required fertility and energy as they can. So far as they can, they must replace purchased goods and services with natural health and diversity and with their own intelligence. To increase production by increasing costs, as farmers have been doing for the last half century, is not only unintelligent; it is crazy. If farmers do not wish to cooperate any longer in their own destruction, then they will have to reduce their dependence on those global economic forces that intend and approve and profit from the destruction of farmers, and they will have to increase their dependence on local nature and local intelligence.

The second effort involves cooperation between local farmers and local consumers. If farmers hope to exercise any control over their markets, in a time when a global economy and global transportation make it possible for the products of any region to be undersold by the products of any other region, then they will have to look to local markets. The long-broken connections between towns and cities and their surrounding landscapes will have to be restored. There is much promise and much hope in such a restoration. But farmers must understand that this requires an economics of cooperation rather than competition. They must understand also that such an economy sooner or later will require some rational means of production control.

If communities of farmers and consumers wish to promote a sustainable, safe, reasonably inexpensive supply of good food, then they must see that the best, the safest, and most dependable source of food for a city is not the global economy, with its extreme vulnerabilities and extravagant transportation costs, but its own surrounding countryside. It is, in every way, in the best interest of urban consumers to be surrounded by productive land, well farmed and well maintained by thriving farm families in thriving farm communities.

If a safe, sustainable local food economy appeals to some of us as a goal that we would like to work for, then we must be careful to recognize not only the great power of the interests arrayed against us but also our own weakness. The hope for such a food economy as we desire is represented by no political party and is spoken for by no national public officials of any consequence. Our national political leaders do not know what we are talking about, and they are without the local affections and allegiances that would permit them to learn what we are talking about.

But we should also understand that our predicament is not without precedent; it is approximately the same as that of the proponents of American independence at the time of the Stamp Act—and with one difference in our favor: in order to do the work that we must do, we do not need a national organization. What we must do is simple: we must shorten the distance that our food is transported so that we are eating more and more from local supplies, more and more to the benefit of local farmers, and more and more to the satisfaction of local consumers. This can be done by cooperation among small organizations: conservation groups, churches, neighborhood associations, consumer co-ops, local merchants, local independent banks, and organizations of small farmers. It also can be done by cooperation between individual producers and consumers. We should not be discouraged to find that local food economies can grow only gradually; it is better that they should grow gradually. But as they grow they will bring about a significant return of power, wealth, and health to the people.

One thing at least should be obvious to us all: the whole human population of the world cannot live on imported food. Some people somewhere are going to have to grow the food. And wherever food is grown the growing of it will raise the same two questions: How do you preserve the land in use? And how do you preserve the people who use the land?

The farther the food is transported, the harder it will be to answer those questions correctly. The correct answers will not come as the inevitable by-products of the aims, policies, and procedures of international trade, free or unfree. They cannot be legislated or imposed by international or national or state agencies. They can only be supplied locally, by skilled and highly motivated local farmers meeting as directly as possible the needs of informed local consumers.

Conserving Forest Communities

I LIVE IN Henry County, near the lower end of the Kentucky River Valley, on a small farm that is half woodland. Starting from my back door, I could walk for days and never leave the woods except to cross the roads. Though Henry County is known as a farming county, 25 percent of it is wooded. From the hillside behind my house I can see thousands of acres of trees in the counties of Henry, Owen, and Carroll.

Most of the trees are standing on steep slopes of the river and creek valleys that were cleared and plowed at intervals from the early years of settlement until about the time of World War II. These are rich woodlands nevertheless. The soil, though not so deep as it once was, is healing from agricultural abuse and, because of the forest cover, is increasing in fertility. The plant communities consist of some cedar and a great diversity of hardwoods, shrubs, and wildflowers.

The history of these now-forested slopes over the last two centuries can be characterized as a cyclic alternation of abuse and neglect. Their best hope, so far, has been neglect—though even neglect has often involved their degradation by livestock grazing. So far, almost nobody has tried to figure out or has even wondered what might be the best use and the best care for such places. Often the trees have been regarded merely as obstructions to row cropping, which, because of the steepness of the terrain, has necessarily caused severe soil losses from water erosion. If an accounting is ever done, we will be shocked to learn how much ecological capital this kind of farming required for an almost negligible economic return: thousands of years of soil building were squandered on a few crops of corn or tobacco.

In my part of Kentucky, as in other parts, we never developed a local forest economy, and I think this was because of our preoccupation with tobacco. In the wintertime when farmers in New England, for example, employed themselves in the woods, our people went to their stripping rooms. Though in the earliest times we depended on the maple groves for syrup and sugar, we did not do so for very long. In this century, the fossil fuels weaned most of our households from firewood. For those reasons and others, we have never very consistently or very competently regarded trees as an economic resource.

And so as I look at my home landscape, I am happy to see that I am to a considerable extent a forest dweller. But I am unhappy to remember every time I look—for the landscape itself reminds me—that I am a dweller in a forest for which there is, properly speaking, no local forest culture and no local forest economy. That is to say that I live in a threatened forest.

Such woodlands as I have been describing are now mostly ignored so long as they are young. After the trees have reached marketable size, especially in a time of agricultural depression, the landowners come under pressure to sell them. And then the old cycle is repeated, as neglect is once more superseded by abuse. The salable trees are marked, and the tract of timber is sold to somebody who may have no connection, economic or otherwise, to the local community. The trees are likely to be felled and dragged from the woods in ways that do more damage than necessary to the land and the young trees. The skidder may take the logs straight upslope, leaving scars that (depending on how they catch the runoff) will be slow to heal or will turn into gullies that will never heal. There is no local *interest* connecting the woods workers to the woods. They do not regard the forest as a permanent resource but rather as a purchased "crop" that must be "harvested" as quickly and as cheaply as possible.

The economy of this kind of forestry is apt to be as deplorable as its ecology. More than likely only the prime log of each tree is taken—that is, the felled tree is cut in two below the first sizable branch, leaving many board feet in short logs (that would be readily usable, say, if there were small local woodworking shops) as well as many cords of firewood. The trees thus carelessly harvested will most likely leave the local community and the state as sawlogs or, at best, rough lumber. The only local economic benefit may well be the single check paid by the timber company to the landowner.

But the small landowners themselves may not receive the optimum benefit, for the prevailing assumptions and economic conditions encourage or require them to sell all their marketable trees at the same time. Unless the landowner is also a logger with the know-how and the means of cutting timber and removing it from the woods, the small, privately owned woodland is not likely to be considered a source of steady income, producing a few trees every year or every few years. For most such landowners in Kentucky, a timber sale may be thinkable only once or twice in a lifetime.

Furthermore, such landowners now must, as a matter of course, sell their timber on a market in which they have no influence, in which the power is held almost exclusively by the buyer. The sellers, of course, may choose not to sell—but only if they can *afford* not to sell. The private owners of Kentucky woodlands are in much the same fix that Kentucky tobacco producers were in before the time of the Burley Tobacco Growers Co-operative Association—and in much the same fix as most American farmers today. They cannot go to market except by putting themselves at the mercy of the market. This is a matter of no little significance and concern in a rural state in which 90.9 percent of the forestland "is owned by approximately 440,000 nonindustrial private owners, whose average holding is 26 acres."

I have been describing one version of present-day commercial forestry in Kentucky—what might be called the casual version.

But we also have in view another version. This is the big-money, large-scale corporate version. It involves the building of a large factory in a forested region, predictably accompanied by political advertisements about "job creation" and "improving the local economy." This factory, instead of sawing trees into boards, will reduce them to pulp for making paper, or it will grind or shred them and make boards or prefabricated architectural components by gluing together the resulting chips or strands.

Obviously, there are some advantages to these methods. Pulping or shredding can certainly use more of a tree than, say, a conventional sawmill. The laminated-strand process can make good building material out of low-quality trees. And there is no denying our society's need for paper and for building materials.

But from the point of view of either the forest or the local human community, there are also a number of problems associated with this kind of operation.

The fundamental problem is that it is costly and large in scale. It is therefore beyond the reach of small rural communities and so will be run inevitably for the benefit not of the local people but of absentee investors. And because of its cost and size, a large wood-products factory establishes in the local forest an enormous appetite for trees.

The very efficiency of a shredding mill—its ability to use small or low-quality trees—necessarily predisposes it to clear-cutting rather than to selective and sustained production. And a well-known inclination of such industries is toward forest monocultures, which do not have the ecological stability of natural forests.

As Kentuckians know from plenty of experience, nonexploitive relationships between large industries and small communities are extremely rare, if they exist at all. A large industrial operation might conceivably be established upon the most generous and forbearing principles of forestry and with the most benevolent intentions toward the local people. But we must remember that this large operation involves a large investment. And experience has taught us that large investments tend to take precedence over ecosystems and communities. In a time of economic adversity, the community and the forest will be sacrificed before the factory will be. The ideal of such operations is maximum profit to the owners or shareholders, who are not likely to be members of the local community. This means what it has always meant: labor and materials must be procured as cheaply as possible, and real human and ecological costs must be "externalized"—charged to taxpayers or to the future.

And so Kentucky forestry, at present, is mainly of two kinds: the casual and careless logging that is hardly more than an afterthought of farming, and the large-scale exploitation of the forest by absentee owners of corporations. Neither kind is satisfactory, by any responsible measure, in a state whose major natural resources will always be its productive soils and whose landscape today is one-half forested.

Kentucky has 12,700,000 acres of forest—almost 20,000 square miles. Very little of this is mature forest; nearly all of the old-growth timber had been cut down by 1940. Kentucky woodlands are nevertheless a valuable economic resource, supporting at present a wood-products industry with an annual payroll of $300 million and employing about 25,000 people. In addition, our forestlands contribute significantly to Kentucky's attractiveness to tourists, hunters, fishermen, and campers. They contribute indirectly to the economy by protecting our watersheds and our health.

But however valuable our forests may be now, they are nothing like so valuable as they can become. If we use the young forests we have now in the best way and if we properly care for them, they will continue to increase in board footage, in health, and in beauty for several more human generations. But already we are running into problems that can severely limit the value and usefulness of this resource to our people, because we have neglected to learn to practice good forest stewardship.

Moreover, we have never understood that the only appropriate human response to a diversified forest ecosystem is a diversified local forest economy. We have failed so far to imagine and put in place the sort of small-scale, locally owned logging and wood-products industries that would be the best guarantors of the long-term good use and good care of our forests. At present, it is estimated that up to 70 percent of the timber production of our forests leaves the state as logs or as raw lumber.

Lest you think that the situation and the problems I have outlined are of interest only to "tree huggers," let me remind you that during most of the history of our state, our rural landscapes and our rural communities have been in bondage to an economic colonialism that has exploited and misused both land and people. This exploitation has tended to become more severe with the growth of industrial technology. It has been most severe and most obvious in the coalfields of eastern Kentucky, but it has been felt and has produced its dire effects everywhere. With few exceptions our country people, generation after generation, have been providers of cheap fuels and raw materials to be used or manufactured in other places and to the profit of other people. They have added no value to what they have produced, and they have gone onto the markets without protection. They have sold their labor, their mineral rights, their crops, their livestock, and their trees with the understanding that the offered price was the price that they must take. Except for the tobacco program and the coal miners' union, rural Kentuckians have generally been a people without an asking price. We have developed the psychology of a subject people, willing to take whatever we have been offered and to believe whatever we have been told by our self-designated "superiors."

Now, with the two staple economies of coal and tobacco in doubt, we ask, "What can we turn to?" This is a question for every Kentuckian, but immediately it is a question for the rural communities. It is a question we may have to hold before ourselves for a long time, because the answer is going to be complex and difficult. If, however, as a part of the answer, we say, "Timber," I believe we will be right.

But we must be careful. In the past we have too often merely trusted that the corporate economy or the government would dispose of natural resources in a way that would be best for the land and the people. I hope we will not do that again. That trust has too often been catastrophically misplaced. From now on we should disbelieve that any corporation ever comes to any rural place to do it good, to "create jobs," or to bring to the local people the benefits of the so-called free market. It will be a tragedy if the members of Kentucky's rural communities ever again allow themselves passively to be sold off as providers of cheap goods and cheap labor. To put the bounty and the health of our land, our only commonwealth, into the hands of people who do not live on it and share its fate will always be an error. For whatever determines the fortune of the land determines also the fortune of the people. If the history of Kentucky teaches anything, it teaches that.

But the peculiarity of our history, so far, is that we have not had to learn the lesson. When the Old World races settled here, they saw a natural abundance so vast they could not imagine that it could be exhausted or ruined. Because it was vast and because virtually a whole continent was opening to the west, many of our forebears felt free to use the land carelessly and to justify their carelessness on the assumption that they could escape what they ruined. That early regardlessness of

consequence infected our character, and so far it has dominated the political and economic life of our state. So far, for
every Kentuckian, like Harry Caudill, willing to speak of the
natural limits within which we have been living all along, there
have been many who have wished only to fill their pockets and
move on, leaving their ecological debts to be paid by somebody else's children.

But by this time, the era of cut-and-run economics *ought* to
be finished. Such an economy cannot be rationally defended or
even apologized for. The proofs of its immense folly, heartlessness, and destructiveness are everywhere. Its failure as a way
of dealing with the natural world and human society can no
longer be sanely denied. That this economic system persists
and grows larger and stronger in spite of its evident failure
has nothing to do with rationality or, for that matter, with
evidence. It persists because, embodied now in multinational
corporations, it has discovered a terrifying truth: If you can
control a people's economy, you don't need to worry about
its politics; its politics have become irrelevant. If you control
people's choices as to whether or not they will work, and where
they will work, and what they will do, and how well they will
do it, and what they will eat and wear, and the genetic makeup
of their crops and animals, and what they will do for amusement, then why should you worry about freedom of speech? In
a totalitarian economy, any "political liberties" that the people
might retain would simply cease to matter. If, as is often the
case already, nobody can be elected who is not wealthy, and if
nobody can be wealthy without dependence on the corporate
economy, then what is your vote worth? The citizen thus becomes an economic subject.

A totalitarian economy might "correct" itself, of course, by
a total catastrophe—total explosion or total contamination
or total ecological exhaustion. A far better correction, however, would be a cumulative process by which states, regions,
communities, households, or even individuals would begin to
work toward economic self-determination and an appropriate
measure of local independence. Such a course of action would
involve us in a renewal of thought about our history and our
predicament. We must ask again whether or not we really want

to be a free people. We must consider again the linkages between land and landownership and land use and liberty. And
we must ask, as we have not very seriously asked before, what
are the best ways to use and to care for our land, our neighbors, and our natural resources.

If economists ever pay attention to such matters, they may
find that as the scale of an enterprise increases, its standards
become more and more simple, and it answers fewer and fewer
needs in the local community. For example, in the summer of
1982, according to an article in *California Forestry Notes*, three
men, using five horses, removed 400,780 board feet from a
35.5-acre tract in Latour State Forest. This was a "thinning operation." Two of the men worked full time as teamsters, using
two horses each; one man felled the trees and did some skidding with a single horse. The job required sixty-four days. It
was profitable both for the state forest and for the operator.
During the sixty-four days the skidders barked a total of eight
trees, only one of which was damaged badly enough to require
removal. Soil disturbance in the course of the operation was
rated as "slight."

At the end of this article the author estimates that a tractor
could have removed the logs two and a half times as fast as
the horses. And thus he implies a question that he does not
attempt to answer: Is it better for two men and four horses to
work sixty-four days, or for one man and one machine to do
the same work in twenty-five and a half days? Assuming that
the workers would all be from the local community, it is clear
that the community, a timber company, and a manufacturer of
mechanical skidders would answer that question in different
ways. The timber company and the manufacturer would answer on the basis of a purely economic efficiency: the need to
produce the greatest volume, hence the greatest profit, in the
shortest time. The community, on the contrary—and just as
much as a matter of self-interest—might reasonably prefer the
way of working that employed the most people for the longest
time and did the least damage to the forest and the soil. The
community might conclude that the machine, in addition to
the ecological costs of its manufacture and use, not only replaced the work of one man but more than halved the working

time of another. From the point of view of the community, it is *not* an improvement when the number of employed workers is reduced by the introduction of labor-saving machinery.

This question of which technology is better is one that our society has almost never thought to ask on behalf of the local community. It is clear nevertheless that the corporate standard of judgment, in this instance as in others, is radically oversimplified, and that the community standard is sufficiently complex. By using more people to do better work, the economic need is met, but so are other needs that are social and ecological, cultural and religious.

We can safely predict that for a long time there are going to be people in places of power who will want to solve our local problems by inviting in some great multinational corporation. They will want to put millions of dollars of public money into an "incentive package" to make it worthwhile for the corporation to pay low wages for our labor and to pay low prices for, let us say, our timber. It is well understood that nothing so excites the glands of a free-market capitalist as the offer of a government subsidy.

But before we agree again to so radical a measure, producing maximum profits to people who live elsewhere and minimal, expensive benefits to ourselves and our neighbors, we ought to ask if we cannot contrive local solutions for our local problems, and if the local solutions might not be the best ones. It is not enough merely to argue against a renewal of the old colonial economy. We must have something else competently in mind.

If we don't want to subject our forests to the rule of absentee exploiters, then we must ask what kind of forest economy we would like to have. By "we" I mean all the people of our state, of course, but I mean also, and especially, the people of our state's rural counties and towns and neighborhoods.

Obviously, I cannot speak for anybody but myself. But as a citizen of this state and a member of one of its rural communities, I would like to offer a description of what I believe would be a good forest economy. The following are not my own ideas, as you will see, but come from the work of many people who have put first in their thoughts the survival and the good health of their communities.

A good forest economy, like any other good land-based economy, would aim to join the local human community and the local natural community or ecosystem together as conservingly and as healthfully as possible.

A good forest economy would therefore be a local economy, and the forest economy of a state or region would therefore be a decentralized economy. The only reason to centralize such an economy is to concentrate its profits into the fewest hands. A good forest economy would be owned locally. It would afford a decent livelihood to local people. And it would propose to serve local needs and fill local demands first, before seeking markets elsewhere.

A good forest economy would preserve the local forest in its native diversity, quality, health, abundance, and beauty. It would recognize no distinction between its own prosperity and the prosperity of the forest ecosystem. A good forest economy would function in part as a sort of lobby for the good use of the forest.

A good forest economy would be properly scaled. Individual enterprises would be no bigger than necessary to ensure the best work and the best livelihood for workers. The ruling purpose would be to do the work with the least possible disturbance to the local ecosystem and the local human community. Keeping the scale reasonably small is good for the forest. Only a local, small-scale forest economy would permit, for example, the timely and selective logging of small woodlots.

Another benefit of smallness of scale is that it preserves economic democracy and the right of private property. Property boundaries, as we should always remember, are human conventions, useful for defining not only privileges but also responsibilities, so that use may always be accompanied by knowledge, affection, care, and skill. Such boundaries exist only because the society as a whole agrees to their existence. If the right of landownership is used only to protect an owner's wish to abuse or destroy the land, upon which the community's welfare ultimately depends, then society's interest in maintaining the convention understandably declines. And so in the interest of democracy and property rights, there is much to be gained by keeping especially the land-based industries small.

A good forest economy would be locally complex. People

in the local community would be employed in forest management, logging, and sawmilling, in a variety of value-adding small factories and shops, and in satellite or supporting industries. The local community, that is, would be enabled by its economy to realize the maximum income from its local resource. This is the opposite of a colonial economy. It would answer unequivocally the question, To *whom* is the value added?

Furthermore, a local forest economy, living by the measure of local economic health, might be led to some surprising alterations of logging technology. For example, it would almost certainly have to look again at the use of draft animals in logging. This would not only be kinder to the forest but would also be another way of elaborating the economy locally, requiring lower investment and less spending outside the community.

A good forest economy would make good forestry attractive to landowners, providing income from recreational uses of their woodlands, markets for forest products other than timber, and so on.

A good forest economy would obviously need to be much interested in local education. It would, of course, need to pass on to its children the large culture's inheritance of book learning. But also, both at home and in school, it would want its children to acquire a competent knowledge of local geography, ecology, history, natural history, and of local songs and stories. And it would want a system of apprenticeships, constantly preparing young people to carry on the local work in the best way.

All along, I have been implying that a good forest economy would be a limited economy. It would be limited in scale and limited by the several things it would not do. But it would be limited also by the necessity to leave some wilderness tracts of significant acreage unused. Because of its inclination to be proud and greedy, human character needs this practical deference toward things greater than itself; this is, I think, a religious deference. Also, for reasons of self-interest and our own survival, we need wilderness as a standard. Wilderness gives us the indispensable pattern and measure of sustainability.

To assure myself that what I have described as a good forest economy is a real possibility, I went to visit the tribal forest of the Menominee Indians in northern Wisconsin. In closing, I want to say what I learned about that forest—from reading; from talking with Marshall Pecore, the forest manager, and others; and from seeing for myself.

The Menominee originally inhabited a territory of perhaps ten million acres in Wisconsin and northern Michigan. By the middle of the nineteenth century, as the country was taken up by white settlers, the tribal holding had been reduced to 235,000 acres, 220,000 acres of which were forested.

The leaders understood that if the Menominee were to live, they would have to give up their old life of hunting and gathering and make timber from their forest a major staple of their livelihood; they understood also that if the Menominee were to survive as a people, they would have to preserve the forest while they lived from it. And so in 1854 they started logging, having first instituted measures to ensure that neither the original nature nor the productive capacity of the forest would be destroyed by their work. Now, 140 years later, Menominee forest management has become technically sophisticated, but it is still rooted in cultural tradition, and its goal has remained exactly the same: to preserve the identification of the human community with the forest, and to give an absolute priority to the forest's ecological integrity. The result, in comparison to the all-too-common results of land use in the United States, is astonishing. In 1854, when logging was begun, the forest contained an estimated billion and a half board feet of standing timber. No records exist for the first thirteen years, but from 1865 to 1988 the forest yielded two billion board feet. And today, after 140 years of continuous logging, the forest still is believed to contain a billion and a half board feet of standing timber. Over those 140 years, the average diameter of the trees has been reduced by only one half of one inch—and that by design, for the foresters want fewer large hemlocks.

About 20 percent of the forest is managed in even-aged stands of aspen and jack pine, which are harvested by clearcutting and which regenerate naturally. The rest of the forest is divided into 109 compartments, to each of which the foresters return every fifteen years to select trees for cutting. Their rule is to cut the worst and leave the best. That is, the loggers remove only those trees that are unlikely to survive for another fifteen years, those that are stunted or otherwise defective, and those that need to be removed in order to improve the stand. Old trees that are healthy and still growing are left uncut. As a result, this is an old forest, containing, for example, 350-year-old hemlocks, as well as cedars that are probably older. The average age of harvested maples is 140 to 180 years.

In support of this highly selective cutting, the forest is kept under constant study and evaluation. And loggers in the forest are strictly regulated and supervised. Even though the topography of the forest is comparatively level, skidders must be small and rubber-tired. Loggers must use permanent skid trails. And all logging contractors must attend training sessions.

The Menominee forest economy currently employs—in forest management, logging, milling, and other work—215 tribe members, or nearly 16 percent of the adult population of the reservation. As the Menominee themselves know, this is not enough; the economy of the forest needs to be more diverse. Its products at present are sawed lumber, logs, veneer logs, pulpwood, and "specialty woods" such as paneling and moldings. More value-adding industries are needed, and the Menominee are working on the problem. One knowledgeable observer has estimated that "they could probably turn twice the profit with half the land under management if they used more secondary processing."

Kentuckians looking for the pattern of a good local forest economy would have to conclude, I think, that the Menominee example is not complex enough, but that in all other ways it is excellent. We have much to learn from it. The paramount lesson undoubtedly is that the Menominee forest economy is as successful as it is because it is not understood primarily as an economy. Everybody I talked to on my visit urged me to understand that the forest is the basis of a culture and that the unrelenting cultural imperative has been to keep the forest intact—to preserve its productivity and the diversity of its trees, both in species and in age. The goal has always been a diverse, old, healthy, beautiful, productive, community-supporting forest that is home not only to its wild inhabitants but also to its human community. To secure this goal, the Menominee, following the dictates of their culture, have always done their work bearing in mind the needs of the seventh generation of their descendants.

And so, to complete my description of a good forest economy, I must add that it would be a long-term economy. Our
modern economy is still essentially a crop-year economy—as
though industrialism had founded itself upon the principles of
the worst sort of agriculture. The ideal of the industrial economy is to shorten as much as possible the interval separating
investment and payoff; it wants to make things fast, especially
money. But even the slightest acquaintance with the vital statistics of trees places us in another kind of world. A forest makes
things slowly; a good forest economy would therefore be a
patient economy. It would also be an unselfish one, for good
foresters must always look toward harvests that they will not
live to reap.

Health Is Membership

FROM OUR CONSTANT and increasing concerns about health, you can tell how seriously diseased we are. Health, as we may remember from at least some of the days of our youth, is at once wholeness and a kind of unconsciousness. Disease (dis-ease), on the contrary, makes us conscious not only of the state of our health but of the division of our bodies and our world into parts.

The word "health," in fact, comes from the same Indo-European root as "heal," "whole," and "holy." To be healthy is literally to be whole; to heal is to make whole. I don't think mortal healers should be credited with the power to make holy. But I have no doubt that such healers are properly obliged to acknowledge and respect the holiness embodied in all creatures, or that our healing involves the preservation in us of the spirit and the breath of God.

If we were lucky enough as children to be surrounded by grown-ups who loved us, then our sense of wholeness is not just the sense of completeness in ourselves but also is the sense of belonging to others and to our place; it is an unconscious awareness of community, of having in common. It may be that this double sense of singular integrity and of communal belonging is our personal standard of health for as long as we live. Anyhow, we seem to know instinctively that health is not divided.

Of course, growing up and growing older as fallen creatures in a fallen world can only instruct us painfully in division and disintegration. This is the stuff of consciousness and experience. But if our culture works in us as it should, then we do not age merely into disintegration and division, but that very experience begins our education, leading us into knowledge of wholeness and of holiness. I am describing here the story of Job, of Lazarus, of the lame man at the pool of Bethesda, of Milton's Samson, of King Lear. If our culture works in us as it should, our experience is balanced by education; we are led out of our lonely suffering and are made whole.

In the present age of the world, disintegration and division, isolation and suffering seem to have overwhelmed us.
The balance between experience and education has been overthrown; we are lost in experience, and so-called education is
leading us nowhere. We have diseases aplenty. As if that were
not enough, we are suffering an almost universal hypochondria. Half the energy of the medical industry, one suspects,
may now be devoted to "examinations" or "tests"—to see if,
though apparently well, we may not be latently or insidiously
diseased.

If you are going to deal with the issue of health in the modern world, you are going to have to deal with much absurdity.
It is not clear, for example, why death should increasingly be
looked upon as a curable disease, an abnormality, by a society that increasingly looks upon life as insupportably painful
and/or meaningless. Even more startling is the realization
that the modern medical industry faithfully imitates disease in
the way that it isolates us and parcels us out. If, for example,
intense and persistent pain causes you to pay attention only
to your stomach, then you must leave home, community, and
family and go to a sometimes distant clinic or hospital, where
you will be cared for by a specialist who will pay attention only
to your stomach.

Or consider the announcement by the Associated Press on
February 9, 1994, that "the incidence of cancer is up among
all ages, and researchers speculated that environmental exposure to cancer-causing substances other than cigarettes may be
partly to blame." This bit of news is offered as a surprise, never
mind that the environment (so called) has been known to be
polluted and toxic for many years. The blame obviously falls on
that idiotic term "the environment," which refers to a world
that surrounds us but is presumably different from us and distant from us. Our laboratories have proved long ago that ciga-
rette smoke gets inside us, but if "the environment" surrounds
us, how does *it* wind up inside us? So much for division as a
working principle of health.

This, plainly, is a view of health that is severely reductive. It
is, to begin with, almost fanatically individualistic. The body is
seen as a defective or potentially defective machine, singular,

solitary, and displaced, without love, solace, or pleasure. Its health excludes unhealthy cigarettes but does not exclude unhealthy food, water, and air. One may presumably be healthy in a disintegrated family or community or in a destroyed or poisoned ecosystem.

So far, I have been implying my beliefs at every turn. Now I had better state them openly.

I take literally the statement in the Gospel of John that God loves the world. I believe that the world was created and approved by love, that it subsists, coheres, and endures by love, and that, insofar as it is redeemable, it can be redeemed only by love. I believe that divine love, incarnate and indwelling in the world, summons the world always toward wholeness, which ultimately is reconciliation and atonement with God.

I believe that health is wholeness. For many years I have returned again and again to the work of the English agriculturist Sir Albert Howard, who said, in *The Soil and Health*, that "the whole problem of health in soil, plant, animal, and man [is] one great subject."

I am moreover a Luddite, in what I take to be the true and appropriate sense. I am not "against technology" so much as I am for community. When the choice is between the health of a community and technological innovation, I choose the health of the community. I would unhesitatingly destroy a machine before I would allow the machine to destroy my community.

I believe that the community—in the fullest sense: a place and all its creatures—is the smallest unit of health and that to speak of the health of an isolated individual is a contradiction in terms.

We speak now of "spirituality and healing" as if the only way to render a proper religious respect to the body is somehow to treat it "spiritually." It could be argued just as appropriately (and perhaps less dangerously) that the way to respect the body fully is to honor fully its materiality. In saying this, I intend no reduction. I do not doubt the reality of the experience and knowledge we call "spiritual" any more than I doubt the reality of so-called physical experience and knowledge; I recognize the rough utility of these terms. But I strongly doubt the advantage, and even the possibility, of separating these two realities.

What I'm arguing against here is not complexity or mystery but dualism. I would like to purge my own mind and language of such terms as "spiritual," "physical," "metaphysical," and "transcendental"—all of which imply that the Creation is divided into "levels" that can readily be peeled apart and judged by human beings. I believe that the Creation is one continuous fabric comprehending simultaneously what we mean by "spirit" and what we mean by "matter."

Our bodies are involved in the world. Their needs and desires and pleasures are physical. Our bodies hunger and thirst, yearn toward other bodies, grow tired and seek rest, rise up rested, eager to exert themselves. All these desires may be satisfied with honor to the body and its maker, but only if much else besides the individual body is brought into consideration. We have long known that individual desires must not be made the standard of their own satisfaction. We must consider the body's manifold connections to other bodies and to the world. The body, "fearfully and wonderfully made," is ultimately mysterious both in itself and in its dependences. Our bodies live, the Bible says, by the spirit and the breath of God, but it does not say how this is so. We are not going to *know* about this.

The distinction between the physical and the spiritual is, I believe, false. A much more valid distinction, and one that we need urgently to learn to make, is that between the organic and the mechanical. To argue this—as I am going to do—puts me in the minority, I know, but it does not make me unique. In *The Idea of a Christian Society*, T. S. Eliot wrote, "We may say that religion, as distinguished from modern paganism, implies a life in conformity with nature. It may be observed that the natural life and the supernatural life have a conformity to each other which neither has with the mechanistic life."

Still, I wonder if our persistent wish to deal spiritually with physical things does not come either from the feeling that physical things are "low" and unworthy or from the fear, especially when speaking of affection, that "physical" will be taken to mean "sexual."

The *New York Review of Books* of February 3, 1994, for example, carried a review of the correspondence of William and

Henry James along with a photograph of the two brothers standing together with William's arm around Henry's shoulders. Apropos of this picture, the reviewer, John Bayley, wrote that "their closeness of affection was undoubted and even took on occasion a quasi-physical form." It is Mr. Bayley's qualifier, "quasi-physical," that sticks in one's mind. What can he have meant by it? Is this prurience masquerading as squeamishness, or vice versa? Does Mr. Bayley feel a need to assure his psychologically sophisticated readers that even though these brothers touched one another familiarly, they were not homosexual lovers?

The phrase involves at least some version of the old dualism of spirit and body or mind and body that has caused us so much suffering and trouble and that raises such troubling questions for anybody who is interested in health. If you love your brother and if you and your brother are living creatures, how could your love for him not be physical? Not spiritual or mental only, not "quasi-physical," but physical. How could you not take a simple pleasure in putting your arm around him?

Out of the same dualism comes our confusion about the body's proper involvement in the world. People seriously interested in health will finally have to question our society's long-standing goals of convenience and effortlessness. What is the point of "labor saving" if by making work effortless we make it poor, and if by doing poor work we weaken our bodies and lose conviviality and health?

We are now pretty clearly involved in a crisis of health, one of the wonders of which is its immense profitability both to those who cause it and to those who propose to cure it. That the illness may prove incurable, except by catastrophe, is suggested by our economic dependence on it. Think, for example, of how readily our solutions become problems and our cures pollutants. To cure one disease, we need another. The causes, of course, are numerous and complicated, but all of them, I think, can be traced back to the old idea that our bodies are not very important except when they give us pleasure (usually, now, to somebody's profit) or when they hurt (now, almost invariably, to somebody's profit).

This dualism inevitably reduces physical reality, and it does

so by removing its mystery from it, by dividing it absolutely
from what dualistic thinkers have understood as spiritual or
mental reality.

A reduction that is merely theoretical might be harmless
enough, I suppose, but theories find ways of getting into action. The theory of the relative unimportance of physical reality has put itself into action by means of a metaphor by which
the body (along with the world itself) is understood as a machine. According to this metaphor—which is now in constant
general use—the human heart, for example, is no longer understood as the center of our emotional life or even as an organ
that pumps; it is understood as "a pump," having somewhat
the same function as a fuel pump in an automobile.

If the body is a machine for living and working, then it must
follow that the mind is a machine for thinking. The "progress"
here is the reduction of mind to brain and then of brain to
computer. This reduction implies and requires the reduction of
knowledge to "information." It requires, in fact, the reduction
of everything to numbers and mathematical operations.

This metaphor of the machine bears heavily upon the question of what we mean by health and by healing. The problem
is that like any metaphor, it is accurate only in some respects.
A girl is only in some respects like a red rose; a heart is only in
some respects like a pump. This means that a metaphor must
be controlled by a sort of humorous intelligence, always mindful of the exact limits within which the comparison is meaningful. When a metaphor begins to control intelligence, as this
one of the machine has done for a long time, then we must
look for costly distortions and absurdities.

Of course, the body in most ways is not at all like a machine.
Like all living creatures and unlike a machine, the body is not
formally self-contained; its boundaries and outlines are not so
exactly fixed. The body alone is not, properly speaking, a body.
Divided from its sources of air, food, drink, clothing, shelter,
and companionship, a body is, properly speaking, a cadaver,
whereas a machine by itself, shut down or out of fuel, is still a
machine. Merely as an organism (leaving aside issues of mind
and spirit) the body lives and moves and has its being, minute
by minute, by an interinvolvement with other bodies and other
creatures, living and unliving, that is too complex to diagram

or describe. It is, moreover, under the influence of thought and feeling. It does not live by "fuel" alone.

A mind, probably, is even less like a computer than a body is like a machine. As far as I am able to understand it, a mind is not even much like a brain. Insofar as it is usable for thought, for the association of thought with feeling, for the association of thoughts and feelings with words, for the connections between words and things, words and acts, thought and memory, a mind seems to be in constant need of reminding. A mind unreminded would be no mind at all. This phenomenon of reminding shows the extensiveness of mind—how intricately it is involved with sensation, emotion, memory, tradition, communal life, known landscapes, and so on. How you could locate a mind within its full extent, among all its subjects and necessities, I don't know, but obviously it cannot be located within a brain or a computer.

To see better what a mind is (or is not), we might consider the difference between what we mean by knowledge and what the computer now requires us to mean by "information." Knowledge refers to the ability to do or say the right thing at the right time; we would not speak of somebody who does the wrong thing at the wrong time as "knowledgeable." People who perform well as musicians, athletes, teachers, or farmers are people of knowledge. And such examples tell us much about the nature of knowledge. Knowledge is formal, and it informs speech and action. It is instantaneous; it is present and available when and where it is needed.

"Information," which once meant that which forms or fashions from within, now means merely "data." However organized this data may be, it is not shapely or formal or in the true sense in-forming. It is not present where it is needed; if you have to "access" it, you don't have it. Whereas knowledge moves and forms acts, information is inert. You cannot imagine a debater or a quarterback or a musician performing by "accessing information." A computer chock full of such information is no more admirable than a head or a book chock full of it.

The difference, then, between information and knowledge is something like the difference between a dictionary and somebody's language.

Where the art and science of healing are concerned, the

machine metaphor works to enforce a division that falsifies the process of healing because it falsifies the nature of the creature needing to be healed. If the body is a machine, then its diseases can be healed by a sort of mechanical tinkering, without reference to anything outside the body itself. This applies, with obvious differences, to the mind; people are assumed to be individually sane or insane. And so we return to the utter anomaly of a creature that is healthy within itself.

The modern hospital, where most of us receive our strictest lessons in the nature of industrial medicine, undoubtedly does well at surgery and other procedures that permit the body and its parts to be treated as separate things. But when you try to think of it as a place of healing—of reconnecting and making whole—then the hospital reveals the disarray of the medical industry's thinking about health.

In healing, the body is restored to itself. It begins to live again by its own powers and instincts, to the extent that it can do so. To the extent that it can do so, it goes free of drugs and mechanical helps. Its appetites return. It relishes food and rest. The patient is restored to family and friends, home and community and work.

This process has a certain naturalness and inevitability, like that by which a child grows up, but industrial medicine seems to grasp it only tentatively and awkwardly. For example, any ordinary person would assume that a place of healing would put a premium upon rest, but hospitals are notoriously difficult to sleep in. They are noisy all night, and the routine interventions go on relentlessly. The body is treated as a machine that does not need to rest.

You would think also that a place dedicated to healing and health would make much of food. But here is where the disconnections of the industrial system and the displacement of industrial humanity are most radical. Sir Albert Howard saw accurately that the issue of human health is inseparable from the health of the soil, and he saw too that we humans must responsibly occupy our place in the cycle of birth, growth, maturity, death, and decay, which is the health of the world. Aside from our own mortal involvement, food is our fundamental connection to that cycle. But probably most of the complaints you hear about hospitals have to do with the food, which, according to the testimony I have heard, tends to range from unappetizing to sickening. Food is treated as another unpleasant substance to inject. And this is a shame. For in addition to the obvious nutritional link between food and health, food can be a pleasure. People who are sick are often troubled or depressed, and mealtimes offer three opportunities a day when patients could easily be offered something to look forward to. Nothing is more pleasing or heartening than a plate of nourishing, tasty, beautiful food artfully and lovingly prepared. Anything less is unhealthy, as well as a desecration.

Why should rest and food and ecological health not be the basic principles of our art and science of healing? Is it because the basic principles already are technology and drugs? Are we confronting some fundamental incompatibility between mechanical efficiency and organic health? I don't know. I only know that sleeping in a hospital is like sleeping in a factory and that the medical industry makes only the most tenuous connection between health and food and no connection between health and the soil. Industrial medicine is as little interested in ecological health as is industrial agriculture.

A further problem, and an equally serious one, is that illness, in addition to being a bodily disaster, is now also an economic disaster. This is so whether or not the patient is insured. It is a disaster for us all, all the time, because we all know that personally or collectively, we cannot continue to pay for cures that continue to get more expensive. The economic disturbance that now inundates the problem of illness may turn out to be the profoundest illness of all. How can we get well if we are worried sick about money?

I wish it were not the fate of this essay to be filled with questions, but questions now seem the inescapable end of any line of thought about health and healing. Here are several more:

1. Can our present medical industry produce an adequate definition of health? My own guess is that it cannot do so. Like industrial agriculture, industrial medicine has depended increasingly on specialist methodology, mechanical technology, and chemicals; thus, its point of reference has become more and more its own technical prowess and less and less the health of creatures and habitats. I don't expect this problem to be

solved in the universities, which have never addressed, much less solved, the problem of health in agriculture. And I don't expect it to be solved by the government.

2. How can cheapness be included in the criteria of medical experimentation and performance? And why has it not been included before now? I believe that the problem here is again that of the medical industry's fixation on specialization, technology, and chemistry. As a result, the modern "health care system" has become a way of marketing industrial products, exactly like modern agriculture, impoverishing those who pay and enriching those who are paid. It is, in other words, an industry such as industries have always been.

3. Why is it that medical strictures and recommendations so often work in favor of food processors and against food producers? Why, for example, do we so strongly favor the pasteurization of milk to health and cleanliness in milk production? (Gene Logsdon correctly says that the motive here "is monopoly, not consumer health.")

4. Why do we so strongly prefer a fat-free or a germ-free diet to a chemical-free diet? Why does the medical industry strenuously oppose the use of tobacco, yet complacently accept the massive use of antibiotics and other drugs in meat animals and of poisons on food crops? How much longer can it cling to the superstition of bodily health in a polluted world?

5. How can adequate medical and health care, including disease prevention, be included in the structure and economy of a community? How, for example, can a community and its doctors be included in the same culture, the same knowledge, and the same fate, so that they will live as fellow citizens, sharers in a common wealth, members of one another?

II

It is clear by now that this essay cannot hope to be complete; the problems are too large and my knowledge too small. What I have to offer is an association of thoughts and questions wandering somewhat at random and somewhat lost within the experience of modern diseases and the often bewildering industry that undertakes to cure them. In my ignorance and bewilderment, I am fairly representative of those who go, or go with loved ones, to doctors' offices and hospitals. What I have written so far comes from my various efforts to make as much sense as I can of that experience. But now I had better turn to the experience itself.

On January 3, 1994, my brother John had a severe heart attack while he was out by himself on his farm, moving a feed trough. He managed to get to the house and telephone a friend, who sent the emergency rescue squad.

The rescue squad and the emergency room staff at a local hospital certainly saved my brother's life. He was later moved to a hospital in Louisville, where a surgeon performed a double-bypass operation on his heart. After three weeks John returned home. He still has a life to live and work to do. He has been restored to himself and to the world.

He and those who love him have a considerable debt to the medical industry, as represented by two hospitals, several doctors and nurses, many drugs and many machines. This is a debt that I cheerfully acknowledge. But I am obliged to say also that my experience of the hospital during John's stay was troubled by much conflict of feeling and a good many unresolved questions, and I know that I am not alone in this.

In the hospital what I will call the world of love meets the world of efficiency—the world, that is, of specialization, machinery, and abstract procedure. Or, rather, I should say that these two worlds come together in the hospital but do not meet. During those weeks when John was in the hospital, it seemed to me that he had come from the world of love and that the family members, neighbors, and friends who at various times were there with him came there to represent that world and to preserve his connection with it. It seemed to me that the hospital was another kind of world altogether.

When I said early in this essay that we live in a world that was created and exists and is redeemable by love, I did not mean to sentimentalize it. For this is also a fallen world. It involves error and disease, ignorance and partiality, sin and death. If this world is a place where we may learn of our involvement in immortal love, as I believe it is, still such learning is only possible here because that love involves us so inescapably in the limits, sufferings, and sorrows of mortality.

*

Like divine love, earthly love seeks plenitude; it longs for the full membership to be present and to be joined. Unlike divine love, earthly love does not have the power, the knowledge, or the will to achieve what it longs for. The story of human love on this earth is a story by which this love reveals and even validates itself by its failures to be complete and comprehensive and effective enough. When this love enters a hospital, it brings with it a terrifying history of defeat, but it comes nevertheless confident of itself, for its existence and the power of its longing have been proved over and over again even by its defeat. In the face of illness, the threat of death, and death itself, it insists unabashedly on its own presence, understanding by its persistence through defeat that it is superior to whatever happens.

The world of efficiency ignores both loves, earthly and divine, because by definition it must reduce experience to computation, particularity to abstraction, and mystery to a small comprehensibility. Efficiency, in our present sense of the word, allies itself inevitably with machinery, as Neil Postman demonstrates in his useful book *Technopoly*. "Machines," he says, "eliminate complexity, doubt, and ambiguity. They work swiftly, they are standardized, and they provide us with numbers that you can see and calculate with." To reason, the advantages are obvious, and probably no reasonable person would wish to reject them out of hand.

And yet love obstinately answers that no loved one is standardized. A body, love insists, is neither a spirit nor a machine; it is not a picture, a diagram, a chart, a graph, an anatomy; it is not an explanation; it is not a law. It is precisely and uniquely what it is. It belongs to the world of love, which is a world of living creatures, natural orders and cycles, many small, fragile lights in the dark.

In dealing with problems of agriculture, I had thought much about the difference between creatures and machines. But I had never so clearly understood and felt that difference as when John was in recovery after his heart surgery, when he was attached to many machines and was dependent for breath on a respirator. It was impossible then not to see that the breathing of a machine, like all machine work, is unvarying, an oblivious regularity, whereas the breathing of a creature is ever changing, exquisitely responsive to events both inside and outside the body, to thoughts and emotions. A machine makes breaths as a machine makes buttons, all the same, but every breath of a creature is itself a creature, like no other, inestimably precious.

Logically, in plenitude some things ought to be expendable. Industrial economics has always believed this: abundance justifies waste. This is one of the dominant superstitions of American history—and of the history of colonialism everywhere. Expendability is also an assumption of the world of efficiency, which is why that world deals so compulsively in percentages of efficacy and safety.

But this sort of logic is absolutely alien to the world of love. To the claim that a certain drug or procedure would save 99 percent of all cancer patients or that a certain pollutant would be safe for 99 percent of a population, love, unembarrassed, would respond, "What about the one percent?"

There is nothing rational or perhaps even defensible about this, but it is nonetheless one of the strongest strands of our religious tradition—it is probably the most essential strand—according to which a shepherd, owning a hundred sheep and having lost one, does not say, "I have saved 99 percent of my sheep," but rather, "I have lost one," and he goes and searches for the one. And if the sheep in that parable may seem to be only a metaphor, then go on to the Gospel of Luke, where the principle is flatly set forth again and where the sparrows stand not for human beings but for all creatures: "Are not five sparrows sold for two farthings, and not one of them is forgotten before God?" And John Donne had in mind a sort of equation and not a mere metaphor when he wrote, "If a clod be washed away by the sea, Europe is the less, as well as if a promontory were, as well as if a manor of thy friend's or of thine own were. Any man's death diminishes me."

It is reassuring to see ecology moving toward a similar idea of the order of things. If an ecosystem loses one of its native species, we now know that we cannot speak of it as itself minus one species. An ecosystem minus one species is a different ecosystem. Just so, each of us is made by—or, one might better say, made as—a set of unique associations with unique persons, places, and things. The world of love does not admit the principle of the interchangeability of parts.

When John was in intensive care after his surgery, his wife, Carol, was standing by his bed, grieving and afraid. Wanting to reassure her, the nurse said, "Nothing is happening to him that doesn't happen to everybody."

And Carol replied, "I'm not everybody's wife."

In the world of love, things separated by efficiency and specialization strive to come back together. And yet love must confront death, and accept it, and learn from it. Only in confronting death can earthly love learn its true extent, its immortality. Any definition of health that is not silly must include death. The world of love includes death, suffers it, and triumphs over it. The world of efficiency is defeated by death; at death, all its instruments and procedures stop. The world of love continues, and of this grief is the proof.

In the hospital, love cannot forget death. But like love, death is in the hospital but not of it. Like love, fear and grief feel out of place in the hospital. How could they be included in its efficient procedures and mechanisms? Where a clear, small order is fervently maintained, fear and grief bring the threat of large disorder.

And so these two incompatible worlds might also be designated by the terms "amateur" and "professional"—amateur, in the literal sense of lover, one who participates for love; and professional in the modern sense of one who performs highly specialized or technical procedures for pay. The amateur is excluded from the professional "field."

For the amateur, in the hospital or in almost any other encounter with the medical industry, the overriding experience is that of being excluded from knowledge—of being unable, in other words, to make or participate in anything resembling an "informed decision." Of course, whether doctors make informed decisions in the hospital is a matter of debate. For in the hospital even the professionals are involved in experience; experimentation has been left far behind. Experience, as all amateurs know, is not predictable, and in experience there are no replications or "controls"; there is nothing with which to compare the result. Once one decision has been made, we have destroyed the opportunity to know what would have happened if another decision had been made. That is to say that medicine is an exact science until applied; application involves intuition, a sense of probability, "gut feeling," guesswork, and error.

In medicine, as in many modern disciplines, the amateur is divided from the professional by perhaps unbridgeable differences of knowledge and of language. An "informed decision" is really not even imaginable for most medical patients and their families, who have no competent understanding of either the patient's illness or the recommended medical or surgical procedure. Moreover, patients and their families are not likely to know the doctor, the surgeon, or any of the other people on whom the patient's life will depend. In the hospital, amateurs are more than likely to be proceeding entirely upon faith—and this is a peculiar and scary faith, for it must be placed not in a god but in mere people, mere procedures, mere chemicals, and mere machines.

It was only after my brother had been taken into surgery, I think, that the family understood the extremity of this deed of faith. We had decided—or John had decided and we had concurred—on the basis of the best advice available. But once he was separated from us, we felt the burden of our ignorance. We had not known what we were doing, and one of our difficulties now was the feeling that we had utterly given him up to what we did not know. John himself spoke out of this sense of abandonment and helplessness in the intensive care unit, when he said, "I don't know what they're going to do to me or for me or with me."

As we waited and reports came at long intervals from the operating room, other realizations followed. We realized that under the circumstances, we could not be told the truth. We would not know, ever, the worries and surprises that came to the surgeon during his work. We would not know the critical moments or the fears. If the surgeon did any part of his work ineptly or made a mistake, we would not know it. We realized, moreover, that if we were told the truth, we would have no way of knowing that the truth was what it was.

We realized that when the emissaries from the operating room assured us that everything was "normal" or "routine," they were referring to the procedure and not the patient. Even as amateurs—perhaps *because* we were amateurs—we knew that what was happening was not normal or routine for John or for us.

*

That these two worlds are so radically divided does not mean
that people cannot cross between them. I do not know how an
amateur can cross over into the professional world; that does
not seem very probable. But that professional people can cross
back into the amateur world, I know from much evidence.
During John's stay in the hospital there were many moments
in which doctors and nurses—especially nurses!—allowed or
caused the professional relationship to become a meeting between two human beings, and these moments were invariably
moving.

The most moving, to me, happened in the waiting room
during John's surgery. From time to time a nurse from the
operating room would come in to tell Carol what was happening. Carol, from politeness or bravery or both, always stood to
receive the news, which always left us somewhat encouraged
and somewhat doubtful. Carol's difficulty was that she had to
suffer the ordeal not only as a wife but as one who had been a
trained nurse. She knew, from her own education and experience, in how limited a sense open-heart surgery could be said
to be normal or routine.

Finally, toward the end of our wait, two nurses came in. The
operation, they said, had been a success. They explained again
what had been done. And then they said that after the completion of the bypasses, the surgeon had found it necessary to
insert a "balloon pump" into the aorta to assist the heart. This
possibility had never been mentioned, nobody was prepared
for it, and Carol was sorely disappointed and upset. The two
young women attempted to reassure her, mainly by repeating things they had already said. And then there was a long
moment when they just looked at her. It was such a look as
parents sometimes give to a sick or suffering child, when they
themselves have begun to need the comfort they are trying to
give.

And then one of the nurses said, "Do you need a hug?"

"Yes," Carol said.

And the nurse gave her a hug.

Which brings us to a starting place.

LIFE IS A MIRACLE

An Essay Against Modern Superstition
(2 0 0 0)

"We are not getting something for nothing.
We are getting nothing for everything."

CONTENTS

Thy life's a miracle. Speak yet again.
King Lear, IV, vi, 55

I. Ignorance

T‍HE EXPRESSED dissatisfaction of some scientists with the dangerous oversimplifications of commercialized science has encouraged me to hope that this dissatisfaction will run its full course. These scientists, I hope, will not stop with some attempt at a merely theoretical or technical "correction," but will press on toward a new, or a renewed, propriety in the study and the use of the living world.

No such change is foreseeable in the terms of the presently dominant mechanical explanations of things. Such a change is imaginable only if we are willing to risk an unfashionable re-course to our cultural tradition. Human hope may always have resided in our ability, in time of need, to return to our cultural landmarks and reorient ourselves.

One of the principal landmarks of the course of my own life is Shakespeare's tragedy of *King Lear*. Over the last forty-five years I have returned to *King Lear* many times. Among the effects of that play—on me, and I think on anybody who reads it closely—is the recognition that in all our attempts to renew or correct ourselves, to shake off despair and have hope, our starting place is always and only our experience. We can begin (and we must always be beginning) only where our history has so far brought us, with what we have done.

Lately my thoughts about the inevitably commercial genetic manipulations already in effect or contemplated have sent me back to *King Lear* again. The whole play is about kindness, both in the usual sense, and in the sense of truth-to-kind, natu-ralness, or knowing the limits of our specifically *human* nature. But this issue is dealt with most explicitly in an episode of the subplot, in which the Earl of Gloucester is recalled from de-spair so that he may die in his full humanity.

The old earl has been blinded in retribution for his loyalty to the king, and in this fate he sees a kind of justice for, as he says, "I stumbled when I saw." He, like Lear, is guilty of hubris or presumption, of treating life as knowable, predictable, and within his control. He has falsely accused and driven away his loyal son, Edgar. Exiled and under sentence of death, Edgar has disguised himself as a madman and beggar. He becomes, in that role, the guide of his blinded father, who asks to be led to Dover where he intends to kill himself by leaping off a cliff. Edgar's task is to save his father from despair, and he succeeds, for Gloucester dies at last " 'Twixt two extremes of passion, joy and grief. . . ." He dies, that is, within the proper bounds of the human estate. Edgar does not want his father to give up on life. To give up on life is to pass beyond the possibility of change or redemption. And so he does not lead his father to the cliff's verge, but only *tells* him he has done so. Gloucester renounces the world, blesses Edgar, his supposedly absent son, and, according to the stage direction, "Falls forward and swoons."

When he returns to consciousness, Edgar now speaks to him in the guise of a passer-by at the bottom of the cliff, from which he pretends to have seen Gloucester fall. Here he assumes explicitly the role of spiritual guide to his father.

Gloucester, dismayed to find himself still alive, attempts to refuse help: "Away, and let me die."

And then Edgar, after an interval of several lines in which he represents himself as a stranger, speaks the filial (and fatherly) line about which my thoughts have gathered:

Thy life's a miracle. Speak yet again.

This is the line that calls Gloucester back—out of hubris, and the damage and despair that invariably follow—into the properly subordinated human life of grief and joy, where change and redemption are possible.

The power of that line read in the welter of innovation and speculation of the bioengineers will no doubt be obvious. One immediately recognizes that suicide is not the only way to give up on life. We know that creatures and kinds of creatures can be killed, deliberately or inadvertently. And most farmers know that any creature that is sold has in a sense been given up on; there is a big difference between selling this year's lamb crop, which is, as such, all that it can be, and selling the breeding flock or the farm, which hold the immanence of a limitless promise.

A little harder to compass is the danger that we can give up on life also by presuming to "understand" it—that is by reducing it to the *terms* of our understanding and by treating it as predictable or mechanical. The most radical influence of reductive science has been the virtually universal adoption of the idea that the world, its creatures, and all the parts of its creatures are machines—that is, that there is no difference between creature and artifice, birth and manufacture, thought and computation. Our language, wherever it is used, is now almost invariably conditioned by the assumption that fleshly bodies are machines full of mechanisms, fully compatible with the mechanisms of medicine, industry, and commerce; and that minds are computers fully compatible with electronic technology.

This may have begun as a metaphor, but in the language as it is used (and as it affects industrial practice) it has evolved from metaphor through equation to identification. And this usage institutionalizes the human wish, or the sin of wishing, that life might be, or might be made to be, predictable.

I have read of Werner Heisenberg's principle that "Wherever one treats living organisms as physicochemical systems they must necessarily behave as such." I am not competent to have an opinion about the truth of that. I do feel able to say that whenever one treats living organisms as machines they must necessarily be *perceived* to behave as such. And I can see that the proposition is reversible: Whenever one perceives living organisms as machines they must necessarily be treated as such. William Blake made the same point earlier in this age of reduction and affliction:

> What seems to Be, Is, To those to whom
> It seems to Be, & is productive of the most dreadful
> Consequences to those to whom it seems to Be . . .

For quite a while it has been possible for a free and thoughtful person to see that to treat life as mechanical or predictable or understandable is to reduce it. Now, almost suddenly, it is becoming clear that to reduce life to the scope of our understanding (whatever "model" we use) is inevitably to enslave it, make property of it, and put it up for sale.

This is to give up on life, to carry it beyond change and redemption, and to increase the proximity of despair.

Cloning—to use the most obvious example—is not a way to improve sheep. On the contrary, it is a way to stall the sheep's lineage and make it unimprovable. No true breeder could consent to it, for true breeders have their farm and their market in mind, and always are trying to breed a better sheep. Cloning, besides being a new method of sheep-stealing, is only a pathetic attempt to make sheep predictable. But this is an affront to reality. As any shepherd would know, the scientist who thinks he has made sheep predictable has only made himself eligible to be outsmarted.

The same sort of limitation and depreciation is involved in the proposed cloning of fetuses for body parts, and in other extreme measures for prolonging individual lives. No individual life is an end in itself. One can live fully only by participating fully in the succession of the generations, in death as well as in life. Some would say (and I am one of them) that we can live fully only by making ourselves as answerable to the claims of eternity as to those of time.

The problem, as it appears to me, is that we are using the wrong language. The language we use to speak of the world and its creatures, including ourselves, has gained a certain analytical power (along with a lot of expertish pomp) but has lost much of its power to designate *what* is being analyzed or to convey any respect or care or affection or devotion toward it. As a result we have a lot of genuinely concerned people calling upon us to "save" a world which their language simultaneously reduces to an assemblage of perfectly featureless and dispirited "ecosystems," "organisms," "environments," "mechanisms," and the like. It is impossible to prefigure the salvation of the world in the same language by which the world has been dismembered and defaced.

By almost any standard, it seems to me, the reclassification of the world from creature to machine must involve at least a perilous reduction of moral complexity. So must the shift in our attitude toward the creation from reverence to understanding. So must the shift in our perceived relationship to nature from that of steward to that of absolute owner, manager, and engineer. So even must our permutation of "holy" to "holistic."

At this point I can only declare myself. I think that the poet and scholar Kathleen Raine was correct in reminding us that life, like holiness, can be known only by being experienced. To experience it is not to "figure it out" or even to understand it, but to suffer it and rejoice in it as it is. In suffering it and rejoicing in it as it is, we know that we do not and cannot understand it completely. We know, moreover, that we do not wish to have it appropriated by somebody's claim to have understood it. Though we have life, it is beyond us. We do not know how we have it, or why. We do not know what is going to happen to it, or to us. It is not predictable; though we can destroy it, we cannot make it. It cannot, except by reduction and the grave risk of damage, be controlled. It is, as Blake said, holy. To think otherwise is to enslave life, and to make, not humanity, but a few humans its predictably inept masters.

We need a new Emancipation Proclamation, not for a specific race or species, but for life itself—and that, I believe, is precisely what Edgar urges upon his once presumptuous and now desperate father:

> Thy life's a miracle. Speak yet again.

Gloucester's attempted suicide is really an attempt to recover control over his life—a control he believes (mistakenly) that he once had and has lost:

> O you mighty gods!
> This world I do renounce, and in your sights
> Shake patiently my great affliction off.

The nature of his despair is delineated in his belief that he can control his life by killing himself, which is a paradox we will meet again three and a half centuries later at the extremity of industrial warfare when we believed that we could "save" by means of destruction.

Later, under the guidance of his son, Gloucester prays a prayer that is exactly opposite to his previous one—

> You ever-gentle gods, take my breath from me;
> Let not my worser spirit tempt me again
> To die before you please

—in which he renounces control over his life. He has given up his life as an understood possession, and has taken it back as miracle and mystery. And his reclamation as a human being is acknowledged in Edgar's response: "Well pray you, father."

It seems clear that humans cannot significantly reduce or mitigate the dangers inherent in their use of life by accumulating more information or better theories or by achieving greater predictability or more caution in their scientific and industrial work. To treat life as less than a miracle is to give up on it.

*

I am aware how brash this commentary will seem, coming from me, who have no competence or learning in science. The issue I am attempting to deal with, however, is not knowledge but ignorance. In ignorance I believe I may pronounce myself a fair expert.

One of our problems is that we humans cannot live without acting; we *have* to act. Moreover, we *have* to act on the basis of what we know, and what we know is incomplete. What we have come to know so far is demonstrably incomplete, since we keep on learning more, and there seems little reason to think that our knowledge will become significantly more complete. The mystery surrounding our life probably is not significantly reducible. And so the question of how to act in ignorance is paramount.

Our history enables us to suppose that it may be all right to act on the basis of incomplete knowledge *if* our culture has an effective way of telling us that our knowledge is incomplete, and also of telling us how to act in our state of ignorance. We may go so far as to say that it is all right to act on the basis of sure knowledge, since our studies and our experience have given us knowledge that seems to be pretty sure. But apparently it is dangerous to act on the assumption that sure knowledge is complete knowledge—or on the assumption that our knowledge will increase fast enough to outrace the bad consequences of the arrogant use of incomplete knowledge. To trust "progress" or our putative "genius" to solve all the problems that we cause is worse than bad science; it is bad religion.

A second human problem is that evil exists and is an ever-present and lively possibility. We know that malevolence is always ready to appropriate the means that we have intended for good. For example, the technical means that have industrialized agriculture, making it (by very limited standards) more efficient and productive and easy, have also made it more toxic, more violent, and more vulnerable—have made it, in fact, far less dependable if not less predictable than it used to be.

One kind of evil certainly is the willingness to destroy what we cannot make—life, for instance—and we have greatly enlarged our means of doing that. And what are we to do? Must we let evil and our implication in it drive us to despair?

The present course of reductive science—as when we allow agriculture to be invaded by the technology of war and the economics of industrialism—*is* driving us to despair, as witness the incidence of suicide among farmers.

If we lack the cultural means to keep incomplete knowledge from becoming the basis of arrogant and dangerous behavior, then the intellectual disciplines themselves become dangerous. What is the point of the further study of nature if that leads to the further destruction of nature? To study the "purpose" of the organ within the organism or of the organism within the ecosystem is *still* reductive if we do so with the assumption that we will or can finally figure it out. This simply captures the world as the subject of present or future "understanding," which will become the basis of further industrial and commercial optimism, which will become the basis of further exploitation and destruction of communities, ecosystems, and local cultures.

I am not of course proposing an end to science and other intellectual disciplines, but rather a change of standards and goals. The standards of our behavior must be derived, not from the capability of technology, but from the nature of places and communities. We must shift the priority from production to local adaptation, from innovation to familiarity, from power to elegance, from costliness to thrift. We must learn to think about propriety in scale and design, as determined by human and ecological health. By such changes we might again make our work an answer to despair.

II. Propriety

MY GENERAL concern is with what I take to be the increasing inability of the scientific, artistic, and religious disciplines to help us address the issue of propriety in our thoughts and acts. "Propriety" is an old term, even an old-fashioned one, and is not much in favor. Its value is in its reference to the fact that we are not alone. The idea of propriety makes an issue of the fittingness of our conduct to our place or circumstances, even to our hopes. It acknowledges the always-pressing realities of context and of influence; we cannot speak or act or live out of context. Our life inescapably affects other lives, which inescapably affect our life. We are being measured, in other words, by a standard that we did not make and cannot destroy. It is by that standard, and only by that standard, that we know we are in a crisis in our relationship to nature. The term "environmental crisis," crude and inexact as it is, acknowledges that we have invoked this standard and have measured ourselves by it. A civilization that is destroying all of its sources in nature has raised starkly the issue of propriety, whether or not it wishes to have done so.

Propriety is the antithesis of individualism. To raise the issue of propriety is to deny that any individual's wish is the ultimate measure of the world. The issue presents itself as a set of questions: Where are we? (This question applies, with as much particularity as human competence will allow, to all of the world's millions of small localities.) Who are we? (The proper answer to this question depends on where we are and where we have been, and it includes history.) What is our condition? (This is a *practical* question.) What are our abilities? (This also is a practical question. It refers to abilities that are *proven*, not to abilities that are theoretical or potential, such as "aptitude" or I.Q.) What appropriately may we do in our own interest *here*? (And this question submits to the standard of the health of the place.) These questions address themselves to all the disciplines, but they do not call for specialized answers. They cannot, I think, be answered by specialists—or not, at least, by specialists in isolation from one another.

To ask such questions seriously now is not quite absurd, for the questions are valid and urgent, but it is nonetheless to risk a sort of comedy, for the questions are as foreign to our sciences and arts as presently practiced, and to our institutions of government, learning, and religion, as they are to the global corporations whose existence depends upon their (and our) willingness to ignore any such questions.

All of the disciplines are increasingly identifiable as professionalisms, which are increasingly conformable to the aims and standards of industrialism. All of the disciplines are failing the test of propriety because they are failing the test of locality. The professionals of the disciplines don't *care* where they are. Though they are inescapably in context, they assume or pretend that they think and work without context. They subscribe to the preeminence of the mind and (logically from that) of the career. The questions of propriety, calling as they must for local answers, call necessarily for *small* answers. But small local answers are now as far beneath the notice of professionalism as of commercialism. Professionalism aspires to *big* answers that will make headlines, money, and promotions. It longs, moreover, for answers that are uniform and universal—the same styles, explanations, routines, tools, methods, models, beliefs, amusements, etc., for everybody everywhere. And like the corporations, whose appetite for "growth" seems now ungovernable, the institutions of government, education, and religion are now all too likely to measure their success in terms of size and number. All the institutions seem to have learned to imitate the organizational structures and to adopt the values and aims of industrial corporations. It is astonishing to realize how quickly and shamelessly doctors and lawyers and even college professors have taken to drumming up trade, and how readily hospitals, once run according to the laws of healing, mercy, and charity, have submitted to the laws of professionalism, industrial methodology, careerism, and profit.

This is happening to all the disciplines, but because science is the most influential category of the disciplines, and increasingly has set the pattern for the rest, we must be concerned first of all with science. Stephen Edelglass, Georg Maier, Hans Gebert, and John Davy in their book, *The Marriage of Sense and Thought*, wrote that "Science now functions in society rather as the Church did in the Middle Ages." What kind of religion science is, and how it works as such, are questions we will have to deal with.

One used to hear a great deal about "pure science." The universities, one was given to understand, were full of scientists who were disinterestedly pursuing truth. "Pure science" did not permit the scientist to ask so crude and pragmatic a question as *why* this or that truth was being pursued; it was just assumed, not only that to know the truth was good, but that, once the truth was discovered, it would somehow be *used* for good. This is a singularly naive view of science (as it would be of any human enterprise), but it survived at least into the early days of space exploration, when a lot of aficionados of so-called high technology assumed that NASA existed to sponsor voyages of pure discovery: to learn whatever might be learned, to take pictures of the earth and other planets, and to provide extremely expensive mystical experiences to astronauts. Some people believed that this enterprise was really a sort of spiritual quest, and would always remain above the gross concerns of, for example, the military-industrial complex. It would promote instead a renewed tenderness toward our "planet" by such devices as pictures of half of said planet, taken at a distance that reduced it to a blue bauble something like a Christmas tree ornament. In our foolish insistence on substituting technology for vision, we forget that we are not the first to have seen "the whole earth" from such a distance. Dante saw it from a higher level of human accomplishment, and at far less economic and ecological cost, several hundred years before NASA.

The possibility of pure science was significantly diminished, surely, by the time early scientists had invented metallurgy and then gunpowder, and it diminished steadily from then on. By now, when the possibilities of application have so enormously multiplied and the greed of corporations has grown so elaborate that they wish to patent discoveries before they have been discovered, it appears safest to assume that all sciences are "applied." Science may at times have been altruistically applied. But even such nominally altruistic sciences as medicine and plant-breeding have now become so deeply interpenetrated with economics and politics that their motives are at best mixed with, and at worst replaced by, the motives of corporations and governments. If nothing else, the increasing costliness of the practice of conventional science, and its consequent dependence on large grants or investments, would mitigate against its purity. One can only assume that pure science now needs to move fast (and beg hard) to keep its skirts from being lifted by the ever randy and handy corporate giants.

As I have already confessed, I am not at all a scientist. And yet, like every human inhabitant of the modern world, I have experienced many of the effects (costs and benefits) of science; I have received a great deal of hearsay of it; and I know that I am always under its influence and at its mercy. Though I am unable to comment on its methods or the truth of its discoveries, I am nonetheless appropriately interested in its motives—in what it thinks it is doing and in how it justifies itself. I agree with the proposition that science (or "science-and-technology") has now become a sort of religion; I am aware also that in many ways it rules over us. I want to know by what power it has crowned and mitered itself.

I believe it is generally agreed that "science" means knowledge of a special kind: factual knowledge that can be proven by measure, that can stand up to empirical testing. Scientific knowledge is the hard cash of the modern economy of thought; its worth is constant, no matter who has it; its value is not derived from belief or opinion or speculation or desire. Once established, it cannot be argued about.

Science has to do, famously, with theory. "Theory," at root, is related to the word "theater"; it has to do with watching, with observation. A scientific theory is an aid to observation. It involves assumptions that appear to be consistent with known facts. It is not proven; it is useful because it may lead to evidence or to proofs.

Science also involves prediction. Prediction is a highly disciplined concept when it is used in relation to the methodology of proof: A thing is true only if it is *predictably* true; a thing is true, not because it is true now, but because it is true always. But in the hands of such "scientists" as meteorologists and economists, whose putative usefulness depends directly upon their ability to predict and whose predictions are frequently wrong, the meaning of prediction begins to slide from science toward journalism. The same slide occurs when scientists, on the basis of early results, predict the success of a course of experimentation. Alert readers of newspapers will certainly have noticed the frequency of reports that scientists "may have" discovered something or other, or that new data "may prove" something or other. Journalists, and apparently some scientists also, are partial to news stories beginning "Scientists foresee" or "Scientists predict."

This seems to come from abuse of faith, which is another essential attribute of science. There is a sort of scientific faith that is legitimate. It is hard to see how the work of science could be done if scientists did not have faith in the workability and soundness of their methods. This is not faith of the highest sort, obviously, but is akin to the unproven confidence with which we non-scientists face the unknowns of our own workdays. But under various suasions of profession and personality, this legitimate faith in scientific methodology seems to veer off into a kind of religious faith in the power of science to know all things and solve all problems, whereupon the scientist may become an evangelist and go forth to save the world.

This religification and evangelizing of science, in defiance of scientific principles, is now commonplace and is widely accepted or tolerated by people who are not scientists. We really seem to have conceded to scientists, to the extent of their own regrettable willingness to occupy it, the place once occupied by the prophets and priests of religion. This can have happened only because of a general abdication of our responsibility to be critical and, above all, self-critical.

Why is there not a robust, profoundly questioning criticism of science within the scientific disciplines? One reason, I assume, is that such self-criticism, especially in public, would be considered "unprofessional." Another reason is that the modern sciences, working always in such proximity to "application," are simply too lucrative or too potentially lucrative to be self-critical. The professions increasingly have adopted the standards and thought patterns of business: If you're making money, what can be wrong? The criticism of science most familiar to ordinary citizens is more than likely to take the form of a public protest against some ruinous local manifestation of applied science. The most ubiquitous and unignorable result of modern chemistry, for example, is pollution, but typically this result is dealt with by ordinary citizens, not by chemists.

In 1959, C. P. Snow spoke of science as having an "automatic corrective." At that time, maybe, one could reasonably suppose that "pure" science, safely withdrawn from application, might by its own processes of experimentation and proof more or less automatically correct itself. By now we know that the applied sciences are subject to no such corrective. The scale of experimentation has become too greatly enlarged, for now science may be said to be conducting many of its experiments on the scale of the world. Among the results are Chernobyl, the ozone hole, the acceleration of species extinction, and universal pollution.

If there are critics of science in the governments and the bureaucracies, they are largely inaudible. In the universities, the scientists generally proceed from promotion to promotion and from grant to grant, leaving few recorded moments of conscience or professional self-doubt; and the professors of the humanities seem for the most part merely to be abashed by the sciences, deferring to their certainties, adopting their values, admiring their wealth, and longing even to imitate their methodology and their jargon. The journalists think it intellectually chic to stand open-mouthed before any wonder of science whatsoever. The media, cultivating their mediocrity, seem quite comfortably unaware that many of the calamities from which science is expected to save the world were caused in the first place by science—which meanwhile is busy propagating further calamities, hailed now as wonders, from which later it will undertake to save the world. Nobody, so far as I have heard, is attempting to figure out how much of the progress resulting from this enterprise is *net*. It is as if a whole population has been genetically deprived of the ability to subtract.

I know that there are some scientists who are speaking and writing sound criticism of science or of scientific abuses of science, but these people seem to have the status of dissidents or heretics; they are not accepted as partners in a necessary dialogue. Typically, their criticisms and objections are not even answered. (If you are making money and have power, why debate?) In short, the scientific critics of science are not effective. That there has been no effective criticism of science is demonstrated, for instance, by science's failure to attend to the possibility of small-scale or cheap or low-energy or ecologically benign technologies. Most applications of science to our problems result in large payments to large corporations and in damages to ecosystems and communities. These eventually will have to be subtracted (but not, if they can help it, by the inventors or manufacturers) from whatever has been gained.

III. *On Edward O. Wilson's* Consilience

APPARENTLY everywhere in the "developed world" human communities and their natural and cultural supports are being destroyed, not by natural calamities or "acts of God" or invasion by foreign enemies, but by a sort of legalized vandalism known as "the economy." The economy now famously depends upon the authority and the applicable knowledge of science. It would therefore be useful to say what is the character of this science that has benefited us in so many ways, and yet has cost us so dearly and exacted from us such deference and such questionable permissions.

Since I am not able to conduct any kind of survey, I will focus my study upon a single book: *Consilience*, by Edward O. Wilson. I am aware of the several objections to treating any one book as representative, but I am encouraged to do so, not only by the advantage of economy, but also by my belief that Mr. Wilson's assumptions are widely shared both by his colleagues and by non-scientists, and that there is no idea in his book that would be surprising to any fairly regular reader of articles on science in a daily newspaper. I think, in short, that despite his pretensions to iconoclasm, Mr. Wilson speaks for a popular scientific orthodoxy. His book reads as though it was written to confirm the popular belief that science is entirely good, that it leads to unlimited progress, and that it has (or will have) all the answers.

I am interested in Mr. Wilson's book also because, like me, he is a conservationist. We have perhaps some other things in common, but we have differences too, and these concern me just as much and (as will presently be evident) more seriously and at greater length. Our fundamental difference may be that he is a university man through and through, and I have always been most comfortable out of school. Whereas Mr. Wilson apparently is satisfied with the modern university's commitment to departmented specialization, professional standards, industry-sponsored research, and a scheme of promotion and tenure based upon publication, I am distrustful of that commitment and think it has done harm, both to learning and to the world.

Obviously, I have no authority from which to question Mr. Wilson's scientific knowledge, which I believe to be great and admirable, as human knowledge goes. My interest, rather, is in his attitudes toward what he knows and (more important) what he doesn't know. It is in these attitudes, I believe, that he is most conventional, and in them he most conforms to the values and the psychology of industrialism.

To show how Mr. Wilson stands on the issue of knowledge and ignorance, I will catalogue here his fundamental biases and assumptions as they appear in his book, observing when he is being properly scientific, and when he is not. *Consilience* is in effect a scientific credo; its opinions are plainly stated.

I. MATERIALISM

Mr. Wilson is, to begin with, a materialist. He believes that this is "a lawful material world," all the laws of which can be explained and understood empirically, and are subject to scientific proofs. He holds that "all tangible phenomena, from the birth of stars to the workings of social institutions, are based on material processes that are ultimately reducible . . . to the laws of physics."

Science is an enterprise of materiality, dealing in empirical proofs, in the tangible, the measurable, and the countable. And so, in terms of procedure, there can be no objection to Mr. Wilson's materialism; he is a scientist, and from a scientist we require truths that are materially verifiable. But as a doctrine of belief, materialism takes him into several kinds of trouble. These troubles show up pretty plainly against his obviously genuine concern for conservation of the natural world.

A minor problem, perhaps, is the tendency of materialism to objectify the world, dividing it from the "objective observer" who studies it. The world thus becomes "the environment," a word which Mr. Wilson uses repeatedly when speaking of conservation, and which means "surroundings," a place that one is *in* but not *of*. The question raised by this objectifying procedure and its vocabulary is whether the problems of conservation can be accurately defined by an objective observer who observes at an intellectual remove, forgetting that he eats, drinks, and breathes the so-called environment.

A more serious problem is this: A theoretical materialism so strictly principled as Mr. Wilson's is inescapably deterministic. We and our works and acts, he holds, are determined by our genes, which are determined by the laws of biology, which are determined ultimately by the laws of physics. He sees that this directly contradicts the idea of free will, which even as a scientist he seems unwilling to give up, and which as a conservationist he cannot afford to give up. He deals with this dilemma oddly and inconsistently.

First, he says that we have, and need, "the illusion of free will" which, he says further, is "biologically adaptive." I have read his sentences several times, hoping to find that I have misunderstood them, but I am afraid that I understand them. He is saying that there is an evolutionary advantage in illusion. The proposition that our ancestors survived because they were foolish enough to believe an illusion is certainly optimistic, but it does not seem very probable. And what are we to think of a materialism that can be used to validate an illusion? Mr. Wilson nevertheless insists upon his point; in another place he speaks of "self-deception" as granting to our species the "adaptive edge."

Later, in discussing the need for conservation, Mr. Wilson affirms the Enlightenment belief that we can "choose wisely." How a wise choice can be made on the basis of an illusory freedom of will is impossible to conceive, and Mr. Wilson wisely chooses not to try to conceive it.

2. MATERIALISM AND MYSTERY

Furthermore, as internally awry and inconsistent as it is, Mr. Wilson makes of his materialism a little platform from which to look down upon (as he thinks) and patronize the opposition. He is a *militant* materialist, with a doctrinaire intolerance for any sort of mystery—or, for that matter, any sort of ambiguity or uncertainty. He understands mystery as attributable entirely to human ignorance, and thereby appropriates it for the future of human science; in his formula, the unknown = the to-be-known. I will have more to say about this presently. For now, it is enough merely to notice that he has no ability to confront mystery (or even the unknown) as such, and therefore has learned none of the lessons that humans have always learned when they have confronted mystery as such. His book is an exercise in a sort of academic hubris.

If modern science is a religion, then one of its presiding deities must be Sherlock Holmes. To the modern scientist as to the great detective, every mystery is a problem, and every problem can be solved. A mystery can exist only because of human ignorance, and human ignorance is always remediable. The appropriate response is not deference or respect, let alone reverence, but pursuit of "the answer."

This pursuit, however, is properly scientific only so long as the mystery is empirically or rationally solvable. When a scientist denies or belittles a mystery that cannot be solved, then he or she is no longer within the bounds of science.

Thus, when Mr. Wilson asserts that *Paradise Lost* owes nothing to "God's guidance of Milton's thoughts, as the poet himself believed," he is talking far beyond the reach of proof. He does not consider that *Paradise Lost* is the poem it is because Milton was a man faithful and humble enough to invoke the assistance of the "Heav'nly Muse." The only empirical truth available here is that Milton, believing, wrote *Paradise Lost*, and that Mr. Wilson, disbelieving, wrote *Consilience*, a book of a different order.

Believing that whatever is intangible does not exist, Mr. Wilson like many materialists, atheists, rationalists, realists, etc., thinks he has struck a killing blow against religious faith when he has asked to see its evidence. But of course religious faith *begins* with the discovery that there is no "evidence." There is no argument or trail of evidence or course of experimentation that can connect unbelief and belief.

By insisting upon so narrow a definition of reality, Mr. Wilson does not defeat religion, but only misunderstands it. He does not appreciate, because he cannot suspect, the possibility that religious faith may be a way of knowing things that cannot otherwise be known. Misunderstanding religion, he often appropriates and misuses such words as *transcendent*, *create*, *archetype*, *reverence*, and *sacred*.

In his chapter on "Ethics and Religion," he has "constructed a debate" between "the transcendentalist" (a straw man) and "the empiricist" (a stuffed shirt). The empiricist, having found the transcendentalist's simple-minded argument (rigged for defeat by Mr. Wilson) to be lacking in "objective evidence" and "statistical proofs," then proceeds to make the same sort of irrational swerve that Mr. Wilson himself made previously in affirming "the illusion of free will." Empiricism, the empiricist concedes, "is bloodless. People need . . . the poetry of affirmation. . . . It would be a sorry day if we abandoned our venerated sacral traditions. It would be a tragic misreading of history to expunge *under God* from the American Pledge of Allegiance. Whether atheists or true believers, let oaths be taken with hand on the Bible. . . . Call upon priests and ministers and rabbis to bless civil ceremony with prayer . . ." All this, Mr. Wilson calls "the presence of poetry."

"But to share reverence," the empiricist continues hopefully, "is not to surrender the precious self. . . ." A language can hardly endure this sort of abuse. It is impossible to tell what Mr. Wilson may mean by "share reverence," but to *feel* reverence, to *be* reverent, is exactly to surrender the "precious self," and is nothing else.

And then the empiricist administers his coup de grâce: "We can be proud as a species because, having discovered that we are alone, we owe the gods very little." This would be a noble blasphemy, like that of Job's wife, if the empiricist or Mr. Wilson believed in the gods, but neither one of them does. It is only a weary little cliché of a too familiar "scientific" iconoclasm—hubris without a bang.

But this materialism raises a question that *Consilience* does not acknowledge. If at last "all tangible phenomena" are empirically reduced to the laws of physics, then we will merely have completed a circle. We will have arrived again at the question that preceded Genesis: Where did the physical world come from? And physics of course can have no answer.

The principal point to be made, so far, is that Mr. Wilson's initially reputable materialism has led him far beyond objective evidence and statistical proofs and into what looks very much like poppycock. More examples will follow.

3. IMPERIALISM

Mr. Wilson's scientific "faith" (as he sometimes calls it) is in the ultimate empirical explainability of everything—that is, the "consilience" of all the disciplines "by the linking of facts and fact-based theory . . . to create a common groundwork of explanation." He concedes from time to time that he may be wrong, but this doubt is a mere gesture, entirely absent from the passages in which he is elatedly confident that he is right. His humility is only etiquette, not a conviction or even an attitude. It is not involved in his thought, and his thought is unaffected by it. His book contains, moreover, several passages in which he writes with candor of the great difficulties and perhaps the impossibilities of the consilience he wishes for, but these also do not dampen or qualify his passionate conviction. Waving aside ignorance and mystery and human limitation as merely illusory or irrelevant, he claims not only all knowledge but all future knowledge and everything unknown as the property of science. So confident is he in his power to know that his book abounds in one-sentence definitions of, for examples, the mind, consciousness, meaning, mood, creativity, insanity, art, science, the self (all of these are in one chapter, "The Mind"). The definitions themselves are mostly jargon, singularly unhelpful—"What we call *meaning* is the linkage among the neural networks created by the spreading excitation that enlarges imagery and engages emotion"—but their purpose is clear enough: It is to take possession of the subject in the name of the idea of consilience.

This idea is explicitly imperialistic, and it is implicitly tyrannical. Mr. Wilson is perfectly frank about his territorial ambitions. He wishes to see all the disciplines linked or unified—but strictly on the basis of science. Non-scientists are not invited to the negotiations, or at least they are not to participate on their own terms: "The key to the exchange between [science and the arts] is . . . reinvigoration of interpretation with the knowledge of science and its proprietary sense of the future." And if you have any doubt of the political and economic implications of modern science and of Mr. Wilson's advocacy, consider the following: "Governmental and private patrons of the brain scientists, like royal geographic commissions of past centuries, are generous. They know that history can be made by a single sighting of coastland, where inland lies virgin land and the future lineaments of empire."

Mr. Wilson's book scarcely acknowledges the existence of politics, and perhaps for good reason, for the putative ability to explain everything along with the denial of religion (or the appropriation of its appearances) is a property of political tyranny. So is the belief that one's explanations will save the world from some great threat. And Mr. Wilson himself, in his conviction that everything that is not a science *ought* to be and *will* be, shows himself to be a man with a fiercely proprietary mind and dire intentions toward the unenlightened.

The logic of his position is clear, and it is most disquieting. Would not, for example, any social theory be almost inescapably totalitarian if it were to be (in Mr. Wilson's terms) general, consilient, and predictive? Or if it *thought* it was, for the supposition would be just as dangerous to freedom as the fact. To suppose that the theory was predictive would prepare the way for a bloody "exogenous shock" that it did not predict. Mr. Wilson, in fact, alludes briefly to such possibilities in his chapter "The Enlightenment," but he does not confront the issue; he races past it and within three pages he is praising the Enlightenment's "new freedom": "It waved aside everything, every form of religious and civil authority, every imaginable fear, to give precedence to the ethic of free inquiry." The direction of Mr. Wilson's consilience, prescribed and announced, is toward an empirical dogma of dead certainty, "a common groundwork of explanation," the terms of which would exclude whatever cannot be empirically explained. Mere "humanitarianism" would do the rest: If you know with certainty what is true, should you not *enforce* the truth? If you see your poor subjects struggling and suffering in their error, how can you rightly forbear to impose the necessary corrections? Having spied out the coast, why should you not extend inland the "lineaments of empire"?

And this, though with some of us it would not be all right, would be incontestable if we had good reason to believe that everything could be explained on the terms proposed, which fortunately we do not. The only science we have or can have is *human* science; it has human limits and is involved always

with human ignorance and human error. It is a fact that the
solutions invented or discovered by science have tended to lead
to new problems or to become problems themselves. Scientists
discovered how to use nuclear energy to solve some problems,
but any use of it is enormously dangerous to us all, and scientists have not discovered what to do with the waste. (They
have not discovered what to do with old tires.) The availability
of antibiotics leads to the overuse of antibiotics. And so on.
Our daily lives are a daily mockery of our scientific pretensions. We are learning to know precisely the location of our
genes, but significant numbers of us don't know the whereabouts of our children. Science does not seem to be lighting
the way; we seem rather to be leapfrogging into the dark along
series of scientific solutions, which become problems, which
call for further solutions, which science is always eager to supply, and which it sometimes cannot supply. Sometimes it fails
us infamously and fearfully. The so-called Y2K problem—the
failure to manufacture computers capable of recognizing the
year 2000—could have been prevented by perfectly ordinary
human foresight, as limited as that is. The coming of the year
2000 could have been foretold by every child old enough to
count. But among all the scientists who helped to develop
"computer science," all who taught it in great universities, all
who used it in their work, all who evangelized it as the answer
to every intellectual problem, apparently no one with authority
foresaw the Y2K fiasco until almost too late.

For a long time we humans have fairly successfully (but not
invariably) avoided error within our systems of thought, but
the systems themselves have often proved to be wrong. That
is, our systems have made it possible (within the limits of the
systems) to be consistent, but they have not preserved us from
error. Our experience suggests that they cannot preserve us
from error. Should we regret this? Probably not, since it is always the errors of our systems that have released us (so far)
from the tyranny of our systems.

The presently dominant system of thought—which we
should call, not "science," but "science-technology-and-
industry"—has produced an unsurprising number of errors
and an unsurprising number of failures. It is hard to see how
our systems of thought could be other than fallible, once we

grant that they cannot be contrived except by fallible creatures; fallibility is an infection in us that we inevitably communicate to our works. Who does not know this? Most of us begin every day with some kind of plan, and every day we see that plan altered or foiled because it necessarily lacks the scope of our nature and character, let alone that of reality. Our view of the world and even of our own experience is always to some degree distorted, oversimplified, or reduced, and so is varyingly liable to be in error. That we are in error means that our plans or systems tend to suffer the interference of bad surprises—and, let us not forget, also of good surprises. It is impossible to argue that we can know empirically anything that is beyond our mental capacity. What we understand has necessarily been limited by the limits of our understanding. Mr. Wilson predicts approvingly that "The world henceforth will be run by synthesizers, people able to put together the right information at the right time, think critically about it, and make important choices wisely." Synthesis, he says, is "holism." He does not acknowledge that the synthesis he is talking about will be neither whole nor holy, but rather an artifact made of parts that we have isolated and in our fashion understood ("We murder to dissect") and put together again in a way we understand.

The fallibility of a human system of thought is always the result of incompleteness. In order to include some things, we invariably exclude others. We can't include everything because we don't know everything; we can't comprehend what comprehends us. The incompleteness of a system is rarely if ever perceptible to those who made it or to those who benefit from it. To those who are excluded from it, the incompleteness of a system is, or eventually becomes, plain enough. One weakness of the present system, which Mr. Wilson does not mind, is that it excludes all inscrutable and ineffable things, including the life history of the human soul.

Another weakness, which Mr. Wilson does mind, is that it excludes the principle or the standard of ecological health. Science-technology-and-industry has enabled us to be precise (apparently) in describing objects that are extremely small and near or extremely large and far away. It has failed utterly to provide us with even adequate descriptions of the places and communities we live in—probably because it *cannot* do so.

There are scientists, one must suppose, who know all about atoms or molecules or genes, or galaxies or planets or stars, but who do not know where they are geographically, historically, or ecologically. Our schools are turning out millions of graduates who do not know, in this sense, where they are. Certain lamentable results predictably follow. Mr. Wilson thinks the present system can correct itself merely by enlarging its present claims. I think the present system can correct itself only by conscientiously trying to include what it has so far excluded —which, of course, would make it an entirely different system with entirely different claims.

If one has a science that is manifestly incomplete—that is surrounded by mystery—and yet one believes with passionate intensity that it should, can, and will complete itself by means of a consilience of all the disciplines, then as an immediate and necessary effect one subjects one's language to a heavy strain. Mr. Wilson's project calls for a language of great assurance, but he is writing inescapably about what he does not know, and so his language often is necessarily tentative. The future that his thesis forces him to try to see into is finally as obscure to him as it is to the rest of us. His writing about consilience is always under the sway of conditional verbs, of protestations of faith, of "if" and "until" and "likely" and "perhaps." And so it is not scientific; it is not theorizing; it is only a fairly ordinary kind of human supposing, guided only by certain popular and professional prejudices.

But the most unscientific and the most disturbing thing about this book is Mr. Wilson's appropriation of whatever is unknown. He does this by variations on the themes of "until" or "not yet." He cannot bring himself to say that scientists do not know something; he must say that they do not know it *yet*; he must say that one thing cannot be known *until* another thing is known. He says repeatedly things like this: "The belief in the possibility of consilience beyond science and across the great branches of learning is not yet science. . . . It cannot be proved with logic from first principles or grounded in any definitive set of empirical tests, at least not by any yet conceived." This "not yet" forthrightly appropriates mystery as future knowledge. It takes possession of life and the future of life in the names of its would-be explainers—and, it follows, of its would-be exploiters. As soon as a mystery is scheduled for solution, it is no longer a mystery; it is a problem. The most tyrannic of all reductions has thus been accomplished; a self-aggrandizing science has thus asserted its "proprietary sense of the future."

The practical result of such language is a sort of moral blindness. We cannot derive sound thinking about propriety of scale and conduct from the proposition that what we need to know we do not know "yet." Such an idea simply overrides the issue of limits. Without a lively recognition of our own limits —chiefly of our knowledge and of our ability to know—we cannot even approach the issue of the limits of nature.

What is the possibility of "consilience beyond science and across the great branches of learning"?
We don't know yet.
Why do the innocent suffer?
We don't know yet.

I am not proposing, of course, that mental work of any sort can do without hypothesis or theory or any other way of articulating one's sense of possibility. Because our knowledge is discontinuous and we are ignorant of the future, we must grant the necessity of some manner of supposing from one point of evidence to another. At the same time it seems only fair to insist that this process should not be extended indefinitely without evidence, and that the points of evidence ought to be reasonably close together. In shallow water one may not risk much by postulating the existence of an as yet invisible stepping stone just beneath the surface. But if the water is deep and swift, one should not start across if some of the stepping stones are hypothetical. It is absurd to accumulate enormous quantities of nuclear waste, telling ourselves that we don't know *yet* how to dispose of it. We might face the future a good deal more confidently now if nuclear scientists had had the humility and the candor to say simply, We don't know.

Clearly, there ought to be a limit beyond which we cease to hedge our ignorance with promises to "continue to study the problem." Scrupulous minds, in this age as in any other, not only must be constrained occasionally to confess ignorance, but also must continue to live with the old proposition that some things are not knowable. *Consilience*, at any rate, shows us a man's mind leaping with exuberant confidence from one merely conjectural stepping stone to another, oblivious of the rushing waters.

4. REDUCTIONISM

Reductionism, like materialism, has uses that are appropriate, and it also can be used inappropriately. It is appropriately used as a way (one way) of understanding what is empirically known or empirically knowable. When it becomes merely an intellectual "position" confronting what is not empirically known or knowable, then it becomes very quickly absurd, and also grossly desensitizing and false. Like materialism, reductionism belongs legitimately to science; as an article of belief, it causes trouble.

According to Mr. Wilson, "Science . . . is the *organized, systematic enterprise that gathers knowledge about the world and condenses the knowledge into testable laws and principles*" [his italics]. He says further that "The cutting edge of science is reductionism, the breaking apart of nature into its natural constituents." And reductionism has "a deeper agenda," which is "to fold the laws and principles of each level of organization into those at more general, hence more fundamental levels. Its strong form is total consilience, which holds that nature is organized by simple universal laws of physics to which all other laws and principles can eventually be reduced." Toward the end of his book, Mr. Wilson adds the following: "There is abundant evidence to support and none absolutely to refute the proposition that consilient explanations are congenial to the entirety of the great branches of learning."

Mr. Wilson's definitions of science and reductionism, granting him his prejudices, seem to me perfectly appropriate. His definition of consilience, however, like his exposition of it, becomes more contestable the farther it goes.

There obviously is a necessary usefulness in the processes of reduction. They are indispensable to scientists—and to the rest of us as well. It is valuable (sometimes) to know the parts of a thing and how they are joined together, to know what things do and do not have in common, and to know the laws or principles by which things cohere, live, and act. Such inquiries are native to human thought and work.

But reductionism also has one inherent limitation that is paramount, and that is abstraction: its tendency to allow the particular to be absorbed or obscured by the general. It is a curious paradox of science that its empirical knowledge of the material world gives rise to abstractions such as statistical averages which have no materiality and exist only as ideas. There is, empirically speaking, no average and no type. Between the species and the specimen the creature itself, the individual creature, is lost. Having been classified, dissected, and explained, the creature has disappeared into its class, anatomy, and explanation. The tendency is to equate the creature (or its habitat) with one's formalized knowledge of it. Mr. Wilson is somewhat aware of this problem for he insists upon the importance of "synthesis and integration." But he does not acknowledge that synthesis and integration are merely parts of an explanation, which is invariably and inevitably less than the thing explained. The synthesizing and integrating scientist is only ordering and making sense of as much as he knows. He is not making whole that which he has taken apart, and he should not claim credit for putting together what was already together.

The uniqueness of an individual creature is inherent, not in its physical or behavioral anomalies, but in its *life*. Its life is not its "life history," the typical cycle of members of its species from conception to reproduction to death. Its life is all that happens to it in its place. Its wholeness is inherent in its life, not in its physiology or biology. This wholeness of creatures and places together is never going to be apparent to an intelligence coldly determined to be empirical or objective. It shows itself to affection and familiarity.

The frequent insultingness of modern (scientific-technological-industrial) medicine is precisely its inclination to regard individual patients apart from their lives, as representatives or specimens of their age, sex, pathology, economic status, or some other category. The specialist to whom you have been "referred" may never have seen you before, may know nothing about you, and may never see you again, and yet he or she presumes to know exactly what is wrong with you. The same insultingness is now also a commonplace of politics, which treats individuals as representatives of racial, sexual, geographic, economic, ideological, and other categories, each with typical faults, complaints, rights, or virtues.

Science speaks properly a language of abstraction and abstract categories when it is properly trying to sort out and put in order the things it knows. But it often assumes improperly that it has said—or known—enough when it has spoken of "the cell" or "the organism," "the genome" or "the ecosystem" and given the correct scientific classification and name. Carried too far, this is a language of false specification and pretentious exactitude, never escaping either abstraction or the cold-heartedness of abstraction.

The giveaway is that even scientists do not speak of their loved ones in categorical terms as "a woman," "a man," "a child," or "a case." Affection requires us to break out of the abstractions, the categories, and confront the creature itself in its life in its place. The importance of this for Mr. Wilson's (and my) cause of conservation can hardly be overstated. For things cannot survive as categories but only as individual creatures living uniquely where they live.

We know enough of our own history by now to be aware that people *exploit* what they have merely concluded to be of value, but they *defend* what they love. To defend what we love we need a particularizing language, for we love what we particularly know. The abstract, "objective," impersonal, dispassionate language of science can, in fact, help us to know certain things, and to know some things with certainty. It can help us, for instance, to know the value of species and of species diversity. But it cannot replace, and it cannot become, the language of familiarity, reverence, and affection by which things of value ultimately are protected.

The abstractions of science are too readily assimilable to the abstractions of industry and commerce, which see everything as interchangeable with or replaceable by something else. There is a kind of egalitarianism which holds that any two things equal in price are equal in value, and that nothing is better than anything that may profitably or fashionably replace it. Forest = field = parking lot; if the price of alteration is right, then there is no point in quibbling over differences. One place is as good as another, one use is as good as another, one

life is as good as another—if the price is right. Thus political
sentimentality metamorphoses into commercial indifference or
aggression. This is the industrial doctrine of the interchangeability of parts, and we apply it to places, to creatures, and to
our fellow humans as if it were the law of the world, using
all the while a sort of middling language, imitated from the
sciences, that cannot speak of heaven or earth, but only of concepts. This is a rhetoric of nowhere, which forbids a passionate
interest in, let alone a love of, anything in particular.

Directly opposed to this reduction or abstraction of things
is the idea of the preciousness of individual lives and places.
This does not come from science, but from our cultural and
religious traditions. It is not derived, and it is not derivable,
from any notion of egalitarianism. If all are equal, none can
be precious. (And perhaps it is necessary to stop here to say
that this ancient delight in the individuality of creatures is not
the same thing as what we now mean by "individualism." It is
the opposite. Individualism, in present practice, refers to the
supposed "right" of an individual to act alone, in disregard of
other individuals.)

We now have the phenomenon of "mitigation banking" by
which a developer may purchase the "right" to spoil one place
by preserving another. Science can measure and balance acreages in this way just as cold-heartedly as commerce; developers
involved in such trading undoubtedly have the assistance of
ecologists. Nothing insists that one place is not interchangeable with another except affection. If the people who live in
such places and love them cannot protect them, nobody can.

It is not quite imaginable that people will exert themselves
greatly to defend creatures and places that they have dispassion-
ately studied. It is altogether imaginable that they will greatly
exert themselves to defend creatures and places that they have
involved in their lives and invested their lives in—and of course
I know that many scientists make this sort of commitment.

I have been working this morning in front of a window where I
have been at work on many mornings for thirty-seven years.
Though I have been busy, today as always I have been aware of
what has been happening beyond the window. The ground is
whitened by patches of melting snow. The river, swollen with

the runoff, is swift and muddy. I saw four wood ducks riding the current, apparently for fun. A great blue heron was fishing, standing in water up to his belly feathers. Through binoculars I saw him stoop forward, catch, and swallow a fish. At the feeder on the window sill, goldfinches, titmice, chickadees, nuthatches, and cardinals have been busy at a heap of free (to them) sunflower seeds. A flock of crows has found something newsworthy in the cornfield across the river. The woodpeckers are at work, and so are the squirrels. Sometimes from this out-look I have seen wonders: deer swimming across, wild turkeys feeding, a pair of newly fledged owls, otters at play, a coyote taking a stroll, a hummingbird feeding her young, a peregrine falcon eating a snake. When the trees are not in leaf, I can see the wooded slopes on both sides of the valley. I have known this place all my life. I long to protect it and the creatures who belong to it. During the thirty-seven years I have been at work here, I have been thinking a good part of the time about how to protect it. This is a small, fragile place, a slender strip of woodland between the river and the road. I know that in two hours a bulldozer could make it unrecognizable to me, and perfectly recognizable to every "developer."

The one thing that I know above all is that even to hope to protect it, I have got to break out of all the categories and confront it as it is; I must be present in its presence. I know at least some of the categories and value them and have found them useful. But here I am in my life, and I know I am not here as a representative white male American human, nor are the birds and animals and plants here as representatives of their sex or species. We all have our ways, forms, and habits. We all are what we are partly because we are here and not in another place. Some of us are mobile; some of us (such as the trees) have to be content merely to be flexible. All of us who are mobile are required by happenstance and circumstance and ac-cident to make choices that are not instinctive, and that force us out of categories into our lives here and now. Even the trees are under this particularizing influence of place and time. Each one, responding to happenstance and circumstance and acci-dent, has assumed a shape not quite like that of any other tree of its kind. The trees stand rooted in their mysteriously deter-mined places, no place quite like any other, in strange finality.

The birds and animals have their nests in holes and burrows and crotches, each one's place a little unlike any other in the world—and so is the nest my mate and I have made.

In all of the thirty-seven years I have worked here, I have been trying to learn a language particular enough to speak of this place as it is and of my being here as I am. My success, as I well know, has been poor enough, and yet I am glad of the effort, for it has helped me to make, and to remember always, the distinction between reduction and the thing reduced. I know the usefulness of reductive language. To know that I am "a white male American human," that a red bird with black wings is "a scarlet tanager," that a tree with white bark is "a sycamore," that this is "a riparian plant community"—all that is helpful to a necessary kind of thought. But when I try to make my language more particular, I see that the life of this place is always emerging beyond expectation or prediction or typicality, that it is unique, given to the world minute by minute, only once, never to be repeated. And then is when I see that this life is a miracle, absolutely worth having, absolutely worth saving.

We are alive within mystery, by miracle. "Life," wrote Erwin Chargaff, "is the continual intervention of the inexplicable." We have more than we can know. We know more than we can say. The constructions of language (which is to say the constructions of thought) are formed *within* experience, not the other way around. Finally we live beyond words, as also we live beyond computation and beyond theory. There is no reason whatever to assume that the languages of science are less limited than other languages. Perhaps we should wish that after the processes of reduction, scientists would return, not to the processes of synthesis and integration, but to the world of our creatureliness and affection, our joy and grief, that precedes and (so far) survives all of our processes.

5. CREATURES AS MACHINES

There is a reduction, by now more formulaic than procedural, that seems endemic to modern science, and from science it has spread everywhere. I mean the definition or identification of the world and all its creatures as "machines." This is one of the fundamental assertions in *Consilience*. The Enlightenment thinkers, Mr. Wilson says, encouraged us to "think of the world as God's machine"—"if you still insist on a divine intervention." A little later he says, "People, after all, are just extremely complicated machines." Further on, he says that we are "organic machines," and that "an organism is a machine."

This machine business may once have had meaning. It may have been a way of asserting belief in the integrity of Creation and the physical coherence of creatures; it may have been a way of insisting on the indispensability of part to whole. The machine, in other words, had a certain usefulness as a *metaphor*. But the legitimacy of a metaphor depends upon our understanding of its limits. A friend of mine remembered an aunt who noted, correctly, that when Jesus said "I am the door," He did not mean that He had hinges and a knob. We must be careful to remember that a profession is not altogether like a field, or a camera altogether like a room, or a pedigree altogether like a crane's foot.

When a metaphor is construed as an equation, it is out of control; when it is construed as an identity, it is preposterous. If we are to assume that our language means anything at all, then the world is not a machine, and neither is an organism. A machine, to state only the greatest and most obvious difference, is a human artifact, and a world or an organism is not.

But Mr. Wilson, like many others, is fond of this error, and he carries it further. He says, first, that "the brain is a machine," and then he says that "the mind . . . is the brain at work." To lock the subject into this definition, and so forestall any difference of opinion or the introduction of any contradictory evidence, he says, "The surest way to grasp complexity in the brain, as in any other biological system, is to think of it as an engineering problem."

The proposed theory of human mentality, then, is a simple formula: mind = brain = machine. That is to say that the mind is singular, material, and altogether what it is in itself—just as a machine (once made) is a machine per se, singular and material. This is pretty close to what I would call the Tarzan theory of the mind, invented by Edgar Rice Burroughs. The Tarzan theory holds that a human, raised entirely by apes, would have a mind nonetheless fully human: a human brain, in (so to speak) a social and cultural vacuum, would still function on its own as a human mind.

But this raises an interesting question: Is there such a thing as a mind which is merely a brain which is a machine? Would one have a mind if one had no body, or no body except for a brain (whether or not it is a machine)—if one had no sense organs, no hands, no ability to move or speak, no sensory pains or pleasures, no appetites, no bodily needs? If we grant (for the sake of argument) that such may be theoretically possible, we must concede at the same time that it is not imaginable, and for the most literal of reasons: Such a mind could contain no image.

And now let us grant the mind a body and all that the body brings with it and implies: sense, imagery, motion, desire. Another question arises: Would such a mind alone be a mind? Would one mind alone, alone and therefore lacking a language and any need for signs or signification, be recognizable as a mind? Again, we may grant that this is theoretically possible (though I see no reason why we should). But if we encountered such a mind, how would we know that it was a mind?

Suppose, then, that we put two embodied minds together, making them male and female, and yet deny them a habitat and any familiarity of place and time. Now we would be approaching something recognizably a mind, because we would have two. There would need to be some sort of language. These two minds would be different, they would have desires toward one another, they would need to negotiate. But without a dwelling place this language would be too poverty-stricken and crude to be called human. We can grant, as before, a theoretical possibility to this sort of mind, at the same time being forced to grant also that it would not be recognizably a human mind. (All of us have seen, and most of us have been, adolescent lovers who had nothing to talk about but themselves, and we know how intelligent *that* is.)

But we have been begging the question all along with our grantings of theoretical possibility, for one can't be a brain without a body, or a body (for very long) without a familiar homeland. To have one mind you have got to have at least two (and undoubtedly many more) and a world. We could call this the Adam and Eve theory of the mind. The correct formula, in fact, is more like this: mind = brain + body + world + local dwelling place + community + history. "History" here would mean not just documented events but the whole heritage of culture, language, memory, tools, and skills. Mind in this definition has become hard to locate in an organ, organism, or place. It has become an immaterial presence or possibility that is capable of being embodied and placed.

And here the difference between organisms and machines becomes clearer. The idea of a mind and the idea of an organism are not separate ideas. Or we could say that they are momentarily separable—but only momentarily—for the purposes of thought. Every living creature embodies enough mind to know how to be itself and survive in its place, else it cannot live. A machine embodies none of the mind that made it, and it has nothing of an organism's dependence on its world and community and place and history. A machine, if shot into outer space never to return, would simply go on and on being a machine; after it ran out of fuel or traveled beyond guidance, it would still be a machine. A human mind, necessarily embodied, if shot into outer space never to return, would die as soon as it went beyond its sustaining connections and references.

How far from home can a mind go and still be a mind? Probably no scientist has yet made this measurement, but we can answer confidently: Not too far. How far can a machine go from home (supposing, for the sake of argument, that a machine has a home) and still be a machine? Theoretically, if it is not destroyed, it can go on forever.

If the mind is incomplete without a home, without familiar associations and points of reference outside itself, then it becomes possible to argue that the longer the mind of an individual or a community is at home the better it may become. But this "better" implies the willingness and the ability to practice the virtues of domestic economy: frugality, continuous household maintenance and repair, neighborliness, good husbandry of soil and water, ecosystem and watershed.

If the mind fails in this earthly housekeeping, it can only get worse. How much can a mind diminish its culture, its community and its geography—how much topsoil, how many species can it lose—and still be a mind?

*

If the mind is as complexly formed as I have suggested, then it seems unlikely that the mind of an individual can be the origin of intelligence or truth, any more than an individual brain can be. And I fail to see how an individual brain alone can have any originating power whatsoever.

Of the material origin of intelligence or truth, or even of mind, any answer given will lead only to another question. To settle the matter, one would have to *see* experimentally the point at which some physical activity or excitement of the brain, alone, transformed itself into an original idea.

But how can an idea, which is not material, have a material origin? "Average," for example, is an idea which partakes of none of the physical properties of the things that are averaged. Materialism itself is an idea, just as immaterial as any other. And Edward O. Wilson, despite his materialism, shows himself to be a man as interested in ideas as in the material world.

If ideas are not material, how can they have a material origin? If they are not material in origin, how can their origin be explained by materialist science? This is the major fault line of Mr. Wilson's book: His interest in explaining the origin of things whose authentic existence is denied by the terms of the proposed investigation.

Anybody who thinks that this "scientific" reduction of creatures to machines is merely an issue to be pondered by academic intellectuals is in need of a second thought. I suppose that there are no religious implications in this reductionism, for if you think creatures are machines, you have no religion. For artists who do not think of themselves as machines, there is one artistic implication: Don't be mechanical. But the implications for politics and conservation are profound.

It is evident to us all by now that modern totalitarian governments become more mechanical as they become more total. Under *any* political system there is always a tendency to expect the government to work with mechanical "efficiency"—that is, with speed and no redundancy. (Mechanical efficiency always "externalizes" inefficiencies, such as exhaust fumes, but still one can understand the temptation.) Our system, however, which claims freedom as its purpose, involves several powerful concepts that tend to retard the speed and efficiency of government and to make it unmechanical: the ideas of government by consent of the governed, of minority rights, of checks and balances, of trial by jury, of appellate courts, and so on. If we were to implement politically the idea that creatures are machines, we would lose all of those precious impediments to mechanical efficiency in government. The basis of our rights and liberties would be undermined. If people are machines, what is wrong, for example, with slavery? Why should a machine wish to be free? Why should a large machine honor a small machine's quaint protestations that it has thoughts or feelings or affections or aspirations?

It is not beside the point to remember that our government at times has seen fit to look upon the prosperity of many small producers and manufacturers as a political and economic good, and so has placed appropriate restraints upon the mechanical efficiencies of monopolists and foreign competitors. It is not mechanically efficient to recognize that unrestrained competition between an individual farmer or storekeeper and a great corporation is neither democratic nor fair. I suppose that our so-called conservatives have at least no inconsistency to apologize for; they have espoused the "freedom" of the corporations and their "global economy," and they have no conflicting inhibitions in favor of democracy and fairness. The "liberals," on the other hand, have made political correctness the measure of their social policy at the same time that they advance the economic determinism of the conservatives. Reconciling these "positions" is not rationally possible; you cannot preserve the traditional rights and liberties of a democracy by the mechanical principles of economic totalitarianism.

But for the time being (may it be short) the corporations thrive, and they are doing so at the expense of everything else. Their dogma of the survival of the wealthiest (i.e. mechanical efficiency) is the dominant intellectual fashion. A letter to the *New York Times* of July 8, 1999 stated it perfectly: "While change is difficult for those affected, the larger, more efficient business organization will eventually emerge and industry consolidation will occur to the benefit of the many." When you read or hear those words "larger" and "more efficient" you may expect soon to encounter the word "inevitable," and this letter writer conformed exactly to the rule: "We should not try to prevent the inevitable consolidation of the farming industry." This way of talking is now commonplace among supposedly intelligent people, and it has only one motive: the avoidance of difficult thought. Or one might as well say that the motive is the avoidance of thought, for that use of the word "inevitable" obviates the need to consider any alternative, and a person confronting only a single possibility is well beyond any need to think. The message is: "The machine is coming. If you are small and in the way, you must lie down and be run over." So high a level of mental activity is readily achieved by terrapins.

The reduction of creatures to machines is in principle directly opposed to the effort of conservation. It is, in the first place, part and parcel of the determinism that derives from materialism. Conservation depends upon our ability to make qualitative choices affecting our influence on the ecosystems we live in or from. Machines can make no such choices, and neither, presumably, can creatures who are machines. If we are machines, we can only do as we are bidden to do by the mechanical laws of our mechanical nature. By what determinism we *regret* our involvement in our mechanical devastations of the natural world has not been explained by Mr. Wilson or (so far as I know) by anybody.

But suppose we *don't* subscribe to this determinism. Suppose we don't believe that creatures are machines. Then we must see the extent to which conservation has been hampered by this idea, whether consciously advocated by scientists or thoughtlessly mouthed about in the media and in classrooms. The widespread belief that creatures *are* machines obviously makes it difficult to form an advocacy for creatures *against* machines. To confuse or conflate creatures with machines not only makes it impossible to see the differences between them; it also masks the conflict between creatures and machines that under industrialism has resulted so far in an almost continuous sequence of victories of machines over creatures.

To say as much puts me on difficult ground, I know. To confess, these days, that you think some things are more important than machines is almost sure to bring you face to face with somebody who will accuse you of being "against technology" —against, that is, "the larger, more efficient business organization" that will emerge inevitably "to the benefit of the many."

And so I would like to be as plain as possible. What I am against—and without a minute's hesitation or apology—is our slovenly willingness to allow machines and the idea of the machine to prescribe the terms and conditions of the lives of creatures, which we have allowed increasingly for the last two centuries, and are still allowing, at an incalculable cost to other creatures *and to ourselves.* If we state the problem that way, then we can see that the way to correct our error, and so deliver ourselves from our own destructiveness, is to quit using our technological capability as the reference point and standard of our economic life. We will instead have to measure our economy by the health of the ecosystems and human communities where we do our work.

It is easy for me to imagine that the next great division of the world will be between people who wish to live as creatures and people who wish to live as machines.

6. ORIGINALITY AND THE "TWO CULTURES"

If one of the deities or mythological prototypes of modern science is Sherlock Holmes, another, surely, is the pioneering navigator or land discoverer: Christopher Columbus or Daniel Boone. Mr. Wilson's book returns to this image again and again. He says that "Original discovery is everything." And he speaks of "new terrain," "the frontier," "the mother lode," "virgin soil," "the growing edge," "the cutting edge," and "virgin land." He speaks of scientists as "prospectors," as navigators who "steer for blue water, abandoning sight of land for a while" and (in several places) as explorers of unmapped territory.

This figure of the heroic discoverer, so prominent in the mind of so eminent a scientist, dominates as well the languages of scientific journalism and propaganda. It defines, one guesses, the ambition or secret hope of most scientists, industrial technologists, and product developers: to go where nobody has previously gone, to do what nobody has ever done.

There is nothing intrinsically wrong with heroic discovery. However, it is as much subject to criticism as anything else. That is to say that it may be either good or bad, depending on what is discovered and what use is made of it. Intelligence minimally requires us to consider the possibility that we might well have done without some discoveries, and that there might be two opinions from different perspectives about any given discovery—for example, the opinion of Cortés, and that of Montezuma. Perhaps intelligence requires us to consider even that some unexplored territory had better be treated as forbidden territory.

As a personal ambition, heroic discovery has obvious risks, even for the heroically gifted. The greatest risk is that one will trade one's life—all the ordinary satisfactions of homeland and family life—for the sake of a hope not ordinarily realizable. William Butler Yeats saw these possibilities as mutually exclusive and the choice between them as inescapable; he was wrong, I think, and yet he was undoubtedly right about the cost of choosing work over life:

> The intellect of man is forced to choose
> Perfection of the life, or of the work,
> And if it take the second must refuse
> A heavenly mansion, raging in the dark.

> When all that story's finished, what's the news?
> In luck or out the toil has left its mark:
> That old perplexity an empty purse,
> Or the day's vanity, the night's remorse.

Mr. Wilson believes that the desire to seek out "virgin land and the future lineaments of empire" is "basic to human nature." Maybe so, but it seems more likely to be basic to the nature of only some humans, among whom have been some of the worst. Its cultural justification is to be found in works of romantic individualism and self-glorification such as Tennyson's "Ulysses," in which the hero (rehabilitated from Dante's *Inferno*) yearns toward "that untraveled world," and desires

> To follow knowledge like a sinking star
> Beyond the utmost bound of human thought.

Dante, as Tennyson did not say, found this Ulysses in Hell among the Evil Counsellors. And we, in making a cultural ideal of the same heroic ambition, see only the good that we believe is inevitably in it, forgetting how much it may partake of adolescent fantasy, adult megalomania, and intellectual snobbery, or how closely allied it is to our continuing history of imperialism and colonialism.

As a *norm* of expectation or ambition, then, heroic discovery is potentially ruinous, and maybe insane. It is one of the versions of our obsession with "getting to the top." Unlike the culture of the European Middle Ages, which honored the vocations of the learned teacher, the country parson, and the plowman as well as that of the knight, or the culture of Japan in the Edo period which ranked the farmer and the craftsman above the merchant, our own culture places an absolute premium upon various kinds of stardom. This degrades and impoverishes ordinary life, ordinary work, and ordinary experience. It depreciates and underpays the work of the primary producers of goods, and of the performers of all kinds of essential but unglamorous jobs and duties. The inevitable practical results are that most work is now poorly done; great cultural and natural resources are neglected, wasted, or abused; the land and its creatures are destroyed; and the citizenry is poorly taught, poorly governed, and poorly served.

Moreover, in education, to place so exclusive an emphasis upon "high achievement" is to lie to one's students. Versions of Mr. Wilson's "original discovery is everything" are now commonly handed out in public schools. The goal of education-as-job-training, which is now the dominant pedagogical idea, is a high professional salary. Young people are being told, "You can be anything you want to be." Every student is given to understand that he or she is being prepared for "leadership." All of this is a lie. Original discovery is *not* everything. You don't, for instance, have to be an original discoverer in order to be a good science teacher. A high professional salary is *not* everything. You *can't* be everything you want to be; nobody can. Everybody *can't* be a leader; not everybody even wants to be. And these lies are not innocent. They lead to disappointment. They lead good young people to think that if they have an ordinary job, if they work with their hands, if they are farmers or house-wives or mechanics or carpenters, they are no good.

C. P. Snow, in his 1959 lecture "The Two Cultures," alluded to this problem. His comment occurs in a footnote, but it is nevertheless maybe the most troubling insight of his lecture. One consequence of industrialism, he said, "is that there are no people left, clever, competent and resigned to a humble job. . . . Postal services, railway services, are likely slowly to deteriorate just because the people who once ran them are now being educated for different things."

Snow's general argument was that in Britain and other coun-tries of the West, the literary and scientific cultures had be-come separated by lack of common knowledge and a common language. His lecture, which I think was never first-rate, is still provocative and useful; it was controversial from the start, and the events of the last forty years certainly have imposed ques-tions and qualifications upon it. But his point, if not his bias, is still valid. The "two cultures" remain divided; they are divided both between and within themselves; and this state of things is still regrettable.

Mr. Wilson appears to accept Snow's argument without qualification, since it justifies his own project for reuniting all the intellectual disciplines by means of consilience. But on this issue of the primacy of originality and innovation, of the cut-ting edge and the unmapped territory, the scientific and the literary cultures appear now to be pretty much in agreement. The two cultures don't exactly meet under this heading, but here at least they overlap.

This agreement is to a considerable extent the result of the absorption of all the disciplines into the organization (and the value system) of the modern, corporatized university, and of the literary culture's envy of the power, wealth, and prestige of the scientific culture within that organization. Given the present structure of incentives and rewards, it is perhaps only natural that non-sciences would aspire to become sciences, and that non-scientists would aspire to be, like scientists, heroes of original discovery (or at least of "the liberation of the hu-man spirit"), scouting the frontiers of human knowledge or experience, wielding the cutting edge of some social science or some critical theory or some "revolutionary" art.

If there is an economy of the life of the mind—as I assume there has to be, for the life of the mind involves the distribution of limited amounts of time, energy, and attention—then that economy, like any other, subsists upon the making of critical choices. You can't think, read, research, study, learn, or teach everything. To choose one thing is to choose against many things. To know some things well is to know other things not so well, or not at all. Knowledge is always surrounded by ignorance. We are, moreover, differently talented and are called by different vocations. All this explains, and to some extent justifies, any system of specialization in work or study. One cannot sensibly choose against specialization because, if for no other reason, all of us by nature are to some degree specialized. There can be no objection in principle to organizing a university as a convocation of specialties and specialists; that is what a university is bound to be.

But some serious questions remain, the most serious of which I would put this way: Can this convocation of specialists, who have been "called together" to learn and teach, actually come together? In other words, can the convocation become a conversation? For that, the convocation would have to have a common purpose, a common standard, and a common language. It would have to understand itself as a part, for better or worse, of the surrounding community. For reasons both selfish and altruistic, it would have to make the good health of its community the primary purpose of all its work. If that were the avowed purpose, then all the members and branches of the university would have to converse with one another, and their various professional standards would have to submit to the one standard of the community's health.

This has not happened in our universities. The opposite, in fact, is happening. Unlike the English agriculturist Sir Albert Howard, who moved from his specialty, mycology, to the "one great subject" of health, the modern university specialist moves ever away from health toward the utter departmentalization and disintegration of the life of the mind and of communities. The various specialties are moving ever outward from any center of interest or common ground, becoming ever farther apart, and ever more unintelligible to one another. Among the causes, I think, none is more prominent than the by now ubiquitous and nearly exclusive emphasis upon originality and innovation. This emphasis, operating within the "channels" of administration, affects in the most direct and practical ways all the lives within the university. It imposes the choice of work over life, exacting not only the personal costs spoken of in Yeats's poem, but very substantial costs to the community as well. And these are costs that can be accounted.

"Over the years," Mr. Wilson writes, "I have been presumptuous enough to counsel new Ph.D.'s in biology as follows: If you choose an academic career you will need forty hours a week to perform teaching and administrative duties, another twenty hours on top of that to conduct respectable research, and still another twenty hours to accomplish really important research." Mr. Wilson is thus prescribing to the young a normative work week of eighty hours. Since he mentions no days off, let us assume that he is speaking of seven workdays of about 11½ hours each, lasting, say, from eight o'clock in the morning until 7:30 at night, or until eight at night if we allow half an hour for lunch. There are 168 hours in a seven-day week. Eighty from 168 leaves 88 hours. If the young Ph.D. sleeps eight hours a night, that takes another 56 hours, leaving 32 hours, or about 4½ hours per day. In that 4½ hours he or she must eat, keep clean, shop, do domestic chores, commute, read, care for his or her (unfortunate) children, etc. The time left over may presumably be used for amusement and for taking part in family and community life.

I suppose we ought to yield a certain admiration to such a dedicated life of work and sacrifice. It is certain that all of us have benefited from such effort on the part of some people. But it is just as certain that we have been damaged and are threatened by similar effort on the part of other people. In fact, most people can be driven to such an extent only by a kind of professional or careerist panic. Young Ph.D.'s or assistant professors find themselves in a man-made evolutionary crisis known in the universities as "publish or perish." To survive they must "produce," for the fundamental academic fact of life is that (as Mr. Wilson puts it) "a discovery does not exist until it is safely reviewed and in print." All "tenure track" professors in all universities, at least until tenure, are under life-or-death pressure to "find their way to a publishable conclusion." If a tree falls in the absence of a refereed journal or a foundation, does it make a sound? The answer, in the opinion of the imitation corporate executives who now run our universities, is no.

This academic Darwinism inflicts severe penalties both upon those who survive and upon those who perish. Both must submit to an absolute economic system which values their lives strictly according to their "productivity"—which is to say that they submit to a form of slavery. Both must submit, at least until tenure, to a university-prescribed regimen of life in which time = work = original discovery = career, thus assuring the ascendancy of professional standards in the minds of the young, and the eclipse of any standard of any other kind. The modern university thus enforces obedience, not to the academic ideal of learning and teaching what is true, as a community of teachers and scholars passing on to the young the knowledge of the old, but obedience rather to the industrial economic ideals of high productivity and constant innovation. The problem here is not that we should object to hard work and exacting study, which any school might appropriately expect, but that we certainly can find reason to object to turning schools into factories, and to making originality or innovation the exclusive goal and measure of so much effort. Mr. Wilson in his counsel to the young is, in fact, helping to perpetuate a system of education that conforms exactly to the demands of the economic system the effects of which, as a conservationist, he so much regrets. The "cutting edge" is not critical or radical or intellectually adventurous. The cutting edge of science is now fundamentally the same as the cutting edge of product development. The university emphasis upon productivity and innovation is inherently conventional and self-protective. It is part and parcel of the status quo. The goal is innovation but not difference. The system exists to prevent "academic freedom" from causing unhappy surprises to corporations, governments, or university administrators.

The present conformity between science and the industrial economy is virtually required by the costliness of the favored kinds of scientific research and the consequent dependence of scientists on patronage. Mr. Wilson writes that "Science, like art, and as always through history, follows patronage." Even I know of some scientists whose work did not "follow patronage," and there certainly have been artists whose work did not, but this statement seems generally true of modern science. Mr. Wilson elaborates; there is, he says, "a cardinal principle in the conduct of scientific research: Find a paradigm for which you can raise money and attack with every method of analysis at your disposal." This principle, in effect, makes the patron the prescriber of the work to be done. It would seem to eliminate the scientist as a person or community member who would judge whether or not the work *ought* to be done. It removes the scientist from the human and ecological circumstances in which the work will have its effect, and which should provide one of the standards by which the work is to be judged; the scientist is thus isolated, by this principle of following patronage, in a career with a budget. What this has to do with the vaunted aim of pursuing truth cannot be determined until one knows where the money comes from and what the donor expects. The donor will determine what truth (and how much) will be pursued, and how far, and to what effect. The scientist, having succeeded or failed with one paradigm, will then presumably be free to find another, and another patron.

The young Ph.D.'s who work eighty hours a week in a system devoted to "really important research" are not going to have time to know their community, let alone to wonder about the possible effects of their work upon its health. They are not going to have time to confront the problems invariably raised by innovation, or to perform the necessary criticism. Nor, in reality, will they have time to know their students very well, or to teach very well. If your educational system gives the preponderance of its rewards (promotions, salary increases, tenure, publication, prizes, grants) to "original discovery" and "really important research," then to the same extent it discourages teaching. It is simply a matter of fact that if teachers know that their careers and their livelihoods depend almost entirely on research, then most will steal time from teaching to give to research—exactly as any rational person would expect.

Teaching, anyhow, cannot do well under the cult of innovation. Devotion to the new enforces a devaluation and dismissal of the old, which is necessarily the subject of teaching. Even if its goal is innovation, science does not *consist* of innovation; it consists of what has been done, what is so far known, what has been thought—just like the so-called humanities. And here we meet a strange and difficult question that may be uniquely modern: Can the past be taught, can it even be known, by people who have no respect for it? If you believe in the absolute superiority of the new, can you learn and teach anything identifiable as old? Here, as before, Mr. Wilson speaks from an entirely conventional point of view. He takes seriously no history before the Enlightenment, which he believes began the era of modern science. Of "prescientific cultures" he makes short work: "they are wrong, always wrong." They know nothing about "the real world," but can only "invent ingenious speculations and myths." And: "Without the instruments and accumulated knowledge of the natural sciences—physics, chemistry, and biology—humans are trapped in a cognitive prison. They are like intelligent fish born in a deep, shadowed pool." I think (or, anyhow, hope) he does not realize how merciless this is—for he has thus flipped away most human history, most human lives, and most of the human cultural inheritance—or how small and dull a world it leaves him in. To escape the "cognitive prison" of religion and mythology, he has consigned himself to the prison of materialist and reductive cognition. *Consilience*, exactly like Genesis, explains only what it is capable of explaining. But, unlike Genesis, it concedes nothing to mystery; it simply rules out or blots out whatever it can't explain or doesn't like. One thing it blots out is the damage that this intellectual complacency and condescension has done and is doing still to prescientific cultures and their homelands around the world.

"I mean no disrespect," Mr. Wilson says, "when I say that prescientific people, regardless of their innate genius, could never guess the nature of physical reality beyond the tiny sphere attainable by unaided common sense. . . . No shaman's spell or fast upon a sacred mountain can summon the electromagnetic spectrum. Prophets of the great religions were kept unaware of its existence, not because of a secretive god but because they lacked the hard-won knowledge of physics." It seems only courteous to inquire at this point if anybody, ever before, has had the originality to propose that the prophets *needed* to know about the electromagnetic spectrum? One may imagine a little play, as follows:

> Isaiah (*finger in the air and somewhat oblivious of the historical superiority of the modern audience*): The voice said, Cry. And he said, What shall I cry? All flesh is grass, and all the goodliness thereof is as the flower of the field . . .
>
> Edward O. Wilson (*somewhat impressed, but nonetheless determined to do his bit for "evolutionary progress"*): But . . . But, sir! Are you aware of the existence of the electromagnetic spectrum?

CURTAIN

Even as a believer in "the potential of indefinite human progress," Mr. Wilson can be properly humble, when he has the notion. For instance, he says that "evolutionary progress is an obvious reality" if we mean by it "the production through time of increasingly complex and controlling organisms and societies, in at least some lines of descent, with regression always a possibility. . . ." One notes with gratitude this consent to the possibility of regression, but in fact Mr. Wilson is not much impressed or detained by any such possibility. Later on, returning to the subject of "preliterate humans," he concedes: "We are all still primitives compared to what we might become." There follows an avowal of the largeness of human ignorance, which, characteristically, he hastens past and quickly forgets: "*Yet* [my emphasis] the great gaps in knowledge are beginning to be filled . . . knowledge continues to expand globally. . . . Any trained person can retrieve and augment any part of it. . . . The explanations can be joined in space from molecule to ecosystem, and in time from microsecond to millennium." And then, speaking as if his program has already been completed, he says, "Now, with science and the arts combined [in consilience], we have it all." Here there is no functioning doubt or question, no live sense of the possibility of regression, no acknowledgment of the possibility that knowledge, if it can be accumulated, can also be lost. There is no hint that knowledge can be misused.

Though his head sometimes tells him that such concessions

should be made, his heart never does. In his heart, he is in agreement with the apparent majority of the public who now believe that the new inevitably replaces or invalidates the old, because the new, coming from an ever-growing fund of data, is inevitably better than the old. The rails of the future have been laid by genetic (or technological or economic) determination, and as we move forward we destroy justly and properly the rails of the past. This is strong, easeful, and reassuring doctrine, so long as one does not count its costs or number its losses.

If under the demands of a university system obsessively concentrated upon originality and innovation, science serves progress, industry, and the corporate economy, then the literary culture (to use the phrase of C. P. Snow) gives its tacit approval to the program of science-technology-and-industry and, itself, serves nothing—except, perhaps, for certain politically correct ideologies that could be as well served anywhere else. It serves less and less even the cause of literacy, which the university system has made the specialty of the English department, and which the English department has made the specialty of the freshman English program. The university as a whole gives no support to the cause of literacy. If technical or workmanly competence in writing is required only by teachers of freshman English, and virtually all other teachers either require no written work or grade what they do require "on content," then the message is unmistakable: Competence in writing does not matter and is not necessary. And that is what most students believe.

The English department, working under the same dire pressure to "produce" publishable books and articles as every other department, must assign the teaching of English composition to graduate students and a few specialists in the methodology and technology of composition-teaching, while most of the regular faculty concentrate on matters exalted far above grammar and punctuation and sentence structure. One result, as I know from my own experience and observation, is the certification of public school English teachers who do not necessarily know how to construct a coherent English sentence, or punctuate it, or make its nouns and verbs agree, or spell its common English words. Nor is it by any means certain that these certified English teachers will have a sense of literary history and tradition. Another result is the virtual languagelessness of many professional journalists and "communicators." Another is the increasing editorial slovenliness of newspapers and publishing companies.

The cult of progress and the new, along with the pressure to originate, innovate, publish, and attract students, has made the English department as nervously susceptible to fashion as a flock of teenagers. The academic "profession" of literature seems now to be merely tumbling from one critical or ideological fad to another, constantly "revolutionizing" itself in pathetic imitation of the "revolutionary" sciences, issuing all the while a series of passionless, jargonizing, "publishable" but hardly readable articles and books, in which a pretentious obscurity and dullness masquerade as profundity. And this, I think, is not easily definable as the fault of anybody in particular. It is the fault of a bad system—howbeit one that most people in it don't to all appearances object to, and one that nobody in it has effectively objected to. The university's convocation of the disciplines is not a conversation; it is incapable of criticizing itself. One of the most dangerous effects of the specialist system is to externalize its critics, and thus deprive them of standing.

Originality and innovation in science may be a danger to the community, because newness is not inherently good, and because the scientific disciplines use only professional standards in judging their work. There is no real criticism. (Ezra Pound has reminded us in *ABC of Reading* that, at root, to criticize is to choose.) Nobody seems able to subtract the negative results of scientific "advances" from the positive. Not many modern scientists would say, with Erwin Chargaff, that "all great scientific discoveries . . . carry . . . an irreversible loss of something that mankind cannot afford to lose." But, then, Chargaff had confronted fully the implication of modern science in the bombing of Hiroshima and Nagasaki and "the German extermination factories." He wrote: "The Nazi experiment in eugenics . . . was the outgrowth of the same kind of mechanistic thinking that, in an outwardly very different form, contributed to what most people would consider the glories of modern science."

In the literary culture, the preponderant aim of originality and innovation strikes directly at the community by granting precedence to intellectual fashion, and so depreciating literacy and literature and the cultural inheritance. The new supposedly dazzles the old out of existence, and people of our era are encouraged to pity their ancestors who had not the good fortune to be as we are. The cult of originality, however, seems to have produced about the expectable amount of bad work, but not much that is truly original. Instead, it has produced fashions and uniforms of originality. Political correctness becomes the intellectual and literary cutting edge.

One worries that the cultists of the new and original think they are doing what Ezra Pound told them to do. In fact, Pound did say that writers should "make it new," but that was probably as traditional an instruction as he ever gave. It is a statement perhaps too easy to understand as a flippant rejection of the old, but that is not what he meant. Pound used, to begin with, the verb "make," and he, like virtually every poet until recently, knew that our word "poet" came from a Greek word meaning "maker." To make, one must know how to make. And how does one learn? By reading. To Pound, how to write was the same question as how to read. To learn to write one must learn both a considerable portion of what has been written and *how* it was written. And so the first reference of the pronoun in "make it new" is the literary inheritance; one must renew the means of literature, which is to say the literary tradition, by making it newly applicable to contemporary needs and occasions. The new must come from the old, for where else would you get it? Not, anyhow, from contempt for the old, or from ambition. Pound's work, at its sanest, was always a testing of the usefulness of what he had read.

But I think he meant much more than that. There are passages in *ABC of Reading* that can be understood as glosses on "make it new." "A classic," Pound wrote there, "is classic not because it conforms to certain structural rules, or fits certain definitions. . . . It is classic because of a certain eternal and irrepressible freshness." And furthermore: "Great literature is simply language charged with meaning to the utmost possible degree." (Note that he does not use the adjective "new" in that sentence.) And furthermore: "Literature is news that STAYS news." The business of literature, then, is to renew not only itself but also our sense of the perennial newness of the world and of our experience; it is to renew our sense of the newness of what is eternally new.

Ananda Coomaraswamy, who was a more systematic student of artistic tradition than Pound, also wrote usefully on the subject of making it new: "There can be no property in ideas, because these are gifts of the Spirit, and not to be confused with talents. . . . No matter how many times [it] may already have been 'applied' by others, whoever conforms himself to an idea and makes it his own, will be working originally, but not so if he is expressing only his own ideals or opinions." And perhaps even more helpfully he wrote that "when there is realization, when the themes are felt and art *lives*, it is of no moment whether . . . the themes are new or old." It should be fairly clear that a culture has taken a downward step when it forsakes the always difficult artistry that renews what is neither new nor old and replaces it with an artistry that merely exploits what is fashionably or adventitiously "new," or merely displays the "originality" of the artist.

Scientists who believe that "original discovery is everything" justify their work by the "freedom of scientific inquiry," just as would-be originators and innovators in the literary culture justify their work by the "freedom of speech" or "academic freedom." Ambition in the arts and the sciences, for several generations now, has conventionally surrounded itself by talk of freedom. But surely it is no dispraise of freedom to point out that it does not exist spontaneously or alone. The hard and binding requirement that freedom must answer, if it is to last, or if in any meaningful sense it is to exist, is that of responsibility. For a long time the originators and innovators of the two cultures have made extravagant use of freedom, and in the process have built up a large debt to responsibility, little of which has been paid, and for most of which there is not even a promissory note.

The debt can be paid only by thought, work, deference, and affection given to the integrity of our ecological and cultural life. The condition which that integrity (or that one-time integrity) imposes on human work and human freedom is that everything we do has an effect or an influence. But it is generally true to say that among the originators of the modern era there has been no flinching before effects, for the purpose of the originators (as understood by themselves) has been the origination of causes only. This is the moral absurdity of specialization driven to the limit. The effects are understood simply as the causes of other original work by other specialists. And thus we have assumed that all problems merely lead to solutions, an article of pathological faith.

All along, the enterprise of science-industry-and-technology has been accompanied by a tradition of objection. Blake's revulsion at the "dark Satanic Mills" and Wordsworth's perception that "we murder to dissect" have been handed down through a succession of lives and works, and among the inheritors have been scientists as well as artists. The worry, I think, has always been that in our ever-accelerating effort to explain, control, use, and sell the world we would destroy the wholeness and the sanctity of all that which it is our highest obligation to "make new."

On the day after Hitler's troops marched into Prague, the Scottish poet Edwin Muir, then living in that city, wrote in his journal a note that recalls a similar lamentation of Montaigne about four hundred years earlier: "So many goodly citties ransacked and razed," Montaigne wrote; "so many nations destroyed and made desolate; so infinite millions of harmelesse people of all sexes, states and ages, massacred, ravaged and put to the sword; and the richest, the fairest and the best part of the world topsiturvied, ruined and defaced for the traffick of Pearles and Pepper: Oh mechanicall victories, oh base conquest." Muir wrote: "Think of all the native tribes and peoples, all the simple indigenous forms of life which Britain trampled upon, corrupted, destroyed . . . in the name of commercial progress. All these things, once valuable, once human, are now dead and rotten. The nineteenth century thought that machinery was a moral force and would make men better. How could the steam-engine make men better? Hitler marching into Prague is connected with all this. If I look back over the last hundred years it seems to me that we have lost more than we have gained, that what we have lost was valuable, and that what we have gained is trifling, for what we have lost was old and what we have gained is merely new."

Laboring in the shadow of the scientific apocalypse of World War II, C. S. Lewis wrote: "Dreams of the far future destiny of man were dragging up from its shallow and unquiet grave the old dream of Man as God. The very experiences of the dissecting room and the pathological laboratory were breeding a conviction that the stifling of all deep-set repugnances was the first essential for progress."

Albert Howard, at about the same time rethinking the role of science in agriculture, wrote: "It is a severe question, but one which imposes itself as a matter of public conscience, whether agricultural research in adopting the esoteric attitude, in putting itself above the public and above the farmer whom it professes to serve, in taking refuge in the abstruse heaven of the higher mathematics, has not subconsciously been trying to cover up what must be regarded as a period of ineptitude and of the most colossal failure. Authority has abandoned the task of illuminating the laws of Nature, has forfeited the position of the friendly judge, scarcely now ventures even to adopt the tone of the earnest advocate: it has sunk to the inferior and petty work of photographing the corpse. . . ."

And in the autobiographical meditation written when he was old, looking back over the course of science from the time of his own crisis of conscience at the revelations of World War II, Erwin Chargaff wrote: "The wonderful, inconceivably intricate tapestry is being taken apart strand by strand; each thread is being pulled out, torn up, and analyzed; and at the end even the memory of the design is lost and can no longer be recalled."

It is not easily dismissable that virtually from the beginning of the progress of science-technology-and-industry that we call the Industrial Revolution, while some have been confidently predicting that science, going ahead as it has gone, would solve all problems and answer all questions, others have been in mourning. Among these mourners have been people of the highest intelligence and education, who were speaking, not from nostalgia or reaction or superstitious dread, but from knowledge, hard thought, and the promptings of culture.

What were they afraid of? What were their "deep-set repugnances"? What did they mourn? Without exception, I think, what they feared, what they found repugnant, was the violation of life by an oversimplifying, feelingless utilitarianism; they feared the destruction of the living integrity of creatures, places, communities, cultures, and human souls; they feared the loss of the old prescriptive definition of humankind, according to which we are neither gods nor beasts, though partaking of the nature of both. What they mourned was the progressive death of the earth.

Wes Jackson of the Land Institute said once, thinking of the nuclear power and genetic engineering industries, "We ought to stay out of the nuclei." I remember that because I felt that he was voicing, not scientific intelligence, but a wise instinct: an intuition, common enough among human beings, that some things are and ought to be forbidden to us, off-limits, unthinkable, foreign, *properly* strange. I remember it furthermore because my own instinctive wish was to "stay out of the nuclei," and, as I well knew, this wish amounted exactly to nothing. One can hardly find a better example of modern science as a public predicament. For modern scientists work with everybody's proxy, whether or not that proxy has been given. A good many people, presumably, would have chosen to "stay out of the nuclei," but that was a choice they did not have. When a few scientists decided to go in, they decided for everybody. This "freedom of scientific inquiry" was immediately transformed into the freedom of corporate and/or governmental exploitation. And so the freedom of the originators and exploiters has become, in effect, the abduction and imprisonment of all the rest of us. Adam was the first, but not the last, to choose for the whole human race.

The specialist system, using only professional standards, thus isolates and overwhelmingly empowers the specialist as the only authorizer of his work—she alone is made the sole moral judge of the need or reason for her work. This solitary assumption of moral authority, of course, must *precede* the acceptance of patronage. Originality as a professional virtue gives far too much importance and power to originators, and at the same time isolates them socially and morally.

The specialist within the literary culture is isolated, it seems to me, in precisely the same way, and in fact *wishes* to be so isolated. The effects of the work of the literary specialists are not, of course, so directly practical as those of the scientists, but they are in the long run a part of the same disintegration, and are equally serious.

That the arts have been envious of the prestige, the drama, and the glamour of innovative science is suggested by the long-enduring vogue of "experimental art." "Experiment" is a word that seems displaced and uncomfortable outside of science; in science, I suppose, a failed experiment is still science, but in art a failed experiment, whatever else it may be, is not art. Misnomer or not, "experimentation" in the arts certainly bespeaks a hankering among artists for the heroism of life on the "cutting edge."

The science closest to art (in the opinion, anyhow, of many artists) is psychology and especially psychoanalysis. The study of the "psyche" is not a very exact science, but its subject matter is indigenous to the arts, and it is not hard to understand how attractive among artists have been the psychological theories of consciousness and "the unconscious." The idea of imitating in writing the "stream of consciousness" occurred early to novelists, and the psychoanalysts carried on and encouraged the artists' age-old fascination with dreams.

Maybe because modern artists took so many promptings from psychology, the scientific goal of "original discovery" became in art, and particularly in literature, the goal of original disclosure. It seems generally true that in the twentieth century writers' interest in personal life and in the inward life of persons became more intense, intimate, and in certain ways more articulate than before. The difference, roughly, is that between Tennyson's "Ulysses" and Eliot's "The Love Song of J. Alfred Prufrock." There is a new keenness, or a new kind of keenness, in understanding how people understand themselves.

Along with this interest in the intimate, inward histories of fictional persons has come (in what relation of cause or effect, I don't know) an interest in disclosing the private lives of real persons. Such disclosures are now conventional and commonplace in biography, in "confessional poetry," and in "fictionalized" accounts of actual lives and events. The most inward life is laid open, the most intimate details are shown, exactly as in dissection (which means cutting apart) or autopsy (which means seeing for yourself). One of the paramount originations of the modern literary culture is the discovery that privacy is

penetrable and publishable, and that publication is not likely
to be legally actionable. In fiction and poetry, in biography,
in journalism and the entertainment industry, and finally in
politics, the cutting edge for most of the twentieth century
has been the dis-covering of the intimate, the secret, the sexual, the private, and the obscene. And this process of exposure
has been carried on in the name of freedom by people priding
themselves on their courage.

Has it required courage? So long as it involved legal or
professional penalties, it most certainly did require courage.
But now that the penalties have been removed, no courage is
necessary. Public sexual revelations and public obscenity are
now merely clichés, part of the uniform behavior of modish
nonconformity and fashionable bad manners, but always performed by people who wish to be thought courageous.

Has it increased freedom? Well, of course people have become more free when they have earned or taken or been given
the right to do what they previously were forbidden to do. But,
as always, the worth of freedom depends upon how it is used.
The value of freedom is probably not intrinsic and is certainly
not limitless. It is generally understood by people who think
about it that freedom can be abused, and that it rests, in the
long run, on a common understanding of fairness: One should
not increase one's freedom by reducing somebody else's.

I would question also the worth of freedom from what
C. S. Lewis called our "deep-set repugnances," among which I
would include our native and proper repugnance against nos-
iness, against having our privacy invaded. This has to do, I
think, with our rightful fear of being misunderstood or too
simply understood, or of having our profoundest experience
misvalued. This, surely, is one of the reasons for Christ's insistence on the privacy of prayer. It is a part of our deepest and
most precious integrity that we should speak (if we wish) for
ourselves. We do not want self-appointed spokesmen for our
souls. Sex and worship especially are inward to us, and they are
especially fragile as possessions. Their nature is to be shared,
and yet it is dangerous to speak of them carelessly. To speak of
them carelessly is to violate yet another nucleus that ought to
be sacrosanct.

Our present idea of freedom in science is too often reducible to thoughtlessness of consequence. Freedom in the arts

frequently looks like mere carelessness in self-exposure or in exposing others. In both science and art there is a principled resistance to any suggestion that the specialist, within his or her work, might be subject or subordinate to anything.

On October 19, 1998, in New York City, the Authors Guild and the Authors Guild Foundation held a panel discussion, "Whose Life Is It, Anyway?"—a transcript of which was published in the *Authors Guild Bulletin*. The panelists, Cynthia Ozick, David Leavitt, Janna Malamud Smith, and Judy Collins, "addressed the moral, ethical and artistic implications of the writer's appropriation of others' lives and experiences." Parts of their conversation are illustrative of the problem of freedom, and I am going to quote from it at some length. Several of my ellipses indicate large omissions.

Cynthia Ozick said: "I could not fathom that fiction might not be an arena of total freedom. . . . I remember sitting on the edge of a bed with my mother-in-law, explaining how I'd been keeping a diary since 1953, and that everyone was in it. She was terrifically disturbed. 'No, no,' she cried, 'erase it, you can't have it, you mustn't do this.' And instantly I realized . . . that no writer could ever take that view. Life becomes real only through having been written. . . . Inevitably, writers are responsible for wounds and hurts—but the writer must say, I don't care, I don't give a damn. . . ."

David Leavitt quoted with approval a writing teacher who once told a class, "For every writer it is a rite of passage to write the story after which a member of your family will no longer speak to you." And later Mr. Leavitt states his credo: "I say anything goes in fiction—anything goes. If you start to take away bit by bit the rights of writers doing what they want, what you end up eroding is your own freedom."

Ms. Ozick agreed that "anything goes." But then she made an exception: "Yet I do have certain lines of limits. . . . I would not admire—I would strenuously object to—a novel which took a Holocaust-denial point of view. . . . But that's an extreme issue. For the writer . . . it's only make-believe, it's the world of enchantment. Make-believe and enchantment can't really harm anyone."

Janna Malamud Smith, politely on the contrary, said this: "When rationalizing their exposure of others, writers tend to claim two values as having overriding worth. One is the aesthetic goal of telling the story well. There's often a feeling that writing beautifully is an ultimate good, that telling a tale very well compensates any harm it might do to its subjects. The second virtue writers tend to honor is outing the truth. We take seriously the job of looking behind hypocrisy and social facade. . . . We like to believe there is a version of the truth that is superior and that we can state it. These are serious premises. . . . But I think they thrive best when they are occasionally pruned by opposing values. . . . The reason people feel betrayed when they find themselves in people's books is this: Intimacy . . . works because you are allowed to do things in a friendship, in a love relationship, that you can't do in public. So when the private things intimacy has allowed you to expose are suddenly made public, that is a legitimate reason for a feeling of profound betrayal. . . . The fact is that betrayals are a real thing."

Later, Ms. Smith speaks of the possibility that people might be "led astray by romantic novels," and she says, "Influence is real, and probably we need to think about that as well."

It certainly is true that writers are burdened with the responsibility to bear witness to such truth as they have seen or think they have seen. The responsibility goes with the trade: There is no value in telling anything if one does not try with all one's might to tell the truth. If human beings could be utterly confident of their ability to tell—or know—the truth, then the problem would be at least smaller, though a writer's insistence upon "total freedom" to tell the truth about other people would still be questionable. Since human beings can be wrong, since even with the best intentions they may know falsely and tell falsehoods, to say that "anything goes" and to leave it at that is far too simple.

My intention here is certainly not to promote any abridgment of the freedoms of speech and inquiry, though I believe that those freedoms are now being pretty severely abused and that the abuse of freedom threatens its survival. But I agree with Janna Malamud Smith that betrayal and influence are real and must be thought about.

I don't believe that the connection between art and life can ever be finally or even very satisfactorily resolved, any more than can be the connection between science and life. We join ourselves to the living world by the artifacts of art and science —by made things. And we are always going to be at least somewhat at fault, because we are ignorant and fallible and small; the living world is larger and more complex than our works. Because we must be always correcting our errors, art and science always need to be free to shift their ground and start again. The unendable, the necessarily ongoing problem of justice to the world and to one another thus enforces practically the requirement of freedom. And this freedom can survive, I believe, only by being well used.

What "well used" may mean was clearly shown by Richard C. Strohman in an article of April 1, 1999 in *The Daily Californian*. Mr. Strohman was writing in defense of "unimpeded science." He was worried about the costs to science of "new initiatives for university-corporate alliances" in the development of biotechnology. The costs, apparently, will be the familiar ones of too much specialization and of a falsifying oversimplification. Mr. Strohman wrote: "The corporate need for technology dedicated to specific products will . . . must . . . subvert the scientific need for unimpeded research. In academic biology the technological need to define complex behavior in terms of simple causality subverts the need for a wider, more complex research context.

"Here then is the real danger of the university-corporate 'merger' . . . a corporate need that must repress new ways of seeing nature."

And so science too must be concerned with "making it new" and with renewing itself, and now it must do so for a reason both new and urgent: to see that nature escapes the corporations which are newly empowered to oversimplify it, commodify it, and put it up for sale. Mr. Strohman's answer, I think, is the correct one: Enlarge the context of the work.

Freedom in both science and art probably depends upon enlarging the context of our work, increasing (rather than decreasing) the number of considerations we allow to bear upon it. This is because the ultimate context of our work is the world, which is always larger than the context of our thought. And so to complicate the consideration of freedom in literature by the considerations of betrayal and influence is not a diminishment of literature or freedom; it is a *just* enlargement of the context of work. If we could faithfully commit ourselves to the principle that nothing whatever can safely be said to lie outside the context of our work, then artists and scientists would have to be ready at any time to see that they have been wrong and to start again, making yet larger the context of the work. *That* is true freedom. It means simply that beyond all error we can begin again; redemption is possible. From this principle also we can make our way to critical judgments of an amplitude beyond specialization and professionalism: Work that diminishes the possibility of a new start, of "making it new," is bad work.

Janna Malamud Smith said, "You don't trade betrayal for writing . . ." That is not a simple statement, because to be a writer is not a simple predicament. There is a constant relationship, though never altogether settled and never altogether clear, between imagination and reality. If you are a fiction writer, you may, at one extreme, tell a story that is almost the story of something that actually happened; at the opposite extreme, you may tell a story that you have almost entirely imagined. But what you have imagined will always be somewhat informed by what you have actually known, and your actual knowing will always be somewhat informed by imagination. The extremes of reality and imagination, within the limits of human experience, are never pure. And so there is always some risk of betrayal. It is possible to allow imagination to abuse reality; it is possible by imagination to violate a real intimacy —and this leaves aside the possibility of deliberately tattling for meanness or revenge or some version of success. It is always possible too that imagination may be debased by a false or too narrow understanding of what is real.

Both imagination and a competent sense of reality are necessary to our life, and they necessarily discipline one another. Only imagination, for example, can give our home landscape and community a presence in our minds that is a sort of vision at once geographical and historical, practical and protective, affectionate and hopeful. But if that vision is not repeatedly corrected by a fairly accurate sense of reality, if the vision becomes fantastical or merely wishful, then both we and the landscape fall into danger; we may destroy the landscape, or the landscape (especially if damaged by us in our illusion) may destroy us.

To speak of betrayal as a possibility in literature is one way of acknowledging this necessary and inescapable tension between imagination and reality. Fiction can abuse reality by violating intimacy or confidence or privacy, or by being wrong. Not least among the offenses of literary artists is their frequent indifference to facts of history or natural history or ways of work.

To be careless of such betrayals is to reduce one's subject to the status of "raw material"—exactly as the ubiquitous enterprise of science-technology-and-industry reduces *its* subjects. The parallels of value and attitude among contemporary arts and sciences and the industrial economy are obvious, are ratified by convention, and are almost unnoticed by the would-be pioneers and heroes of the cutting edges. But if one lives, as I do, in a rural place, which is to say in the midst of other people's "raw material," then one does notice. And if one notices, then one knows that artistic and scientific betrayals are real and are serious. They are an affront to one's subject, and they endanger it.

Too much disclosure of the intimate, the secret, the sexual, the private, and the obscene is accomplished by mentioning or representing or picturing but not imagining. To represent the intimacy of desire or of grief without the art that compels one to imagine these things as the events of lives and of shared lives is actually to misrepresent them. This is the "objectivity" of the schools and the professions, which allows a university or a corporation to look at the community—its *own* community —as one looks at a distant landscape through fog. This sort of objectivity functions in art much the same as in science; it obstructs compassion; it obscures the particularity of creatures and places. In both, it is a failure of imagination.

Journalism and the electronic media, for example, routinely exhibit representations or disclosures of intimate emotion as objects of curiosity, as intrinsically interesting, or as proofs of artistic or journalistic courage. The perennial act of cutting-edge enterprise in reporting is to shove a camera or a microphone into the face of a grieving woman. But what is the qualitative difference between the man who cold-heartedly shoots another and the photographer who cold-heartedly photographs the corpse or the grieving widow? Are they not simply two parts of the same epidemic failure

of imagination, which is to say a failure of compassion and
of community life?

Such exposures do not make us free, and they do not increase our knowledge. They only compound human cruelty by
a self-induced numbness to the suffering of others and to our
common suffering.

To be indifferent to hurts given by one's writing to its human subjects, which exactly parallels the scientific-industrial
indifference to the suffering of animal or human subjects of
exploitation or experimentation—to say "I don't care, I don't
give a damn"—is a betrayal not only of the subject of writing,
which is invariably our common life, our neighborhood, but
also of imagination itself. It is a refusal to be compassionate, a
denial of the vital link between imagination and compassion.
How can such a betrayal not impair one's ability to know the
truth and to make art?

The world and its neighborhoods, natural and human, are
not passively the subjects of art, any more than they are passively the subjects of science-industry-and-technology. They
are affected by all that we do. And they respond. The world
does not exist merely to be written about, any more than it
exists merely to be studied. It is real, before and after human
work. What we write is finally to be measured by the health of
what we write about. What we think we know affects the health
of the thing we think we know.

The problem of influence also is real, and it is inescapable.
Ms. Ozick acknowledges as much when she wishes to exempt
the Holocaust from her credo of not giving a damn, and so she
undermines all her affirmation of artistic superiority and autonomy. To say that writing about the Holocaust may be influential is to say that writing may be influential, period. Who can
deny that writing about Jews with contempt may cause them
to be treated with contempt, or with violence? But the history
of oppression forbids us to limit that liability to the Jews. We
treat people, places, and things in accordance with the way we
perceive them, and literature influences our perceptions. To
leave aside more fashionable examples, who can deny that the
history of coal mining in the southern Appalachians has been
under the influence of writers who have written of the mountaineers as "briars" or "hillbillies"? And who wishes to say that

our long exploitation and finally our virtual destruction of our farm population has not been influenced by generations of writers who have represented farmers as "yokels" who live in the "sticks" and do "mind-numbing work"? We can't deny that writing has an influence unless we can also deny that stereotyping and character assassination have an influence.

The question for art, then, is exactly the same as the question for science: Can it properly subordinate itself to concerns that are larger than its own? Can it judge itself by standards that are higher and more comprehensive than professional standards? The issue is the old one of propriety. Is every artist and every scientist to be "free" to work as if his or her discipline were the only one, or the dominant one? Or is it possible still to see one's work as occurring within a larger and ultimately a mysterious pattern of causes and influences? If we can see that we are mutually dependent upon one another and upon that mosaic of natural and human neighborhoods we call "the world," then it should not be too hard to see that there ought to be responsible connections between science and the knowledge of how to live, and between art and the art of living, and that there is always, inescapably, acknowledged or not, a complex connection between art and science.

7. PROGRESS WITHOUT SUBTRACTION

The task of thinking about Mr. Wilson's book is made difficult at every point by his adherence to the rather simple-minded popular doctrine of mechanical or automatic progress. His book perhaps was written as a defense of that doctrine. He believes, with the Enlightenment thinkers, in "the potential of indefinite human progress." He affirms the necessity to speak of "evolutionary progress." In spite of his perfunctory acknowledgment of the possibility of regression he speaks twice of the "Ratchet of Progress." He says that "humanity accepted the Ratchet of Progress" as if to suggest that we had a choice, but he doesn't say what we might have accepted instead. The practical effect of his belief in the inevitability of progress is to make him a poor critic of his own thought. In fact, for all his enthusiasm, he is a rather passive consumer of scientific platitudes. His idea of progress, for example, is both starkly deterministic (it is "evolutionary" and a "ratchet") and hazily romantic: Modern science, he says, is "driven by the faith that if we dream, press to discover, explain, and dream again, thereby plunging repeatedly into new terrain, the world will somehow come clearer and we will grasp the true strangeness of the universe. And the strangeness will all prove to be connected and make sense." Later, in his very sobering appraisal of our destruction of "the environment," he says, "We must plunge ahead and make the best of it, worried but confident of success. . . ."

This is utterly baffling. If our future is already determined by "evolutionary progress" and the "ratchet" is in place, there is no use in "plunging" anywhere—unless it is to exercise our "biologically adaptive" "illusion of free will." But if free will is an illusion, to what purpose do we make the world clearer? What Mr. Wilson evidently means by "plunging" is merely going ahead as we are going with our "really important research" and "following patronage." It is hard to see how any of this could be encouraging or useful to a conservationist.

If regression really is a possibility, then should we not watch for the signs of it? And should we not attempt to subtract regression from progression to get at least an approximate notion of net gain or net loss? Mr. Wilson concedes that people forget and die, but he says that "knowledge continues to expand globally while passing from one generation to the next." But in fact as knowledge expands globally it is being lost locally. This is the paramount truth of the modern history of rural places everywhere in the world. And it is the gravest problem of land use: Modern humans typically are using places whose nature they have never known and whose history they have forgotten; thus ignorant, they almost necessarily abuse what they use. If science has sponsored both an immensity of knowledge and an immensity of violence, what is the gain? If we "grasp the true strangeness of the universe" but forget how to farm, what is the gain?

Such questions, seriously asked and intelligently answered, lead directly to choices that people have the ability to make, but no such possibility is suggested in *Consilience*.

In Edward O. Wilson's view, the world is not a place where we all make in our daily lives intelligent or unintelligent choices affecting the future of the world. It is, rather, a place where the most genetically favored and the most richly subsidized scientists determine the future by "plunging ahead," each isolated in his or her vision of "new terrain," and each cut off from any restraining affection for old terrain.

Why should we trust them?

IV. Reduction and Religion

I T IS CLEARLY bad for the sciences and the arts to be divided into "two cultures." It is bad for scientists to be working without a sense of obligation to cultural tradition. It is bad for artists and scholars in the humanities to be working without a sense of obligation to the world beyond the artifacts of culture. It is bad for both of these cultures to be operating strictly according to "professional standards," without local affection or community responsibility, much less any vision of an eternal order to which we all are subordinate and under obligation. It is even worse that we are actually confronting, not just "two cultures," but a whole ragbag of disciplines and professions, each with its own jargon more or less unintelligible to the others, and all saying of the rest of the world, "That is not my field."

The badness of all this is manifested first in the loss even of the pretense of intellectual or academic community. This is a loss increasingly ominous because intellectual engagement among the disciplines, across the lines of the specializations —that is to say *real* conversation—would enlarge the context of work; it would press thought toward a just complexity; it would work as a system of checks and balances, introducing criticism that would reach beyond the professional standards. Without such a vigorous conversation originating in the universities and emanating from them, we get what we've got: sciences that spread their effects upon the world as if the world were no more than an experimental laboratory; arts and "humanities" as unmindful of their influence as if the world did not exist; institutions of learning whose chief purpose is to acquire funds and be administered by administrators; governments whose chief purpose is to provide offices to members of political parties.

The ultimate manifestation of this incoherence is loss of trust—loss, moreover, of the entire cultural pattern by which we understand what it means to give and receive trust. The general assumption now is that everybody is working in his or her own interest and will continue to do so until checked by somebody whose self-interest is more powerful. That nobody now trusts the politicians or their governments is probably the noisiest of present facts. More quietly, people are withdrawing their trust from the professions, the corporations, the education system, the religious institutions, the medical industry. Perhaps no expert has yet assigned a quantitative value to trust; it is nonetheless certain that when we have finished subtracting trust from all we think we have gained, not much will be left.

And so it certainly is desirable—it probably is necessary—that the arts and the sciences should cease to be "two cultures" and become fully communicating, if not always fully cooperating, parts of one culture. (I believe, as I will show, that this culture when it comes will be in fact a mosaic of cultures, based upon every community's recognition that all its members have a common ground, and that this ground is the ground under their feet.) I have, therefore, not the slightest inclination to disagree with Mr. Wilson's wish for a "linkage of the arts and humanities." With his goal of "consilience," though I sympathize, I do not agree.

I do not agree because I do not think it is possible. I do not think it is possible because, as he defines it, it would impose the scientific methodology of reductionism upon cultural properties, such as religion and the arts, that are inherently alien to it, and that are often expressly resistant to reduction of any kind. Consilience, Mr. Wilson says, is "literally a 'jumping together' of knowledge by the linking of facts and fact-based theory across disciplines to create a common groundwork of explanation." And: "The only way either to establish or to refute consilience is by methods developed in the natural sciences—not . . . an effort led by scientists, or frozen in mathematical abstraction, but rather one allegiant to the habits of thought that have worked so well in exploring the material universe." The project of consilience, then, is not for scientists only, but it is only for science.

Whether or not science, religion, and the arts can be linked on "a common groundwork of explanation" depends upon a further question: Can religion and the arts be explained in the same way that science can be, or can they, in any comprehensive way, be explained at all? And this, it seems to me, depends upon another question that is even more important: Is knowledge by definition explainable, or is there such a thing as unexplainable knowledge?

I have in mind three statements that seem to me to test this issue of knowledge and explainability:

At the end of *King Lear*, the broken-hearted old king comes in with his faithful daughter Cordelia dead in his arms. He says: "Thou'lt come no more, / Never never never never never."

In II Samuel 18:33, David the king has just been told that his son, who has been his enemy, is dead. The King says: "O my son Absalom, my son, my son Absalom! would God I had died for thee, O Absalom, my son, my son!"

After the battle of Gettysburg, General Lee was overheard saying to himself, "Too bad! Too bad! Oh, too bad!"

These outcries "out of the depths" certainly express knowledge, and precisely too. They communicate knowledge. But the knowledge they convey cannot be proved, demonstrated, or explained; it cannot be taught or learned. These utterances are not "self-explanatory." They are as far as possible unlike what we now call "information." One either does or does not know what they mean. The idea of explaining them to someone who does not know is merely laughable.

Statements of religious faith seem to me to be of the same general kind. Job says: "I know that my redeemer liveth, and that he shall stand at the latter day upon the earth: And though . . . worms destroy this body, yet in my flesh shall I see God. . . ." This statement rests upon no evidence, no proof. It is not in any respectable sense a theory. Job calls it knowledge: He "knows" that what he says is true. A great many people who have read these verses have agreed; they too have known that this is so.

"The empiricist" in Mr. Wilson's chapter on "Ethics and Religion" would find Job's knowledge readily explainable as a "beneficent" falsehood, supported by no "objective evidence" or "statistical proofs." Mr. Wilson himself understands it as a genetically implanted "urge": "Perhaps . . . it can all eventually be explained as brain circuitry and deep, genetic history." People follow religion, he says, because it is "easier" than empiricism, the lab evidently being harder to bear than the cross. Mr. Wilson forgets, in calling attention to religion's

want of statistical proofs, that empiricism can supply no statistical disproofs. His explanation of religion rather tends to
prove that it is not explainable. God and the devices of human
understanding are not the same subject.

Suppose, granting the hopelessness of empirical proof, that
you took Job's statement of faith as seriously as Mr. Wilson
wishes you to take empiricism; how, then, could you explain it
to Mr. Wilson? It seems to me that you would have to concede
—and here empirical evidence is available—that it could not
be done.

His statement of his own "position" brings no clarification;
though it is a statement of a faith somewhat less than scientific,
for it has no proofs, it carefully does not touch the issue of
religious faith: "I am an empiricist. On religion I lean toward
deism but consider its proof largely *a problem* in astrophysics.
The existence of a cosmological God who created the universe
(as envisioned by deism) is *possible*, and *may eventually* be settled, *perhaps* by forms of material evidence *not yet* imagined.
Or the matter may be forever beyond human reach. In contrast . . . the existence of a biological God, one who directs
organic evolution and intervenes in human affairs (as envisioned by theism) is *increasingly* contravened by biology and
the brain sciences." My emphases call attention to the extreme
tentativeness of the thought. Mr. Wilson concedes on the same
page, "I may be wrong," but that very concession exposes the
hopelessness of the argument that he is proposing to settle by
consilience. How could he be "proven" wrong? The faith of
an empirical deist will probably have to wait a good while for
proof or disproof by astrophysics. About as long, I imagine, as
it will take the "increasing" evidence of biology and the brain
sciences to culminate in empirical disproof of theism.

What is the difference between an "empirical" faith so
hedged about and religious faith? One difference, to use Edwin
Muir's terms, is that whereas religious faith is old, the empirical
faith is merely new. A second difference is that religious faith
has lived to grow old because to hundreds of generations it
has appeared to rest upon a knowledge that is not empirical,
whereas the empirical faith, as its language shows, rests only
upon speculation.

There is no reason, as I hope and believe, that science and

religion might not live together in amity and peace, so long as they both acknowledge their real differences and each remains within its own competence. Religion, that is, should not attempt to dispute what science has actually proved; and science should not claim to know what it does not know, it should not confuse theory and knowledge, and it should disavow any claim on what is empirically unknowable.

The two cannot be reconciled by Mr. Wilson's consilience because consilience requires the acceptance of empiricism as a ruling dogma or orthodoxy, denying standing or consideration to any thought not subject to empirical proof. His proposed consilience, by attempting to impose on art and religion the methods and values of reductive science, would prolong the disunity and disintegration it is meant to heal. Like a naive politician, Mr. Wilson thinks he has found a way to reconcile two sides without realizing that his way is one of the sides. There is simply no reason for any person of faith to discuss consilience with Mr. Wilson. One cannot, in honesty, propose to reconcile Heaven and Earth by denying the existence of Heaven.

The danger of this sort of reconciliation, as twentieth-century politics has shown, is that whatever proposes to invalidate or abolish religion (and this is what consilience pretty openly proposes) is in fact attempting to put itself in religion's place. Science-as-religion is clearly a potent threat to freedom. Beyond that, it endangers real science. Science can function as religion only by making two unscientific claims: that it will *eventually* know everything, and that it will *eventually* solve all human problems. And here it is enough to note that at times Mr. Wilson allows the term "science" to become altogether too elastic.

Religion, as empiricists must finally grant, deals with a reality beyond the reach of empiricism. This larger reality does not manifest itself in the manner of laboratory results or in the manner of a newspaper front page. Christ does not come down from the cross and confound his tormentors, as good a movie as that would make. God does not speak loudly from Heaven in the most popular modern languages for all to hear. (If He did, we would have no need for science, or religion either.) It is nevertheless true that people believe in the existence of this larger reality, and accept religious truth as knowledge, because

of their *experience*. John Milton, to whom Mr. Wilson so easily condescends, is only one of many poets in our tradition
who wrote of an unevident reality, and who invoked the muse
for aid in so great a task. The walls of the rational, empirical
world are famously porous. What come through are dreams,
imaginings, inspirations, visions, revelations. There is no use
in stooping over these with a magnifying lens. Beyond any
earthly reason we experience beauty in excess of use, justice in
excess of anger, mercy in excess of justice, love in excess of deserving or fulfillment. We have known evil beyond imagining
and seemingly beyond intention. We have known compassion
and forgiveness beyond measure. And all of this is in excess of
what Mr. Wilson means by "religion" and of what he means
by "ethics."

Religion, it seems to me, has dealt with this reality clumsily enough, and that is why the history of a religion and its
organizations is so frequently a blight on its teachings. But
religion at least attempts to deal with religious experience on
its own terms; it does not try to explain it by terms that are
fundamentally alien to it. For thousands of years, for example,
people (who were not dummies) have supposed that dreams
come from outside the waking world, speaking to us at least
some of the time, and however unclearly, of a reality beyond
that world. Hamlet speaks for a lot of people, and very much
to my point, when he says, "I could be bounded in a nutshell
and count myself a king of infinite space were it not that I have
bad dreams." The same, of course, is true of good dreams. Mr.
Wilson says, typically, that "dreaming is a kind of insanity, a
rush of visions, largely unconnected to reality . . . arbitrary
in content . . . very likely a side effect of the reorganization
and editing of information in the memory banks of the brain."
Something of the sort, of course, may be said of inspiration,
imagination, beauty, justice, mercy, and love—which consilience would require us to understand as mere strategies of
survival encoded in our genes. But this kind of reduction is
sufficiently answered by the fact that these things, thus explained, are no longer even conceptually what they were. Reduction does not necessarily limit itself to compacting and
organizing knowledge; it also has the power to change what
is known.

But biblical religion (which is the only religion that Mr. Wilson talks about) is also explicitly against reductionism. Mr. Wilson's spokesman "the empiricist" hauls out, as if he had thought of it himself, the most popular "environmental" cliché about Christianity: "With a second life waiting, suffering can be endured—especially in other people. The natural environment can be used up." This little platitude has passed from mouth to mouth for years, chewable but not swallowable. It is untrue. Nobody who has actually read the Gospels could believe it. It ignores the very point of the Incarnation. It ignores Christ's unfailing compassion for sufferers, whom He healed, one by one, as they came or were carried to Him. And there is nowhere in the Bible a single line that gives or implies a permission to "use up" the "natural environment."

On the contrary, the Bible says that between all creatures and God there is an absolute intimacy. All flesh lives by the spirit and breath of God. We "live, and move, and have our being" in God. In the Gospels it is a principle of faith that God's love for the world includes *every* creature individually, not just races or species. God knows of the fall of every sparrow; He has numbered "the very hairs of your head." Edgar was being perfectly scriptural when he said to his father, "Thy life's a miracle," and so was William Blake when he said that "everything that lives is holy." Julian of Norwich also was following scripture when she said that God "wants us to know that not only does he care for great and noble things, but equally for little and small, lowly and simple things as well." Stephanie Mills is witness to the survival of this tradition when she writes: "*A Sand County Almanac* is suffused with affection for distinct beings. . . ."

No attentive reader of the Bible can fail to see the writers' alertness to the individuality of things. The characters of humans are sharply observed and are appreciated for their unique qualities. And surely nobody, having read of him once, can forget the warhorse in Job 39:25, who "saith among the trumpets, Ha, ha." I don't know where you could find characterizations more deft and astute than those in the story of the resurrection in John 20:1–17. And again and again the biblical writers write of their pleasure and wonder in the "manifold" works of God, all keenly observed.

People who blame the Bible for the modern destruction of nature have failed to see its delight in the variety and individuality of creatures and its insistence upon their holiness. But that delight—in, say, the final chapters of Job or the 104th psalm—is far more useful to the cause of conservation than the undifferentiating abstractions of science. Empiricists fail to see how the language of religion (and I mean such language as I have quoted, not pulpit clichés) can speak of a non-empirical reality and convey knowledge, and how it can instruct those who use it in good faith. Reverence gives standing to creatures, and to our perception of them, just as the law gives standing to a citizen. Certain things appear only in certain lights. "The gods' presence in the world," Herakleitos said, "goes unnoticed by men who do not believe in the gods." To define knowledge as merely empirical is to limit one's ability to know; it enfeebles one's ability to feel and think.

We have come face to face with a paradox that we had better notice. Mr. Wilson's materialism is theoretical and reductionistic, tending, in his idea of consilience, toward "unity." People of faith, on the other hand, have always believed in the unity of truth in God, whose works are endlessly and countlessly various. There is a world of difference between this humanly unknowable unity of truth and Mr. Wilson's theoretical unity of knowledge, which supposes that mere humans can know, in some definitive or final way, the truth. And the results are wonderfully different: Acceptance of the mystery of unitary truth in God leads to glorification of the multiplicity of His works, whereas Mr. Wilson's goal of a cognitive unity produced by science leads to abstraction and reduction, the opposite of which is not synthesis. The principle that is opposite to reduction—and, when necessary, its sufficient answer —is God's love for all things, for each thing for its own sake and not for its category.

V. Reduction and Art

B<small>Y "THE ARTS"</small> Mr. Wilson means "the creative arts, the personal productions of literature, visual arts, drama, music, and dance marked by those qualities which . . . we call the true and the beautiful." He says further that "The defining quality of the arts is the expression of the human condition by mood and feeling, calling into play all the senses, evoking both order and disorder." And he makes a strict distinction between science and the arts: "While biology has an important part to play in scholarly interpretation, the creative arts themselves can never be locked in by this or any other discipline of science. The reason is that the exclusive role of the arts is the transmission of the intricate details of human experience by artifice to intensify aesthetic and emotional response. Works of art communicate feeling directly from mind to mind, with no intent to explain why the impact occurs. In this defining quality, the arts are the antithesis of science.

"When addressing human nature, science is coarse-grained and encompassing, as opposed to the arts, which are fine-grained and interstitial. That is, science aims to create principles and use them in human biology to define the diagnostic qualities of the species; the arts use fine details to flesh out and make strikingly clear by implication those same qualities."

These proposed differences notwithstanding, Mr. Wilson argues that science and the arts can be brought into alignment or unity by "consilient explanation." The means of consilience is to be interpretation, which is "the logical channel of consilient explanation between science and the arts." Two questions about the arts are "the central concern of interpretation": "where they come from in both history and personal experience, and how their essential qualities of truth and beauty are to be described through ordinary language." Interpretation of the arts needs to be reinvigorated "with the knowledge of science and its proprietary sense of the future." Mr. Wilson expects that, thus reinvigorated, interpretation will finally show (whether theoretically or by proof is not clear to me) that the arts originate in "an inborn human nature"—that is, in "the material processes of the human mind." Again, the mind is equated with the brain; the consilient explanation of the arts depends upon the explanation of the brain:

"If the brain is ever to be charted and an enduring theory of the arts created as a part of the enterprise, it will be by stepwise and consilient contributions from the brain sciences, psychology, and evolutionary biology. And if during this process the creative mind is to be understood, it will need collaboration between scientists and humanities scholars.

"The collaboration, now in its early stages, is likely to conclude that innovation is a concrete biological process founded upon an intricacy of nerve circuitry and neuro-transmitter release."

Great artists are genetically gifted, not by "singular neurobiological traits," but rather "by a quantitative edge in powers shared in smaller degree with those less gifted," and this quantitative edge produces works that are "qualitatively new." Art is to be accounted for both by genetic evolution and by cultural evolution, but cultural evolution is under the sway of "epigenetic rules of human nature" that draw creative minds toward "certain thoughts and behavior," which, in turn, "bias cultural evolution toward the invention of archetypes, the widely recurring abstractions and core narratives that are dominant themes in the arts." The most enduring works of art are those that are truest to their origins in human nature: "It follows that even the greatest works of art might be understood fundamentally with knowledge of the biologically evolved epigenetic rules that guided them."

Having tried conscientiously to summarize Mr. Wilson's "working hypothesis" of "the biological origin of the arts," I find a residue of statements that I don't understand well enough to include in my summary. For example, he says, "*The arts are innately focused toward certain forms and themes but are otherwise freely constructed*" [his italics]. If the forms and themes, especially the forms, are determined by innate predisposition, then it is not clear how much latitude there can be for freedom of construction. What is called for, apparently, is a sample analysis of a work of art, showing what is "innate" and what is "freely constructed," and how the innate and the free can be conjoined in a work that is "qualitatively new" when innovation has already been described as "a concrete biological process."

Nor am I able to understand the statement that the quality of the arts "is measured by . . . the precision of their adherence to human nature." If the forms and themes of the arts are determined by "an inborn human nature" ("the material processes of the human mind"), then how could they not adhere to it? In a naturally determined system, how can anything happen, or how is anything conceivable, that is unnatural? We need now an example of a work of art that does not adhere to human nature—which, if produced, would testify to the authenticity of that freedom of will which Mr. Wilson has said is illusory. But in a system of biological determinism, how does the issue of quality arise in the first place? If everything is originated biologically and free will is an illusion, then what we get is what we've got, qualitative standards are irrelevant, and critical judgment also is an illusion.

But even the parts of Mr. Wilson's "working hypothesis" that I am able to comprehend are frequently in error.

He is much mistaken, to begin with, in his wish to limit the arts to "expression of the human condition by mood and feeling" and to "aesthetic and emotional response." The arts, of course, "express" by their native means: words, colors, shapes, sounds, etc. They also include knowledge. They can instruct. Literature, at least, can convey facts, adduce evidence, and make arguments. *Paradise Lost*, which is the only work of literature that Mr. Wilson discusses at length, is for his thesis particularly unfortunate. Milton's purpose in that poem was avowedly *not* to express the human condition by mood and feeling. His purpose was, as he said, to "assert Eternal Providence, / And justifie the wayes of God to men." His poem is, among much else, a great argument. If you read *Paradise Lost*, you will certainly be obliged to feel and to experience moods and to respond aesthetically and emotionally, but you will also have to employ all of your mind to think and comprehend and to make critical judgments. Milton would have been indignant at the suggestion that his art was in any exclusive way "the antithesis of science."

Mr. Wilson would like to exclude science from art, which is easy to do, maybe, in theory, but harder in practice, when one considers how much the arts have been influenced by science and how often science has provided the subject matter of art. It would be a daunting critical exercise to subtract astronomy from *The Divine Comedy*, or biology from *Walden*.

But he would also like to exclude art from science. Though he speaks of the need for "collaboration between scientists and humanities scholars," it is hard to see what use he would have for the humanities scholars, except maybe to provide a little bibliography. His "working hypothesis" of "the biological origin of the arts" is strictly a scientific hypothesis, and it proposes only scientific tasks. Mr. Wilson's councils obviously could not include any humanities scholars who might, for example, take seriously Milton's faith, or his poetic purpose, or his invocation to the Heavenly Muse. The humanities scholars of choice would be those who would affirm Mr. Wilson's materialism, in which case Milton (and a host of other artists) would not be represented or would be misrepresented.

Since Mr. Wilson sees the arts as products of "gene-culture coevolution," he naturally sees them as serving the cause of "survival and reproduction." I am happy to concede him this point. Though I am not much impressed by evolution as the ultimate explanation of life, I am altogether convinced that the arts have helped us to survive and reproduce; to believe otherwise, I would have to deny the existence and the efficacy of love songs. But species survival alone does not adequately account for the existence of the arts, and (if quality is an issue) it does not provide an adequate standard of art criticism. "Survival value," it seems to me, must deal in minimums, since any species dependent upon maximums would be too vulnerable to survive. The human race has survived because of its ability to survive famine, not because of its ability to survive feasts. Survival is possible at minimal levels—in poverty, exile, concentration camps—and this ability merely to persist and endure undoubtedly owes much to instinct, to "inborn human nature," unlearned. But surviving is not the same thing, it is not as high an accomplishment, as the desire to go on living one's own life after surviving, say, defeat or famine or poverty or illness or grief. To live at a high level, desiring and aspiring throughout a human lifetime with its inevitable griefs and troubles, requires culture that, beyond any genetic determination or epigenetic rules, must be deliberately taught and learned. Obviously, the desire to live at a high level can have "survival value" also. Nevertheless, the desire to survive and the desire to live are two different desires, and the second is more conscious, more deliberate, more a matter of education and cultural choice than the first.

Mr. Wilson speaks of human nature as if it were *only* inborn, a product only of evolution. And so he has little choice but to speak of art in the same way. His fundamental error, in proposing his consilience of science and art, is his assumption that works of art are properties of nature in the same way that organisms are. (He thus extends his reductive formula to read: work of art = organism = machine.) He understands works of art as the products of "talent," not as artifacts, not as things made by arts which exist by being taught and learned. Once, he says that the masters of the arts have "exceptional knowledge" and "technical skill," but nowhere does he speak of the cultural continuum by which such knowledge and skill are kept alive and handed down. He is interested almost exclusively in the artists' "talent" and their "intuitive grasp of inborn human nature." He does not understand the arts as ways of making or works of art as made things. He asks two questions about the arts: where they come from and how their qualities can be described; he does not ask how they are made. And so he can think of the arts and human nature merely as "natural." He thinks of human nature as "inborn," not as both inborn and to be learned from (among other things) works of art. If human nature (and therefore all its manifestations, such as the arts) is merely natural or inborn, then it is merely a subject of study; no standards of judgment are necessary. If human nature is also the product of learning and is to some extent made by art, then critical judgment is both possible and necessary, and we must deal with issues of will and choice. We are ready to ask, for example, what may be the effect of our cultural and artistic choices upon the natural world. And at this point we can see the error of segregating the "fine" or "creative" arts from the arts that are practical or economic. Why should our universities sponsor an active criticism of the fine arts (by specialized or professional standards, ignoring their effect on the world) but no criticism of farming or forestry or mining or manufacturing? This question, of course, can be answered by a crude evolutionism—those who survive do not bite the corporations that feed them—but it ought to give some anxiety to a conservationist.

Finally, if innovation (the "qualitatively new") is a primary requirement for art, then why are we still interested in works that are no longer new?

Can science and the arts be "linked" by "a common groundwork of explanation"? The answer depends upon the extent to which the arts are reducible to explanation. Mr. Wilson's project of consilience depends upon his assumption that works of art can be rendered into "interpretations" that can then be aligned with the laws of biology and ultimately with the laws of physics. He assumes, in other words, that a sufficient response to a work of art is to "interpret" it, and moreover that the resulting interpretation is as good as or is equal to the work of art. He says that "criticism can be as inspired and idiosyncratic as the work it addresses." (It is consistent with his view of things that he should both deny the possibility of inspiration and use "inspired" as a term of praise, but how he reconciles the supposed intellectual virtue of idiosyncrasy with his zeal for reduction to laws and principles is not clear.) To propose that the value of a work of art lies in its interpretation is to propose further that it is of interest only as an instance or specimen and that it can be not only explained but explained away.

And of course this would be all right if works of art were so constructed as to have extractable meanings or principles or laws. The problem is that they are not so constructed, and in this way they are in fact much like organisms. A chickadee is not constructed to exemplify the principles of its anatomy or the laws of aerodynamics or the life history of its species, and it has not been explained when these things have been extracted (or subtracted) from it.

For a while, in thinking of this question, I proposed to myself that the only things really explainable are explanations. That is not quite true, but it is near enough to the truth that I am unwilling to forget it.

What can be explained? Experiments, ideas, patterns, cause-effect relationships and connections *within defined limits*, anything that can be calculated, graphed, or diagrammed. And yet explanation changes whatever is explained into something explainable. Explanation is reductive, not comprehensive; most of the time, when you have explained something, you discover leftovers. An explanation is a bucket, not a well.

What can't be explained? I don't think creatures can be explained. I don't think lives can be explained. What we know about creatures and lives must be pictured or told or sung or danced. And I don't think pictures or stories or songs or dances can be explained. The arts are indispensable precisely because they are so nearly antithetical to explanation.

The arts are constitutionally resistant to the reduction that Mr. Wilson wishes to subject them to. This resistance manifests itself in two ways: Art insists upon the irreducibility of its subjects; and works of art, as objects, are by nature not reducible.

The power of art tends to be an individuating power, and that tendency is itself an affirmation of the value of individuals and of individuality. It is true that in our literature we have some allegories such as the play *Everyman* and *Pilgrim's Progress*, in which the characters represent abstractions, but this genre, though it contains important works, is a minor one. The dominant flow of our artistic tradition rises from the Bible and from Homer's epics, great works of individualization, which pause to delineate the characters not just of heroes and seers but also of children, housewives, bureaucrats, prostitutes, and tax collectors, of swineherds and old nurses and animals. This tradition, both sacred and democratic, has given us Odysseus and Penelope, Eumaios and Eurykleia, King David and Mary Magdalene, the Wife of Bath, Dante and Virgil, King Lear and Rosalind, Corin the shepherd and Falstaff, Tom Jones, Emma Woodhouse and Mr. Knightley, Captain Ahab, Huckleberry Finn, Tess of the D'Urbervilles, Leopold and Molly Bloom, Joe Christmas and Lena Grove. However much these characters may "stand for" us humans in our quests, flights, trials, and follies, they are each also intransigently themselves, and are valued as such. They all come out of the common fund of human experience, and so we recognize them, but not one of them is the same as anybody else. This tradition gives us the true-to-life portraits of shepherds in Flemish nativity scenes. It is realized pointedly in many a painting of the Virgin, in which she is represented both as the mother of Christ and as whatever ordinary girl posed for the artist.

The truest tendency of art is toward the exaltation, not the reduction, of its subjects. The highest art, as William Blake said, is able

> To see a World in a Grain of Sand
> And a Heaven in a Wild Flower

To paint a convincing portrait of the Virgin is to realize that for Christ to be born into this world He had to have a human mother. To write believably of a pilgrimage from Hell to Heaven, or of the transfiguring destitution of Gloucester and Lear, is to require time to remember eternity. Mr. Wilson's science, on the contrary, cannot see a world in a grain of sand. It can classify, name, and (within limits) explain a grain of sand, and divide it into ever smaller parts. There is no reason to say that this work is not admirable, valuable, or useful. But there is reason to say that it is not equivalent to, and it does not replace, the imagination of William Blake. Blake's lines remind us again of the miraculousness of life. This news has been delivered to us time after time in our long tradition. It cannot be proved. It only can be told or shown.

All art that rises above competence insists upon the irreducibility of its subjects, its materials, and its finished works. It makes things that are inherently valuable in themselves and are not interchangeable with other things. To a merely competent carpenter, one sound board may be pretty much the same as another. But to a fine carpenter or cabinetmaker, every board is unique. The better artist a woodworker becomes, the more aware he or she becomes of the individuality of boards and of the differences between them. The increase of art accounts for the increase of perception.

In the same way, to be competent a farmer must know the nature of species and breeds of animals. But the better the farmer, the more aware he or she is of the animals' individuality. "Every one is different," you hear the good stockmen say. "No two are alike." The ideal of livestock breeding over the centuries has not been to produce clones. Recognition of "type" is certainly important. But paramount is the ability to recognize the outstanding individual.

The plainest and most emphatic denunciation of critical reductionism, and one that is generally ignored by critics, is the "Notice" posted at the beginning of *Huckleberry Finn*: "Persons attempting to find a motive in this narrative will be prosecuted; persons attempting to find a moral in it will be banished; persons attempting to find a plot in it will be shot." Mark Twain's point, I think, is not that his book had no motive or moral or plot, but rather that its motive, its moral, and its plot were peculiar to itself as a whole, and could be conveyed only by itself as a whole. The motive, the moral, and the plot were not to be extracted and studied piecemeal like the organs of a laboratory frog. And the reason for this is plain: The value of *Huckleberry Finn* is not in its motive or moral or plot, but in its language. The book is valuable because it is a story *told*, not a story explained.

Or the problem of reduction in art may be illustrated by the problem of translating a poem from one language to another. The problem is that the poetry is in the language, or is the language. We can certainly translate the "sense" of one language to another, but the question of how to translate a poem from one language to another is the same question as how to translate a language from one language to another, which cannot be done. And so translators of poetry must accept failure as the primary condition of their work. They must settle, at best, for second best: Their translations must succeed or fail as new poems in their own language which at the same time serve as approximations or shadows of the original poems. Nobody, I think, has ever believed that there was an equation between a translation and the original.

You cannot translate a poem into an explanation, any more than you can translate a poem into a painting or a painting into a piece of music or a piece of music into a walking stick. A work of art says what it says in the only way it can be said. Beauty, for example, cannot be interpreted. It is not an empirically verifiable fact; it is not a quantity. Artists and critics and teachers and students certainly ought to notice that some things are beautiful and some are not; they ought to ask, and learn if they can, the difference between beauty and ugliness; they should learn how beautiful things are made and how things are made beautiful; but they might as well not ask what are the equivalents of beauty in ideas or pulse rates or dollars or "ordinary language." To believe that the arts can be interpreted so as to make them consilient with biology or physics is about equivalent to the belief that literary classics can survive as comic books or movies.

The truth too, as it appears in art, cannot be extracted as an idea or paraphrase. If we didn't, to start with, feel that a work of art was true, we wouldn't bother with it, or not for long. I don't think we stand before Gerard David's *Annunciation*, in the Metropolitan Museum, as speculators of the truth. We don't say, Can this be true? or, Might it have happened this way? Either we see that in the painting it is happening, or we don't. If we don't, we pass on by. Of course, if we assent to the painting, and if we are responsible people, we finally must ask if it is true. We must measure it against our knowledge of other paintings and other visions of holy things. We must ask if we are being fooled, or are fooling ourselves. Nevertheless, the painting must be accepted or rejected as itself, not on the basis of interpretation or our opinion of Luke 1:28–35. The painting says what it says in the only way it can be said.

I don't mean at all to say that criticism is impossible, or that it cannot be useful. Obviously, we need to talk about works of art. We must test our ways of knowing about them. We must learn them and teach them and describe them and study the ways they are made. We must compare them with one another, and evaluate them by whatever standards we can make applicable. But a work of criticism is not equivalent to a work of art and cannot replace it. The English departments and the biology departments and all other would-be consilient departments can spend the next millennium interpreting *King Lear*, and at the end of all that work the interpretation will still be one kind of thing and *King Lear* will still be a thing of a different kind. And so Mr. Wilson's idea that the arts' "essential qualities of truth and beauty [can be] described through ordinary language" is not merely off the subject, as might first appear; it is a violation of sense. It is saying, in effect, that extraordinary language can be described in ordinary language. This is too flimsy a scaffold to hold up much in the way of art criticism.

The question remains, Is it science? If "science" means proven knowledge or a methodology for proving knowledge, then Mr. Wilson's chapter on "The Arts and Their Interpretation" is no closer to science than it is to art criticism. In the chapter, he makes this startling confession: "Gene-culture co-evolution is, I believe, the underlying process by which the brain evolved and the arts originated. It is the conceivable means most consistent with the joint findings of the brain sciences, psychology, and evolutionary biology. Still, *direct* evidence with reference to the arts is slender." On this "I believe," this "conceivable," and this "slender" evidence, Mr. Wilson's enormous speculation teeters, like the Balanced Rock. It is not a reassuring place for a picnic.

VI. A Conversation Out of School

THE DISCIPLINES are different from one another, each distinct in itself, and rightly so. Science and art are neither fundamental nor immutable. They are not life or the world. They are tools. The arts and the sciences are our kit of cultural tools. Science cannot replace art or religion for the same reason that you cannot loosen a nut with a saw or cut a board in two with a wrench. The first question about the disciplines is not how they originated but how and for what they are to be used.

But if the sciences and the arts are divided into "two cultures," or into many subcultures, they are nobody's kit of tools. They are not the subjects of one conversation. They cannot be used in collaboration. And if they cannot be gathered together in one culture by consilience—which, on the evidence of Mr. Wilson's book, is not probable—then what can gather them together?

The only reason, really, that we need this kit of tools is to build and maintain our dwelling here on earth. (Those who wish to live or do business in other worlds should be free to depart, but not to return.) Our dwelling here is the proper work of culture. If the tools can be used collaboratively, then maybe we can find what are the appropriate standards for our work and can then build a good and lasting dwelling—which actually would be a diversity of dwellings suited to the diversity of homelands. If the tools cannot be so used, then they will be used to destroy such dwellings as we have accomplished so far, and our homelands as well.

To begin to think of the possibility of collaboration among the disciplines, we must realize that the "two cultures" exist as such because both of them belong to the one culture of division and dislocation, opposition and competition, which is to say the culture of colonialism and industrialism. This culture has steadily increased the dependence of individuals, regions, and nations upon larger and larger collective economies at the same time that it has thrown individuals, regions, and nations into a competitiveness with one another that is limitlessly destructive and demeaning. This state of universal competition understands the world as an anti-pattern in which each thing is opposed to every other thing, and it destroys the self-sufficiency of all places—households, farms, communities, regions, nations—even as it destroys the self-sufficiency of the world.

The collective economy is run for the benefit of a decreasing number of increasingly wealthy corporations. These corporations understand their "global economy" as a producer of money, not of goods. The goods of the world such as topsoil or forests must decline so that the money may increase. To facilitate this process, the corporations patronize the disciplines, chiefly the sciences, but some of the money, as "philanthropy," trickles down upon the arts. The brokers of this patronage are the universities, which are the organizers of the disciplines in our time. Since the universities are always a-building and are always in need of money, they accept the economy's fundamental principle of the opposition of money to goods. Having thus accepted as real the world as an anti-pattern of competing opposites, it is merely inevitable that they should organize learning, not as a conversation of collaborating disciplines, but as an anti-system of opposed and competing divisions. They have departmented our one great responsibility to live ably and generously into a nest of irresponsibilities. The sciences are sectioned like a stockyard the better to serve the corporations. The so-called humanities, which might have supplied at least a corrective or chastening remembrance of the good that humans have sometimes accomplished, have been dismembered into utter fecklessness, turning out "communicators" who have nothing to say and "educators" who have nothing to teach.

Must we reconcile ourselves to this cultural disintegration, this cacophony of the disciplines? Is it possible, failing consilience, to bring the arts and the sciences into healthful coherence and community of purpose? Edward O. Wilson would like to consiliate art and science on the terms of science—wrongly, I believe. The correct response is not to substitute the terms of art, or to look about for some hardly imaginable compromise.

The correct response, I think, is to ask if science and art are inherently at odds with one another. It seems obvious that they are not. To see that they are not may require extracurricular thought, but once we have cracked the crust of academic convention we can see that "science" means knowing and that "art" means doing, and that one is meaningless without the other. Out of school, the two are commonly inter-involved and naturally cooperative in the same person—a farmer, say, or a woodworker—who knows and does, both at the same time. It may be more or less possible to know and do nothing, but it is not possible to do and know nothing. One does as one knows. It is not possible to imagine a farmer who does not use both science and art.

It is also obvious that there is no insuperable natural or inherent division between scientists and artists—at least there is none outside of the academic pigeonholes. It is possible for a scientist and an artist to take part in the same conversation. On this subject I can speak from experience. I have been for the greater part of my life an artist of sorts, a cottage industrialist of literature, and for the past nineteen years I have been involved in a conversation with Wes Jackson, who is a scientist, a plant geneticist, and co-founder of The Land Institute in Salina, Kansas. This conversation has been, from the beginning, an alliance and a friendship. To me, it has been an indispensable source of instruction and a continuous testing of my thoughts. I can't speak for Wes, of course, and I make no claims as to the quality of our talk; the point here is only that we have been able to talk to each other out of our supposedly estranged disciplines, making our disciplines in the process useful to one another. In many meetings, telephone calls, notes, and letters during nineteen years, we have almost always had questions, and sometimes have had answers, for each other.

One of the most interesting facts about our conversation, from the standpoint of this essay, is that we were not prepared for it by our schooling. Wes's Ph.D. in genetics and my M.A. in English were not designed to give us things to say to each other. Because of his knowledge of the Bible and various works of literature, Wes was better prepared to talk to me than I to talk to him. Before I met him, I had been for perhaps fifteen years under the influence of the writings of the English agricultural scientist, Sir Albert Howard, and this was all that enabled me to understand Wes's germinal idea that, to be enduring, agriculture must imitate the local processes of nature. Though Wes is a writer (we both are essayists), I am not a scientist. I am perfectly ignorant of some things that Wes knows perfectly. While I have been writing, in addition to essays about agriculture, a series of fragments of the history of an imagined rural community, Wes has been at work on a project to renew agriculture by the development of perennial grain crops, in imitation of the native prairie plant communities, thereby reducing the amount of plowing necessary for food production, thereby reducing our presently ruinous rates of soil erosion. Obviously, two men so divergently occupied, and so divided by education, cannot talk together by any notion of the unification or consilience of their disciplines. And so what has made our conversation possible? The list of reasons amounts in implication to a fairly complete criticism of the present organization of the disciplines and the assumptions of that organization:

1. Though Wes and I were specialized, and maybe too much so, by our formal schooling, that schooling was superimposed upon an earlier, older education that we have in common. We both were raised in agrarian families. We were taught as children to know, respect, and love farming. From childhood until now, our thoughts about agriculture have been informed and conditioned by the actual work of farming, and this is work that we *like*.

2. Though we both have taught in universities, I more than Wes, neither of us has made a life in a university or in a "university community." We have lived in the countryside, among farming people, and have been involved in farming. In our minds, the problems we have talked about have always had the aspect of particular places and people, intimately known and cared about.

3. In ways sometimes different and sometimes the same, we have been at work on the same problem: how to change from a culture and a system of agriculture that destroy land and people to a culture and a system able to conserve both.

4. Because good land use involves both science and art (knowing and doing) and cannot be understood or practiced as either alone, we have had no illusions about the self-sufficiency or the adequacy of either of our disciplines. Our conversation has been between two parts of an always uncompleted whole.

It has lasted so long partly, of course, because it has been enjoyable, but also because it has been necessary.

5. The questions that have concerned us have been the same, and all of them raise the most practical issues of propriety: How can land and people be well used? What is good use? What is good knowledge, good thought, good work? How can one become genuinely and honorably native to one's place? This last question (the terms of which are set forth in Wes's book *Becoming Native to This Place*) is paramount. Asking it removes one permanently from the "two cultures" of careerist artists and scientists.

6. Neither of us believes that either art or science can be "neutral." Influence and consequence are inescapable. History continues. You cannot serve both God and Mammon, and you cannot work without serving one or the other.

7. Though each of us possesses the specialized vocabulary of his discipline, our conversation uses such talk only when necessary. We both can speak common English. Each of us, moreover, can speak a local English that is a source both of pleasure and exactitude. Our conversation is always striving to be local and particular. It is full of proper nouns, names of places and people. This subject of language is of the greatest importance, and I will have to return to it.

VII. Toward a Change of Standards

I HAVE JUST implied that a scientist and an artist may have to live and work outside the university if they are to have a sustained, mutually instructive, and effective conversation. Some will argue with this, and so be it. I will only point out that the modern university is organized to divide the disciplines; that universities pay little or no attention to the local and earthly effects of the work that is done in them; and that in the universities one discipline is rarely called upon to answer questions that might be asked of it by another discipline. If the universities sponsored an authentic conversation among the disciplines, then, for example, the colleges of agriculture would long ago have been brought under questioning by the college of arts and sciences or of medicine. A vital, functioning intellectual community *could* not sponsor patterns of land use that are increasingly toxic, violent, and destructive of rural communities.

I don't at all mean to suggest that I know how to reorganize the disciplines. I don't know how to do that, and I doubt that anybody does. But I feel no hesitation in saying that the standards and goals of the disciplines need to be changed. It used to be that we thought of the disciplines as ways of being useful to ourselves, for we needed to earn a living, but also and more importantly we thought of them as ways of being useful to one another. As long as the idea of vocation was still viable among us, I don't believe it was ever understood that a person was "called" to be rich or powerful or even successful. People were taught the disciplines at home or in school for two reasons: to enable them to live and work both as self-sustaining individuals and as useful members of their communities, and to see that the disciplines themselves survived the passing of the generations.

Now we seem to have replaced the ideas of responsible community membership, of cultural survival, and even of usefulness, with the idea of professionalism. Professional education proceeds according to ideas of professional competence and according to professional standards, and this explains the decline in education from ideals of service and good work, citizenship and membership, to mere "job training" or "career preparation." The context of professionalism is not a place or a community but a career, and this explains the phenomenon of "social mobility" and all the evils that proceed from it. The religion of professionalism is progress, and this means that, in spite of its vocal bias in favor of practicality and realism, professionalism forsakes both past and present in favor of the future, which is never present or practical or real. Professionalism is always offering up the past and the present as sacrifices to the future, in which all our problems will be solved and our tears wiped away—and which, being the future, never arrives. The future is always free of past limitations and present demands, always stocked with newer merchandise than any presently available, always promising that what we are going to have is better than what we have. The future is the utopia of academic thought, for virtually anything is hypothetically possible there; and it is the always-expanding frontier of the industrial economy, the fictive real estate against which losses are debited and to which failures are exiled. The future is not anticipated or provided for, but is only bought or sold. The present is ever diminished by this buying and selling of shares in the future that rightfully are owned by the unborn.

Wallace Stegner knew, both from his personal experience and from his long study of his region, that the two cultures of the American West are not those of the sciences and the arts, but rather those of the two human kinds that he called "boomers" and "stickers," the boomers being "those who pillage and run," and the stickers "those who settle, and love the life they have made and the place they have made it in." This applies to our country as a whole, and maybe to all of Western civilization in modern times. The first boomers were the oceanic navigators of the European Renaissance. They were gold seekers. All boomers have been gold seekers. They are would-be Midases who want to turn all things into gold: plants and animals, trees, food and drink, soil and water and air, life itself, even the future.

The sticker theme has so far managed to survive, and to preserve in memory and even in practice the ancient human gifts of reverence, fidelity, neighborliness, and stewardship. But un-questionably the dominant theme of modern history has been that of the boomer. It is no surprise that the predominant arts and sciences of the modern era have been boomer arts and boomer sciences.

The collaboration of boomer science with the boomer mentality of the industrial corporations has imposed upon us a state of virtually total economy in which it is the destiny of every creature (humans not excepted) to have a price and to be sold. In a total economy, all materials, creatures, and ideas become commodities, interchangeable and disposable. People become commodities along with everything else. Only such an economy could seek to impose upon the world's abounding geographic and creaturely diversity the tyranny of technological and genetic monoculture. Only in such an economy could "life forms" be patented, or the renewability of nature and culture be destroyed. Monsanto's aptly named "terminator gene"—which, implanted in seed sold by Monsanto, would cause the next generation of seed to be sterile—is as grave an indicator of totalitarian purpose as a concentration camp.

The complicity of the arts and humanities in this conquest is readily apparent in the enthusiasm with which the disciplines, schools, and libraries have accepted their ever-growing dependence (at public expense) on electronic technologies that are, in fact, as all of history shows, not necessary to learning or teaching, and which have produced no perceptible improvement in either. This was accomplished virtually without a dissenting voice, without criticism, without regard even for the economic cost. It is the clearest demonstration so far that the cult of originality and innovation is in fact a crowd of conformists, tramping on one another's heels for fear of being the last to buy whatever is for sale.

With the same ardor, and in more or less the same stampede, this crowd has mastered a slang of personal "liberation" that has done little for real freedom (which requires perception of authentic differences and distinctions), but has set many free from their rightful obligations and responsibilities—to, for example, their spouses and their children. The arts, especially in their well-paying popular versions, have become adept as permission givers for this sort of freedom. But there is too close a kinship between the personal freedom from reverence, fidelity, neighborliness, and stewardship and the corporate freedom to pollute and exterminate. When, if ever, the accounting is properly done, many of our present "liberties" and "necessities" will be seen to owe too much to the exploitation of "cheap" labor, raw materials, energy, and food.

The dominant story of our age, undoubtedly, is that of adultery and divorce. This is true both literally and figuratively: The dominant *tendency* of our age is the breaking of faith and the making of divisions among things that once were joined. This story obviously must be told by somebody. Perhaps, in one form or another, it must be told (because it must be experienced) by everybody. But how has it been told, and how ought it be told? This is a critical question, but not a question merely for art criticism. The story can be told in a way that clarifies, that makes imaginable and compassionable, the suffering and the costs; or it can be told in a way that seems to grant an easy permission and absolution to adultery and divorce. Can literature, for example, be written according to standards that are not merely literary? Obviously it can. And it had better be.

Suppose, then, that we should change the standards, as in fact some scientists and some artists already are attempting to do. Suppose that the ultimate standard of our work were to be, not professionalism and profitability, but the health and durability of human and natural communities. Suppose we learned to ask of any proposed innovation the question that so far only the Amish have been wise enough to ask: What will this do to our community? Suppose we attempted the authentic multiculturalism of adapting our ways of life to the nature of the places where we live. Suppose, in short, that we should take seriously the proposition that our arts and sciences have the power to help us adapt and survive. What then?

Well, we certainly would have a healthier, prettier, more diverse and interesting world, a world less toxic and explosive, than we have now.

And how might this come about? Again, I have to say that I don't know. I don't like or trust large, official programs of improvement, and I don't want to appear to be inviting any such thing. But perhaps there is no harm in making suggestions, if I acknowledge that the suggestions are only mine, and if I make sure that my suggestions apply primarily to the thinking, work, and conduct of individuals. Here is my list:

1. Rather than the present economic hierarchy of the professions, which results in the denigration and undercompensation of essential jobs of work, particularly in the economies of land use, we should think and work toward an appropriate subordination of all the disciplines to the health of creatures, places, and communities. A science or an art, for example, that served settlement rather than the exploitation of "frontiers" would be subordinate to reverence, fidelity, neighborliness, and stewardship, to affection and delight. It would aim to keep our creatureliness intact.

2. We should banish from our speech and writing any use of the word "machine" as an explanation or definition of anything that is not a machine. Our understanding of creatures and our use of them are *not* improved by calling them machines.

3. We should abandon the idea that this world and our human life in it can be brought by science to some sort of mechanical perfection or predictability. We are creatures whose intelligence and knowledge are not invariably equal to our circumstances. The radii of knowledge have only pushed back —and enlarged—the circumference of mystery. We live in a world famous for its ability both to surprise us and to deceive us. We are prone to err, ignorantly or foolishly or intentionally or maliciously. One of the oddest things about us is the interdependency of our virtues and our faults. Our moral code depends on our shortcomings as much as our knowledge. It is only when we confess our ignorance that we can see our need for "the law and the prophets." It is only because we err and are ignorant that we make promises, which we keep, not because we are smart, but because we are faithful.

4. We should give up the frontier and its boomer "ethics" of greed, cunning, and violence, and, so near too late, accept settlement as our goal. Wes Jackson says that our schools now have only one major, upward mobility, and that we need to

offer a major in homecoming. I agree, and would only add that a part of the sense of "homecoming" must be home*making*, for we now must begin sometimes with remnants, sometimes with ruins.

5. We need to require from our teachers, researchers, and leaders—and attempt for ourselves—a responsible accounting of technological progress. What have we gained by computers, for example, *after* we have subtracted the ecological costs of making them, using them, and throwing them away, the value of lost time and work when "the computers are down," and the enormous economic cost of the "Y2K" correction?

6. We ought conscientiously to reduce our tolerance for ugliness. Why, if we are in fact "progressing," should so much expense and effort have resulted in so much ugliness? We ought to begin to ask ourselves what are the limits—of scale, speed, and probably expense as well—beyond which human work is bound to be ugly.

7. We should recognize the insufficiency, to our life here among living creatures, of the abstract categories of reductionist thought. Resist classification! Without some use of abstraction, thought is incoherent or unintelligible, perhaps unthinkable. But abstraction alone is merely dead. And here we return again to the crucial issue of language.

In our public dialogue (such as it is) we are now using many valuable words that are losing their power of reference, and have as a consequence become abstract, merely gestures. I have in mind words such as "patriotism," "freedom," "equality," and "rights," or "nature," "human," "wild," and "sustainable." We could make a longish list of words such as these, which we often use without thought or feeling, just to show which side we suppose we want to be on. This situation calls for language that is not sloganish and rhetorical, but rather is capable of reference, specification, precision, and refinement —a language never far from experience and example. In the work of the great poets, the heavenly and the earthly are not abstract, but are *present*; the language of those poets is whole and precise. We are mistaken to think that we can increase our earthly knowledge by ignoring Heaven, or become more intelligent by giving up Dante or condescending to Milton. The middling, politically correct language of the professions is incapable either of reverence or familiarity; it is headless and footless, loveless, a language of nowhere.

I believe that this need for a whole, vital, particularizing language applies just as strongly to the sciences as to the arts and humanities. For the human necessity is not just to know, but also to cherish and protect the things that are known, and to know the things that can be known only by cherishing. If we are to protect the world's multitude of places and creatures, then we must know them, not just conceptually but imaginatively as well. They must be pictured in the mind and in memory; they must be known with affection, "by heart," so that in seeing or remembering them the heart may be said to "sing," to make a music peculiar to its recognition of each particular place or creature that it knows well. I am remembering here the importance to Confucius of "the tones given off by the heart." To know imaginatively is to know intimately, particularly, precisely, gratefully, reverently, and with affection.

In *Consilience*, Edward O. Wilson says, "Today the entire planet has become home ground," but that is a conceptual statement only, and doubtful as such. No human has ever known, let alone imagined, the entire planet. And even in an age of "world travel," none of us lives on the entire planet; in fact, owing to so much mobility, a lot of people (as some of them will tell you) don't live anywhere. But if we are to know any part of the planet intimately, particularly, precisely, and with affection, then we must live somewhere in particular for a long time. We must be able to call up to the mind's eye by name a lot of local places, people, creatures, and things.

One of the most significant costs of the economic destruction of farm populations is the loss of local memory, local history, and local names. Field names, for instance, even such colorless names as "the front field" and "the back field," are vital signs of a culture. If the arts and the sciences ever waken from their rapture of academic specialization, they will make themselves at home in places they have helped to spoil, and set about reconstructing histories and remembering names.

8. We should value familiarity above innovation. Boomer scientists and artists want to discover (so to speak) a place where they have not been. Sticker scientists and artists want to know where they are. There is no reason that familiarity cannot be a goal just as worthy, demanding, and exciting as innovation —or, as I would argue, much more so. It would certainly give worthwhile employment to more people. And in fact its boundaries are much larger. Innovation is limited always by human ingenuity and human means; familiarity is limited only by the limits of life. The real infinitude of experience is in familiarity.

My own experience has shown me that it is possible to live in and attentively study the same small place decade after decade, and find that it ceaselessly evades and exceeds comprehension. There is nothing that it can be reduced to, because "it" is always, and not predictably, changing. It is never the same two days running, and the better one pays attention the more aware one becomes of these differences. Living and working in the place day by day, one is continuously revising one's knowledge of it, continuously being surprised by it and in error about it. And even if the place stayed the same, one would be getting older and growing in memory and experience, and would need for that reason alone to work from revision to revision. One knows one's place, that is to say, only within limits, and the limits are in one's mind, not in the place. This is a description of life in time in the world. A place, apart from our now always possible destruction of it, is inexhaustible. It *cannot* be altogether known, seen, understood, or appreciated.

That Cézanne returned many times to paint Mont Sainte-Victoire, or that William Carlos Williams spent a long life writing about Rutherford, New Jersey, does not mean that those places are now exhausted as subjects. It means that they are inexhaustible. There are many examples of this. One that I have kept in mind for nearly forty years, to define a hope and a consolation, is that of the French entomologist Jean Henri Fabre. Too poor to travel, Fabre spent the last thirty-odd years of his life studying the insects and other creatures of his small *harmas* near Sérignan, "a patch of pebbles enclosed within four walls." And surely his enthusiasm lasted so long because he studied the living creatures in their—and his—dwelling place.

There is nothing intrinsically wrong with an interest in discovery and innovation. It only becomes wrong when it is thought to be the norm of culture and of intellectual life. As such it is in the first place misleading. As I have already suggested, the effort of familiarity is always leading to discovery and the new, just as do the quests of explorers and "original researchers." The difference is that innovation for its own sake, and especially now when it so directly serves the market, is disruptive of human settlement, whereas the revelations of familiarity elaborate the local cultural pattern and tend toward settlement, which they also prevent from becoming static.

If local adaptation is important, as I believe it unquestionably is, then we must undertake, in both science and art, the effort of familiarity. In doing so, we will confront the endlessness of human knowledge, work, and experience. But we should not mislead ourselves: We will confront mystery too. There is more to the world, and to our own work in it, than we are going to know.

One of the best studies of local adaptation that I know is a book by George Sturt, *The Wheelwright's Shop*, which looks at traditional farm carts and wagons as products of "the age-long effort of Englishmen to fit themselves close and ever closer into England." Sturt understands these vehicles as distinctly local products, whose form and fabric evolved in a long, only partly conscious give and take between the people and the landscape. The accommodation inherent in the design was elegant, though not in any sure sense explainable:

> But where begin to describe so efficient an organism, in which all the parts interacted until it is hard to say which was modified first, to meet which other? Was it to suit the horses or the ruts, the loading or the turning, that the front wheels had to have a diameter of about four feet? Or was there something in the average height of a carter, or in the skill of wheel-makers, that fixed these dimensions? One only knew that, by a wonderful compromise, all these points had been provided for in the country tradition of fore-wheels for a waggon. And so all through. Was it to suit these same wheels that the front of a waggon was slightly curved up, or was that

done in consideration of the loads, and the circumstance merely taken advantage of for the wheels? I could not tell. I cannot tell. I only know that in these and a hundred details every well-built farm-waggon (of whatever variety) was like an organism, reflecting in every curve and dimension some special need of its own countryside, or, perhaps, some special difficulty attending wheelwrights with the local timber.

This is the way a locally adapted culture works. Over a long time it learns to conform its artifacts to the local landscape, local circumstances, and local needs. This is exactly opposite to the way of industrialism, which forces the locality to conform to industrial artifacts, always with the most dreadful consequences to the locality. Having in the industrial age exchanged, as Sturt says, "local needs . . . for cosmopolitan wishes," we are a long way now from the saving elegance that his book recalls. And the most resolute and expensive projects of discovery and innovation on the part of science-technology-and-industry cannot take us there. Only a long, patient, loving effort of familiarity can do that.

Hanging his project from an "if," as usual, Edward O. Wilson looks forward in *Consilience* to "a Magellanic voyage that eventually encircles the whole of reality." This presumably will be the long end run that will carry us and "the environment" over the goal line of survival.

But in an earlier book, *Biophilia*, Mr. Wilson set forth a far different conclusion, one which he has now evidently repudiated, but which I wish to affirm: "That the naturalist's journey has only begun and for all intents and purposes will go on forever. That it is possible to spend a lifetime in a magellanic voyage around the trunk of a single tree."

A single tree? Well, life is a miracle and therefore infinitely of interest everywhere. We have perhaps sufficient testimony, from artists and scientists both, that if we watch, refine our intelligence and our attention, curb our greed and our pride, work with care, have faith, a single tree might be enough.

VIII. Some Notes in Conclusion

IN THE PROCESS that carries knowledge from the laboratory to the market there is not enough fear. And in the history of that process there has not been adequate accounting.

Richard Strohman has made clear what is objectionable about the infusion of money from biotechnology corporations into the universities: These grants will press university scientists in the direction of product development; an interest in product development increases the emphasis upon predictability (you cannot market, or not for long, an unpredictable product); but an undue emphasis upon predictability will tend to narrow the context of experimentation, making the product, in the context of the world, unpredictable in effect and influence.

This process has a historical analogue in the introduction of the internal combustion engine into agriculture. In the commercial workshops tractors had only to pass the test of mechanical correctness: They had to start and run more or less predictably. In the context of the world, however, these machines had effects and exerted influences that far surpassed their merely mechanical limits. They replaced agriculture's old dependence on the free energy of the sun with a dependence on purchased energy; in general, they increased farming's dependence on a supply economy that farmers cannot control or influence; over the years, these dependences have radically oversimplified the patterns of farming, replacing diversity with monoculture, crop rotation with continuous tillage, and human labor with machines and chemicals; they have replaced (in Wes Jackson's words) nature's wisdom with human cleverness; they have caused widespread, profound social and cultural disruption. All these changes are still in progress. Whatever the technological or quantitative gains, this industrialization of farming has been costly, and it will continue to be. Most of the costs have been "externalized"—that is, charged to nature or the public or the future.

The response to this in the land grant universities has been applause.

*

The faculties and administrations of universities are inexcusably bewildered between the superstition that knowledge is invariably good and the fact that it can be monetarily valuable and also dangerous.

There is in fact no reason to think that the professions are self-correcting, or that new knowledge necessarily compensates for old error.

*

The time is past, if ever there was such a time, when you can just discover knowledge and turn it loose in the world and assume that you have done good.

This, to me, is a sign of the incompleteness of science in itself—which is a sign of the need for a strenuous conversation among all the branches of learning. This is a conversation that the universities have failed to produce, and in fact have obstructed.

*

If we were as fearful of our knowledge and our power as in our ignorance we ought to be—and as our cultural and religious traditions instruct us to be—then we would be trying to reconnect the disciplines both within the universities and in the conduct of the professions.

*

In our present economic predicament, ethics, ecology, environmental law, etc. won't *as specialties* have much corrective force. They will be used to rationalize what is wrong.

*

The anti-smoking campaign, by its insistent reference to the expensiveness to government and society of death by smoking, has raised a question that it has not answered: What is the best and cheapest disease to die from, and how can the best and cheapest disease best be promoted?

*

An idea of health that does not generously and gracefully accommodate the fact of death is obviously incomplete. The crudest manifestation of modern medicine is its routine, stubborn, and finally cruel resistance to death. This comes of the refusal to accept death not only as part of health, which it demonstrably is, but also as a great mystery both in itself and as a part of the mystery that surrounds us all our lives. The medical industry's resistance is only sometimes an instance of scientific heroism; sometimes it is the fear of what we don't know anything about.

*

Science can teach us and help us to resist death, but it can't teach us to prepare for death or to die well.

The question of how you want to die is somewhat fantastical but nonetheless it is one that all the living need to consider, one that belongs to the issue of health, and one that health science can't answer.

Do you want to die at home with your people "in blessed peace around you," which is the death Tiresias foresaw for Odysseus and the one Homer seems to recommend?

Or do you want to die in the hands of the best medical professionals wherever they are?

Such questions may seem irrelevant until you realize that they define two very different lives.

*

The refusal of modern medicine to confront the deaths of its patients is only a function of its refusal to confront the unique and unempirically precious lives of its patients.

Analogous to that is modern agriculture's refusal to confront the life of a good or healthy farm as a cultural artifact, unique in place and character, complex in form, mysterious in its sources, and many times more fascinating and precious than the "unit of production" to which it has been reduced by the economics of "agribusiness" and the colleges of agriculture.

*

My worry is only partly about science-as-Pandora, an activity of questioning or curiosity which cannot undo the harm that it may do and sometimes does. (Science has armed us but it cannot disarm us.)

I am worried also about the *application* of science, which I think is generally crude. This cannot be solved merely by keeping the context of research as large as it actually is in application —which, of course, would be a sensible precaution. However large the context, however generous the acknowledgment of context, the results of the research still cannot be applied *both* generally and sensitively. Finally it is "brought home" to a specific community of persons and creatures in a specific place. If it is then applied in its abstract or generalized or marketable form, it will obscure the uniquenesses of the subject persons or creatures or places, or of their community, and this sort of application is almost invariably destructive.

The only remedy I can see is for scientists (and artists also) to understand and imagine themselves as members of, and sharers in, the fate of affected communities. Our schools now encourage people to regard as mere privileges the power and influence that they call leadership. But leadership without membership is a terrible thing.

*

The issue of the application of science is a political issue. If the science is applied only by, or can only be applied by, a government or a large corporation, then it is tyrannical. If it is to be applied, it should be applied locally by local people on a local scale, using the health of the locality as the standard of application and judgment.

The use of science by or upon people who do not understand it is always potentially tyrannical, and it is always dangerous.

*

Applying knowledge—scientific or otherwise—is an art. An artist is somebody who knows what to put where, and when to put it. A good artist is one who applies knowledge skillfully and sensitively to the particular creatures and places of the world. Good farmers and good architects and good doctors are the most obvious examples, but the same potential is in all the arts. This is why it is such a shame to see the so-called fine arts elaborating themselves as academic or professional specialties without reference to much of anything, perhaps in imitation of the supposedly pure sciences.

*

The modern scientific enterprise apparently is directed toward the goal of complete knowledge. But if you had complete knowledge, if you knew everything, could you then act? Could you apply what you knew, or would you be paralyzed by a surplus of considerations? If you were to map within a circle all possible relationships among all the points along its circumference, you would end up with a black circle—an engorgement of "information" that would not be knowledge, but rather the practical equivalent of the blank circle you began with.

Thus the proposition that it would be good to know every-thing is probably false. The real question that is always to be addressed is the one that arises from our state of ignorance: How does one act well—sensitively, compassionately, without irreparable damage—on the basis of *partial* knowledge?

Perhaps the most proper, and the most natural, response to our state of ignorance is not haste to increase the amount of available information, or even to increase knowledge, but rather a lively and convivial engagement with the issues of form, elegance, and kindness. These issues of "sustainability" are both scientific and artistic.

*

The problem that we confront in our sciences and arts, as I understand it, is not a problem of information but a problem of ignorance. Or, to put it another way, the problem is not primarily one of mass; it is a problem of form.

It is out of our confrontation with our ignorance that we come to the problem of form. The ignorant must hope, and with study they may come to know, that it is possible to achieve forms that partake of wholeness and even holiness and make sense, even though one does not know everything.

*

Good artists are people who can stick things together so that they stay stuck. They know how to gather things into formal arrangements that are intelligible, memorable, and lasting. Good forms confer health upon the things that they gather to-gether. Farms, families, and communities are forms of art just as are poems, paintings, and symphonies. None of these things would exist if we did not make them. We can make them either well or poorly; this choice is another thing that we make.

*

Always informing our practice of artistic form is our sense of the formality of creation. This "sense" is not knowledge of the empirical or scientific sort. It does not tend toward any sort of description. It is the perfectly assimilated, perfectly forgotten knowledge by which all creatures live in their places.

*

Overhanging all our thought and work is the question of how certain of itself human knowledge can be.

I learned from the theologian Philip Sherrard to ask this question: If things are evolving, and if human consciousness is evolving along with everything else, where do we find a standpoint from which to understand the whole process?

To make the same point in a more practical way, let us take that ubiquitous and misleading word "environment"—which, as used, proposes that reality is composed of a creature and its surroundings. But if, as in fact we know, the creature is not only *in* its environment but *of* it, and if the relationship between creature and environment is mutually formative, and if this relationship is a process that cannot be stopped short of the creature's death, then how can we get outside the relationship in order to predict with certainty the effects of our participation?

Religion begins with such questions. But even reason can see that they define the issues of propriety and scale. If we can't know with final certainty what we are doing, then reason cautions us to be humble and patient, to keep the scale small, to be careful, to go slow.

*

In speaking of the reductionism of modern science, we should not forget that the primary reductionism is in the assumption that human experience or human meaning can be adequately represented in any human language. This assumption is false.

To show what I mean, I will give the example that is most immediate to my mind:

My grandson, who is four years old, is now following his father and me over some of the same countryside that I followed my father and grandfather over. When his time comes, my grandson will choose as he must, but so far all of us have been farmers. I know from my grandfather that when he was a child he too followed his father in this way, hearing and seeing, not knowing yet that the most essential part of his education had begun.

And so in this familiar spectacle of a small boy tagging along behind his father across the fields, we are part of a long procession, five generations of which I have seen, issuing out of generations lost to memory, going back, for all I know, across previous landscapes and the whole history of farming.

Modern humans tend to believe that whatever is known can be recorded in books or on tapes or on computer discs and then again learned by those artificial means.

But it is increasingly plain to me that the meaning, the cultural significance, even the practical value, of this sort of family procession across a landscape can be known but not told. These things, though they have a public value, do not have a public meaning; they are too specific to a particular small place and its history. This is exactly the tragedy in the modern displacement of people and cultures.

That such things can be known but not told can be shown by answering a simple question: *Who* knows the meaning, the cultural significance, and the practical value of this rural family's generational procession across its native landscape? The answer is not so simple as the question: No one person ever will know all the answer. My grandson certainly does not know it. And my son does not, though he has positioned himself to learn some of it, should he be so blessed.

I am the one who (to some extent) knows, though I know also that I cannot tell it to anyone living. I am in the middle now between my grandfather and my father, who are alive in my memory, and my son and my grandson, who are alive in my sight.

If my son, after thirty more years have passed, has the good pleasure of seeing his own child and grandchild in that procession, then he will know something like what I now know.

This living procession through time in a place *is* the record by which such knowledge survives and is conveyed. When the procession ends, so does the knowledge.

CITIZENSHIP PAPERS

(2003)

A Citizen's Response to "The National Security Strategy of the United States of America"

> The constituent parts of a state are obliged to hold their public faith with each other, and with all those who derive any serious interest under their engagements, as much as the whole state is bound to keep its faith with separate communities. Otherwise competence and power would soon be confounded, and no law be left but the will of a prevailing force.
> —Edmund Burke, *On the Revolution in France*

> America! America!
> God mend thine every flaw,
> Confirm thy soul in self control,
> Thy liberty in law.
> —Katharine Lee Bates, "America the Beautiful"

> To seek for peace by way of war is the same as to seek for chastity by way of fornication.
> —Anonymous theologian of the first century

THE NEW "National Security Strategy" published by the White House in September 2002, if carried out, would amount to a radical revision of the political character of our nation. This document was conceived in reaction to the terrorist attacks of September 11, 2001. Its central and most significant statement is this:

> While the United States will constantly strive to enlist the support of the international community, we will not hesitate to act alone, if necessary, to exercise our right of self-defense by acting preemptively against such terrorists . . .

A democratic citizen, properly uneasy, must deal here first of all with the question, Who is this "we"? It is not the "we" of the Declaration of Independence, which referred to a small group

239

of signatories bound by the conviction that "governments [derive] their just powers from the consent of the governed." And it is not the "we" of the Constitution, which refers to "*the people* [my emphasis] of the United States."

Because of what is implied by the commitment to act alone and preemptively, this "we" of the new strategy can refer only to the president. It is a royal "we." A head of state, preparing to act alone in starting a preemptive war, will need to justify his intention by secret information, and will need to plan in secret and execute his plan without forewarning. A preemptive attack widely known and discussed, as in a democratic polity, would risk being preempted by a preemptive attack by the other side. The idea of a government acting alone in preemptive war is inherently undemocratic, for it does not require or even permit the president to obtain the consent of the governed. As a policy, this new strategy depends on the acquiescence of a public kept fearful and ignorant, subject to manipulation by the executive power, and on the compliance of an intimidated and office-dependent legislature. Even within the narrow logic of warfare, there is a substantial difference between a defensive action, for which the reason would be publicly known, and a preemptive or aggressive action, for which the reason would be known only by a few at the center of power. The responsibilities of the president obviously are not mine, and so I hesitate to doubt absolutely the necessity of governmental secrecy. But I feel no hesitation in saying that to the extent that a government is secret, it cannot be democratic or its people free. By this new doctrine, the president alone may start a war against any nation at any time, and with no more forewarning than preceded the Japanese attack on Pearl Harbor.

Would-be participating citizens of a democratic nation, unwilling to have their consent coerced or taken for granted, therefore have no choice but to remove themselves from the illegitimate constraints of this "we" in as immediate and public a way as possible.

But as this document and its supporters insist, we have now entered a new era when acts of war may be carried out not only by nations and "rogue nations," but also by individuals using weapons of mass destruction, and this requires us to give up some measure of freedom in return for some increase of security. The lives of every one of us may at any time be in jeopardy.

Even so, we need to ask: What does real security require of us? What does true patriotism require of us? What does freedom require of us?

The alleged justification for this new strategy is the recent emergence in the United States of international terrorism. But why the events of September 11, 2001, horrifying as they were, should have called for a radical new investiture of power in the executive branch is not clear.

"The National Security Strategy" defines terrorism as "premeditated, politically motivated violence perpetrated against innocents." This is truly a distinct kind of violence, but it is a kind old and familiar, even in the United States. All that was really new about the events of September 11, 2001, was that they raised the scale of such violence to that of "legitimate" modern warfare.

To imply by the word "terrorism" that this sort of terror is the work exclusively of "terrorists" is misleading. The "legitimate" warfare of technologically advanced nations likewise is premeditated, politically motivated violence perpetrated against innocents. The distinction between the *intention* to perpetrate violence against innocents, as in "terrorism," and the *willingness* to do so, as in "war," is not a source of comfort. We know also that modern war, like ancient war, often involves intentional violence against innocents.

A more correct definition of "terrorism" would be this: violence perpetrated unexpectedly without the authorization of a national government. Violence perpetrated unexpectedly *with* such authorization is not "terrorism" but "war." If a nation perpetrates violence officially—whether to bomb an enemy airfield or a hospital—it is not guilty of "terrorism." But there is no need to hesitate over the difference between "terrorism" and any violence or threat of violence that is terrifying. "The National Security Strategy" wishes to cause "terrorism" to be seen "in the same light as slavery, piracy, or genocide"—but not in the same light as war. It accepts and affirms the legitimacy of war.

This document concedes that "we are menaced less by fleets and armies than by catastrophic technologies in the hands of the embittered few." And yet our government, with our permission, continues to manufacture, stockpile, and trade in these catastrophic technologies, including nuclear, chemical, and biological weapons. In nuclear or biological warfare, in which we know we cannot limit effects, how do we distinguish our enemies from our friends—or our enemies from ourselves? Does this not bring us exactly to the madness of terrorists who kill themselves in order to kill others?

The official definition of "terrorism," then, is far too exclusive if we seriously wish to free the world of the terrors induced by human violence. But let us suppose that our opposition to terror could be justly or wisely limited to a "war against terrorism." How effective might such a war be?

The war against terrorism is not, strictly speaking, a war against nations, even though it has already involved international war in Afghanistan and presidential threats against other nations. This is a war against "the embittered few"—"thousands of trained terrorists"—who are "at large" among many thousands and even millions of others who are, in the language of this document, "innocents," and thus deserving of our protection.

Hunting these terrorists down will be like combing lice out of a head of hair. Unless we are willing to kill innocents in order to kill the guilty—unless we are willing to blow our neighbor's head off, or blow our own head off, to get rid of the lice —the need to be lethal will be impeded constantly by the need to be careful. Because of the inherent difficulties and because we must suppose a new supply of villains to be always in the making, we can expect the war on terrorism to be more or less endless, endlessly costly and endlessly supportive of a thriving bureaucracy.

Unless, that is, we should become willing to ask why, and to do something about the causes. Why do people become terrorists? Such a question is often dismissed as evidence of "liberal softness" toward malefactors. But that is not necessarily the case. Such a question may also arise from the recognition that problems have causes. There is, however, no acknowledgment in "The National Security Strategy" that terrorism might have a cause that could possibly be discovered and possibly remedied. "The embittered few," it seems, are merely "evil."

II

Much of the obscurity of our effort so far against terrorism originates in the now official idea that the enemy is evil and that we are (therefore) good, which is the precise mirror image of the official idea of the terrorists.

The epigraph of Part III of "The National Security Strategy" contains this sentence from President Bush's speech at the National Cathedral on September 14, 2001: "But our responsibility to history is already clear: to answer these attacks and rid the world of evil." A government, committing its nation to rid the world of evil, is assuming necessarily that it and its nation are good.

But the proposition that anything so multiple and large as a nation can be good is an insult to common sense. It is also dangerous, because it precludes any attempt at self-criticism or self-correction; it precludes public dialogue. It leads us far indeed from the traditions of religion and democracy that are intended to measure and so to sustain our efforts to be good. "There is none good but one, that is, God," Christ said. Also: "He that is without sin among you, let him first cast a stone at her." And Thomas Jefferson justified general education by the obligation of citizens to be critical of their government: "for nothing can keep it right but their own vigilant *and distrustful* [my emphasis] superintendence." An inescapable requirement of true patriotism, love for one's land, is a vigilant distrust of any determinative power, elected or unelected, that may preside over it.

And so it is not without reason or precedent that a would-be participating citizen should point out that in addition to evils originating abroad and supposedly correctable by catastrophic technologies in "legitimate" hands, we have an agenda of domestic evils, not only those that properly self-aware humans can find in their own hearts, but also several that are indigenous to our history as a nation: issues of economic and social justice, and issues related to the continuing and worsening maladjustment between our economy and our land.

There are kinds of violence that have nothing directly to do with unofficial or official warfare but are accepted as normal to our economic life. I mean such things as toxic pollution, land destruction, soil erosion, and the destruction of biological diversity, and of the degradation of ecological supports of agriculture. To anybody with a normal concern for health and sanity, these "externalized costs" are terrible and are terrifying.

I don't wish to make light of the threats and dangers that now confront us. There can be no doubt of the reality of terrorism as defined and understood by "The National Security Strategy," or of the seriousness of our situation, or of our need for security. But frightening as all this is, it does not relieve us of the responsibility to be as intelligent, principled, and practical as we can be. To rouse the public's anxiety about foreign terror while ignoring domestic terror, and to fail to ask if these terrors are in any way related, is wrong.

It is understandable that we should have reacted to the attacks of September 11, 2001, by curtailment of civil rights, by defiance of laws, and by resort to overwhelming force, for those actions are the ready products of fear and hasty thought. But they cannot protect us against the destruction of our own land by ourselves. They cannot protect us against the selfishness, wastefulness, and greed that we have legitimized here as economic virtues, and have taught to the world. They cannot protect us against our government's long-standing disdain for any form of self-sufficiency or thrift, or against the consequent dependence, which for the present at least is inescapable, on foreign supplies such as oil from the Middle East.

And they cannot protect us from what may prove to be the greatest danger of all: the estrangement of our people from one another and from our land. Increasingly, Americans—including, notoriously, their politicians—are not *from* anywhere. And so they have in this "homeland," which their government now seeks to make secure on their behalf, no home *place* that they are strongly moved to know or love or use well or protect.

*

It is no wonder that "The National Security Strategy," growing as it does out of unresolved contradictions in our domestic life, should attempt to compound a foreign policy out of contradictory principles.

There is, first of all, the contradiction of peace and war, or of war as the means of achieving and preserving peace. This document affirms peace; it also affirms peace as the justification of war and war as the means of peace—and thus it perpetuates a hallowed absurdity. But implicit in its assertion of this (and, by implication, any other) nation's right to act alone in its own interest is an acceptance of war as a permanent condition. Either way, it is cynical to invoke the ideas of cooperation, community, peace, freedom, justice, dignity, and the rule of law (as this document repeatedly does), and then proceed to assert one's intention to act alone in making war. One cannot reduce terror by holding over the world the threat of what it most fears.

All the things we supposedly want to secure are thus subverted by our proposed means of securing them. Edmund Burke recognized this contradiction and was wary of it: "Laws are commanded to hold their tongues amongst arms; and tribunals fall to the ground with the peace they are no longer able to uphold."

This is a contradiction not reconcilable except by a self-righteousness almost inconceivably naive. The authors write that "We will . . . use our foreign aid to promote freedom and support those who struggle *non-violently* [my emphasis] for it"; and they observe that

> In pursuing advanced military capabilities that can threaten its neighbors . . . China is following an out-dated path that, in the end, will hamper its own pursuit of national greatness. In time, China will find that social and political freedom is the only source of that greatness.

Thus we come to the authors' implicit definition of "rogue state": any nation pursuing national greatness by advanced military capabilities that can threaten its neighbors—any nation, that is, except *our* nation.

If you think our displeasure with "rogue states" might have any underpinning in international law, then you will be disappointed to learn that

> We will take the actions necessary to ensure that our efforts to meet our global security commitments and protect Americans are not impaired by the potential for investigations, inquiry, or prosecution by the International Criminal Court (ICC), whose jurisdiction does not extend to Americans and which we do not accept.

The rule of law in the world, then, is to be upheld by a nation that has declared itself to be above the law. A childish hypocrisy here assumes the dignity of a nation's foreign policy. But if we perceive an illegitimacy in the catastrophic weapons and *ad lib* warfare of other nations, how can we not perceive the same illegitimacy in our own?

The contradiction between peace and war implies, of course, at every point a contradiction between security and war. We wish, this document says, to be cooperative with other nations, but the authors seem not to realize how rigidly our diplomacy and our offers to cooperate will be qualified by this new threat of overwhelming force to be used merely at the president's pleasure. We cannot hope to be secure when our government has declared, by its announced readiness "to act alone," its willingness to be everybody's enemy.

III

A further contradiction is that between war and commerce. This issue arises first of all in the war economy, which unsurprisingly regards war as a business and weapons as merchandise. However nationalistic may be the doctrine of "The National Security Strategy," the fact is that the business of warfare and the weapons trade have been thoroughly internationalized. Saddam Hussein possesses weapons of mass destruction, for example, partly because we sold him such weapons and the means of making them back when (madman or not) he was our "friend." But the internationalization of the weapons trade is a result inherent in international trade itself. It is a part of globalization. Mr. Bush's addition of this Security Strategy to the previous bipartisan commitment to globalization exposes an American dementia that has not been so plainly displayed before.

The America Whose Business Is Business has been internationalizing its economy in haste (for bad reasons, and with little foresight), looking everywhere for "trading partners," cheap labor, and tax shelters. Meanwhile, the America Whose Business Is National Defense is withdrawing from the world in haste (for bad reasons, with little foresight), threatening left and right, repudiating agreements, and angering friends. The problem of participating in the Global Economy for the benefit of Washington's corporate sponsors while maintaining a nationalist belligerence and an isolationist morality calls for superhuman intelligence in the Secretary of Commerce. The problem of "acting alone" in an international war while maintaining simultaneously our ability to import the foreign goods (for instance, oil) on which we have become dependent even militarily will call, likewise, for overtopping genius in the Secretary of Defense.

"The National Security Strategy" devotes a whole chapter to the president's resolve to "ignite a new era of global economic growth through free markets and free trade." But such a project cannot be wedded, even theoretically, to his commitment to a militarist nationalism ever prepared "to act alone." One must wonder when the government's corporate sponsors will see the contradiction and require the nation to assume a more humble posture in the presence of the global economy.

The conflict in future-abuse between this document and the sales talk of the corporations is stark, and it is pretty absurd. On the one hand, we have the future as a consumer's paradise in which everybody will be able to buy comfort, convenience, and happiness. On the other hand, we have this government's new future in which terrible things are bound to happen if we don't do terrible things in the present—which, of course, will make terrible things even more likely to happen in the future.

After World War II, we hoped the world might be united for the sake of peacemaking. Now the world is being "globalized" for the sake of trade and the so-called free market—for the sake, that is, of plundering the world for cheap labor, cheap energy, and cheap materials. How nations, let alone regions and communities, are to shape and protect themselves within this "global economy" is far from clear. Nor is it clear how the global economy can hope to survive the wars of nations.

If a nation cannot be "good" in any simple or incontestable way, then what can it reasonably be that is better than bad?

A nation can be charitable, as we can say with some confidence, for we need not look beyond our own for an example. Our nation, sometimes, has been charitable toward its own people; it has been kind to the elderly, the sick, the unemployed, and others unable to help themselves. And sometimes it has been charitable toward other nations, as when we helped even our onetime enemies to recover from World War II. But "charity" does not refer only to the institutional or governmental help we give to the "less fortunate." The word means "love." The commandment to "Love your enemies" suggests that charity must be without limit; it must include everything. A nation's charity must come from the heart and the imagination of its people. It requires us ultimately to see the world as a community of all the creatures, a community which, to be possessed by any, must be shared by all.

Perhaps that is only a better way of saying that a nation can be civilized. To be civil is to conduct oneself as a responsible citizen, honoring the lives and the rights of others. Our courts and jails are filled with the uncivil, who have presumed to act alone in their own interest. And we well know that incivility is now almost conventionally in business among us.

A nation can be independent, as our founders instructed us. If a nation cannot within reasonable measure be independent, it is hard to see how its existence might be justified. Though independence may at times require some sort of self-defense, it cannot be maintained by defiance of other nations or by making war against them. It can be maintained only by the most practical economic self-reliance. At the very least, a nation should be able sustainably to feed, clothe, and shelter its citizens, using its own sources and by its own work.

And of course that requires a nation to be, in the truest sense, patriotic: Its citizens must love their land with a knowing, intelligent, sustaining, and protective love. They must not, for any price, destroy its beauty, its health, or its productivity.

And they must not allow their patriotism to be degraded to a mere loyalty to symbols or any present set of officials.

A nation also can abide under the rule of law. Since the Alien and Sedition Laws of 1798, in times of national stress or emergency there have been arguments for the abridgement of citizenship under the Constitution. But the weakness of those arguments is in their invariable implication that a democracy such as ours can work only in the most favorable circumstances. If constitutional guarantees of rights and immunities cannot be maintained in unfavorable circumstances, what is their point or value? Their value in fact originates in the acknowledgment of their usefulness in the times of greatest difficulty and to those in greatest need, as does the value of international law.

It is impossible to think that constitutional government can be suspended in a time of danger, in deference to the greater "efficiency" of centralized power, and then easily or quickly restored. Efficiency may be a political virtue, but only if strictly limited. Our Constitution, by its separation of powers and its system of checks and balances, acts as a restraint upon efficiency by denying exclusive power to any branch of the government. The logic of governmental efficiency, unchecked, runs straight on, not only to dictatorship, but also to torture, assassination, and other abominations.

Such aims as charity, civility, independence, true patriotism, and lawfulness a nation of imperfect human beings may reasonably adopt as its standards. And we may conclude reasonably, rightly, and with no touch of self-contempt, that by those standards we are less charitable, less civil, less independent, less patriotic, and less law-abiding than we might be, and than we need to be. And do these shortcomings relate to the president's perception that we are less secure than we need to be? We would be extremely foolish to suppose otherwise.

One might reasonably assume that a policy of national security would advocate from the start various practical measures to conserve and to use frugally the nation's resources, the objects of this husbandry being a reduction in the nation's dependence on imports and a reduction in the competition between nations for necessary goods. One might reasonably expect the virtues of stewardship, thrift, self-sufficiency, and neighborliness to

receive a certain precedence in the advocacy of political leaders. Since the country, to make itself secure, may be required
to rely on itself, one might reasonably expect a due concern for
the health and longevity of its soils, forests, and watersheds,
its natural and its human communities, its domestic economy,
and the natural systems on which that economy inescapably
depends.

Such a concern could come only from perceiving the contradiction between national security and the present global
economy, but there is no such perception in "The National
Security Strategy." This document, ignoring all conflicts, proposes to go straight ahead with both projects: national security, which it defines forthrightly as isolationist, domineering,
and violent; and the global economy, which it defines as
international humanitarianism. It does allow that there is a
connection between our national security and the economic
condition of "the rest of the world," and that the extreme
poverty of much of the rest of the world is "neither just nor
stable." And of course one can only agree. But the authors
assume that economic wrongs can be righted merely by "economic development" and the "free market," that it is the nature of these things to cure poverty, and that they will not be
impeded by terrorism or a war against terrorism or a preemptive war for the security of one nation. These are articles of a
faith available only to a politically sheltered economic elite.

As for conservation here and elsewhere, the authors provide
a list of proposals that is short, vague or ambiguous, incomplete, and rather wildly miscellaneous:

> We will incorporate labor and environmental concerns
> into U.S. trade negotiations . . .

> We will . . . expand the sources and types of global
> energy . . .

> We will . . . develop cleaner and more energy efficient
> technologies.

> Economic growth should be accompanied by global ef-
> forts to *stabilize* [my emphasis] greenhouse gas concen-
> trations . . .

> Our overall objective is to reduce America's greenhouse
> gas emissions relative to the size of our economy, cutting
> such emissions per unit of economic activity by 18 per-
> cent over the next 10 years . . .

But the only energy technologies specifically promoted here
are those of "clean coal" and nuclear power—for both of which
there is a strong corporate advocacy and a strong conservationist criticism or opposition. There is no mention of land loss, of
soil erosion, of pollution of land, air, and water, or of the various threats to biological diversity—all problems of generally
(and scientifically) recognized gravity.

Agriculture, which is involved with all the problems listed
above, and with several others—and which is the economic activity most clearly and directly related to national security, if
one grants that we all must eat—receives scant and superficial
treatment amounting to dismissal. The document proposes
only:

1. "a global effort to address new technology, science, and
   health regulations that needlessly impede farm exports
   and improved agriculture." This refers, without saying
   so, to the growing consumer resistance to genetically
   modified food. A global effort to overcome this resistance would help, not farmers and not consumers, but
   global agribusiness corporations such as Monsanto.
2. "transitional safeguards which we have used in the agricultural sector." This refers to government subsidies,
   which ultimately help the agribusiness corporations, not
   farmers.
3. promotion of "new technologies, including biotechnology, [which] have enormous potential to improve
   crop yields in developing countries while using fewer
   pesticides and less water." This is offered (as usual and
   questionably) as the solution to hunger, but its immediate benefit would be to the corporate suppliers of the
   technologies.

This is not an agriculture policy, let alone a national security
strategy. It has the blindness, arrogance, and foolishness that

are characteristic of top-down thinking by politicians and academic experts, assuming that "improved agriculture" would
inevitably be the result of catering to the agribusiness corporations, and that national food security can be achieved merely by
going on as before. It does not address any agricultural problem as such, and it ignores the vulnerability of our present food
system—dependent as it is on genetically impoverished monocultures, cheap petroleum, cheap long-distance transportation,
and cheap farm labor—to many kinds of disruption by "the
embittered few," who, in the event of such disruption, would
quickly become the embittered many. On eroding, ecologically
degraded, increasingly toxic landscapes, worked by failing or
subsidy-dependent farmers and by the cheap labor of migrants,
we have erected the tottering tower of "agribusiness," which
prospers and "feeds the world" (incompletely and temporarily)
by undermining its own foundations.

But all our military strength, all our police, all our technologies and strategies of suspicion and surveillance cannot make
us secure if we lose our ability to farm, or if we squander our
forests, or if we exhaust or poison our water sources.

A policy of preemptive war that rests, as this one does, on such
flimsy domestic underpinnings and on too great a concentration of power in the presidency, obviously risks or invites correction. And, as we know, correction can come by three means:

1. By strenuous public debate. But this requires a strong,
   independent political opposition, which at present we do
   not have. The country now contains many individuals
   and groups seriously troubled by issues of civil rights,
   food, health, agriculture, economy, peace, and conservation. These people have much in common, but they have
   no strong political voice, because few politicians have
   seen fit to speak for them.

2. By failure, as by some serious disruption of our food or
   transportation or energy systems. This might cause a
   principled and serious public debate, which might place
   us on a more stable economic and political footing. But
   we are a large nation, highly centralized in almost every
   way, and without much self-sufficiency, either national

or regional. Any failure, therefore, will be large and not nearly so easy to correct as to prevent.

3. By citizens' initiative. Responsibilities abandoned by the government properly are assumed by the people. An example of this is the already well-established movement for local economies, typically beginning with food. There is much hope in this effort, provided that it continues to grow. And we have, surviving from the Vietnam War, a surprisingly strong and numerous peace movement. As we learned from the Vietnam experience, the only effective answer to a secretive and unresponsive government is a citizens' revolt. The revolt against the war in Vietnam was nonviolent, effective, and finally successful. Such a correction is better than no correction, but it is far from ideal—far less to be preferred than correction by public debate. A citizens' revolt necessarily comes too late. And if it is not peaceable and responsibly led, it could easily destroy the things it is meant to save.

IV

The present administration has adopted a sort of official Christianity, and it obviously wishes to be regarded as Christian. But "Christian" war has always been a problem, best solved by avoiding any attempt to reconcile policies of national or imperial militarism with anything Christ said or did. The Christian gospel is a summons to peace, calling for justice beyond anger, mercy beyond justice, forgiveness beyond mercy, love beyond forgiveness. It would require a most agile interpreter to justify hatred and war by means of the Gospels, in which we are bidden to love our enemies, bless those who curse us, do good to those who hate us, and pray for those who despise and persecute us.

This peaceability has grown more practical—it has gained "survival value"—as industrial warfare has developed increasingly catastrophic weapons, which are abominable to our government, so far, only when other governments possess them. But since the end of World War II, when the terrors of industrial war had been fully revealed, many people and, by fits and starts, many governments have recognized that peace is not just a desirable condition, as was thought before, but is a practical necessity. It has become less and less thinkable that we might have a living and a livable world, or that we might have livable lives or any lives at all, if we do not make the world capable of peace.

And yet we have not learned to think of peace apart from war. We have received many teachings about peace and peace-ability in biblical and other religious traditions, but we have marginalized those teachings, have made them abnormal, in deference to the great norm of violence and conflict. We wait, still, until we face terrifying dangers and the necessity to choose among bad alternatives, and then we think again of peace, and again we fight a war to secure it.

At the end of the war, if we have won it, we declare peace; we congratulate ourselves on our victory; we marvel at the newly proved efficiency of our latest, most "sophisticated" weapons; we ignore the cost in lives, materials, and property, in suffering and disease, in damage to the natural world; we ignore the in-evitable residue of resentment and hatred; and we go on as be-fore, having, as we think, successfully defended our way of life.

That is pretty much the story of our victory in the Gulf War of 1991. In the years between that victory and September 11, 2001, we did not alter our thinking about peace and war, which is to say that we thought much about war and little about peace; we continued to punish the defeated people of Iraq and their children; we made no effort to reduce our dependence on the oil we import from other, potentially belligerent countries; we made no improvement in our charity toward the rest of the world; we made no motion toward greater economic self-reliance; and we continued our extensive and often irreversible damages to our own land. We appear to have assumed merely that our victory confirmed our manifest destiny to be the rich-est, most powerful, most wasteful nation in the world. After the catastrophe of September 11, it again became clear to us how good it would be to be at peace, to have no enemies, to have no needless death to mourn. And then, our need for war following with the customary swift and deadly logic our need for peace, we took up the customary obsession with the evil of other people.

And now we are stirring up the question whether or not Islam is a warlike religion, ignoring the question, much more urgent for us, whether or not Christianity is a warlike religion. There is no hope in this. Islam, Judaism, Christianity—all have been warlike religions. All have tried to make peace and rid the world of evil by fighting wars. This has not worked. It is never going to work. The failure belongs inescapably to all of these religions insofar as they have been warlike, and to acknowledge this failure is the duty of all of them. It is the duty of all of them to see that it is wrong to destroy the world, or risk destroying it, to get rid of its evil.

It is useless to try to adjudicate a long-standing animosity by asking who started it or who is the most wrong. The only sufficient answer is to give up the animosity and try forgiveness, to try to love our enemies and to talk to them and (if we pray) to pray for them. If we can't do any of that, then we must begin again by trying to imagine our enemies' children, who, like *our* children, are in mortal danger because of enmity that they did not cause.

We can no longer afford to confuse peaceability with passivity. Authentic peace is no more passive than war. Like war, it calls for discipline and intelligence and strength of character, though it calls also for higher principles and aims. If we are serious about peace, then we must work for it as ardently, seriously, continuously, carefully, and bravely as we have ever prepared for war.

Thoughts in the Presence of Fear

I. The time will soon come when we will not be able to remember the horrors of September 11 without remembering also the unquestioning technological and economic optimism that ended on that day.

II. This optimism rested on the proposition that we were living in a "new world order" and a "new economy" that would "grow" on and on, bringing a prosperity of which every new increment would be "unprecedented."

III. The dominant politicians, corporate officers, and investors who believed this proposition did not acknowledge that the prosperity was limited to a tiny percentage of the world's people, and to an ever smaller number of people even in the United States; that it was founded upon the oppressive labor of poor people all over the world; and that its ecological costs increasingly threatened all life, including the lives of the supposedly prosperous.

IV. The "developed" nations had given to the "free market" the status of a religion, and were sacrificing to it their farmers, farmlands, and rural communities, their forests, wetlands, and prairies, their ecosystems and watersheds. They had accepted universal pollution and global warming as normal costs of doing business.

V. There was, as a consequence, a growing worldwide effort on behalf of economic decentralization, economic justice, and ecological responsibility. We must recognize that the events of September 11 make this effort more necessary than ever. We citizens of the industrial countries must continue the labor of self-criticism and self-correction. We must recognize our mistakes.

VI. The paramount doctrine of the economic and technological euphoria of recent decades has been that everything depends on innovation. It was understood as desirable, and even as necessary, that we should go on and on from one technological innovation to the next, which would cause the economy to "grow" and make everything better and better. This of course implied at every point a hatred of the past, of all things inherited and free. All things superseded in our progress of innovations, whatever their value might have been, were discounted as of no value at all.

VII. We did not anticipate anything like what has now happened. We did not foresee that all our sequence of innovations might be at once overridden by a greater one: the invention of a new kind of war that would turn our previous innovations against us, discovering and exploiting the debits and the dangers that we had ignored. We never considered the possibility that we might be trapped in the webwork of communication and transport that was supposed to make us free.

VIII. Nor did we foresee that the weaponry and the war science that we marketed and taught to the world would become available, not just to recognized national governments which possess so uncannily the power to legitimate large-scale violence, but also to "rogue nations," dissident or fanatical groups, and individuals—whose violence, though never worse than that of nations, is judged by the nations to be illegitimate.

IX. We had apparently accepted the belief that technology is only good; that it cannot serve evil as well as good; that it cannot serve our enemies as well as ourselves; that it cannot be used to destroy what is good, including our homelands and our lives.

X. We had accepted too the corollary belief that an economy (either as a money economy or as a life-support system) that is global in extent, technologically complex, and centralized is invulnerable to terrorism, sabotage, or war, and that it is protectable by "national defense."

XI. We now have a clear, inescapable choice that we must make. We can continue to promote a global economic system of unlimited "free trade" among corporations, held together by long and highly vulnerable lines of communication and supply, but *now* recognizing that such a system will have to be protected by a hugely expensive police force that will be worldwide, whether maintained by one nation or several or all, and that such a police force will be effective precisely to the extent that it oversways the freedom and privacy of the citizens of every nation.

XII. Or we can promote a decentralized world economy that would have the aim of assuring to every nation and region a *local* self-sufficiency in life-supporting goods. This would not eliminate international trade, but it would tend toward a trade in surpluses after local needs have been met.

XIII. One of the gravest dangers to us now, second only to further terrorist attacks against our people, is that we will attempt to go on as before with the corporate program of global "free trade," whatever the cost in freedom and civil rights, without self-questioning or self-criticism or public debate.

XIV. This is why the substitution of rhetoric for thought, always a temptation in a national crisis, must be resisted by officials and citizens alike. It is hard for ordinary citizens to know what is actually happening in Washington in a time of such great trouble; for all we know, serious and difficult thought may be taking place there. But the talk that we are hearing from politicians, bureaucrats, and commentators has so far tended to reduce the complex problems now facing us to issues of unity, security, normality, and retaliation.

XV. National self-righteousness, like personal self-righteousness, is a mistake. It is misleading. It is a sign of weakness. Any war that we may make now against terrorism will come as a new installment in a history of war in which we have fully participated. We are not innocent of making war against civilian populations. The modern doctrine of such warfare was set forth and enacted by General William Tecumseh Sherman, who held that a civilian population could be declared guilty and rightly subjected to military punishment. We have never repudiated that doctrine.

XVI. It is a mistake also—as events since September 11 have shown—to suppose that a government can promote and participate in a global economy and at the same time act exclusively in its own interest by abrogating its international treaties and standing aloof from international cooperation on moral issues.

XVII. And surely, in our country, under our Constitution, it is a fundamental error to suppose that any crisis or emergency can justify any form of political oppression. Since September 11, far too many public voices have presumed to "speak for us" in saying that Americans will gladly accept a reduction of freedom in exchange for greater "security." Some would, maybe. But some others would accept a reduction in security (and in global trade) far more willingly than they would accept any abridgement of our constitutional rights.

XVIII. In a time such as this, when we have been seriously and most cruelly hurt by those who hate us, and when we must consider ourselves to be gravely threatened by those same people, it is hard to speak of the ways of peace and to remember that Christ enjoined us to love our enemies, but this is no less necessary for being difficult.

XIX. Even now we dare not forget that since the attack on Pearl Harbor—to which the present attack has been often and not usefully compared—we humans have suffered an almost uninterrupted sequence of wars, none of which have brought peace or made us more peaceable.

XX. The aim and result of war necessarily are not peace but victory, and any victory won by violence necessarily justifies the violence that won it and leads to further violence. If we are serious about innovation, must we not conclude that we need something new to replace our perpetual "war to end war"?

XXI. What leads to peace is not violence but peaceableness, which is not passivity, but an alert, informed, practiced, and active state of being. We should recognize that while we have extravagantly subsidized the means of war, we have almost totally neglected the ways of peaceableness. We have, for example, several national military academies, but not one peace academy. We have ignored the teachings and the examples of Christ, Gandhi, Martin Luther King, and other peaceable leaders. And here we have an inescapable duty to notice also that war is profitable, whereas the means of peaceableness, being cheap or free, make no money.

XXII. The key to peaceableness is continuous practice. It is wrong to suppose that we can exploit and impoverish the poorer countries, while arming them and instructing them in the newest means of war, and then reasonably expect them to be peaceable.

XXIII. We must not again allow public emotion or the public media to caricature our enemies. If our enemies are now to be some nations of Islam, then we should undertake to *know* those enemies. Our schools should begin to teach the histories, cultures, arts, and languages of the Islamic nations. And our leaders should have the humility and the wisdom to ask the reasons some of those people have for hating us.

XXIV. Starting with the economies of food and farming, we should promote at home and encourage abroad the ideal of local self-sufficiency. We should recognize that this is the surest, the safest, and the cheapest way for the world to live. We should not countenance the loss or destruction of any local capacity to produce necessary goods.

XXV. We should reconsider and renew and extend our efforts to protect the natural foundations of the human economy: soil, water, and air. We should protect every intact ecosystem and watershed that we have left, and begin restoration of those that have been damaged.

XXVI. The complexity of our present trouble suggests as never before that we need to change our present concept of education. Education is not properly an industry, and its proper use is not to serve industries, either by job-training or by industry-subsidized research. Its proper use is to enable citizens to live lives that are economically, politically, socially, and culturally responsible. This cannot be done by gathering or "accessing" what we now call "information"—which is to say facts without context and therefore without priority. A proper education enables young people to put their lives in order, which means knowing what things are more important than other things; it means putting first things first.

XXVII. The first thing we must begin to teach our children (and learn ourselves) is that we cannot spend and consume endlessly. We have got to learn to save and conserve. We do need a "new economy," but one that is founded on thrift and care, on saving and conserving, not on excess and waste. An economy based on waste is inherently and hopelessly violent, and war is its inevitable by-product. We need a peaceable economy.

The Failure of War

IF YOU KNOW even as little history as I know, it is hard not to doubt the efficacy of modern war as a solution to any problem except that of retribution—the "justice" of exchanging one damage for another, which results only in doubling (and continuing) the damage and the suffering.

Apologists for war will immediately insist that war answers the problem of national self-defense. But the doubter, in reply, will ask to what extent the *cost* even of a successful war of national defense—in life, money, material goods, health, and (inevitably) freedom—may amount to a national defeat. And national defense in war always involves *some* degree of national defeat. This paradox has been with us from the very beginning of our republic. Militarization in defense of freedom reduces the freedom of the defenders. There is a fundamental inconsistency between war and freedom.

In asking such a question, the doubter will be mindful also that in a modern war, fought with modern weapons and on the modern scale, neither side can limit to "the enemy" or "the enemy country" the damage that it does. These wars damage the world. We know enough by now to know that you cannot damage a part of the world without damaging all of it. Modern war has not only made it impossible to kill "combatants" without killing "noncombatants," it has made it impossible to damage your enemy without damaging yourself. You cannot kill your enemy's women and children without offering your own women and children to the selfsame possibility. We (and, inevitably, others) have prepared ourselves to destroy our enemy by destroying the entire world—including, of course, ourselves.

That many have considered the increasing unacceptability of modern warfare is shown by the language of the propaganda surrounding modern war. Modern wars have characteristically been fought to end war. They have been fought in the name of peace. Our most terrible weapons have been made, ostensibly, to preserve and assure the peace of the world. "All we want is peace," we say, as we increase relentlessly our capacity to make war.

And yet at the end of a century in which we have fought two wars to end war and several more to prevent war and preserve peace, and in which scientific and technological progress has made war ever more terrible and less controllable in its effects, we still, by policy, give no consideration, and no countenance, to nonviolent means of national defense. We do indeed make much of diplomacy and diplomatic relations, but by diplomacy we mean invariably ultimatums for peace backed by the threat of war. It is always understood that we stand ready to kill those with whom we are "peacefully negotiating."

Our century of war, militarism, and political terror has unsurprisingly produced great—and successful!—advocates of true peace, among whom Mohandas K. Gandhi and Martin Luther King, Jr., are paramount examples. The considerable success that they achieved testifies to the presence, in the midst of violence, of an authentic and powerful desire for peace and, more important, of the proven will to make the necessary sacrifices. But so far as our government is concerned, these men and their great and authenticating accomplishments might as well never have existed. To achieve peace by peaceable means is not yet our goal. We cling to the hopeless paradox of making peace by making war.

Which is to say that we cling, in our public life, to a brutal hypocrisy. In our century of almost universal violence of humans against fellow humans, and against our natural and cultural commonwealth, hypocrisy has been inescapable because our opposition to violence has been selective or merely fashionable. Some of us who approve of our monstrous military budget and our peacekeeping wars nonetheless deplore "domestic violence" and think that our society can be pacified by "gun control." Some of us are against capital punishment but for abortion. Some of us are against abortion but for capital punishment. Most of us, whatever our stand on preserving the lives of the thoughtlessly conceived unborn, thoughtlessly participate in an economy that steals from all the unborn.

One does not have to know very much or think very far in order to see the moral absurdity upon which we have erected our sanctioned enterprises of violence. Abortion-as-birth-control is justified as a "right," which can establish itself only by denying all the rights of another person, which is the most primitive intent of warfare. Capital punishment sinks us all to the same level of primal belligerence, at which an act of violence is avenged by another act of violence. What the justifiers of these wrongs ignore is the fact—as well established by the history of feuds or the history of anger as by the history of war —that violence breeds violence. Acts of violence committed in "justice" or in affirmation of "rights" or in defense of "peace" do not end violence. They prepare and justify its continuation.

The most dangerous superstition of the parties of violence is the idea that sanctioned violence can prevent or control unsanctioned violence. But if violence is "just" in one instance, as determined by the state, why, by a merely logical extension, might it not also be "just" in another instance, as determined by an individual? How can a society that justifies capital punishment and warfare prevent its justifications from being extended to assassination and terrorism? If a government perceives that some causes are so important as to justify the killing of children, how can it hope to prevent the contagion of its logic from spreading to its citizens—or to its citizens' children? If you so devalue human life that the accidentally conceived unborn may be permissibly killed, how do you keep that permission from being assumed by someone who has made the same judgment against the born?

I am aware of the difficulty of assigning psychological causes to acts of violence. Psychological causes abound. But here I am talking about the power of example, precedent, and reason.

If we give to these small absurdities the magnitude of international relations, we produce, unsurprisingly, some much larger absurdities. What could be more absurd than our attitude of high moral outrage against other nations for manufacturing the selfsame weapons that we manufacture? The difference, as our leaders say, is that we will use these weapons virtuously whereas our enemies will use them maliciously—a proposition that too readily conforms to a proposition of much less dignity: We will use them in *our* interest, whereas our enemies will use them in *theirs*. The issue of virtue in war is as obscure, ambiguous, and troubling as Abraham Lincoln found to be the issue of prayer in war: "Both [the North and the South] read the same Bible, and pray to the same God; and each invokes His aid against the other. . . . The prayers of both could not be answered—that of neither could be answered fully."

But recent American wars, having been both "foreign" and "limited," have been fought under a second illusion even more dangerous than the illusion of perfect virtue: We are assuming, and are encouraged by our leaders to assume, that, aside from the sacrifice of life, no personal sacrifice is required. In "foreign" wars, we do not directly experience the damage that we inflict upon the enemy. We hear and see this damage reported in the news, but we are not affected, and we don't mind. These limited, "foreign" wars require that some of our young people will be killed or crippled, and that some families will grieve, but these "casualties" are so widely distributed among our population as hardly to be noticed. Otherwise, we do not feel ourselves to be involved. We pay taxes to support the war, but that is nothing new, for we pay war taxes also in time of "peace." We experience no shortages, we suffer no rationing, we endure no limitations. We earn, borrow, spend, and consume in wartime as in peacetime.

And of course no sacrifice is required of those large economic interests that now principally constitute our "economy." No corporation will be required to submit to any limitation or to sacrifice a dollar. On the contrary, war is understood by some as the great cure-all and opportunity of our corporate economy. War ended the Great Depression of the 1930s, and we have maintained a war economy—an economy, one might justly say, of general violence—ever since, sacrificing to it an enormous economic and ecological wealth, including, as designated victims, the farmers and the industrial working class.

And so great costs are involved in our fixation on war, but the costs are "externalized" as "acceptable losses." And here we see how progress in war, progress in technology, and progress in the industrial economy are parallel to one another—or, very often, are merely identical.

Romantic nationalists, which is to say most apologists for war, always imply in their public speeches a mathematics or an accounting that cannot be performed. Thus by its suffering in the Civil War, the North is said to have "paid for" the emancipation of the slaves and the preservation of the Union. Thus we may speak of our liberty as having been "bought" by the bloodshed of patriots. I am fully aware of the truth in such statements. I know that I am one of many who have benefited from painful sacrifices made by other people, and I would not like to be ungrateful. Moreover, I am a patriot myself, and I know that the time may come for any of us when we must make extreme sacrifices for the sake of liberty.

But still I am suspicious of this kind of accounting. For one reason, it is necessarily done by the living on behalf of the dead. And I think we must be careful about too easily accepting, or being too easily grateful for, sacrifices made by others, especially if we have made none ourselves. For another reason, though our leaders in war always assume that there is an acceptable price, there is never a previously stated level of acceptability. The acceptable price, finally, is whatever is paid.

It is easy to see the similarity between this accounting of the price of war and our usual accounting of "the price of progress." We seem to have agreed that whatever has been (or will be) paid for so-called progress is an acceptable price. If that price includes the diminishment of privacy and the increase of government secrecy, so be it. If it means a radical reduction in the number of small businesses and the virtual destruction of the farm population, so be it. If it means cultural and ecological impoverishment, so be it. If it means the devastation of whole regions by extractive industries, so be it. If it means that a mere handful of people should own more billions of wealth than is owned by all of the world's poor, so be it.

But let us have the candor to acknowledge that what we call "the economy" or "the free market" is less and less distinguishable from warfare. For about half of this century we worried about world conquest by international communism. Now with less worry (so far) we are witnessing world conquest by international capitalism. Though its political means are milder (so far) than those of communism, this newly internationalized capitalism may prove even more destructive of human cultures and communities, of freedom, and of nature. Its tendency is just as much toward total dominance and control. Confronting this conquest, ratified and licensed by the new international trade agreements, no place and no community in the world may consider itself safe from some form of plunder. More and more people all over the world are recognizing that this is so, and they are saying that world conquest of any kind is wrong, period.

They are doing more than that. They are saying that *local* conquest also is wrong, and wherever it is taking place local people are joining together to oppose it. All over my own state of Kentucky this opposition is growing—from the west, where the exiled people of the Land Between the Lakes are struggling to save their confiscated homeland from bureaucratic degradation, to the east, where the native people of the mountains are still struggling to preserve their land from destruction by absentee corporations.

To have an economy that is warlike, that aims at conquest, and that destroys virtually everything that it is dependent on, placing no value on the health of nature or of human communities, is absurd enough. It is even more absurd that this economy, that in some respects is so much at one with our military industries and programs, is in other respects directly in conflict with our professed aim of national defense.

It seems only reasonable, only sane, to suppose that a gigantic program of preparedness for national defense would be founded, first of all, upon a principle of national and even regional economic independence. A nation determined to defend itself and its freedoms should be prepared, and always preparing, to live from its own resources and from the work and the skills of its own people. It should carefully husband and conserve those resources, justly compensate that work, and rigorously cultivate and nurture those skills. But that is not what we are doing in the United States today. What we are doing, as we prepare for and prosecute wars allegedly for national defense, is squandering in the most prodigal manner the natural and human resources of the nation.

At present, in the face of declining finite sources of fossil fuel energy, we have virtually no energy policy, either for conservation or for the development of safe and clean alternative sources. At present, our energy policy simply is to use all that we have. At present, moreover, in the face of a growing population needing to be fed, we have virtually no policy for land conservation, and *no* policy of just compensation to the primary producers of food. At present, our agricultural policy is to use up everything that we have, while depending increasingly on imported food, energy, technology, and labor.

Those are just two examples of our general indifference to our own needs. We thus are elaborating a direct and surely a dangerous contradiction between our militant nationalism and our espousal of the international "free market" ideology. How are we going to defend our freedoms (this is a question both for militarists and for pacifists) when we must import our necessities from international suppliers who have no concern or respect for our freedoms? What would happen if in the course of a war of national defense we were to be cut off from our foreign sources of supply? What would happen if, in a war of national defense, military necessity required us to attack or blockade our foreign suppliers? We have already fought one energy war allegedly in national defense. If our present policies of economic indifference continue, we may face wars for other commodities: food or water or shoes or steel or textiles.

What can we do to free ourselves of this absurdity?

With that question my difficulty declares itself, for I do not ask it as a teacher knowing the answer. I ask it knowing that by doing so I describe my own dilemma. The news media, the industrial economy, perhaps human nature as well, prompt us to want quick, neat answers to our questions, but I don't think my question has a quick, neat answer.

Obviously, we would be less absurd if we took better care of one another and of all our fellow creatures. We would be less absurd if we founded our public policies upon an honest description of our needs and our predicament, rather than upon fantastical descriptions of our wishes. We would be less absurd if our leaders would consider in good faith the proven alternatives to violence.

Such things are easy to say. But finally we must face this daunting question, not as a nation or a group, but as individual persons—as ourselves. We are disposed, somewhat by culture and somewhat by nature, to solve our problems by violence —by maximum force relentlessly applied—and even to enjoy doing so. And yet by now all of us must at least have suspected that our right to live, to be free, and to be at peace is not guaranteed by any act of violence. It can be guaranteed only by our willingness that all other persons should live, be free, and be at peace—and by our willingness to use or give our own lives to make that possible. To be incapable of such willingness is merely to resign ourselves to the absurdity we are in; and yet, if you are like me, you are unsure to what extent you are capable of it.

It appears then that the answer to my question may be only another question. But maybe we can take some encouragement from that. Maybe, if our questions lead to other questions, that is a sign that we are asking the right ones.

Here is the other question that the predicament of modern warfare forces upon us: How many deaths of other people's children by bombing or starvation are we willing to accept in order that we may be free, affluent, and (supposedly) at peace? To that question I answer pretty quickly: *None*. And I know that I am not the only one who would give that answer: Please. No children. Don't kill any children for *my* benefit.

If that is our answer, then we must know that we have not come to rest. Far from it. For now surely we must feel ourselves swarmed about with more questions that are urgent, personal, and intimidating. But perhaps also we feel ourselves beginning to be free, facing at last in our own lives the greatest challenge ever laid before us, the most comprehensive vision of human progress, the best advice and the least obeyed:

"Love your enemies, bless them that curse you, do good to them that hate you, and pray for them which despitefully use you, and persecute you;

"That ye may be the children of your Father which is in Heaven: for He maketh His sun to rise on the evil and the good, and sendeth rain on the just and on the unjust."

(1999)

Postscript

I have been advised that "most readers" will object to my treatment of abortion in this essay. I don't know that most readers will object, but I am sure that some will. And so I will deal with this subject more plainly.

The issue of abortion, so far as I understand it, involves two questions: Is it killing? And what is killed?

It is killing, of course. To kill is the express purpose of the procedure.

What is killed is usually described by apologists for abortion as "a fetus," as if that term names a distinct kind or species of being. But what this being might be, if it is not a human being, is not clear. Generally, pregnant women have thought and spoken of the beings in their wombs as "babies." The attempt to make a categorical distinction between a baby living in the womb and a baby living in the world is as tenuous as would be an attempt to make such a distinction between a living child and a living adult. No living creature is "viable" independently of an enveloping life-support system.

If the creature in the womb is a living human being, and so far also an innocent one, then it is wrong to treat it as an enemy. If we are worried about the effects of treating fellow humans as enemies or enemies of society eligible to be killed, how do we justify treating an innocent fellow human as an enemy-in-the-womb?

As for the "right to control one's own body," I am all for that. But implicit in that right is the responsibility to control one's body in such a way as to avoid dealing irresponsibly or violently or murderously with other bodies.

Women and men generally have understood that when they have conceived a child they have relinquished a significant measure of their independence, and that henceforth they must control their bodies in the interest of the child.

(2003)

In Distrust of Movements

I MUST BURDEN MY READERS as I have burdened myself with the knowledge that I speak from a local, some might say a provincial, point of view. When I try to identify myself to myself I realize that, in my most immediate reasons and affections, I am less than an American, less than a Kentuckian, less even than a Henry Countian, but am a man most involved with and concerned about my family, my neighbors, and the land that is daily under my feet. It is this involvement that defines my citizenship in the larger entities. And so I will remember, and I ask you to remember, that I am not trying to say what is thinkable everywhere, but rather what it is possible to think on the westward bank of the lower Kentucky River in the summer of 1998.

Over the last twenty-five or thirty years I have been making and remaking different versions of the same argument. It is not "my" argument, really, but rather one that I inherited from a long line of familial, neighborly, literary, and scientific ancestors. We could call it "the agrarian argument." This argument can be summed up in as many ways as it can be made. One way to sum it up is to say that we humans can escape neither our dependence on nature nor our responsibility to nature—and that, precisely because of this condition of dependence *and* responsibility, we are also dependent upon and responsible for human culture.

Food, as I have argued at length, is both a natural (which is to say a divine) gift and a cultural product. Because we must *use* land and water and plants and animals to produce food, we are at once dependent on and responsible to what we use. We must know both how to use and how to care for what we use. This knowledge is the basis of human culture. If we do not know how to adapt our desires, our methods, and our technology to the nature of the places in which we are working, so as to make them productive *and to keep them so*, that is a cultural failure of the grossest and most dangerous kind. Poverty and starvation also can be cultural products—if the culture is wrong.

271

Though this argument, in my keeping, has lengthened and acquired branches, in its main assumptions it has stayed the same. What has changed—and I say this with a good deal of wonder and with much thankfulness—is the audience. Perhaps the audience will always include people who are not listening, or people who think the agrarian argument is merely an anachronism, a form of entertainment, or a nuisance to be waved away. But increasingly the audience also includes people who take this argument seriously, because they are involved in one or more of the tasks of agrarianism. They are trying to maintain a practical foothold on the earth for themselves or their families or their communities. They are trying to preserve and develop local land-based economies. They are trying to preserve or restore the health of local communities and ecosystems and watersheds. They are opposing the attempt of the great corporations to own and control all of Creation.

In short, the agrarian argument now has a significant number of friends. As the political and ecological abuses of the so-called global economy become more noticeable and more threatening, the agrarian argument is going to have more friends than it has now. This being so, maybe the advocate's task needs to change. Maybe now, instead of merely propounding (and repeating) the agrarian argument, the advocate must also try to see that this argument does not win friends too easily. I think, myself, that this is the case. The tasks of agrarianism that we have undertaken are not going to be finished for a long time. To preserve the remnants of agrarian life, to oppose the abuses of industrial land use and finally correct them, and to develop the locally adapted economies and cultures that are necessary to our survival will require many lifetimes of dedicated work. This work does not need friends with illusions. And so I would like to speak—in a friendly way, of course—out of my distrust of "movements."

I have had with my friend Wes Jackson a number of useful conversations about the necessity of getting out of movements —even movements that have seemed necessary and dear to us—when they have lapsed into self-righteousness and self-betrayal, as movements seem almost invariably to do. People in movements too readily learn to deny to others the rights and privileges they demand for themselves. They too easily become unable to mean their own language, as when a "peace move-ment" becomes violent. They often become too specialized, as if they cannot help taking refuge in the pinhole vision of the industrial intellectuals. They almost always fail to be radical enough, dealing finally in effects rather than causes. Or they deal with single issues or single solutions, as if to assure them-selves that they will not be radical enough.

And so I must declare my dissatisfaction with movements to promote soil conservation or clean water or clean air or wil-derness preservation or sustainable agriculture or community health or the welfare of children. Worthy as these and other goals may be, they cannot be achieved alone. They cannot be responsibly advocated alone. I am dissatisfied with such efforts because they are too specialized, they are not comprehensive enough, they are not radical enough, they virtually predict their own failure by implying that we can remedy or control ef-fects while leaving the causes in place. Ultimately, I think, they are insincere; they propose that the trouble is caused by *other* people; they would like to change policy but not behavior.

The worst danger may be that a movement will lose its lan-guage either to its own confusion about meaning and practice, or to preemption by its enemies. I remember, for example, my naïve confusion at learning that it was possible for advocates of organic agriculture to look upon the "organic method" as an end in itself. To me, organic farming was attractive both as a way of conserving nature and as a strategy of survival for small farmers. Imagine my surprise in discovering that there could be huge "organic" monocultures. And so I was somewhat pre-pared for the recent attempt of the United States Department of Agriculture to appropriate the "organic" label for food ir-radiation, genetic engineering, and other desecrations by the corporate food economy. Once we allow our language to mean anything that anybody wants it to mean, it becomes impossi-ble to mean what we say. When "homemade" ceases to mean neither more nor less than "made at home," then it means any-thing, which is to say that it means nothing. The same decay is at work on words such as "conservation," "sustainable," "safe," "natural," "healthful," "sanitary," and "organic." The use of such words now requires the most exacting control of context and the use immediately of illustrative examples.

Real organic gardeners and farmers who market their produce locally are finding that, to a lot of people, "organic"
means something like "trustworthy." And so, for a while, it will
be useful for us to talk about the meaning and the economic
usefulness of trust and trustworthiness. But we must be careful. Sooner or later, Trust Us Global Foods, Inc., will be upon
us, advertising safe, sanitary, natural food irradiation. And then
we must be prepared to raise another standard and move on.

As you see, I have good reasons for declining to name the
movement I think I am a part of. I call it The Nameless Movement for Better Ways of Doing—which I hope is too long and
uncute to be used as a bumper sticker. I know that movements
tend to die with their names and slogans, and I believe that this
Nameless Movement needs to live on and on. I am reconciled
to the likelihood that from time to time it will name itself and
have slogans, but I am not going to use its slogans or call it
by any of its names. After this, I intend to stop calling it The
Nameless Movement for Better Ways of Doing, for fear it will
become the NMBWD and acquire a headquarters and a budget and an inventory of T-shirts covered with language that in
a few years will be mere spelling.

Let us suppose, then, that we have a Nameless Movement
for Better Land Use and that we know we must try to keep it
active, responsive, and intelligent for a long time. What must
we do?

What we must do above all, I think, is try to see the problem in its full size and difficulty. If we are concerned about
land abuse, then we must see that this is an economic problem.
Every economy is, by definition, a land-using economy. If we
are using our land wrong, then something is wrong with our
economy. This is difficult. It becomes more difficult when we
recognize that, in modern times, every one of us is a member
of the economy of everybody else. Every one of us has given
many proxies to the economy to use the land (and the air,
the water, and other natural gifts) on our behalf. Adequately
supervising those proxies is at present impossible; withdrawing
them is for virtually all of us, as things now stand, unthinkable.

But if we are concerned about land abuse, we have begun an
extensive work of economic criticism. Study of the history of
land use (and any local history will do) informs us that we have

had for a long time an economy that thrives by undermining its own foundations. Industrialism, which is the name of our economy, and which is now virtually the only economy of the world, has been from its beginnings in a state of riot. It is based squarely upon the principle of violence toward everything on which it depends, and it has not mattered whether the form of industrialism was communist or capitalist; the violence toward nature, human communities, traditional agricultures, and local economies has been constant. The bad news is coming in from all over the world. Can such an economy somehow be fixed without being radically changed? I don't think it can.

The Captains of Industry have always counseled the rest of us to "be realistic." Let us, therefore, be realistic. Is it realistic to assume that the present economy would be just fine if only it would stop poisoning the earth, air, and water, or if only it would stop soil erosion, or if only it would stop degrading watersheds and forest ecosystems, or if only it would stop seducing children, or if only it would quit buying politicians, or if only it would give women and favored minorities an equitable share of the loot? Realism, I think, is a very limited program, but it informs us at least that we should not look for bird eggs in a cuckoo clock.

Or we can show the hopelessness of single-issue causes and single-issue movements by following a line of thought such as this: We need a continuous supply of uncontaminated water. Therefore, we need (among other things) soil-and-water-conserving ways of agriculture and forestry that are not dependent on monoculture, toxic chemicals, or the indifference and violence that always accompany big-scale industrial enterprises on the land. Therefore, we need diversified, small-scale land economies that are dependent on people. Therefore, we need people with the knowledge, skills, motives, and attitudes required by diversified, small-scale land economies. And all this is clear and comfortable enough, until we recognize the question we have come to: *Where are the people?*

Well, all of us who live in the suffering rural landscapes of the United States know that most people are available to those landscapes only recreationally. We see them bicycling or boating or hiking or camping or hunting or fishing or driving along and looking around. They do not, in Mary Austin's phrase, "summer and winter with the land." They are unacquainted with the land's human and natural economies. Though people have not progressed beyond the need to eat food and drink water and wear clothes and live in houses, most people have progressed beyond the domestic arts—the husbandry and wifery of the world—by which those needful things are produced and conserved. In fact, the comparative few who still practice that necessary husbandry and wifery often are inclined to apologize for doing so, having been carefully taught in our education system that those arts are degrading and unworthy of people's talents. Educated minds, in the modern era, are unlikely to know anything about food and drink or clothing and shelter. In merely taking these things for granted, the modern educated mind reveals itself also to be as superstitious a mind as ever has existed in the world. What could be more superstitious than the idea that money brings forth food?

I am not suggesting, of course, that everybody ought to be a farmer or a forester. Heaven forbid! I *am* suggesting that most people now are living on the far side of a broken connection, and that this is potentially catastrophic. Most people are now fed, clothed, and sheltered from sources, in nature and in the work of other people, toward which they feel no gratitude and exercise no responsibility.

We are involved now in a profound failure of imagination. Most of us cannot imagine the wheat beyond the bread, or the farmer beyond the wheat, or the farm beyond the farmer, or the history (human or natural) beyond the farm. Most people cannot imagine the forest and the forest economy that produced their houses and furniture and paper; or the landscapes, the streams, and the weather that fill their pitchers and bathtubs and swimming pools with water. Most people appear to assume that when they have paid their money for these things they have entirely met their obligations. And that is, in fact, the conventional economic assumption. The problem is that it is possible to starve under the rule of the conventional economic assumption; some people are starving now under the rule of that assumption.

Money does not bring forth food. Neither does the technology of the food system. Food comes from nature and from the work of people. If the supply of food is to be continuous for a long time, then people must work in harmony with nature. That means that people must find the right answers to a lot of questions. The same rules apply to forestry and the possibility of a continuous supply of forest products.

People grow the food that people eat. People produce the lumber that people use. People care properly or improperly for the forests and the farms that are the sources of those goods. People are necessarily at both ends of the process. The economy, always obsessed with its need to sell products, thinks obsessively and exclusively of the consumer. It mostly takes for granted or ignores those who do the damaging or the restorative and preserving work of agriculture and forestry. The economy pays poorly for this work, with the unsurprising result that the work is mostly done poorly. But here we must ask a very realistic economic question: Can we afford to have this work done poorly? Those of us who know something about land stewardship know that we cannot afford to pay poorly for it, because that means simply that we will not get it. And we know that we cannot afford land use without land stewardship.

One way we could describe the task ahead of us is by saying that we need to enlarge the consciousness and the conscience of the economy. Our economy needs to know—and care— what it is doing. This is revolutionary, of course, if you have a taste for revolution, but it is also merely a matter of common sense. How could anybody seriously object to the possibility that the economy might eventually come to know what it is doing?

Undoubtedly some people will want to start a movement to bring this about. They probably will call it the Movement to Teach the Economy What It Is Doing—the MTEWIID. Despite my very considerable uneasiness, I will agree to participate, but on three conditions.

My first condition is that this movement should begin by giving up all hope and belief in piecemeal, one-shot solutions. The present scientific quest for odorless hog manure should give us sufficient proof that the specialist is no longer with us. Even now, after centuries of reductionist propaganda, the world is still intricate and vast, as dark as it is light, a place of mystery, where we cannot do one thing without doing many things, or put two things together without putting many things together.

Water quality, for example, cannot be improved without improving farming and forestry, but farming and forestry cannot be improved without improving the education of consumers —and so on.

The proper business of a human economy is to make one whole thing of ourselves and this world. To make ourselves into a practical wholeness with the land under our feet is maybe not altogether possible—how would *we* know?—but, as a goal, it at least carries us beyond *hubris*, beyond the utterly groundless assumption that we can subdivide our present great failure into a thousand separate problems that can be fixed by a thousand task forces of academic and bureaucratic specialists. That program has been given more than a fair chance to prove itself, and we ought to know by now that it won't work.

My second condition is that the people in this movement (the MTEWIID) should take full responsibility for themselves as members of the economy. If we are going to teach the economy what it is doing, then we need to learn what *we* are doing. This is going to have to be a private movement as well as a public one. If it is unrealistic to expect wasteful industries to be conservers, then obviously we must lead in part the public life of complainers, petitioners, protesters, advocates and supporters of stricter regulations and saner policies. But that is not enough. If it is unrealistic to expect a bad economy to try to become a good one, then *we* must go to work to build a good economy. It is appropriate that this duty should fall to us, for good economic behavior is more possible for us than it is for the great corporations with their miseducated managers and their greedy and oblivious stockholders. Because it is possible for us, we must try in every way we can to make good economic sense in our own lives, in our households, and in our communities. We must do more for ourselves and our neighbors. We must learn to spend our money with our friends and not with our enemies. But to do this, it is necessary to renew local economies, and revive the domestic arts. In seeking to change our economic use of the world, we are seeking inescapably to change our lives. The outward harmony that we desire between our economy and the world depends finally upon an inward harmony between our own hearts and the creative spirit that is the life of all creatures, a spirit as near us as our flesh and yet forever beyond the measures of this obsessively measuring age. We can grow good wheat and make good bread only if we understand that we do not live by bread alone.

My third condition is that this movement should content itself to be poor. We need to find cheap solutions, solutions within the reach of everybody, and the availability of a lot of money prevents the discovery of cheap solutions. The solutions of modern medicine and modern agriculture are all staggeringly expensive, and this is caused in part, and maybe altogether, by the availability of huge sums of money for medical and agricultural research.

Too much money, moreover, attracts administrators and experts as sugar attracts ants—look at what is happening in our universities. We should not envy rich movements that are organized and led by an alternative bureaucracy living on the problems it is supposed to solve. We want a movement that is a movement because it is advanced by all its members in their daily lives.

Now, having completed this very formidable list of the problems and difficulties, fears and fearful hopes that lie ahead of us, I am relieved to see that I have been preparing myself all along to end by saying something cheerful. What I have been talking about is the possibility of renewing human respect for this earth and all the good, useful, and beautiful things that come from it. I have made it clear, I hope, that I don't think this respect can be adequately enacted or conveyed by tipping our hats to nature or by representing natural loveliness in art or by prayers of thanksgiving or by preserving tracts of wilderness —though I recommend all those things. The respect I mean can be given only by using well the world's goods that are given to us. This good use, which renews respect—which is the only currency, so to speak, of respect—also renews our pleasure. The callings and disciplines that I have spoken of as the domestic arts are stationed all along the way from the farm to the prepared dinner, from the forest to the dinner table, from stewardship of the land to hospitality to friends and strangers. These arts are as demanding and gratifying, as instructive and as pleasing as the so-called fine arts. To learn them, to practice them, to honor and reward them is, I believe, our profoundest calling. Our reward is that they will enrich our lives and make us glad.

The Total Economy

L ET US BEGIN by assuming what appears to be true: that the so-called environmental crisis is now pretty well established as a fact of our age. The problems of pollution, species extinction, loss of wilderness, loss of farmland, and loss of topsoil may still be ignored or scoffed at, but they are not denied. Concern for these problems has acquired a certain standing, a measure of discussability, in the media and in some scientific, academic, and religious institutions.

This is good, of course; obviously, we can't hope to solve these problems without an increase of public awareness and concern. But in an age burdened with "publicity," we have to be aware also that as issues rise into popularity they rise also into the danger of oversimplification. To speak of this danger is especially necessary in confronting the destructiveness of our relationship to nature, which is the result, in the first place, of gross oversimplification.

The "environmental crisis" has happened because the human household or economy is in conflict at almost every point with the household of nature. We have built our household on the assumption that the natural household is simple and can be simply used. We have assumed increasingly over the last five hundred years that nature is merely a supply of "raw materials," and that we may safely possess those materials merely by taking them. This taking, as our technical means have increased, has involved always less reverence or respect, less gratitude, less local knowledge, and less skill. Our methodologies of land use have strayed from our old sympathetic attempts to imitate natural processes, and have come more and more to resemble the methodology of mining, even as mining itself has become more powerful technologically and more brutal.

And so we will be wrong if we attempt to correct what we perceive as "environmental" problems without correcting the economic oversimplification that caused them. This oversimplification is now either a matter of corporate behavior or of behavior under the influence of corporate behavior. This is sufficiently clear to many of us. What is not sufficiently clear,

perhaps to any of us, is the extent of our complicity, as individuals and especially as individual consumers, in the behavior of
the corporations.

What has happened is that most people in our country, and
apparently most people in the "developed" world, have given
proxies to the corporations to produce and provide *all* of their
food, clothing, and shelter. Moreover, they are rapidly increasing their proxies to corporations or governments to provide
entertainment, education, child care, care of the sick and the
elderly, and many other kinds of "service" that once were carried on informally and inexpensively by individuals or households or communities. Our major economic practice, in short,
is to delegate the practice to others.

The danger now is that those who are concerned will believe
that the solution to the "environmental crisis" can be merely
political—that the problems, being large, can be solved by
large solutions generated by a few people to whom we will
give our proxies to police the economic proxies that we have
already given. The danger, in other words, is that people will
think they have made a sufficient change if they have altered
their "values," or had a "change of heart," or experienced a
"spiritual awakening," and that such a change in passive consumers will necessarily cause appropriate changes in the public
experts, politicians, and corporate executives to whom they
have granted their political and economic proxies.

The trouble with this is that a proper concern for nature
and our use of nature must be practiced, not by our proxy-
holders, but by ourselves. A change of heart or of values without a practice is only another pointless luxury of a passively
consumptive way of life. The "environmental crisis," in fact,
can be solved only if people, individually and in their communities, recover responsibility for their thoughtlessly given
proxies. If people begin the effort to take back into their own
power a significant portion of their economic responsibility,
then their inevitable first discovery is that the "environmental
crisis" is no such thing; it is not a crisis of our environs or
surroundings; it is a crisis of our lives as individuals, as family
members, as community members, and as citizens. We have an
"environmental crisis" because *we* have consented to an economy in which by eating, drinking, working, resting, traveling,

and enjoying ourselves we are destroying the natural, the God-given, world.

We live, as we must sooner or later recognize, in an era of sentimental economics and, consequently, of sentimental politics. Sentimental communism holds in effect that everybody and everything should suffer for the good of "the many" who, though miserable in the present, will be happy in the future for exactly the same reasons that they are miserable in the present.

Sentimental capitalism is not so different from sentimental communism as the corporate and political powers claim to suppose. Sentimental capitalism holds in effect that everything small, local, private, personal, natural, good, and beautiful must be sacrificed in the interest of the "free market" and the great corporations, which will bring unprecedented security and happiness to "the many"—in, of course, the future.

These forms of political economy may be described as sentimental because they depend absolutely upon a political faith for which there is no justification. They seek to preserve the gullibility of the people by issuing a cold check on a fund of political virtue that does not exist. Communism and "free-market" capitalism both are modern versions of oligarchy. In their propaganda, both justify violent means by good ends, which always are put beyond reach by the violence of the means. The trick is to define the end vaguely—"the greatest good of the greatest number" or "the benefit of the many" —and keep it at a distance. For example, the United States government's agricultural policy, or non-policy, since 1952 has merely consented to the farmers' predicament of high costs and low prices; it has never envisioned or advocated in particular the prosperity of farmers or of farmland, but has only promised "cheap food" to consumers and "survival" to the "larger and more efficient" farmers who supposedly could adapt to and endure the attrition of high costs and low prices. And after each inevitable wave of farm failures and the inevitable enlargement of the destitution and degradation of the countryside, there have been the inevitable reassurances from government propagandists and university experts that American agriculture was now more efficient and that everybody would be better off in the future.

The fraudulence of these oligarchic forms of economy is in their principle of displacing whatever good they recognize (as well as their debts) from the present to the future. Their success depends upon persuading people, first, that whatever they have now is no good, and, second, that the promised good is certain to be achieved in the future. This obviously contradicts the principle—common, I believe, to all the religious traditions —that if ever we are going to do good to one another, then the time to do it is now; we are to receive no reward for promising to do it in the future. And both communism and capitalism have found such principles to be a great embarrassment. If you are presently occupied in destroying every good thing in sight in order to do good in the future, it is inconvenient to have people saying things like "Love thy neighbor as thyself" or "Sentient beings are numberless, I vow to save them." Communists and capitalists alike, "liberal" capitalists and "conservative" capitalists alike, have needed to replace religion with some form of determinism, so that they can say to their victims, "I'm doing this because I can't do otherwise. It is not my fault. It is inevitable." This is a lie, obviously, and religious organizations have too often consented to it.

The idea of an economy based upon several kinds of ruin may seem a contradiction in terms, but in fact such an economy is possible, as we see. It is possible, however, on one implacable condition: The only future good that it assuredly leads to is that it will destroy itself. And how does it disguise this outcome from its subjects, its short-term beneficiaries, and its victims? It does so by false accounting. It substitutes for the real economy, by which we build and maintain (or do not maintain) our household, a symbolic economy of money, which in the long run, because of the self-interested manipulations of the "controlling interests," cannot symbolize or account for anything but itself. And so we have before us the spectacle of unprecedented "prosperity" and "economic growth" in a land of degraded farms, forests, ecosystems, and watersheds, polluted air, failing families, and perishing communities.

This moral and economic absurdity exists for the sake of the allegedly "free" market, the single principle of which is this: Commodities will be produced wherever they can be produced at the lowest cost and consumed wherever they will bring the highest price. To make too cheap and sell too high has always been the program of industrial capitalism. The global "free market" is merely capitalism's so far successful attempt to enlarge the geographic scope of its greed, and moreover to give to its greed the status of a "right" within its presumptive territory. The global "free market" is free to the corporations precisely because it dissolves the boundaries of the old national colonialisms, and replaces them with a new colonialism without restraints or boundaries. It is pretty much as if all the rabbits have now been forbidden to have holes, thereby "freeing" the hounds.

The "right" of a corporation to exercise its economic power without restraint is construed, by the partisans of the "free market," as a form of freedom, a political liberty implied presumably by the right of individual citizens to own and use property.

But the "free market" idea introduces into government a sanction of an inequality that is not implicit in any idea of democratic liberty: namely that the "free market" is freest to those who have the most money, and is not free at all to those with little or no money. Wal-Mart, for example, as a large corporation "freely" competing against local, privately owned businesses, has virtually all the freedom, and its small competitors virtually none.

To make too cheap and sell too high, there are two requirements. One is that you must have a lot of consumers with surplus money and unlimited wants. For the time being, there are plenty of these consumers in the "developed" countries. The problem, for the time being easily solved, is simply to keep them relatively affluent and dependent on purchased supplies.

The other requirement is that the market for labor and raw materials should remain depressed relative to the market for retail commodities. This means that the supply of workers should exceed demand, and that the land-using economies should be allowed or encouraged to overproduce.

To keep the cost of labor low, it is necessary first to entice or force country people everywhere in the world to move into the cities—in the manner prescribed by the Committee for Economic Development after World War II—and, second, to continue to introduce labor-replacing technology. In this way it is possible to maintain a "pool" of people who are in the threatful position of being mere consumers, landless and poor, and who therefore are eager to go to work for low wages—precisely the condition of migrant farm workers in the United States.

To cause the land-using economies to overproduce is even simpler. The farmers and other workers in the world's land-using economies, by and large, are not organized. They are therefore unable to control production in order to secure just prices. Individual producers must go individually to the market and take for their produce simply whatever they are paid. They have no power to bargain or to make demands. Increasingly, they must sell, not to neighbors or to neighboring towns and cities, but to large and remote corporations. There is no competition among the buyers (supposing there is more than one), who *are* organized and are "free" to exploit the advantage of low prices. Low prices encourage overproduction, as producers attempt to make up their losses "on volume," and overproduction inevitably makes for low prices. The land-using economies thus spiral downward as the money economy of the exploiters spirals upward. If economic attrition in the land-using population becomes so severe as to threaten production, then governments can subsidize production without production controls, which necessarily will encourage overproduction, which will lower prices—and so the subsidy to rural producers becomes, in effect, a subsidy to the purchasing corporations. In the land-using economies, production is further cheapened by destroying, with low prices and low standards of quality, the cultural imperatives for good work and land stewardship.

This sort of exploitation, long familiar in the foreign and domestic colonialism of modern nations, has now become "the global economy," which is the property of a few supranational corporations. The economic theory used to justify the global economy in its "free market" version is, again, perfectly groundless and sentimental. The idea is that what is good for the corporations will sooner or later—though not of course immediately—be good for everybody.

That sentimentality is based, in turn, upon a fantasy: the proposition that the great corporations, in "freely" competing with one another for raw materials, labor, and market share, will drive each other indefinitely, not only toward greater "efficiencies" of manufacture, but also toward higher bids for raw materials and labor and lower prices to consumers. As a result, all the world's people will be economically secure—in the future. It would be hard to object to such a proposition if only it were true.

But one knows, in the first place, that "efficiency" in manufacture always means reducing labor costs by replacing workers with cheaper workers or with machines.

In the second place, the "law of competition" does *not* imply that many competitors will compete indefinitely. The law of competition is a simple paradox: Competition destroys competition. The law of competition implies that many competitors, competing on the "free market" without restraint, will ultimately and inevitably reduce the number of competitors to one. The law of competition, in short, is the law of war.

In the third place, the global economy is based upon cheap long-distance transportation, without which it is not possible to move goods from the point of cheapest origin to the point of highest sale. And cheap long-distance transportation is the basis of the idea that regions and nations should abandon any measure of economic self-sufficiency in order to specialize in production for export of the few commodities, or the single commodity, that can be most cheaply produced. Whatever may be said for the "efficiency" of such a system, its result (and, I assume, its purpose) is to destroy local production capacities, local diversity, and local economic independence. It destroys the economic security that it promises to make.

This idea of a global "free market" economy, despite its obvious moral flaws and its dangerous practical weaknesses, is now the ruling orthodoxy of the age. Its propaganda is subscribed to and distributed by most political leaders, editorial writers, and other "opinion makers." The powers that be, while continuing to budget huge sums for "national defense," have apparently abandoned any idea of national or local self-sufficiency, even in food. They also have given up the idea that a national or local government might justly place restraints upon economic activity in order to protect its land and its people.

The global economy is now institutionalized in the World Trade Organization, which was set up, without election anywhere, to rule international trade on behalf of the "free market"—which is to say on behalf of the supranational corporations—and to *over*rule, in secret sessions, any national or regional law that conflicts with the "free market." The corporate program of global "free trade" and the presence of the World Trade Organization have legitimized extreme forms of expert thought. We are told confidently that if Kentucky loses its milk-producing capacity to Wisconsin (and if Wisconsin's is lost to California), that will be a "success story." Experts such as Stephen C. Blank, of the University of California, Davis, have recommended that "developed" countries, such as the United States and the United Kingdom, where food can no longer be produced cheaply enough, should give up agriculture altogether.

The folly at the root of this foolish economy began with the idea that a corporation should be regarded, legally, as "a person." But the limitless destructiveness of this economy comes about precisely because a corporation is *not* a person. A corporation, essentially, is a pile of money to which a number of persons have sold their moral allegiance. Unlike a person, a corporation does not age. It does not arrive, as most persons finally do, at a realization of the shortness and smallness of human lives; it does not come to see the future as the lifetime of the children and grandchildren of anybody in particular. It can experience no personal hope or remorse, no change of heart. It cannot humble itself. It goes about its business as if it were immortal, with the single purpose of becoming a bigger pile of money. The stockholders essentially are usurers, people who "let their money work for them," expecting high pay in return for causing others to work for low pay. The World Trade Organization enlarges the old idea of the corporation-as-person by giving the global corporate economy the status of a supergovernment with the power to overrule nations.

I don't mean to say, of course, that all corporate executives and stockholders are bad people. I am only saying that all of them are very seriously implicated in a bad economy.

Unsurprisingly, among people who wish to preserve things other than money—for instance, every region's native capacity to produce essential goods—there is a growing perception that the global "free market" economy is inherently an enemy to the natural world, to human health and freedom, to industrial workers, and to farmers and others in the land-use economies; and, furthermore, that it is inherently an enemy to good work and good economic practice.

I believe that this perception is correct and that it can be shown to be correct merely by listing the assumptions implicit in the idea that corporations should be "free" to buy low and sell high in the world at large. These assumptions, so far as I can make them out, are as follows:

1. That there is no conflict between the "free market" and political freedom, and no connection between political democracy and economic democracy.

2. That there can be no conflict between economic advantage and economic justice.

3. That there is no conflict between greed and ecological or bodily health.

4. That there is no conflict between self-interest and public service.

5. That it is all right for a nation's or a region's subsistence to be foreign-based, dependent on long-distance transport, and entirely controlled by corporations.

6. That the loss or destruction of the capacity anywhere to produce necessary goods does not matter and involves no cost.

7. That, therefore, wars over commodities—our recent Gulf War, for example—are legitimate and permanent economic functions.

8. That this sort of sanctioned violence is justified also by the predominance of centralized systems of production, supply, communications, and transportation which are extremely vulnerable not only to acts of war between nations, but also to sabotage and terrorism.

9. That it is all right for poor people in poor countries to work at poor wages to produce goods for export to affluent people in rich countries.

10. That there is no danger and no cost in the proliferation of exotic pests, vermin, weeds, and diseases that accompany international trade, and that increase with the volume of trade.

11. That an economy is a machine, of which people are merely the interchangeable parts. One has no choice but to do the work (if any) that the economy prescribes, and to accept the prescribed wage.

12. That, therefore, vocation is a dead issue. One does not do the work that one chooses to do because one is called to it by Heaven or by one's natural abilities, but does instead the work that is determined and imposed by the economy. Any work is all right as long as one gets paid for it. (This assumption explains the prevailing "liberal" and "conservative" indifference toward displaced workers, farmers, and small-business people.)

13. That stable and preserving relationships among people, places, and things do not matter and are of no worth.

14. That cultures and religions have no legitimate practical or economic concerns.

These assumptions clearly prefigure a condition of total economy. A total economy is one in which everything—"life forms," for instance, or the "right to pollute"—is "private property" and has a price and is for sale. In a total economy, significant and sometimes critical choices that once belonged to individuals or communities become the property of corporations. A total economy, operating internationally, necessarily shrinks the powers of state and national governments, not only because those governments have signed over significant powers to an international bureaucracy or because political leaders become the paid hacks of the corporations, but also because political processes—and especially democratic processes—are too slow to react to unrestrained economic and technological development on a global scale. And when state and national governments begin to act in effect as agents of the global economy, selling their people for low wages and their people's products for low prices, then the rights and liberties of citizenship must necessarily shrink. A total economy is an unrestrained taking of profits from the disintegration of nations, communities, households, landscapes, and ecosystems. It licenses symbolic or artificial wealth to "grow" by means of the destruction of the real wealth of all the world.

Among the many costs of the total economy, the loss of the

principle of vocation is probably the most symptomatic and, from a cultural standpoint, the most critical. It is by the replacement of vocation with economic determinism that the exterior workings of a total economy destroy human character and culture also from the inside.

In an essay on the origin of civilization in traditional cultures, Ananda Coomaraswamy wrote that "the principle of justice is the same throughout. . . . [It is] that each member of the community should perform the task for which he is fitted by nature. . . ." The two ideas, justice and vocation, are inseparable. That is why Coomaraswamy spoke of industrialism as "the mammon of injustice," incompatible with civilization. It is by way of the practice of vocation that sanctity and reverence enter into the human economy. It was thus possible for traditional cultures to conceive that "to work is to pray."

Aware of industrialism's potential for destruction, as well as the considerable political danger of great concentrations of wealth and power in industrial corporations, American leaders developed, and for a while used, certain means of limiting and restraining such concentrations, and of somewhat equitably distributing wealth and property. The means were: laws against trusts and monopolies, the principle of collective bargaining, the concept of 100 percent parity between the land-using and the manufacturing economies, and the progressive income tax. And to protect domestic producers and production capacities, it is possible for governments to impose tariffs on cheap imported goods. These means are justified by the government's obligation to protect the lives, livelihoods, and freedoms of its citizens. There is, then, no necessity that requires our government to sacrifice the livelihoods of our small farmers, small-business people, and workers, along with our domestic economic independence, to the global "free market." But now all of these means are either weakened or in disuse. The global economy is intended as a means of subverting them.

In default of government protections against the total economy of the supranational corporations, people are where they have been many times before: in danger of losing their economic security and their freedom, both at once. But at the same time the means of defending themselves belongs to them in the form of a venerable principle: Powers not exercised by government return to the people. If the government does not propose to protect the lives, the livelihoods, and the freedoms of its people, then the people must think about protecting themselves.

How are they to protect themselves? There seems, really, to be only one way, and that is to develop and put into practice the idea of a local economy—something that growing numbers of people are now doing. For several good reasons, they are beginning with the idea of a local food economy. People are trying to find ways to shorten the distance between producers and consumers, to make the connections between the two more direct, and to make this local economic activity a benefit to the local community. They are trying to learn to use the consumer economies of local towns and cities to preserve the livelihoods of local farm families and farm communities. They want to use the local economy to give consumers an influence over the kind and quality of their food, and to preserve and enhance the local landscapes. They want to give everybody in the local community a direct, long-term interest in the prosperity, health, and beauty of their homeland. This is the only way presently available to make the total economy less total. It was once the only way to make a national or a colonial economy less total, but now the necessity is greater.

I am assuming that there is a valid line of thought leading from the idea of the total economy to the idea of a local economy. I assume that the first thought may be a recognition of one's ignorance and vulnerability as a consumer in the total economy. As such a consumer, one does not know the history of the products one uses. Where, exactly, did they come from? Who produced them? What toxins were used in their production? What were the human and ecological costs of producing and then of disposing of them? One sees that such questions cannot be answered easily, and perhaps not at all. Though one is shopping amid an astonishing variety of products, one is denied certain significant choices. In such a state of economic ignorance it is not possible to choose products that were produced locally or with reasonable kindness toward people and toward nature. Nor is it possible for such consumers to influence production for the better. Consumers who feel

a prompting toward land stewardship find that in this economy they can have no stewardly practice. To be a consumer
in the total economy, one must agree to be totally ignorant,
totally passive, and totally dependent on distant supplies and
self-interested suppliers.

And then, perhaps, one begins to *see* from a local point of
view. One begins to ask, What is here, what is in my neighborhood, what is in me, that can lead to something better? From
a local point of view, one can see that a global "free market"
economy is possible only if nations and localities accept or ignore the inherent weakness of a production economy based
on exports and a consumer economy based on imports. An
export economy is beyond local influence, and so is an import
economy. And cheap long-distance transport is possible only if
granted cheap fuel, international peace, control of terrorism,
prevention of sabotage, and the solvency of the international
economy.

Perhaps also one begins to see the difference between a small
local business that must share the fate of the local community
and a large absentee corporation that is set up to escape the
fate of the local community by ruining the local community.

So far as I can see, the idea of a local economy rests upon only
two principles: neighborhood and subsistence.

In a viable neighborhood, neighbors ask themselves what
they can do or provide for one another, and they find answers
that they and their place can afford. This, and nothing else, is
the *practice* of neighborhood. This practice must be, in part,
charitable, but it must also be economic, and the economic part
must be equitable; there is a significant charity in just prices.

Of course, everything needed locally cannot be produced
locally. But a viable neighborhood is a community, and a viable
community is made up of neighbors who cherish and protect
what they have in common. This is the principle of subsistence.
A viable community, like a viable farm, protects its own production capacities. It does not import products that it can produce for itself. And it does not export local products until local
needs have been met. The economic products of a viable community are understood either as belonging to the community's
subsistence or as surplus, and only the surplus is considered to

be marketable abroad. A community, if it is to be viable, cannot think of producing solely for export, and it cannot permit importers to use cheaper labor and goods from other places to destroy the local capacity to produce goods that are needed locally. In charity, moreover, it must refuse to import goods that are produced at the cost of human or ecological degradation elsewhere. This principle of subsistence applies not just to localities, but to regions and nations as well.

The principles of neighborhood and subsistence will be disparaged by the globalists as "protectionism"—and that is exactly what it is. It is a protectionism that is just and sound, because it protects local producers and is the best assurance of adequate supplies to local consumers. And the idea that local needs should be met first and only surpluses exported does *not* imply any prejudice against charity toward people in other places or trade with them. The principle of neighborhood at home always implies the principle of charity abroad. And the principle of subsistence is in fact the best guarantee of giveable or marketable surpluses. This kind of protection is not "isolationism."

Albert Schweitzer, who knew well the economic situation in the colonies of Africa, wrote about seventy years ago: "Whenever the timber trade is good, permanent famine reigns in the Ogowe region, because the villagers abandon their farms to fell as many trees as possible." We should notice especially that the goal of production was "as many . . . as possible." And Schweitzer made my point exactly: "These people could achieve true wealth if they could develop their agriculture and trade to meet their own needs." Instead they produced timber for export to "the world market," which made them dependent upon imported goods that they bought with money earned from their exports. They gave up their local means of subsistence, and imposed the false standard of a foreign demand ("as many trees as possible") upon their forests. They thus became helplessly dependent on an economy over which they had no control.

Such was the fate of the native people under the African colonialism of Schweitzer's time. Such is, and can only be, the fate of everybody under the global colonialism of our time. Schweitzer's description of the colonial economy of the Ogowe region is in principle not different from the rural economy in Kentucky or Iowa or Wyoming now. A total economy, for all practical purposes, is a total government. The "free trade," which from the standpoint of the corporate economy brings "unprecedented economic growth," from the standpoint of the land and its local populations, and ultimately from the standpoint of the cities, is destruction and slavery. Without prosperous local economies, the people have no power and the land no voice.

Two Minds

HUMAN ORDERS—scientific, artistic, social, economic, and political—are fictions. They are untrue, not because they necessarily are false, but because they necessarily are incomplete. All of our human orders, however inclusive we may try to make them, turn out to be to some degree exclusive. And so we are always being surprised by something we find, too late, that we have excluded. Think of almost any political revolution or freedom movement or the ozone hole or mad cow disease or the events of September 11, 2001.

The present order, thus surprised, is then required to accommodate new knowledge and thus to be reordered. Thomas Kuhn described this process in *The Structure of Scientific Revolutions.*

But these surprises and changes obviously have their effect also on individual lives and on whole cultures. All of our fictions labor under an ever-failing need to be true. And this means that they labor under an obligation to be continuously revised.

Or, to put it another way, we humans necessarily make pictures in our minds of our places and our world. But we can do this only by selection, putting some things into the picture and leaving the rest out. And so we live in two landscapes, one superimposed upon the other.

First there is the cultural landscape made up of our own knowledge of where we are, of landmarks and memories, of patterns of use and travel, of remindings and meanings. The cultural landscape, among other things, is a pattern of exchanges of work, goods, and comforts among neighbors. It is the country we have in mind.

And then there is the actual landscape, which we can never fully know, which is always going to be to some degree a mystery, from time to time surprising us. These two landscapes are necessarily and irremediably different from each other. But there is danger in their difference; they can become *too* different. If the cultural landscape becomes too different from the actual landscape, then we will make practical errors that will be destructive of the actual landscape or of ourselves or of both.

You can learn this from the study of any landscape that is inhabited by humans, or from teachers such as Barry Lopez and Gary Nabhan, who have written on the traditional economies of the Arctic and the American Southwest. It is easier to understand, perhaps, when thinking about extreme landscapes: If the cultural landscapes come to be too much at odds with the actual landscapes of the Arctic or the desert, the penalties are apt to be swift and lethal. In more forgiving landscapes they will (perhaps) be slower, but finally just as dangerous.

In some temperate and well-watered areas, we humans have applied the most extreme industrial methods of landscape destruction. By disregarding the cultural landscape, and all values and protections that accrue therefrom to the actual landscape, the strip miners entirely destroy the actual landscape. Cropland erosion, caused by a serious incongruity between the cultural and the actual landscapes, is a slower form of destruction than strip mining, but given enough time, it too can be entirely destructive.

At present, in the United States and in much of the rest of the world, most of the cultural landscapes that still exist are hodgepodges of failing local memories, money-making schemes, ignorant plans, bucolic fantasies, misinformation, and the random facts that we now call "information." This is compounded by the outright destruction of innumerable burial sites and other sacred places, of natural and historical landmarks, and of entire actual landscapes. Moreover, we have enormous and increasing numbers of people who have no home landscape, though in every one of their economic acts they are affecting the actual landscapes of the world, mostly for the worse. This is a situation unprecedentedly disorderly and dangerous.

To be disconnected from any actual landscape is to be, in the practical or economic sense, without a home. To have no country carefully and practically in mind is to be without a culture. In such a situation, culture becomes purposeless and arbitrary, dividing into "popular culture," determined by commerce, advertising, and fashion, and "high culture," which is either social affectation, displaced cultural memory, or the merely aesthetic pursuits of artists and art lovers.

We are thus involved in a kind of lostness in which most people are participating more or less unconsciously in the destruction of the natural world, which is to say, the sources of their own lives. They are doing this unconsciously because they see or do very little of the actual destruction themselves, and they don't know, because they have no way to learn, how they are involved. At the same time, many of the same people fear and mourn the destruction, which they can't stop because they have no practical understanding of its causes.

Conservationists, scientists, philosophers, and others are telling us daily and hourly that our species is now behaving with colossal irrationality and that we had better become more rational. I agree as to the dimensions and danger of our irrationality. As to the possibility of curing it by rationality, or at least by the rationality of the rationalists, I have some doubts.

The trouble is not just in the way we are thinking; it is also in the way we, or anyhow we in the affluent parts of the world, are living. And it is going to be hard to define anybody's living as a series of simple choices between irrationality and rationality. Moreover, this is supposedly an age of reason; we are encouraged to believe that the governments and corporations of the affluent parts of the world are run by rational people using rational processes to make rational decisions. The dominant faith of the world in our time is in rationality. That in an age of reason, the human race, or the most wealthy and powerful parts of it, should be behaving with colossal irrationality ought to make us wonder if reason alone can lead us to do what is right.

It is often proposed, nowadays, that if we would only get rid of religion and other leftovers from our primitive past and become enlightened by scientific rationalism, we could invent the new values and ethics that are needed to preserve the natural world. This proposal is perfectly reasonable, and perfectly doubtful. It supposes that we can empirically know and rationally understand everything involved, which is exactly the supposition that has underwritten our transgressions against the natural world in the first place.

*

Obviously we need to use our intelligence. But how much intelligence have we got? And what sort of intelligence is it that we have? And how, at its best, does human intelligence work? In order to try to answer these questions I am going to suppose for a while that there are two different kinds of human mind: the Rational Mind and another, which, for want of a better term, I will call the Sympathetic Mind. I will say now, and try to keep myself reminded, that these terms are going to appear to be allegorical, too neat and too separate —though I need to say also that their separation was not invented by me.

The Rational Mind, without being anywhere perfectly embodied, is the mind all of us are supposed to be trying to have. It is the mind that the most powerful and influential people *think* they have. Our schools exist mainly to educate and propagate and authorize the Rational Mind. The Rational Mind is objective, analytical, and empirical; it makes itself up only by considering facts; it pursues truth by experimentation; it is uncorrupted by preconception, received authority, religious belief, or feeling. Its ideal products are the proven fact, the accurate prediction, and the "informed decision." It is, you might say, the official mind of science, industry, and government.

The Sympathetic Mind differs from the Rational Mind, not by being unreasonable, but by refusing to limit knowledge or reality to the scope of reason or factuality or experimentation, and by making reason the servant of things it considers precedent and higher.

The Rational Mind is motivated by the fear of being misled, of being wrong. Its purpose is to exclude everything that cannot empirically or experimentally be proven to be a fact.

The Sympathetic Mind is motivated by fear of error of a very different kind: the error of carelessness, of being unloving. Its purpose is to be considerate of whatever is present, to leave nothing out.

The Rational Mind is exclusive; the Sympathetic Mind, however failingly, wishes to be inclusive.

These two types certainly don't exhaust the taxonomy of minds. They are merely the two that the intellectual fashions of our age have most deliberately separated and thrown into opposition.

My purpose here is to argue in defense of the Sympathetic Mind. But my objection is not to the use of reason or to reasonability. I am objecting to the exclusiveness of the Rational Mind, which has limited itself to a selection of mental functions such as the empirical methodologies of analysis and experimentation and the attitudes of objectivity and realism. In order to go into business on its own, it has in effect withdrawn from all of human life that involves feeling, affection, familiarity, reverence, faith, and loyalty. The separability of the Rational Mind is not only the dominant fiction but also the master superstition of the modern age.

The Sympathetic Mind is under the influence of certain inborn or at least fundamental likes and dislikes. Its impulse is toward wholeness. It is moved by affection for its home place, the local topography, the local memories, and the local creatures. It hates estrangement, dismemberment, and disfigurement. The Rational Mind tolerates all these things "in pursuit of truth" or in pursuit of money—which, in modern practice, have become nearly the same pursuit.

I am objecting to the failure of the rationalist enterprise of "objective science" or "pure science" or "the disinterested pursuit of truth" to prevent massive damage both to nature and to human economy. The Rational Mind does not confess its complicity in the equation: knowledge = power = money = damage. Even so, the alliance of academic science, government, and the corporate economy, and their unifying pattern of sanctions and rewards, is obvious enough. We have resisted, so far, a state religion, but we are in danger of having both a corporate state and a state science, which some people, in both the sciences and the arts, would like to establish as a state religion.

The Rational Mind is the lowest common denominator of the government–corporation–university axis. It is the fiction that makes high intellectual ability the unquestioning servant of bad work and bad law.

Under the reign of the Rational Mind, there is no firewall between contemporary science and contemporary industry or economic development. It is entirely imaginable, for instance, that a young person might go into biology because of love for plants and animals. But such a young person had better be careful, for there is nothing to prevent knowledge gained for love of the creatures from being used to destroy them for the love of money.

Now some biologists, who have striven all their lives to embody perfectly the Rational Mind, have become concerned, even passionately concerned, about the loss of "biological diversity," and they are determined to do something about it. This is usually presented as a merely logical development from ignorance to realization to action. But so far it is only comedy. The Rational Mind, which has been destroying biological diversity by "figuring out" some things, now proposes to save what is left of biological diversity by "figuring out" some more things. It does what it has always done before: It defines the problem as a big problem calling for a big solution; it calls in the world-class experts; it invokes science, technology, and large grants of money; it propagandizes and organizes and "gears up for a major effort." The comedy here is in the failure of these rationalists to see that as soon as they have become passionately concerned they have stepped outside the dry, objective, geometrical territory claimed by the Rational Mind, and have entered the still mysterious homeland of the Sympathetic Mind, watered by unpredictable rains and by real sweat and real tears.

The Sympathetic Mind would not forget that so-called environmental problems have causes that are in part political and therefore have remedies that are in part political. But it would not try to solve these problems merely by large-scale political protections of "the environment." It knows that they must be solved ultimately by correcting the way people use their home places and local landscapes. Politically, but also by local economic improvements, it would stop colonialism in all its forms, domestic and foreign, corporate and governmental. Its first political principle is that landscapes should not be used by people who do not live in them and share their fate. If that principle were strictly applied, we would have far less need for the principle of "environmental protection."

The Sympathetic Mind understands the vital importance of the cultural landscape. The Rational Mind, by contrast, honors no cultural landscape, and therefore has no protective loyalty or affection for any actual landscape.

The definitive practical aim of the Sympathetic Mind is to adapt local economies to local landscapes. This is necessarily the work of local cultures. It cannot be done as a world-scale *feat* of science, industry, and government. This will seem a bitter bite to the optimists of scientific rationalism, which is scornful of limits and proud of its usurpations. But the science of the Sympathetic Mind is occupied precisely with the study of limits, both natural and human.

The Rational Mind does not work from any sense of geographical whereabouts or social connection or from any basis in cultural tradition or principle or character. It does not see itself as existing or working within a context. The Rational Mind doesn't think there is a context until it gets there. Its principle is to be "objective"—which is to say, unremembering and disloyal. It works within narrow mental boundaries that it draws for itself, as directed by the requirements of its profession or academic specialty or its ambition or its desire for power or profit, thus allowing for the "trade-off" and the "externalization" of costs and effects. Even when working outdoors, it is an indoor mind.

The Sympathetic Mind, even when working indoors, is an outdoor mind. It lives within an abounding and unbounded reality, always partly mysterious, in which everything matters, in which we humans are therefore returned to our ancient need for thanksgiving, prayer, and propitiation, in which we meet again and again the ancient question: How does one become worthy to use what must be used?

Whereas the Rational Mind is the mind of analysis, explanation, and manipulation, the Sympathetic Mind is the mind of our creatureliness.

Creatureliness denotes what Wallace Stevens called "the instinctive integrations which are the reasons for living." In our creatureliness we forget the little or much that we know about the optic nerve and the light-sensitive cell, and *we see*; we forget whatever we know about the physiology of the brain, and *we think*; we forget what we know of anatomy, the nervous system, the gastrointestinal tract, and *we work, eat, and sleep.*

We forget the theories and therapies of "human relationships," and we merely love the people we love, and even try to love the others. If we have any sense, we forget the fashionable determinisms, and we tell our children, "Be good. Be careful. Mind your manners. Be kind."

The Sympathetic Mind leaves the world whole, or it attempts always to do so. It looks upon people and other creatures as whole beings. It does not parcel them out into functions and uses.

The Rational Mind, by contrast, has rested its work for a long time on the proposition that all creatures are machines. This works as a sort of strainer to eliminate impurities such as affection, familiarity, and loyalty from the pursuit of knowledge, power, and profit. This machine-system assures the objectivity of the Rational Mind, which is itself understood as a machine, but it fails to account for a number of things, including the Rational Mind's own worries and enthusiasms. Why should a machine be bothered by the extinction of other machines? Would even an "intelligent" computer grieve over the disappearance of the Carolina parakeet?

The Rational Mind is preoccupied with the search for a sure way to avoid risk, loss, and suffering. For the Rational Mind, experience is likely to consist of a sequence of bad surprises and therefore must be booked as a "loss." That is why, to rationalists, the past and the present are so readily expendable or destructible in favor of the future, the era of no loss.

But the Sympathetic Mind accepts loss and suffering as the price, willingly paid, of its sympathy and affection—its wholeness.

The Rational Mind attempts endlessly to inform itself against its ruin by facts, experiments, projections, scoutings of "alternatives," hedgings against the unknown.

The Sympathetic Mind is informed by experience, by tradition-borne stories of the experiences of others, by familiarity, by compassion, by commitment, by faith.

The Sympathetic Mind is preeminently a faithful mind, taking knowingly and willingly the risks required by faith. The Rational Mind, ever in need of certainty, is always in doubt, always looking for a better way, asking, testing, disbelieving in everything but its own sufficiency to its own needs, which its experience and its own methods continually disprove. It is a skeptical, fearful, suspicious mind, and always a disappointed one, awaiting the supreme truth or discovery it expects of itself, which of itself it cannot provide.

To show how these two minds work, let us place them within the dilemma of a familiar story. Here is the parable of the lost sheep from the Gospel of St. Matthew: "If a man have an hundred sheep, and one of them be gone astray, doth he not leave the ninety and nine, and goeth into the mountains, and seeketh that which is gone astray? And if so be that he find it, verily I say unto you, he rejoiceth more of that sheep, than of the ninety and nine which went not astray."

This parable is the product of an eminently sympathetic mind, but for the moment that need not distract us. The dilemma is practical enough, and we can see readily how the two kinds of mind would deal with it.

The rationalist, we may be sure, has a hundred sheep because he has a plan for that many. The one who has gone astray has escaped not only from the flock but also from the plan. That this particular sheep should stray off in this particular place at this particular time, though it is perfectly in keeping with the nature of sheep and the nature of the world, is not at all in keeping with a rational plan. What is to be done? Well, it certainly would not be rational to leave the ninety and nine, exposed as they would then be to further whims of nature, in order to search for the one. Wouldn't it be best to consider the lost sheep a "trade-off" for the safety of the ninety-nine? Having thus agreed to his loss, the doctrinaire rationalist would then work his way through a series of reasonable questions. What would be an "acceptable risk"? What would be an "acceptable loss"? Would it not be good to do some experiments to determine how often sheep may be expected to get lost? If one sheep is likely to get lost every so often, then would it not be better to have perhaps 110 sheep? Or should one insure the flock against such expectable losses? The annual insurance premium would equal the market value of how many sheep? What is likely to be the cost of the labor of looking for one lost sheep after quitting time? How much time spent looking would equal the market value of the lost sheep? Should not one think of splicing a few firefly genes into one's sheep so that strayed sheep would glow in the dark? And so on.

But (leaving aside the theological import of the parable) the shepherd is a shepherd because he embodies the Sympathetic Mind. Because he is a man of sympathy, a man devoted to the care of sheep, a man who knows the nature of sheep and of the world, the shepherd of the parable is not surprised or baffled by his problem. He does not hang back to argue over risks, trade-offs, actuarial data, or market values. He does not quibble over fractions. He goes without hesitating to hunt for the lost sheep because he has committed himself to the care of the whole hundred, because he understands his work as the fulfillment of his whole trust, because he loves the sheep, and because he knows or imagines what it is to be lost. He does what he does on behalf of the whole flock because he wants to preserve himself as a whole shepherd.

He also does what he does because he has a particular affection for that particular sheep. To the Rational Mind, all sheep are the same; any one is the same as any other. They are interchangeable, like coins or clones or machine parts or members of "the work force." To the Sympathetic Mind, each one is different from every other. Each one is an individual whose value is never entirely reducible to market value.

The Rational Mind can and will rationalize any trade-off. The Sympathetic Mind can rationalize none. Thus we have not only the parable of the ninety and nine, but also the Buddhist vow to save all sentient beings. The parable and the vow are utterly alien to the rationalism of modern science, politics, and industry. To the Rational Mind, they "don't make sense" because they deal with hardship and risk merely by acknowledgment and acceptance. Their very point is to require a human being's suffering to involve itself in the suffering of other creatures, including that of other human beings.

The Rational Mind conceives of itself as eminently practical, and is given to boasting about its competence in dealing with "reality." But if you want to hire somebody to take care of your hundred sheep, I think you had better look past the "animal scientist" and hire the shepherd of the parable, if you can still find him anywhere. For it will continue to be more reasonable,

from the point of view of the Rational Mind, to trade off the
lost sheep for the sake of the sheep you have left—until only
one is left.

If you think I have allowed my argument to carry me entirely
into fantasy and irrelevance, then let me quote an up-to-date
story that follows pretty closely the outline of Christ's parable.
This is from an article by Bernard E. Rollin in *Christian Century*, December 19–26, 2001, p. 26:

> A young man was working for a company that operated
> a large, total-confinement swine farm. One day he de-
> tected symptoms of a disease among some of the feeder
> pigs. As a teen, he had raised pigs himself . . . so he
> knew how to treat the animals. But the company's policy
> was to kill any diseased animals with a blow to the head
> —the profit margin was considered too low to allow for
> treatment of individual animals. So the employee decided
> to come in on his own time, with his own medicine, and
> he cured the animals. The management's response was to
> fire him on the spot for violating company policy.

The young worker in the hog factory is a direct cultural
descendant of the shepherd in the parable, just about opposite and perhaps incomprehensible to the "practical" rationalist. But the practical implications are still the same. Would
you rather have your pigs cared for by a young man who
had compassion for them or by one who would indifferently
knock them in the head? Which of the two would be most
likely to prevent the disease in the first place? Compassion, of
course, is the crux of the issue. For "company policy" must
exclude compassion; if compassion were to be admitted to
consideration, such a "farm" could not exist. And yet one
imagines that even the hardheaded realists of "management"
must occasionally violate company policy by wondering at
night what they would do if *all* the pigs got sick. (I suppose
they would kill them all, collect the insurance, and move on.
Perhaps all that has been foreseen and prepared for in the
business plan, and there is no need, after all, to lie awake and
worry.)

But what of the compassionate young man? The next sentence of Mr. Rollin's account says: "Soon the young man left agriculture for good. . . ." We need to pause here to try to understand the significance of his departure.

Like a strip mine, a hog factory exists in utter indifference to the landscape. Its purpose, as an animal *factory*, is to exclude from consideration both the nature of the place where it is and the nature of hogs. That it is a factory means that it could be in *any* place, and that the hog is a "unit of production." But the young man evidently was farm-raised. He evidently had in his mind at least the memory of an actual place and at least the remnants of its cultural landscape. In that landscape, things were respected according to their nature, which made compassion possible when their nature was violated. That this young man was fired from his job for showing compassion is strictly logical, for the explicit purpose of the hog factory is to violate nature. And then, logically enough, the young man "left agriculture for good." But when you exclude compassion from agriculture, what have you done? Have you not removed something ultimately of the greatest practical worth? I believe so. But this is one of the Rational Mind's world-scale experiments that has not yet been completed.

The Sympathetic Mind is a freedom-loving mind because it knows, given the inevitable discrepancies between the cultural and the actual landscapes, that everybody involved must be free to change. The idea that science and industry and government can discover for the rest of us the ultimate truths of nature and human nature, which then can be infallibly used to regulate our life, is wrong. The true work of the sciences and the arts is to keep all of us moving, in our own lives in our own places, between the cultural and the actual landscapes, making the always necessary and the forever unfinished corrections.

When the Rational Mind establishes a "farm," the result is bad farming. There is a remarkable difference between a hog factory, which exists only for the sake of its economic product, and a good farm, which exists for many reasons, including the pleasure of the farm family, their affection for their home, their satisfaction in their good work—in short, their patriotism. Such a farm yields its economic product as a sort of side effect of the health of a flourishing place in which things live according to their nature. The hog factory attempts to be a totally rational, which is to say a totally economic, enterprise. It strips away from animal life and human work every purpose, every benefit too, that is not economic. It comes about as the result of a long effort on the part of "scientific agriculture" to remove the Sympathetic Mind from all agricultural landscapes and replace it with the Rational Mind. And so good-bye to the shepherd of the parable, and to compassionate young men who leave agriculture for good. Good-bye to the cultural land-scape. Good-bye to the actual landscape. These have all been dispensed with by the Rational Mind, to be replaced by a to-talitarian economy with its neat, logical concepts of world-as-factory and life-as-commodity. This is an economy excluding all decisions but "informed decisions," purporting to reduce the possibility of loss.

Nothing so entices and burdens the Rational Mind as its need, and its self-imposed responsibility, to make "informed deci-sions." It is certainly possible for a mind to be informed—in several ways, too. And it is certainly possible for an informed mind to make decisions on the basis of all that has informed it. But that such decisions are "informed decisions"—in the sense that "informed decisions" are predictably right, or even that they are reliably better than uninformed decisions—is open to doubt.

The ideal of the "informed decision" forces "decision mak-ers" into a thicket of facts, figures, studies, tests, and "projec-tions." It requires long and uneasy pondering of "cost-benefit ratios"—the costs and benefits, often, of abominations. The problem is that decisions all have to do with the future, and all the actual knowledge we have is of the past. It is impossible to make a decision, however well-informed it may be, that is assuredly right, because it is impossible to know what will hap-pen. It is only possible to know or guess that some things *may* happen, and many things that have happened have not been foreseen.

Moreover, having made an "informed decision," even one that turns out well, there is no way absolutely to determine whether or not it was a better decision than another deci-sion that one might have made instead. It is not possible to compare a decision that one made with a decision that one did not make. There are no "controls," no "replication plots," in experience.

The Rational Mind is under relentless pressure to justify governmental and corporate acts on an ever-increasing scale of power, extent, and influence. Given that pressure, it may be not so very surprising that the Rational Mind should have a remarkable tendency toward superstition. Reaching their mental limits, which, as humans, they must do soon enough, the rationalists begin to base their thinking on principles that are sometimes astonishingly unsound: "Creatures are machines" or "Knowledge is good" or "Growth is good" or "Science will find the answer" or "A rising tide lifts all boats." Or they approach the future with a stupefying array of computers, models, statistics, projections, calculations, cost-benefit analyses, experts, and even better computers—which of course cannot foretell the end of a horse race any better than Bertie Wooster.

And so the great weakness of the Rational Mind, contrary to its protestations, is a sort of carelessness or abandonment that takes the form of high-stakes gambling—as when, with optimism and fanfare, without foreknowledge or self-doubt or caution, nuclear physicists or chemists or genetic engineers release their products into the whole world, making the whole world their laboratory.

Or the great innovators and decision makers build huge airplanes whose loads of fuel make them, in effect, flying bombs. And they build the World Trade Center, forgetting apparently the B-25 bomber that crashed into the seventy-ninth floor of the Empire State Building in 1945. And then on September 11, 2001, some enemies—of a kind we well knew we had and evidently had decided to ignore—captured two huge airplanes and flew them, as bombs, into the two towers of the World Trade Center. In retrospect, we may doubt that these shaping decisions were properly informed, just as we may doubt that the expensive "intelligence" that is supposed to foresee and prevent such disasters is sufficiently intelligent.

The decisions, if the great innovators and decision makers were given to reading poetry, might have been informed by

James Laughlin's poem "Above the City," which was written
soon after the B-25 crashed into the Empire State Building:

> You know our office on the 18th
> floor of the Salmon Tower looks
> right out on the
>
> Empire State & it just happened
> we were finishing up some
> late invoices on
>
> a new book that Saturday morning
> when a bomber roared through the
> mist and crashed
>
> flames poured from the windows
> into the drifting clouds & sirens
> screamed down in
>
> the streets below it was unearthly
> but you know the strangest thing
> we realized that
>
> none of us were much surprised be-
> cause we'd always known that those
> two Paragons of
>
> Progress sooner or later would per-
> form before our eyes this demon-
> stration of their
> true relationship

It is tempting now to call this poem "prophetic." But it is so
only in the sense that it is insightful; it perceives the implicit
contradiction between tall buildings and airplanes. This contradiction was readily apparent also to the terrorists of September 11, but evidently invisible within the mist of technological
euphoria that had surrounded the great innovators and decision makers.

    In the several dimensions of its horror the destruction of
the World Trade Center exceeds imagination, and that tells us
something. But as a physical event it is as comprehensible as
1 + 1, and that tells us something else.

*

Now that terrorism has established itself among us as an inescapable consideration, even the great decision makers are beginning to see that we are surrounded by the results of great decisions not adequately informed. We have built many nuclear power plants, each one a potential catastrophe, that will have to be protected, not only against their inherent liabilities and dangers, but against terrorist attack. And we have made, in effect, one thing of our food supply system, and that will have to be protected (if possible) from bioterrorism. These are by no means the only examples of the way we have exposed ourselves to catastrophic harm and great expense by our informed, rational acceptance of the normalcy of bigness and centralization.

After September 11, it can no longer be believed that science, technology, and industry are only good or that they serve only one "side." That never has been more than a progressivist and commercial superstition. Any power that belongs to one side belongs, for worse as well as better, to all sides, as indifferent as the sun that rises "on the evil and on the good." Only in the narrowest view of history can the scientists who worked on the nuclear bomb be said to have worked for democracy and freedom. They worked inescapably also for the enemies of democracy and freedom. If terrorists get possession of a nuclear bomb and use it, then the scientists of the bomb will be seen to have worked also for terrorism. There is (so far as I can now see) nothing at all that the Rational Mind can do, after the fact, to make this truth less true or less frightening. This predicament cries out for a different kind of mind before and after the fact: a mind faithful and compassionate that will not rationalize about the "good use" of destructive power, but will repudiate *any* use of it.

Freedom also is neutral, of course, and serves evil as well as good. But freedom rests on the power of good—by free speech, for instance—to correct evil. A great destructive power simply prevents this small decency of freedom. There is no way to correct a nuclear explosion.

In the midst of the dangers of the Rational Mind's achievement of bigness and centralization, the Sympathetic Mind is as hardpressed as a pacifist in the midst of a war. There is no greater violence that ends violence, and no greater bigness with which to solve the problems of bigness. All that the Sympathetic Mind can do is maintain its difference, preserve its own integrity, and attempt to see the possibility of something better.

The Sympathetic Mind, as the mind of our creatureliness, accepts life in this world for what it is: mortal, partial, fallible, complexly dependent, entailing many responsibilities toward ourselves, our places, and our fellow beings. Above all, it understands itself as limited. It knows without embarrassment its own irreducible ignorance, especially of the future. It deals with the issue of the future, not by knowing what is going to happen, but by knowing—within limits—what to expect, and what should be required, of itself, of its neighbors, and of its place. Its decisions are informed by its culture, its experience, its understanding of nature. Because it is aware of its limits and its ignorance, it is alert to issues of scale. The Sympathetic Mind knows from experience—not with the brain only, but with the body—that danger increases with height, temperature, speed, and power. It knows by common sense and instinct that the way to protect a building from being hit by an airplane is to make it shorter; that the way to keep a nuclear power plant from becoming a weapon is not to build it; that the way to increase the security of a national food supply is to increase the agricultural self-sufficiency of states, regions, and local communities.

Because it is the mind of our wholeness, our involvement with all things beyond ourselves, the Sympathetic Mind is alert as well to the issues of propriety, of the fittingness of our artifacts to their places and to our own circumstances, needs, and hopes. It is preoccupied, in other words, with the fidelity or the truthfulness of the cultural landscape to the actual landscape.

I know I am not the only one reminded by the World Trade Center of the Tower of Babel: "let us build . . . a tower, whose top may reach unto heaven; and let us make us a name, lest we be scattered abroad upon the face of the whole earth." All extremely tall buildings have made me think of the Tower of Babel, and this started a long time before September 11, 2001, for reasons that have become much clearer to me now that "those two Paragons of / progress" have demonstrated again "their true relationship."

Like all such gigantic buildings, from Babel onward, the
World Trade Center was built without reference to its own
landscape or to any other. And the reason in this instance is not
far to find. The World Trade Center had no reference to a landscape because world trade, as now practiced, has none. World
trade now exists to exploit indifferently the landscapes of the
world, and to gather the profits to centers whence they may
be distributed to the world's wealthiest people. World trade
needs centers precisely to prevent the world's wealth from being "scattered abroad upon the face of the whole earth." Such
centers, like the "global free market" and the "global village,"
are utopias, "no-places." They need to be no-places, because
they respect no places and are loyal to no place.

As for the problem of building on Manhattan Island, the Rational Mind has reduced that also to a simple economic principle: Land is expensive but air is cheap; therefore, build in the
air. In the early 1960s my family and I lived on the Lower West
Side, not far uptown from what would become the site of the
World Trade Center. The area was run down, already under
the judgment of "development." But once, obviously, it had
been a coherent, thriving local neighborhood of residential
apartments and flats, small shops and stores, where merchants
and customers knew one another and neighbors were known
to neighbors. Walking from our building to the Battery was a
pleasant thing to do because one had the sense of being in a
real place that kept both the signs of its old human history and
the memory of its geographical identity. The last time I went
there, the place had been utterly dis-placed by the World Trade
Center.

Exactly the same feat of displacement is characteristic of the
air transportation industry, which exists to free travel from all
considerations of place. Air travel reduces place to space in order to traverse it in the shortest possible time. And like gigantic
buildings, gigantic airports must destroy their places and become no-places in order to exist.

People of the modern world, who have accepted the dominance and the value system of the Rational Mind, do not
object, it seems, to this displacement, or to the consequent
disconnection of themselves from neighborhoods and from

the landscapes that support them, or to their own anonymity within crowds of strangers. These things, according to cliché, free one from the suffocating intimacies of rural or small town life. And yet we now are obliged to notice that placelessness, centralization, gigantic scale, crowdedness, and anonymity are conditions virtually made to order for terrorists.

It is wrong to say, as some always do, that catastrophes are "acts of God" or divine punishments. But it is not wrong to ask if they may not be the result of our misreading of reality or our own nature, and if some correction may not be needed. My own belief is that the Rational Mind has been performing impressively within the narrowly drawn boundaries of what it provably knows, but it has been doing badly in dealing with the things of which it is ignorant: the future, the mysterious wholeness and multiplicity of the natural world, the needs of human souls, and even the real bases of the human economy in nature, skill, kindness, and trust. Increasingly, it seeks to justify itself with intellectual superstitions, public falsehoods, secrecy, and mistaken hopes, responding to its failures and bad surprises with (as the terrorists intend) terror and with ever grosser applications of power.

But the Rational Mind is caught, nevertheless, in cross-purposes that are becoming harder to ignore. It is altogether probable that there is an executive of an air-polluting industry who has a beloved child who suffers from asthma caused by air pollution. In such a situation the Sympathetic Mind cries, "Stop! Change your life! Quit your job! At least try to discover the cause of the harm and *do* something about it!" And here the Rational Mind must either give way to the Sympathetic Mind, or it must recite the conventional excuse that is a confession of its failure: "There is nothing to be done. This is the way things are. It is inevitable."

The same sort of contradiction now exists between national security and the global economy. Our government, having long ago abandoned any thought of economic self-sufficiency, having ceded a significant measure of national sovereignty to the World Trade Organization, and now terrified by terrorism, is obliged to police the global economy against the transportation of contraband weapons, which can be detected if the meshes of the surveillance network are fine enough, and also against the transportation of diseases, which cannot be detected. This too will be excused, at least for a while, by the plea of inevitability, never mind that this is the result of a conflict of policies and of "informed decisions." Meanwhile, there is probably no landscape in the world that is not threatened with abuse or destruction as a result of somebody's notion of trade or somebody's notion of security.

When the Rational Mind undertakes to work on a large scale, it works clumsily. It inevitably does damage, and it cannot exempt even itself or its own from the damage it does. You cannot help to pollute the world's only atmosphere and exempt your asthmatic child. You cannot make allies and enemies of the same people at the same time. Finally the idea of the trade-off fails. When the proposed trade-off is on the scale of the whole world—the natural world for world trade, world peace for national security—it can fail only into world disaster.

The Rational Mind, while spectacularly succeeding in some things, fails completely when it tries to deal in materialist terms with the part of reality that is spiritual. Religion and the language of religion deal approximately and awkwardly enough with this reality, but the Rational Mind, though it apparently cannot resist the attempt, cannot deal with it at all.

But most of the most important laws for the conduct of human life probably are religious in origin—laws such as these: Be merciful, be forgiving, love your neighbors, be hospitable to strangers, be kind to other creatures, take care of the helpless, love your enemies. We must, in short, love and care for one another and the other creatures. We are allowed to make no exceptions. Every person's obligation toward the Creation is summed up in two words from Genesis 2:15: "Keep it."

It is impossible, I believe, to make a neat thing of this set of instructions. It is impossible to disentangle its various obligations into a list of discrete items. Selfishness, or even "enlightened self-interest," cannot find a place to poke in its awl. One's obligation to oneself cannot be isolated from one's obligation to everything else. The whole thing is balanced on the verb *to love*. Love for oneself finds its only efficacy in love for everything else. Even loving one's enemy has become a strategy of self-love as the technology of death has grown greater. And this the terrorists have discovered and have accepted: The death of your enemy is your own death. The whole network of interdependence and obligation is a neatly set trap. Love does not let us escape from it; it turns the trap itself into the means and fact of our only freedom.

This condition of lawfulness and this set of laws did not originate in the Rational Mind, and could not have done so. The Rational Mind reduces our complex obligation to care for one another to issues of justice, forgetting the readiness with which we and our governments reduce justice, in turn, to revenge; and forgetting that even justice is intolerable without mercy, forgiveness, and love.

Justice is a rational procedure. Mercy is not a procedure and it is not rational. It is a kind of freedom that comes from sympathy, which is to say imagination—the *felt* knowledge of what it is to be another person or another creature. It is free because it does not have to be just. Justice is desirable, of course, but it is virtually the opposite of mercy. Mercy, says the Epistle of James, "rejoiceth against judgment."

As for the law requiring us to "keep" the given or the natural world (to go in search of the lost sheep, to save all sentient beings), the Rational Mind, despite the reasonable arguments made by some ecologists and biologists, cannot cover that distance either. In response to the proposition that we are responsible for the health of all the world, the Rational Mind begins to insist upon exceptions and trade-offs. It begins to designate the profit-yielding parts of the world that may "safely" be destroyed, and those unprofitable parts that may be preserved as "natural" or "wild." It divides the domestic from the wild, the human from the natural. It conceives of a natural place as a place where no humans live. Places where humans live are not natural, and the nature of such places must be reduced to comprehensibility, which is to say destroyed as they naturally are. The Rational Mind, convinced of the need to preserve "biological diversity," wants to preserve it in "nature preserves." It cannot conceive or tolerate the possibility of preserving biological diversity in the whole world, or of an economic harmony between humans and a world that by nature exceeds human comprehension.

It is because of the world's ultimately indecipherable web-work of vital connections, dependences, and obligations, and because ultimately our response to it must be loving beyond knowing, that the works of the Rational Mind are ultimately disappointing even to some rationalists.

When the Rational Mind fails not only into bewilderment but into irrationality and catastrophe, as it repeatedly does, that is because it has so isolated itself within its exclusive terms that it goes beyond its limits without knowing it.

Finally the human mind must accept the limits of sympathy, which paradoxically will enlarge it beyond the limits of rationality, but nevertheless will limit it. It must find its freedom and its satisfaction by working within its limits, on a scale much smaller than the Rational Mind will easily accept, for the Rational Mind continually longs to extend its limits by technology. But the safe competence of human work extends no further, ever, than our ability to think and love at the same time.

Obviously, we *can* work on a gigantic scale, but just as obviously we cannot foresee the gigantic catastrophes to which gigantic works are vulnerable, any more than we can foresee the natural and human consequences of such work. We can develop a global economy, but only on the conditions that it will not be loving in its effects on its human and natural sources, and that it will risk global economic collapse. We can build gigantic works of architecture too, but only with the likelihood that the gathering of the economic means to do so will generate somewhere the will to destroy what we have built.

The efficacy of a law is in the ability of people to obey it. The larger the scale of work, the smaller will be the number of people who can obey the law that we should be loving toward the world, even those places and creatures that we must use. You will see the problem if you imagine that you are one of the many, or if you *are* one of the many, who can find no work except in a destructive industry. Whether or not it is economic slavery to have no choice of jobs, it certainly is moral slavery to have no choice but to do what is wrong.

And so conservationists have not done enough when they conserve wilderness or biological diversity. They also must conserve the possibilities of peace and good work, and to do that they must help to make a good economy. To succeed, they must help to give more and more people everywhere in the world the opportunity to do work that is both a living and a loving. This, I think, cannot be accomplished by the Rational Mind. It will require the full employment of the Sympathetic Mind—*all* the little intelligence we have.

The Whole Horse

This modern mind sees only half of the horse—that half which may become a dynamo, or an automobile, or any other horsepowered machine. If this mind had much respect for the full-dimensioned, grass-eating horse, it would never have invented the engine which represents only half of him. The religious mind, on the other hand, has this respect; it wants the whole horse, and it will be satisfied with nothing less. I should say a religious mind that requires more than a half-religion.
　　—Allen Tate, "Remarks on the Southern Religion," in *I'll Take My Stand*

O NE OF THE PRIMARY RESULTS—and one of the primary needs—of industrialism is the separation of people and places and products from their histories. To the extent that we participate in the industrial economy, we do not know the histories of our meals or of our habitats or of our families. This is an economy, and in fact a culture, of the one-night stand. "I had a good time," says the industrial lover, "but don't ask me my last name." Just so, the industrial eater says to the svelte industrial hog, "We'll be together at breakfast. I don't want to see you before then, and I won't care to remember you afterwards."

In this condition, we have many commodities, but little satisfaction, little sense of the sufficiency of anything. The scarcity of satisfaction makes of our many commodities, in fact, an infinite series of commodities, the new commodities invariably promising greater satisfaction than the older ones. And so we can say that the industrial economy's most-marketed commodity is satisfaction, and that this commodity, which is repeatedly promised, bought, and paid for, is never delivered. On the other hand, people who have much satisfaction do not need many commodities.

The persistent want of satisfaction is directly and complexly related to the dissociation of ourselves and all our goods from our and their histories. If things do not last, are not made to last, they can have no histories, and we who use these things can have no memories. We buy new stuff on the promise of satisfaction because we have forgot the promised satisfaction for which we bought our old stuff. One of the procedures of the industrial economy is to reduce the longevity of materials. For example, wood, which well made into buildings and furniture and well cared for can last hundreds of years, is now routinely manufactured into products that last twenty-five years. We do not cherish the memory of shoddy and transitory objects, and so we do not remember them. That is to say that we do not invest in them the lasting respect and admiration that make for satisfaction.

The problem of our dissatisfaction with all the things that we use is not correctable within the terms of the economy that produces those things. At present, it is virtually impossible for us to know the economic history or the ecological cost of the products we buy; the origins of the products are typically too distant and too scattered and the processes of trade, manufacture, transportation, and marketing too complicated. There are, moreover, too many reasons for the industrial suppliers of these products not to want their histories to be known.

When there is no reliable accounting and therefore no competent knowledge of the economic and ecological effects of our lives, we cannot live lives that are economically and ecologically responsible. This is the problem that has frustrated, and to a considerable extent undermined, the American conservation effort from the beginning. It is ultimately futile to plead and protest and lobby in favor of public ecological responsibility while, in virtually every act of our private lives, we endorse and support an economic system that is by intention, and perhaps by necessity, ecologically irresponsible.

If the industrial economy is not correctable within or by its own terms, then obviously what is required for correction is a countervailing economic idea. And the most significant weakness of the conservation movement is its failure to produce or espouse an economic idea capable of correcting the economic idea of the industrialists. Somewhere near the heart of the conservation effort as we have known it is the romantic assumption that, if we have become alienated from nature, we can become unalienated by making nature the subject of contemplation or art, ignoring the fact that we live necessarily in and from nature—ignoring, in other words, all the economic issues that are involved. Walt Whitman could say, "I think I could turn and live with animals" as if he did not know that, in fact, we do live with animals, and that the terms of our relation to them are inescapably established by our economic use of their and our world. So long as we live, we are going to be living with skylarks, nightingales, daffodils, waterfowl, streams, forests, mountains, and all the other creatures that romantic poets and artists have yearned toward. And by the way we live we will determine whether or not those creatures will live.

That this nature-romanticism of the nineteenth century ignores economic facts and relationships has not prevented it from setting the agenda for modern conservation groups. This agenda has rarely included the economics of land use, without which the conservation effort becomes almost inevitably long on sentiment and short on practicality. The giveaway is that when conservationists try to be practical they are likely to defend the "sustainable use of natural resources" with the argument that this will make the industrial economy sustainable. A further giveaway is that the longer the industrial economy lasts in its present form, the further it will demonstrate its ultimate impossibility: Every human in the world cannot, now or ever, own the whole catalogue of shoddy, high-energy industrial products, which cannot be sustainably made or used. Moreover, the longer the industrial economy lasts, the more it will eat away the possibility of a better economy.

The conservation effort has at least brought under suspicion the general relativism of our age. Anybody who has studied with care the issues of conservation knows that our acts are being measured by a real, absolute, and unyielding standard that was invented by no human. Our acts that are not in harmony with nature are inevitably and sometimes irremediably destructive. The standard exists. But having no opposing economic idea, conservationists have had great difficulty in applying the standard.

What, then, is the countervailing idea by which we might correct the industrial idea? We will not have to look hard to find it, for there is only one, and that is agrarianism. Our major difficulty (and danger) will be in attempting to deal with agrarianism as "an idea"—agrarianism is primarily a practice, a set of attitudes, a loyalty, and a passion; it is an idea only secondarily and at a remove. To use merely the handiest example: I was raised by agrarians, my bias and point of view from my earliest childhood were agrarian, and yet I never heard agrarianism defined, or even so much as named, until I was a sophomore in college. I am well aware of the danger in defining things, but if I am going to talk about agrarianism, I am going to have to define it. The definition that follows is derived both from agrarian writers, ancient and modern, and from the unliterary and sometimes illiterate agrarians who have been my teachers.

The fundamental difference between industrialism and agrarianism is this: Whereas industrialism is a way of thought based on monetary capital and technology, agrarianism is a way of thought based on land.

Agrarianism, furthermore, is a culture at the same time that it is an economy. Industrialism is an economy before it is a culture. Industrial culture is an accidental by-product of the ubiquitous effort to sell unnecessary products for more than they are worth.

An agrarian economy rises up from the fields, woods, and streams—from the complex of soils, slopes, weathers, connections, influences, and exchanges that we mean when we speak, for example, of the local community or the local watershed. The agrarian mind is therefore not regional or national, let alone global, but local. It must know on intimate terms the local plants and animals and local soils; it must know local possibilities and impossibilities, opportunities and hazards. It depends and insists on knowing very particular local histories and biographies.

Because a mind so placed meets again and again the necessity for work to be good, the agrarian mind is less interested in abstract quantities than in particular qualities. It feels threatened and sickened when it hears people and creatures and places spoken of as labor, management, capital, and raw material. It is not at all impressed by the industrial legendry of gross national products, or of the numbers sold and dollars earned by gigantic corporations. It is interested—and forever fascinated—by questions leading toward the accomplishment of good work: What is the best location for a particular building or fence?

What is the best way to plow *this* field? What is the best course for a skid road in *this* woodland? Should *this* tree be cut or spared? What are the best breeds and types of livestock for *this* farm?—questions which cannot be answered in the abstract, and which yearn not toward quantity but toward elegance. Agrarianism can never become abstract because it has to be practiced in order to exist.

And though this mind is local, almost absolutely placed, little attracted to mobility either upward or lateral, it is not provincial; it is too taken up and fascinated by its work to feel inferior to any other mind in any other place.

An agrarian economy is always a subsistence economy before it is a market economy. The center of an agrarian farm is the household. The function of the household economy is to assure that the farm family lives from the farm so far as possible. It is the subsistence part of the agrarian economy that assures its stability and its survival. A subsistence economy necessarily is highly diversified, and it characteristically has involved hunting and gathering as well as farming and gardening. These activities bind people to their local landscape by close, complex interests and economic ties. The industrial economy alienates people from the native landscape precisely by breaking these direct practical ties and introducing distant dependences.

Agrarian people of the present, knowing that the land must be well cared for if anything is to last, understand the need for a settled connection, not just between farmers and their farms, but between urban people and their surrounding and tributary landscapes. Because the knowledge and know-how of good caretaking must be handed down to children, agrarians recognize the necessity of preserving the coherence of families and communities.

The stability, coherence, and longevity of human occupation require that the land should be divided among many owners and users. The central figure of agrarian thought has invariably been the small owner or smallholder who maintains a significant measure of economic self-determination on a small acreage. The scale and independence of such holdings imply two things that agrarians see as desirable: intimate care in the use of the land, and political democracy resting upon the indispensable foundation of economic democracy.

A major characteristic of the agrarian mind is a longing for independence—that is, for an appropriate degree of personal and local self-sufficiency. Agrarians wish to earn and deserve what they have. They do not wish to live by piracy, beggary, charity, or luck.

In the written record of agrarianism, there is a continually recurring affirmation of nature as the final judge, lawgiver, and pattern-maker of and for the human use of the earth. We can trace the lineage of this thought in the West through the writings of Virgil, Spenser, Shakespeare, Pope, Jefferson, and on into the work of the twentieth-century agriculturists and scientists J. Russell Smith, Liberty Hyde Bailey, Albert Howard, Wes Jackson, John Todd, and others. The idea is variously stated: We should not work until we have looked and seen where we are; we should honor Nature not only as our mother or grandmother, but as our teacher and judge; we should "let the forest judge"; we should "consult the Genius of the Place"; we should make the farming fit the farm; we should carry over into the cultivated field the diversity and coherence of the native forest or prairie. And this way of thinking is surely allied to that of the medieval scholars and architects who saw the building of a cathedral as a symbol or analogue of the creation of the world. The agrarian mind is, at bottom, a religious mind. It subscribes to Allen Tate's doctrine of "the whole horse." It prefers the Creation itself to the powers and quantities to which it can be reduced. And this is a mind completely different from that which sees creatures as machines, minds as computers, soil fertility as chemistry, or agrarianism as an idea. John Haines has written that "the eternal task of the artist and the poet, the historian and the scholar . . . is to find the means to reconcile what are two separate and yet inseparable histories, Nature and Culture. To the extent that we can do this, the 'world' makes sense to us and can be lived in." I would add only that this applies also to the farmer, the forester, the scientist, and others.

The agrarian mind begins with the love of fields and ramifies in good farming, good cooking, good eating, and gratitude to God. Exactly analogous to the agrarian mind is the sylvan mind that begins with the love of forests and ramifies in good forestry, good woodworking, good carpentry, and gratitude to God. These two kinds of mind readily intersect; neither ever intersects with the industrial-economic mind. The industrial-economic mind begins with ingratitude, and ramifies in the destruction of farms and forests. The "lowly" and "menial" arts of farm and forest are mostly taken for granted or ignored by the culture of the "fine arts" and by "spiritual" religions; they are taken for granted or ignored or held in contempt by the powers of the industrial economy. But in fact they are inescapably the foundation of human life and culture, and their adepts are capable of as deep satisfactions and as high attainments as anybody else.

Having, so to speak, laid industrialism and agrarianism side by side, implying a preference for the latter, I will be confronted by two questions that I had better go ahead and answer.

The first is whether or not agrarianism is simply a "phase" that we humans had to go through and then leave behind in order to get onto the track of technological progress toward ever greater happiness. The answer is that although industrialism has certainly conquered agrarianism, and has very nearly destroyed it altogether, it is also true that in every one of its uses of the natural world industrialism is in the process of catastrophic failure. Industry is now desperately shifting—by means of genetic engineering, global colonialism, and other contrivances—to prolong its control of our farms and forests, but the failure nonetheless continues. It is not possible to argue sanely in favor of soil erosion, water pollution, genetic impoverishment, and the destruction of rural communities and local economies. Industrialism, unchecked by the affections and concerns of agrarianism, becomes monstrous. And this is because of a weakness identified by the Twelve Southerners of *I'll Take My Stand* in their "Statement of Principles": Under the rule of industrialism "the remedies proposed . . . are always homeopathic." That is to say that industrialism always proposes to correct its errors and excesses by more industrialization.

The second question is whether or not by espousing the revival of agrarianism we will commit the famous sin of "turning back the clock." The answer to that, for present-day North Americans, is fairly simple. The overriding impulse of agrarianism is toward the local adaptation of economies and cultures. Agrarian people wish to adapt the farming to the farm and the forestry to the forest. At times and in places we latter-day

Americans have come close to accomplishing this goal, and we have a few surviving examples, but it is generally true that we are much further from local adaptation now than we were fifty years ago. We never yet have developed stable, sustainable, locally adapted land-based economies. The good rural enterprises and communities that we will find in our past have been almost constantly under threat from the colonialism, first foreign and then domestic, and now "global," which has so far dominated our history, and which has been institutionalized for a long time in the industrial economy. The possibility of an authentically settled country still lies ahead of us.

If we wish to look ahead, we will see not only in the United States but in the world two economic programs that conform pretty exactly to the aims of industrialism and agrarianism as I have described them.

The first is the effort to globalize the industrial economy, not merely by the expansionist programs of supra-national corporations within themselves, but also by means of government-sponsored international trade agreements, the most prominent of which is the World Trade Organization Agreement, which institutionalizes the industrial ambition to use, sell, or destroy every acre and every creature of the world.

The World Trade Organization gives the lie to the industrialist conservatives' professed abhorrence of big government. The cause of big government, after all, is big business. The power to do large-scale damage, which is gladly assumed by every large-scale industrial enterprise, calls naturally and logically for government regulation, which of course the corporations object to. But we have a good deal of evidence also that the leaders of big business actively desire and promote big government. They and their political allies, while ostensibly working to "downsize" government, continue to promote government helps and "incentives" to large corporations; and, however absurdly, they adhere to their notion that a small government, taxing only the working people, can maintain a big highway system, a big military establishment, a big space program, and big government contracts.

But the most damaging evidence is the World Trade Organization itself, which is in effect a global government, with power to enforce the decisions of the collective against national laws that conflict with it. The coming of the World Trade Organization was foretold seventy years ago in the "Statement of Principles" of *I'll Take My Stand*, which said that "the true Sovietists or Communists . . . are the industrialists themselves. They would have the government set up an economic super-organization, which in turn would become the government." The agrarians of *I'll Take My Stand* did not foresee this because they were fortune-tellers, but because they had perceived accurately the character and motive of the industrial economy.

The second program, counter to the first, is composed of many small efforts to preserve or improve or establish local economies. These efforts on the part of nonindustrial or agrarian conservatives, local patriots, are taking place in countries both affluent and poor all over the world.

Whereas the corporate sponsors of the World Trade Organization, in order to promote their ambitions, have required only the hazy glamour of such phrases as "the global economy," "the global context," and "globalization," the local economists use a much more diverse and particularizing vocabulary that you can actually think with: "community," "ecosystem," "watershed," "place," "homeland," "family," "household."

And whereas the global economists advocate a world-government-by-economic-bureaucracy, which would destroy local adaptation everywhere by ignoring the uniqueness of every place, the local economists found their work upon respect for such uniqueness. Places differ from one another, the local economists say, therefore we must behave with unique consideration in each one; the ability to tender an appropriate practical regard and respect to each place in its difference is a kind of freedom; the inability to do so is a kind of tyranny. The global economists are the great centralizers of our time. The local economists, who have so far attracted the support of no prominent politician, are the true decentralizers and downsizers, for they seek an appropriate degree of self-determination and independence for localities. They seem to be moving toward a radical and necessary revision of our idea of a city. They are learning to see the city, not just as a built and paved municipality set apart by "city limits" to live by trade and transportation from the world at large, but rather as a part of a local community which includes also the city's rural neighbors, its surrounding landscape and its watershed, on which it might depend for at least some of its necessities, and for the health of which it might exercise a competent concern and responsibility.

At this point, I want to say point blank what I hope is already clear: Though agrarianism proposes that everybody has agrarian responsibilities, it does not propose that everybody should be a farmer or that we do not need cities. Nor does it propose that every product should be a necessity. Furthermore, any thinkable human economy would have to grant to manufacturing an appropriate and honorable place. Agrarians would insist only that any manufacturing enterprise should be formed and scaled to fit the local landscape, the local ecosystem, and the local community, and that it should be locally owned and employ local people. They would insist, in other words, that the shop or factory owner should not be an outsider, but rather a sharer in the fate of the place and its community. The deciders should have to live with the results of their decisions.

Between these two programs—the industrial and the agrarian, the global and the local—the most critical difference is that of knowledge. The global economy institutionalizes a global ignorance, in which producers and consumers cannot know or care about one another, and in which the histories of all products will be lost. In such a circumstance, the degradation of products and places, producers and consumers is inevitable.

But in a sound local economy, in which producers and consumers are neighbors, nature will become the standard of work and production. Consumers who understand their economy will not tolerate the destruction of the local soil or ecosystem or watershed as a cost of production. Only a healthy local economy can keep nature and work together in the consciousness of the community. Only such a community can restore history to economics.

I will not be altogether surprised to be told that I have set forth here a line of thought that is attractive but hopeless. A number of critics have advised me of this, out of their charity, as if I might have written of my hopes for forty years without giving a thought to hopelessness. Hope, of course, is always accompanied by the fear of hopelessness, which is a legitimate fear.

And so I would like to conclude by confronting directly the issue of hope. My hope is most seriously challenged by the fact of decline, of loss. The things that I have tried to defend are less numerous and worse off now than when I started, but in this I am only like all other conservationists. All of us have been fighting a battle that on average we are losing, and I doubt that there is any use in reviewing the statistical proofs. The point —the only interesting point—is that we have not quit. Ours is not a fight that you can stay in very long if you look on victory as a sign of triumph or on loss as a sign of defeat. We have not quit because we are not hopeless.

My own aim is not hopelessness. I am not looking for reasons to give up. I am looking for reasons to keep on. In outlining here the concerns of agrarianism, I have intended to show how the effort of conservation could be enlarged and strengthened.

What agrarian principles implicitly propose—and what I explicitly propose in advocating those principles at this time—is a revolt of local small producers and local consumers against the global industrialism of the corporations. Do I think that there is a hope that such a revolt can survive and succeed, and that it can have a significant influence upon our lives and our world?

Yes, I do. And to be as plain as possible, let me just say what I know. I know from friends and neighbors and from my own family that it is now possible for farmers to sell at a premium to local customers such products as organic vegetables, organic beef and lamb, and pasture-raised chickens. This market is being made by the exceptional goodness and freshness of the food, by the wish of urban consumers to support their farming neighbors, and by the excesses and abuses of the corporate food industry.

This is the pattern of an economic revolt that is not only possible but is happening. It is happening for two reasons: First, as the scale of industrial agriculture increases, so does the scale of its abuses, and it is hard to hide large-scale abuses from consumers. It is virtually impossible now for intelligent consumers to be ignorant of the heartlessness and nastiness of animal confinement operations and their excessive use of antibiotics, of the use of hormones in meat and milk production, of the stenches and pollutants of pig and poultry factories, of the use of toxic chemicals and the waste of soil and soil health in industrial row-cropping, of the mysterious or disturbing or threatening practices associated with industrial food storage, preservation, and processing. Second, as the food industries focus more and more on gigantic global opportunities, they cannot help but overlook small local opportunities, as is made plain by the proliferation of "community-supported agriculture," farmers markets, health food stores, and so on. In fact, there are some markets that the great corporations by definition cannot supply. The market for so-called organic food, for example, is really a market for good, fresh, trustworthy food, food from producers known and trusted by consumers, and such food cannot be produced by a global corporation.

But the food economy is only one example. It is also possible to think of good local forest economies. And in the face of much neglect, it is possible to think of local small business economies—some of them related to the local economies of farm and forest—supported by locally owned, community-oriented banks.

What do these efforts of local economy have to do with conservation as we know it? The answer, I believe, is *everything*. The conservation movement, as I said earlier, has a conservation program; it has a preservation program; it has a rather sporadic health-protection program; but it has no economic program, and because it has no economic program it has the status of something exterior to daily life, surviving by emergency, like an ambulance service. In saying this, I do not mean to belittle the importance of protest, litigation, lobbying, legislation, large-scale organization—all of which I believe in and support. I am saying simply that we must do more. We must confront, on the ground, and each of us at home, the economic assumptions in which the problems of conservation originate.

We have got to remember that the great destructiveness of the industrial age comes from a division, a sort of divorce, in our economy, and therefore in our consciousness, between production and consumption. Of this radical division of functions we can say, without much fear of oversimplifying, that the aim of industrial producers is to sell as much as possible and that the aim of industrial consumers is to buy as much as possible. We need only to add that the aim of both producer and consumer is to be so far as possible carefree. Because of various pressures, governments have learned to coerce from producers some grudging concern for the health and solvency of consumers. No way has been found to coerce from consumers any consideration for the methods and sources of production.

What alerts consumers to the outrages of producers is typically some kind of loss or threat of loss. We see that in dividing consumption from production we have lost the function of conserving. Conserving is no longer an integral part of the economy of the producer or the consumer. Neither the producer nor the consumer any longer says, "I must be careful of this so that it will last." The working assumption of both is that where there is some, there must be more. If they can't get what they need in one place, they will find it in another. That is why conservation is now a separate concern and a separate effort.

But experience seems increasingly to be driving us out of the categories of producer and consumer and into the categories of citizen, family member, and community member, in all of which we have an inescapable interest in making things last. And here is where I think the conservation movement (I mean that movement that has defined itself as the defender of wilderness and the natural world) can involve itself in the fundamental issues of economy and land use, and in the process gain strength for its original causes.

I would like my fellow conservationists to notice how many people and organizations are now working to save something of value—not just wilderness places, wild rivers, wildlife habitat, species diversity, water quality, and air quality, but also agricultural land, family farms and ranches, communities, children and childhood, local schools, local economies, local food markets, livestock breeds and domestic plant varieties, fine old buildings, scenic roads, and so on. I would like my fellow conservationists to understand also that there is hardly a small farm or ranch or locally owned restaurant or store or shop or business anywhere that is not struggling to conserve itself.

All of these people, who are fighting sometimes lonely battles to keep things of value that they cannot bear to lose, are the conservation movement's natural allies. Most of them have the same enemies as the conservation movement. There is no necessary conflict among them. Thinking of them, in their great variety, in the essential likeness of their motives and concerns, one thinks of the possibility of a defined community of interest among them all, a shared stewardship of all the diversity of good things that are needed for the health and abundance of the world.

I don't suppose that this will be easy, given especially the history of conflict between conservationists and land users. I only suppose that it is necessary. Conservationists can't conserve everything that needs conserving without joining the effort to use well the agricultural lands, the forests, and the waters that we must use. To enlarge the areas protected from use without at the same time enlarging the areas of *good* use is a mistake. To have no large areas of protected old-growth forest would be folly, as most of us would agree. But it is also folly to have come this far in our history without a single working model of a thoroughly diversified and integrated, ecologically sound, local forest economy. That such an economy is possible is indicated by many imperfect or incomplete examples, but we need desperately to put the pieces together in one place—and then in every place.

The most tragic conflict in the history of conservation is that between the conservationists and the farmers and ranchers. It is tragic because it is unnecessary. There is no irresolvable conflict here, but the conflict that exists can be resolved only on the basis of a common understanding of good practice. Here again we need to foster and study working models: farms and ranches that are knowledgeably striving to bring economic practice into line with ecological reality, and local food economies in which consumers conscientiously support the best land stewardship.

We know better than to expect very soon a working model of a conserving global corporation. But we must begin to expect —and we must, as conservationists, begin working for, and in—working models of nature-conserving local economies. These are possible now. Good and able people are working hard to develop them now. They need the full support of the conservation movement now. Conservationists need to go to these people, ask what they can do to help, and then help. A little later, having helped, they can in turn ask for help.

The Agrarian Standard

THE UNSETTLING OF AMERICA was published twenty-five years ago; it is still in print and is still being read. As its author, I am tempted to be glad of this, and yet, if I believe what I said in that book, and I still do, then I should be anything but glad. The book would have had a far happier fate if it could have been disproved or made obsolete years ago.

It remains true because the conditions it describes and opposes, the abuses of farmland and farming people, have persisted and become worse over the last twenty-five years. In 2002 we have less than half the number of farmers in the United States that we had in 1977. Our farm communities are far worse off now than they were then. Our soil erosion rates continue to be unsustainably high. We continue to pollute our soils and streams with agricultural poisons. We continue to lose farmland to urban development of the most wasteful sort. The large agribusiness corporations that were mainly national in 1977 are now global, and are replacing the world's agricultural diversity, useful primarily to farmers and local consumers, with bioengineered and patented monocultures that are merely profitable to corporations. The purpose of this new global economy, as Vandana Shiva has rightly said, is to replace "food democracy" with a worldwide "food dictatorship."

To be an agrarian writer in such a time is an odd experience. One keeps writing essays and speeches that one would prefer not to write, that one wishes would prove unnecessary, that one hopes nobody will have any need for in twenty-five years. My life as an agrarian writer has certainly involved me in such confusions, but I have never doubted for a minute the importance of the hope I have tried to serve: the hope that we might become a healthy people in a healthy land.

We agrarians are involved in a hard, long, momentous contest, in which we are so far, and by a considerable margin, the losers. What we have undertaken to defend is the complex accomplishment of knowledge, cultural memory, skill, self-mastery, good sense, and fundamental decency—the high and indispensable art—for which we probably can find no better name than "good farming." I mean farming as defined by agrarianism as opposed to farming as defined by industrialism: farming as the proper use and care of an immeasurable gift.

I believe that this contest between industrialism and agrarianism now defines the most fundamental human difference, for it divides not just two nearly opposite concepts of agriculture and land use, but also two nearly opposite ways of understanding ourselves, our fellow creatures, and our world.

The way of industrialism is the way of the machine. To the industrial mind, a machine is not merely an instrument for doing work or amusing ourselves or making war; it is an explanation of the world and of life. The machine's entirely comprehensible articulation of parts defines the acceptable meanings of our experience, and it prescribes the kinds of meanings the industrial scientists and scholars expect to discover. These meanings have to do with nomenclature, classification, and rather short lineages of causation. Because industrialism cannot understand living things except as machines, and can grant them no value that is not utilitarian, it conceives of farming and forestry as forms of mining; it cannot use the land without abusing it.

Industrialism prescribes an economy that is placeless and displacing. It does not distinguish one place from another. It applies its methods and technologies indiscriminately in the American East and the American West, in the United States and in India. It thus continues the economy of colonialism. The shift of colonial power from European monarchy to global corporation is perhaps the dominant theme of modern history. All along—from the European colonization of Africa, Asia, and the New World, to the domestic colonialism of American industries, to the colonization of the entire rural world by the global corporations—it has been the same story of the gathering of an exploitive economic power into the hands of a few people who are alien to the places and the people they exploit. Such an economy is bound to destroy locally adapted agrarian economies everywhere it goes, simply because it is too ignorant not to do so. And it has succeeded precisely to the extent that it has been able to inculcate the same ignorance in workers and consumers. A part of the function of industrial education is to preserve and protect this ignorance.

To the corporate and political and academic servants of global industrialism, the small family farm and the small farming community are not known, are not imaginable, and are therefore unthinkable, except as damaging stereotypes. The people of "the cutting edge" in science, business, education, and politics have no patience with the local love, local loyalty, and local knowledge that make people truly native to their places and therefore good caretakers of their places. This is why one of the primary principles in industrialism has always been to get the worker away from home. From the beginning it has been destructive of home employment and home economies. The office or the factory or the institution is the place for work. The economic function of the household has been increasingly the consumption of purchased goods. Under industrialism, the farm too has become increasingly consumptive, and farms fail as the costs of consumption overpower the income from production.

The idea of people working at home, as family members, as neighbors, as natives and citizens of their places, is as repugnant to the industrial mind as the idea of self-employment. The industrial mind is an organizational mind, and I think this mind is deeply disturbed and threatened by the existence of people who have no boss. This may be why people with such minds, as they approach the top of the political hierarchy, so readily sell themselves to "special interests." They cannot bear to be unbossed. They cannot stand the lonely work of making up their own minds.

The industrial contempt for anything small, rural, or natural translates into contempt for uncentralized economic systems, any sort of local self-sufficiency in food or other necessities. The industrial "solution" for such systems is to increase the scale of work and trade. It is to bring Big Ideas, Big Money, and Big Technology into small rural communities, economies, and ecosystems—the brought-in industry and the experts being invariably alien to and contemptuous of the places to which they are brought in. There is never any question of propriety, of adapting the thought or the purpose or the technology to the place.

The result is that problems correctable on a small scale are replaced by large-scale problems for which there are no large-scale corrections. Meanwhile, the large-scale enterprise has reduced or destroyed the possibility of small-scale corrections. This exactly describes our present agriculture. Forcing all agricultural localities to conform to economic conditions imposed from afar by a few large corporations has caused problems of the largest possible scale, such as soil loss, genetic impoverishment, and ground-water pollution, which are correctable only by an agriculture of locally adapted, solar-powered, diversified small farms—a correction that, after a half century of industrial agriculture, will be difficult to achieve.

The industrial economy thus is inherently violent. It impoverishes one place in order to be extravagant in another, true to its colonialist ambition. A part of the "externalized" cost of this is war after war.

Industrialism begins with technological invention. But agrarianism begins with givens: land, plants, animals, weather, hunger, and the birthright knowledge of agriculture. Industrialists are always ready to ignore, sell, or destroy the past in order to gain the entirely unprecedented wealth, comfort, and happiness supposedly to be found in the future. Agrarian farmers know that their very identity depends on their willingness to receive gratefully, use responsibly, and hand down intact an inheritance, both natural and cultural, from the past. Agrarians understand themselves as the users and caretakers of some things they did not make, and of some things that they cannot make.

I said a while ago that to agrarianism farming is the proper use and care of an immeasurable gift. The shortest way to understand this, I suppose, is the religious way. Among the commonplaces of the Bible, for example, are the admonitions that the world was made and approved by God, that it belongs to Him, and that its good things come to us from Him as gifts. Beyond those ideas is the idea that the whole Creation exists only by participating in the life of God, sharing in His being, breathing His breath. "The world," Gerard Manley Hopkins said, "is charged with the grandeur of God." Such thoughts seem strange to us now, and what has

estranged us from them is our economy. The industrial economy could not have been derived from such thoughts any
more than it could have been derived from the golden rule.

If we believed that the existence of the world is rooted in
mystery and in sanctity, then we would have a different economy. It would still be an economy of use, necessarily, but it
would be an economy also of return. The economy would have
to accommodate the need to be worthy of the gifts we receive
and use, and this would involve a return of propitiation, praise,
gratitude, responsibility, good use, good care, and a proper regard for future generations. What is most conspicuously absent
from the industrial economy and industrial culture is this idea
of return. Industrial humans relate themselves to the world
and its creatures by fairly direct acts of violence. Mostly we
take without asking, use without respect or gratitude, and give
nothing in return. Our economy's most voluminous product
is waste—valuable materials irrecoverably misplaced, or ran-
domly discharged as poisons.

To perceive the world and our life in it as gifts originating
in sanctity is to see our human economy as a continuing moral
crisis. Our life of need and work forces us inescapably to use in
time things belonging to eternity, and to assign finite values to
things already recognized as infinitely valuable. This is a fearful predicament. It calls for prudence, humility, good work,
propriety of scale. It calls for the complex responsibilities of
caretaking and giving-back that we mean by "stewardship." To
all of this the idea of the immeasurable value of the resource
is central.

We can get to the same idea by a way a little more economic
and practical, and this is by following through our literature
the ancient theme of the small farmer or husbandman who
leads an abundant life on a scrap of land often described as
cast-off or poor. This figure makes his first literary appearance,
so far as I know, in Virgil's Fourth Georgic:

> I saw a man,
> An old Cilician, who occupied
> An acre or two of land that no one wanted,
> A patch not worth the ploughing, unrewarding

> For flocks, unfit for vineyards; he however
> By planting here and there among the scrub
> Cabbages or white lilies and verbena
> And flimsy poppies, fancied himself a king
> In wealth, and coming home late in the evening
> Loaded his board with unbought delicacies.

Virgil's old squatter, I am sure, is a literary outcropping of an agrarian theme that has been carried from earliest times until now mostly in family or folk tradition, not in writing, though other such people can be found in books. Wherever found, they don't vary by much from Virgil's prototype. They don't have or require a lot of land, and the land they have is often marginal. They practice subsistence agriculture, which has been much derided by agricultural economists and other learned people of the industrial age, and they always associate frugality with abundance.

In my various travels, I have seen a number of small homesteads like that of Virgil's old farmer, situated on "land that no one wanted" and yet abundantly productive of food, pleasure, and other goods. And especially in my younger days, I was used to hearing farmers of a certain kind say, "They may run me out, but they won't starve me out" or "I may get shot, but I'm not going to starve." Even now, if they cared, I think agricultural economists could find small farmers who have prospered, not by "getting big," but by practicing the ancient rules of thrift and subsistence, by accepting the limits of their small farms, and by knowing well the value of having a little land.

How do we come at the value of a little land? We do so, following this strand of agrarian thought, by reference to the value of *no* land. Agrarians value land because somewhere back in the history of their consciousness is the memory of being landless. This memory is implicit, in Virgil's poem, in the old farmer's happy acceptance of "an acre or two of land that no one wanted." If you have no land you have nothing: no food, no shelter, no warmth, no freedom, no life. If we remember this, we know that all economies begin to lie as soon as they assign a fixed value to land. People who have been landless know that the land is invaluable; it is worth everything. Preagricultural humans, of course, knew this too. And so, evidently, do the animals. It is a fearful thing to be without a "territory." Whatever the market may say, the worth of the land is what it always was: It is worth what food, clothing, shelter, and freedom are worth; it is worth what life is worth. This perception moved the settlers from the Old World into the New. Most of our American ancestors came here because they knew what it was to be landless; to be landless was to be threatened by want and also by enslavement. Coming here, they bore the ancestral memory of serfdom. Under feudalism, the few who owned the land owned also, by an inescapable political logic, the people who worked the land.

Thomas Jefferson, who knew all these things, obviously was thinking of them when he wrote in 1785 that "it is not too soon to provide by every possible means that as few as possible shall be without a little portion of land. The small landholders are the most precious part of a state. . . ." He was saying, two years before the adoption of our Constitution, that a democratic state and democratic liberties depend upon democratic ownership of the land. He was already anticipating and fearing the division of our people into settlers, the people who wanted "a little portion of land" as a home, and, virtually opposite to those, the consolidators and exploiters of the land and the land's wealth, who would not be restrained by what Jefferson called "the natural affection of the human mind." He wrote as he did in 1785 because he feared exactly the political theory that we now have: the idea that government exists to guarantee the right of the most wealthy to own or control the land without limit.

In any consideration of agrarianism, this issue of limitation is critical. Agrarian farmers see, accept, and live within their limits. They understand and agree to the proposition that there is "this much and no more." Everything that happens on an agrarian farm is determined or conditioned by the understanding that there is only so much land, so much water in the cistern, so much hay in the barn, so much corn in the crib, so much firewood in the shed, so much food in the cellar or freezer, so much strength in the back and arms—and no more. This is the understanding that induces thrift, family coherence, neighborliness, local economies. Within accepted limits, these virtues become necessities. The agrarian sense of abundance comes from the experienced possibility of frugality and renewal within limits.

This is exactly opposite to the industrial idea that abundance comes from the violation of limits by personal mobility, extractive machinery, long-distance transport, and scientific or technological breakthroughs. If we use up the good possibilities in this place, we will import goods from some other place, or we will go to some other place. If nature releases her wealth too slowly, we will take it by force. If we make the world too toxic for honeybees, some compound brain, Monsanto perhaps, will invent tiny robots that will fly about, pollinating flowers and making honey.

To be landless in an industrial society obviously is not at all times to be jobless and homeless. But the ability of the industrial economy to provide jobs and homes depends on prosperity, and on a very shaky kind of prosperity too. It depends on "growth" of the wrong things such as roads and dumps and poisons—on what Edward Abbey called "the ideology of the cancer cell"—and on greed with purchasing power. In the absence of growth, greed, and affluence, the dependents of an industrial economy too easily suffer the consequences of having no land: joblessness, homelessness, and want. This is not a theory. We have seen it happen.

I don't think that being landed necessarily means owning land. It does mean being connected to a home landscape from which one may live by the interactions of a local economy and without the routine intervention of governments, corporations, or charities.

In our time it is useless and probably wrong to suppose that a great many urban people ought to go out into the countryside and become homesteaders or farmers. But it is not useless or wrong to suppose that urban people have agricultural responsibilities that they should try to meet. And in fact this is happening. The agrarian population among us is growing, and by no means is it made up merely of some farmers and some country people. It includes urban gardeners, urban consumers who are buying food from local farmers, organizers of local food economies, consumers who have grown doubtful of the healthfulness, the trustworthiness, and the dependability of the corporate food system—people, in other words, who understand what it means to be landless.

Apologists for industrial agriculture rely on two arguments. In one of them, they say that the industrialization of agriculture, and its dominance by corporations, has been "inevitable." It has come about and it continues by the agency of economic and technological determinism. There has been simply nothing that anybody could do about it.

The other argument is that industrial agriculture has come about by choice, inspired by compassion and generosity. Seeing the shadow of mass starvation looming over the world, the food conglomerates, the machinery companies, the chemical companies, the seed companies, and the other suppliers of "purchased inputs" have done all that they have done in order to solve "the problem of hunger" and to "feed the world."

We need to notice, first, that these two arguments, often used and perhaps believed by the same people, exactly contradict each other. Second, though supposedly it has been imposed upon the world by economic and technological forces beyond human control, industrial agriculture has been pretty consistently devastating to nature, to farmers, and to rural communities, at the same time that it has been highly profitable to the agribusiness corporations, which have submitted not quite reluctantly to its "inevitability." And, third, tearful over human suffering as they always have been, the agribusiness corporations have maintained a religious faith in the profitability of their charity. They have instructed the world that it is better for people to buy food from the corporate global economy than to raise it for themselves. What is the proper solution to hunger? Not food from the local landscape, but industrial development. After decades of such innovative thought, hunger is still a worldwide calamity.

The primary question for the corporations, and so necessarily for us, is not how the world will be fed, but who will control the land, and therefore the wealth, of the world. If the world's people accept the industrial premises that favor bigness, centralization, and (for a few people) high profitability, then the corporations will control all of the world's land and all of its wealth. If, on the contrary, the world's people might again see the advantages of local economies, in which people live, so far as they are able to do so, from their home landscapes, and work patiently toward that end, eliminating waste and the cruelties of landlessness and homelessness, then I think they might reasonably hope to solve "the problem of hunger," and several other problems as well.

But do the people of the world, allured by TV, supermarkets, and big cars, or by dreams thereof, *want* to live from their home landscapes? *Could* they do so, if they wanted to? Those are hard questions, not readily answerable by anybody. Throughout the industrial decades, people have become increasingly and more numerously ignorant of the issues of land use, of food, clothing, and shelter. What would they do, and what *could* they do, if they were forced by war or some other calamity to live from their home landscapes?

It is a fact, well attested but little noticed, that our extensive, mobile, highly centralized system of industrial agriculture is extremely vulnerable to acts of terrorism. It will be hard to protect an agriculture of genetically impoverished monocultures that is entirely dependent on cheap petroleum and long-distance transportation. We know too that the great corporations, which grow and act so far beyond the restraint of "the natural affections of the human mind," are vulnerable to the natural depravities of the human mind, such as greed, arrogance, and fraud.

The agricultural industrialists like to say that their agrarian opponents are merely sentimental defenders of ways of farming that are hopelessly old-fashioned, justly dying out. Or they say that their opponents are the victims, as Richard Lewontin put it, of "a false nostalgia for a way of life that never existed." But these are not criticisms. They are insults.

For agrarians, the correct response is to stand confidently on our fundamental premise, which is both democratic and ecological: The land is a gift of immeasurable value. If it is a gift, then it is a gift to all the living in all time. To withhold it from some is finally to destroy it for all. For a few powerful people to own or control it all, or decide its fate, is wrong.

From that premise we go directly to the question that begins the agrarian agenda and is the discipline of all agrarian practice: What is the best way to use land? Agrarians know that this question necessarily has many answers, not just one. We are not asking what is the best way to farm everywhere in the world, or everywhere in the United States, or everywhere in Kentucky or Iowa. We are asking what is the best way to farm in each one of the world's numberless places, as defined by topography, soil type, climate, ecology, history, culture, and local need. And we know that the standard cannot be determined only by market demand or productivity or profitability or technological capability, or by any other single measure, however important it may be. The agrarian standard, inescapably, is local adaptation, which requires bringing local nature, local people, local economy, and local culture into a practical and enduring harmony.

Conservationist and Agrarian

I AM A CONSERVATIONIST and a farmer, a wilderness advocate and an agrarian. I am in favor of the world's wildness, not only because I like it, but also because I think it is necessary to the world's life and to our own. For the same reason, I want to preserve the natural health and integrity of the world's economic landscapes, which is to say that I want the world's farmers, ranchers, and foresters to live in stable, locally adapted, resource-preserving communities, and I want them to thrive.

One thing that this means is that I have spent my life on two losing sides. As long as I have been conscious, the great causes of agrarianism and conservation, despite local victories, have suffered an accumulation of losses, some of them probably irreparable—while the third side, that of the land-exploiting corporations, has appeared to grow ever richer. I say "appeared" because I think their wealth is illusory. Their capitalism is based, finally, not on the resources of nature, which it is recklessly destroying, but on fantasy. Not long ago I heard an economist say, "If the consumer ever stops living beyond his means, we'll have a recession." And so the two sides of nature and the rural communities are being defeated by a third side that will eventually be found to have defeated itself.

Perhaps in order to survive its inherent absurdity, the third side is asserting its power as never before: by its control of politics, of public education, and of the news media; by its dominance of science; and by biotechnology, which it is commercializing with unprecedented haste and aggression in order to control totally the world's land-using economies and its food supply. This massive ascendancy of corporate power over democratic process is probably the most ominous development since the end of World War II, and for the most part "the free world" seems to be regarding it as merely normal.

My sorrow in having been for so long on two losing sides has been compounded by knowing that those two sides have been in conflict, not only with their common enemy, the third side, but also, and by now almost conventionally, with each other. And I am further aggrieved in understanding that every-body on my two sides is deeply implicated in the sins and in the fate of the self-destructive third side.

As a part of my own effort to think better, I decided not long ago that I would not endorse any more wilderness pres-ervation projects that do not seek also to improve the health of the surrounding economic landscapes and human commu-nities. One of my reasons is that I don't think we can preserve either wildness or wilderness areas if we can't preserve the eco-nomic landscapes and the people who use them. This has put me into discomfort with some of my conservation friends, but that discomfort only balances the discomfort I feel when farm-ers or ranchers identify me as an "environmentalist," both be-cause I dislike the term and because I sympathize with farmers and ranchers.

Whatever its difficulties, my decision to cooperate no longer in the separation of the wild and the domestic has helped me to see more clearly the compatibility and even the coherence of my two allegiances. The dualism of domestic and wild is, after all, mostly false, and it is misleading. It has obscured for us the domesticity of the wild creatures. More important, it has obscured the absolute dependence of human domestic-ity upon the wildness that supports it and in fact permeates it. In suffering the now-common accusation that humans are "anthropocentric" (ugly word), we forget that the wild sheep and the wild wolves are respectively ovicentric and lupocen-tric. The world, we may say, is wild, and all the creatures are homemakers within it, practicing domesticity: mating, raising young, seeking food and comfort. Likewise, though the wild sheep and the farm-bred sheep are in some ways unlike in their domesticities, we forget too easily that if the "domestic" sheep become too unwild, as some occasionally do, they become uneconomic and useless: They have reproductive problems, conformation problems, and so on. Domesticity and wildness are in fact intimately connected. What is utterly alien to both is corporate industrialism—a dis-placed economic life that is without affection for the places where it is lived and without respect for the materials it uses.

The question we must deal with is not whether the domestic and the wild are separate or can be separated; it is how, in the human economy, their indissoluble and necessary connection can be properly maintained.

But to say that wildness and domesticity are not separate, and that we humans are to a large extent responsible for the proper maintenance of their relationship, is to come under a heavy responsibility to be practical. I have two thoroughly practical questions on my mind.

The first is: Why should conservationists have a positive interest in, for example, farming? There are lots of reasons, but the plainest is: Conservationists eat. To be interested in food but not in food production is clearly absurd. Urban conservationists may feel entitled to be unconcerned about food production because they are not farmers. But they can't be let off so easily, for they all are farming by proxy. They can eat only if land is farmed on their behalf by somebody somewhere in some fashion. If conservationists will attempt to resume responsibility for their need to eat, they will be led back fairly directly to all their previous concerns for the welfare of nature.

Do conservationists, then, wish to eat well or poorly? Would they like their food supply to be secure from one year to the next? Would they like their food to be free of poisons, antibiotics, alien genes, and other contaminants? Would they like a significant portion of it to be fresh? Would they like it to come to them at the lowest possible ecological cost? The answers, if responsibly given, will influence production, will influence land use, will determine the configuration and the health of landscapes.

If conservationists merely eat whatever the supermarket provides and the government allows, they are giving economic support to all-out industrial food production: to animal factories; to the depletion of soil, rivers, and aquifers; to crop monocultures and the consequent losses of biological and genetic diversity; to the pollution, toxicity, and overmedication that are the inevitable accompaniments of all-out industrial food production; to a food system based on long-distance transportation and the consequent waste of petroleum and the spread of pests and diseases; and to the division of the countryside into ever-larger farms and ever-larger fields receiving always less human affection and human care.

If, on the other hand, conservationists are willing to insist on having the best food, produced in the best way, as close to their homes as possible, and if they are willing to learn to judge the quality of food and food production, then they are going to give economic support to an entirely different kind of land use in an entirely different landscape. This landscape will have a higher ratio of caretakers to acres, of care to use. It will be at once more domestic and more wild than the industrial landscape. Can increasing the number of farms and farmers in an agricultural landscape enhance the quality of that landscape as wildlife habitat? Can it increase what we might call the wilderness value of that landscape? It *can* do so, and the determining factor would be diversity. Don't forget that we are talking about a landscape that is changing in response to an increase in local consumer demand for local food. Imagine a modern agricultural landscape devoted mainly to corn and soybeans and to animal factories. And then imagine its neighboring city developing a demand for good, locally grown food. To meet that demand, local farming would have to diversify.

If that demand is serious, if it is taken seriously, if it comes from informed and permanently committed consumers, if it promises the necessary economic support, then that radically oversimplified landscape will change. The crop monocultures and animal factories will give way to the mixed farming of plants and animals. Pastured flocks and herds of meat animals, dairy herds, and poultry flocks will return, requiring, of course, pastures and hayfields. If the urban consumers would extend their competent concern for the farming economy to include the forest economy and its diversity of products, that would improve the quality and care, and increase the acreage, of farm woodlands. And we should not forget the possibility that good farmers might, for their own instruction and pleasure, preserve patches of woodland unused. As the meadows and woodlands flourished in the landscape, so would the wild birds and animals. The acreages devoted to corn and soybeans, grown principally as livestock feed or as raw materials for industry, would diminish in favor of the fruits and vegetables required by human dinner tables.

As the acreage under perennial cover increased, soil erosion would decrease and the water-holding capacity of the soil would increase. Creeks and rivers would grow cleaner and their flow more constant. As farms diversified, they would tend to become smaller because complexity and work increase with diversity, and so the landscape would acquire more owners. As the number of farmers and the diversity of their farms increased, the toxicity of agriculture would decrease—insofar as agricultural chemicals are used to replace labor and to defray the biological costs of monoculture. As food production became decentralized, animal wastes would be dispersed, and would be absorbed and retained in the soil as nutrients rather than flowing away as waste and as pollutants. The details of such a transformation could be elaborated almost endlessly. To make short work of it here, we could just say that a dangerously oversimplified landscape would become healthfully complex, both economically and ecologically.

Moreover, since we are talking about a city that would be living in large measure from its local fields and forests, we are talking also about a local economy of decentralized, small, nonpolluting value-adding factories and shops that would be scaled to fit into the landscape with the least ecological or social disruption. And thus we can also credit to this economy an increase in independent small businesses, in self-employment, and a decrease in the combustible fuel needed for transportation and (I believe) for production.

Such an economy is technically possible, there can be no doubt of that; we have the necessary methods and equipment. The capacity of nature to accommodate, and even to cooperate in, such an economy is also undoubtable; we have the necessary historical examples. This is not, from nature's point of view, a pipe dream.

What *is* doubtable, or at least unproven, is the capacity of modern humans to choose, make, and maintain such an economy. For at least half a century we have taken for granted that the methods of farming could safely be determined by the mechanisms of industry, and that the economies of farming could safely be determined by the economic interests of industrial corporations. We are now running rapidly to the end of the possibility of that assumption. The social, ecological, and even the economic costs have become too great, and the costs are still increasing, all over the world.

Now we must try to envision an agriculture founded, not on mechanical principles, but on the principles of biology and ecology. Sir Albert Howard and Wes Jackson have argued at length for such a change of standards. If you want to farm sustainably, they have told us, then you have got to make your farming conform to the natural laws that govern the local ecosystem. You have got to farm with both plants and animals in as great a diversity as possible, you have got to conserve fertility, recycle wastes, keep the ground covered, and so on. Or, as J. Russell Smith put it seventy years ago, you have got to "fit the farming to the land"—not to the available technology or the market, as important as those considerations are, but to the land. It is necessary, in short, to maintain a proper connection between the domestic and the wild. The paramount standard by which the work is to be judged is the health of the place where the work is done.

But this is not a transformation that we can just drift into, as we drift in and out of fashions, and it is not one that we should wait to be forced into by large-scale ecological breakdown. It won't happen if a lot of people—consumers and producers, city people and country people, conservationists and land users—don't get together deliberately to make it happen.

Those are some of the reasons why conservationists should take an interest in farming and make common cause with good farmers. Now I must get on to the second of my practical questions.

Why should farmers be conservationists? Or maybe I had better ask why *are* good farmers conservationists? The farmer lives and works in the meeting place of nature and the human economy, the place where the need for conservation is most obvious and most urgent. Farmers either fit their farming to their farms, conform to the laws of nature, and keep the natural powers and services intact—or they do not. If they do not, then they increase the ecological deficit that is being charged to the future. (I had better admit that some farmers do increase the ecological deficit, but they are not the farmers I am talking about. I am not asking conservationists to support destructive ways of farming.)

Good farmers, who take seriously their duties as stewards of Creation and of their land's inheritors, contribute to the welfare of society in more ways than society usually acknowledges, or even knows. These farmers produce valuable goods, of course; but they also conserve soil, they conserve water, they conserve wildlife, they conserve open space, they conserve scenery.

All that is merely what farmers *ought* to do. But since our present society's first standard in all things is profit and it loves to dwell on "economic reality," I can't resist a glance at these good farmers in their economic circumstances, for these farmers will be poorly paid for the goods they produce, and for the services they render to conservation they will not be paid at all. Good farmers today may market products of high quality and perform well all the services I have listed, and *still* be unable to afford health insurance, and *still* find themselves mercilessly caricatured in the public media as rural simpletons, hicks, or rednecks. And then they hear the voices of the "economic realists": "Get big or get out. Sell out and go to town. Adapt or die." We have had fifty years of such realism in agriculture, and the result has been more and more large-scale monocultures and factory farms, with their ever larger social and ecological —and ultimately economic—costs.

Why do good farmers farm well for poor pay and work as good stewards of nature for no pay, many of them, moreover, having no hope that their farms will be farmed by their children (for the reasons given) or that they will be farmed by anybody?

Well, I was raised by farmers, have farmed myself, and have in turn raised two farmers—which suggests to me that I may know something about farmers, and also that I don't know very much. But over the years I along with a lot of other people have wondered, "Why do they do it?" Why do farmers farm, given their economic adversities on top of the many frustrations and difficulties normal to farming? And always the answer is: "Love. They must do it for love." Farmers farm for the love of farming. They love to watch and nurture the growth of plants. They love to live in the presence of animals. They love to work outdoors. They love the weather, maybe even when it is making them miserable. They love to live where they work and to work where they live. If the scale of their farming is small enough, they like to work in the company of their children and with the help of their children. They love the measure of independence that farm life can still provide. I have an idea that a lot of farmers have gone to a lot of trouble merely to be self-employed, to live at least a part of their lives without a boss.

And so the first thing farmers as conservationists must try to conserve is their love of farming and their love of independence. Of course they can conserve these things only by handing them down, by passing them on to their children, or to *somebody's* children. Perhaps the most urgent task for all of us who want to eat well and to keep eating is to encourage farm-raised children to take up farming. And we must recognize that this only can be done economically. Farm children are not encouraged by watching their parents take their products to market only to have them stolen at prices less than the cost of production.

But farmers obviously are responsible for conserving much more than agrarian skills and attitudes. I have already told why farmers should be, as much as any conservationists, conservers of the wildness of the world—and that is their inescapable dependence on nature. Good farmers, I believe, recognize a difference that is fundamental between what is natural and what is man-made. They know that if you treat a farm as a factory and living creatures as machines, or if you tolerate the idea of "engineering" organisms, then you are on your way to something destructive and, sooner or later, too expensive. To treat creatures as machines is an error with large practical implications.

Good farmers know too that nature can be an economic ally. Natural fertility is cheaper, often in the short run, always in the long run, than purchased fertility. Natural health, inbred and nurtured, is cheaper than pharmaceuticals and chemicals. Solar energy—if you know how to capture and use it: in grass, say, and the bodies of animals—is cheaper than petroleum. The highly industrialized factory farm is entirely dependent on "purchased inputs." The agrarian farm, well integrated into the natural systems that support it, runs to an economically significant extent on resources and supplies that are free.

It is now commonly assumed that when humans took to agriculture they gave up hunting and gathering. But hunting and gathering remained until recently an integral and lively part of my own region's traditional farming life. People hunted for wild game; they fished the ponds and streams; they gathered wild greens in the spring, hickory nuts and walnuts in the fall; they picked wild berries and other fruits; they prospected for wild honey. Some of the most memorable, and least regrettable, nights of my own youth were spent in coon hunting with farmers. There is no denying that these activities contributed to the economy of farm households, but a further fact is that they were pleasures; they were wilderness pleasures, not greatly different from the pleasures pursued by conservationists and wilderness lovers. As I was always aware, my friends the coon hunters were not motivated just by the wish to tree coons and listen to hounds and listen to each other, all of which were sufficiently attractive; they were coon hunters also because they wanted to be afoot in the woods at night. Most of the farmers I have known, and certainly the most interesting ones, have had the capacity to ramble about outdoors for the mere happiness of it, alert to the doings of the creatures, amused by the sight of a fox catching grasshoppers, or by the puzzle of wild tracks in the snow.

As the countryside has depopulated and the remaining farmers have come under greater stress, these wilderness pleasures have fallen away. But they have not yet been altogether abandoned; they represent something probably essential to the character of the best farming, and they should be remembered and revived.

Those, then, are some reasons why good farmers are conservationists, and why all farmers ought to be.

What I have been trying to do is to define a congruity or community of interest between farmers and conservationists who are not farmers. To name the interests that these two groups have in common, and to observe, as I did at the beginning, that they also have common enemies, is to raise a question that is becoming increasingly urgent: Why don't the two groups publicly and forcefully agree on the things they agree on, and make an effort to cooperate? I don't mean to belittle their disagreements, which I acknowledge to be important. Nevertheless, cooperation is now necessary, and it is possible. If Kentucky tobacco farmers can meet with antismoking groups, draw up a set of "core principles" to which they all agree, and then support those principles, something of the sort surely could happen between conservationists and certain land-using enterprises: family farms and ranches, small-scale, locally owned forestry and forest products industries, and perhaps others. Something of the sort, in fact, is beginning to happen, but so far the efforts are too small and too scattered. The larger organizations on both sides need to take an interest and get involved.

If these two sides, which need to cooperate, have so far been at odds, what is the problem? The problem, I think, is economic. The small land users, on the one hand, are struggling so hard to survive in an economy controlled by the corporations that they are distracted from their own economy's actual basis in nature. They also have not paid enough attention to the difference between their always threatened local economies and the apparently thriving corporate economy that is exploiting them.

On the other hand, the mostly urban conservationists, who mostly are ignorant of the economic adversities of, say, family-scale farming or ranching, have paid far too little attention to the connection between *their* economic life and the despoliation of nature. They have trouble seeing that the bad farming and forestry practices that they oppose as conservationists are done on their behalf, and with their consent implied in the economic proxies they have given as consumers.

These clearly are serious problems. Both of them indicate that the industrial economy is not a true description of economic reality, and moreover that this economy has been wonderfully successful in getting its falsehoods believed. Too many land users and too many conservationists seem to have accepted the doctrine that the availability of goods is determined by the availability of cash, or credit, and by the market. In other words, they have accepted the idea always implicit in the arguments of the land-exploiting corporations: that there can be, and that there is, a safe disconnection between economy and ecology, between human domesticity and the wild world. Industrializing farmers have too readily assumed that the nature of their land could safely be subordinated to the capability of their technology, and that conservation could safely be left to conservationists. Conservationists have too readily assumed that the integrity of the natural world could be preserved mainly by preserving tracts of wilderness, and that the nature and nurture of the economic landscapes could safely be left to agribusiness, the timber industry, debt-ridden farmers and ranchers, and migrant laborers.

To me, it appears that these two sides are as divided as they are because each is clinging to its own version of a common economic error. How can this be corrected? I don't think it can be, so long as each of the two sides remains closed up in its own conversation. I think the two sides need to enter into *one* conversation. They have got to talk to one another. Conservationists have got to know and deal competently with the methods and economics of land use. Land users have got to recognize the urgency, even the economic urgency, of the requirements of conservation.

Failing this, these two sides will simply concede an easy victory to their common enemy, the third side, the corporate totalitarianism which is now rapidly consolidating as "the global economy" and which will utterly dominate both the natural world and its human communities.

THE WAY OF IGNORANCE

Secrecy vs. Rights

O N JANUARY 12, 2004, according to David Stout of the *New York Times*, the Supreme Court refused to reconsider a lower court's ruling "that the Justice Department was within its rights in refusing to identify more than 700 people, most of those Arabs or Muslims, arrested for immigration violations in connection with the attacks" of September 11, 2001. Mr. Stout usefully reduced the case to its gist by writing that it "pitted two fundamental values against each other—the right of the public to know details of how its government operates and the government's need to keep some information secret to protect national security."

The Supreme Court thus decided, in effect, to support the Justice Department's policy of holding prisoners in secret for the sake of national security, but it did not resolve the conflict between "the right of the public to know" and "the government's need" for secrecy.

The first thing we notice, of course, is that secret imprisonment necessarily denies to the prisoner any semblance of due process. But more is involved than that, and I want to try to say what more is involved.

The rights of the people are openly declared (not granted, but affirmed) in the founding documents of our nation. These rights were not understood as given, and therefore retractable, by the government at its discretion; they were understood, rather, as entitlements originating in "the laws of nature and of nature's God." The government's need for secrecy, by contrast, is a need that can be defined only by the government, and only in secret.

A government's wish to rule in secret on its own initiative and authority is perfectly understandable; this is a merely human weakness. *Of course* government officials would like to keep some information secret for reasons of national security, as well as for many other reasons that may readily be imagined. It is nevertheless true that a government's wish to govern in secret is the same as the wish to govern tyrannically, as we are shown by the secret imprisonment of the seven hundred.

Perhaps it is always the tendency of those in power to wish to rule autocratically, and this is what our founders feared and provided against. Aside from any issue of faith or theory, there is a practical reason for ascribing human rights to "the laws of nature and of nature's God." Rights that originate beyond and above governmental power cannot justly be abridged or revoked by a government.

In the United States the major political dialogue has been about rights: Can the fundamental human rights, set forth in the Declaration of Independence and the Constitution as belonging equally to all, be justifiably withheld from some? Over the course of our history until now, we have decided that they cannot be. We have not decided this without reason. The most persuasive argument of the Civil Rights Movement, for example, was that you cannot withhold civil rights from some people and still confidently guarantee the same rights to others; the denial turned against one group at one time can, at another time, be turned against another group. Injustices condoned against a native racial minority may with the same impunity be used against Arab and Muslim immigrants, or against dissident white protestants. Governments, the Declaration says, derive "their just powers from the consent of the governed." If the governed accept or allow a government that is arbitrary, self-authorized, and self-justified, then, whatever may be the fate of national security, mere citizens will no longer be secure in their rights.

There is no reason, now or ever, to make light of what we are now calling "national security." People want, naturally enough, "to be secure in their persons, houses, papers, and effects," not only against "unreasonable searches and seizures," but also against violence, whether domestic or foreign in origin. That is why we threw off the arbitrary rule of England and instituted a government of our own; we wished to be secure, which is to say that we wished not to be governed except by our own consent.

We may say then that national security has been our concern from the beginning, and that from the beginning the designated purpose of the nation was to secure the rights to which its citizens individually were entitled by the laws of nature and of nature's God. The two kinds of security were understood as one. If the security of the nation ceases to imply the security of all its inhabitants in their God-given rights, then, according to the Declaration of Independence, that nation's government will have delegitimized itself and should be replaced. That is not the wild idea of somebody in the current crop of left-wingers or right-wingers, but merely what the Declaration says, and we have been living with it unobjectingly for going on 228 years.

People's wish to be safe is undoubtedly one of the paramount concerns of politics and government. We want to be safe because we have perceived accurately that we live under threat of many dangers. We expect, rightly, that the government should give us reasonable protections against at least some of those dangers, especially the ones against which we can only be protected collectively. And we have to reckon with the likelihood of circumstances in which these protections may be supplied only selectively and incompletely.

The Bill of Rights was written in anticipation both of the tendency of government to usurp the rights of individuals and of circumstances in which those rights will be hard to preserve. It would always be reasonable to foresee times of stress between the government's obligation to safeguard the lives of people and its obligation to respect their rights. This is the reason for appending a Bill of Rights to the Constitution.

However, the Bill of Rights insists that the rights of individual persons must be unfailingly respected in all circumstances excepting only one, which is set forth in Article V: "No person shall be held to answer for a capital, or otherwise infamous crime, unless on a presentment or indictment of a grand jury, except in cases arising in the land or naval forces, or in the militia, when in actual service in time of war or public danger . . ." This exception is made, apparently, to validate military courts. But Article V expressly disdains to limit its application merely to citizens; it says "No person," not "No citizen." And having stated the exception, it returns promptly to the inclusive language it began with: "nor shall any person . . . be deprived of life, liberty, or property, without due process of law . . ."

It is clear, then, that secret imprisonment by the government of *any person*, citizen or immigrant or alien or enemy, is necessarily a denial of due process of law, insofar as no person can at the same time be secretly imprisoned and publicly tried. (We now have only the government's word that the number of prisoners is, or was, seven hundred.)

A further point is somewhat harder to state, but I think it is no less obvious. Terrorism, against which the government without formally declaring war says we are "at war," at the same time that it seeks to circumvent the legal conventions of war, has certainly precipitated a time of public danger. How badly frightened the general public may be by this state of affairs is a question hard to answer. But the federal government and the courts have given evidence that they, as the terrorists intended, are badly frightened. They are so badly frightened as to believe that they have no choice but to sacrifice the rights of persons in deference to the government's need for secrecy. But it is an error to believe that these two "fundamental values" can somehow be justly "balanced" by the government or the courts, or that the people can judge responsibly between their rights, which they can easily know, and a proclaimed "need," which the government so far forbids them to know. The Constitution, anyhow, does not provide for its own suspension by the fearful in a time of war and public danger.

Contempt for Small Places

NEWSPAPER EDITORIALS deplore such human-caused degradations of the oceans as the Gulf of Mexico's "dead zone," and reporters describe practices like "mountain removal" mining in eastern Kentucky. Some day we may finally understand the connections.

The health of the oceans depends on the health of rivers; the health of rivers depends on the health of small streams; the health of small streams depends on the health of their watersheds. The health of the water is exactly the same as the health of the land; the health of small places is exactly the same as the health of large places. As we know, disease is hard to confine. Because natural law is in force everywhere, infections move.

We cannot immunize the continents and the oceans against our contempt for small places and small streams. Small destructions add up, and finally they are understood collectively as large destructions. Excessive nutrient runoff from farms and animal factories in the Mississippi watershed has caused, in the Gulf of Mexico, a hypoxic or "dead zone" of five or six thousand square miles. In forty-odd years, strip mining in the Appalachian coal fields, culminating in mountain removal, has gone far toward the destruction of a whole region, with untold damage to the region's people, to watersheds, and to the waters downstream.

There is not a more exemplary history of our contempt for small places than that of Eastern Kentucky coal mining, which has enriched many absentee corporate shareholders and left the region impoverished and defaced. Coal industry representatives are now defending mountain removal—and its attendant damage to forests, streams, wells, dwellings, roads, and community life—by saying that in "10, 15, 20 years" the land will be restored, and that such mining has "created the [level] land" needed for further industrial development.

But when you remove a mountain you also remove the topsoil and the forest, and you do immeasurable violence to the ecosystem and the watershed. These things are not to be restored in ten or twenty years, or in ten or twenty hundred

361

years. As for the manufacture of level places for industrial development, the supply has already far exceeded any foreseeable demand. And the devastation continues.

The contradictions in the state's effort "to balance the competing interests" were stated as follows by Ewell Balltrip, director of the Kentucky Appalachian Commission: "If you don't have mining, you don't have an economy, and if you don't have an economy you don't have a way for the people to live. But if you don't have environmental quality, you won't create the kind of place where people want to live."

Yes. And if the clearly foreseeable result is a region of flat industrial sites where nobody wants to live, we need a better economy.

Compromise, Hell!

W<small>E ARE DESTROYING OUR COUNTRY</small>—I mean our country itself, our land. This is a terrible thing to know, but it is not a reason for despair unless we decide to continue the destruction. If we decide to continue the destruction, that will not be because we have no other choice. This destruction is not necessary. It is not inevitable, except that by our submissiveness we make it so.

We Americans are not usually thought to be a submissive people, but of course we are. Why else would we allow our country to be destroyed? Why else would we be rewarding its destroyers? Why else would we all—by proxies we have given to greedy corporations and corrupt politicians—be participating in its destruction? Most of us are still too sane to piss in our own cistern, but we allow others to do so, and we reward them for it. We reward them so well, in fact, that those who piss in our cistern are wealthier than the rest of us.

How do we submit? By not being radical enough. Or by not being thorough enough, which is the same thing.

Since the beginning of the conservation effort in our country, conservationists have too often believed that we could protect the land without protecting the people. This has begun to change, but for a while yet we will have to reckon with the old assumption that we can preserve the natural world by protecting wilderness areas while we neglect or destroy the economic landscapes—the farms and ranches and working forests—and the people who use them. That assumption is understandable in view of the worsening threats to wilderness areas, but it is wrong. If conservationists hope to save even the wild lands and wild creatures, they are going to have to address issues of economy, which is to say issues of the health of the landscapes and the towns and cities where we do our work, and the quality of that work, and the well-being of the people who do the work.

Governments seem to be making the opposite error, believing that the people can be adequately protected without protecting the land. And here I am not talking about parties or party doctrines, but about the dominant political assumption.

Sooner or later, governments will have to recognize that if the land does not prosper, nothing else can prosper for very long. We can have no industry or trade or wealth or security if we don't uphold the health of the land and the people and the people's work.

It is merely a fact that the land, here and everywhere, is suffering. We have the "dead zone" in the Gulf of Mexico and undrinkable water to attest to the toxicity of our agriculture. We know that we are carelessly and wastefully logging our forests. We know that soil erosion, air and water pollution, urban sprawl, the proliferation of highways and garbage are making our lives always less pleasant, less healthful, less sustainable, and our dwelling places more ugly.

Nearly forty years ago my state of Kentucky, like other coal-producing states, began an effort to regulate strip mining. While that effort has continued, and has imposed certain requirements of "reclamation," strip mining has become steadily more destructive of the land and the land's future. We are now permitting the destruction of entire mountains and entire watersheds. No war, so far, has done such extensive or such permanent damage. If we know that coal is an exhaustible resource, whereas the forests over it are with proper use inexhaustible, and that strip mining destroys the forest virtually forever, how can we permit this destruction? If we honor at all that fragile creature the topsoil, so long in the making, so miraculously made, so indispensable to all life, how can we destroy it? If we believe, as so many of us profess to do, that the Earth is God's property and is full of His glory, how can we do harm to any part of it?

In Kentucky, as in other unfortunate states, and again at great public cost, we have allowed—in fact we have officially encouraged—the establishment of the confined animal-feeding industry, which exploits and abuses everything involved: the land, its people, the animals, and the consumers. If we love our country, as so many of us profess to do, how can we so desecrate it?

But the economic damage is not confined just to our farms and forests. For the sake of "job creation" in Kentucky, and in other backward states, we have lavished public money on corporations that come in and stay only so long as they can exploit people here more cheaply than elsewhere. The general purpose of the present economy is to exploit, not to foster or conserve.

Look carefully, if you doubt me, at the centers of the larger towns in virtually every part of our country. You will find that they are economically dead or dying. Good buildings that used to house needful, useful, locally owned small businesses of all kinds are now empty or have evolved into junk stores or antique shops. But look at the houses, the churches, the commercial buildings, the courthouse, and you will see that more often than not they are comely and well made. And then go look at the corporate outskirts: the chain stores, the fast-food joints, the food-and-fuel stores that no longer can be called service stations, the motels. Try to find something comely or well made there.

What is the difference? The difference is that the old town centers were built by people who were proud of their place and who realized a particular value in living there. The old buildings look good because they were built by people who respected themselves and wanted the respect of their neighbors. The corporate outskirts, on the contrary, were built by people who manifestly take no pride in the place, see no value in lives lived there, and recognize no neighbors. The only value they see in the place is the money that can be siphoned out of it to more fortunate places—that is, to the wealthier suburbs of the larger cities.

Can we actually suppose that we are wasting, polluting, and making ugly this beautiful land for the sake of patriotism and the love of God? Perhaps some of us would like to think so, but in fact this destruction is taking place because we have allowed ourselves to believe, and to live, a mated pair of economic lies: that nothing has a value that is not assigned to it by the market, and that the economic life of our communities can safely be handed over to the great corporations.

We citizens have a large responsibility for our delusion and our destructiveness, and I don't want to minimize that. But I don't want to minimize, either, the large responsibility that is borne by government.

It is commonly understood that governments are instituted to provide certain protections that citizens individually cannot provide for themselves. But governments have tended to assume that this responsibility can be fulfilled mainly by the police and the military services. They have used their regulatory powers reluctantly and often poorly. Our governments have only occasionally recognized the need of land and people to be protected against economic violence. It is true that economic violence is not always as swift, and is rarely as bloody, as the violence of war, but it can be devastating nonetheless. Acts of economic aggression can destroy a landscape or a community or the center of a town or city, and they routinely do so.

Such damage is justified by its corporate perpetrators and their political abettors in the name of the "free market" and "free enterprise," but this is a freedom that makes greed the dominant economic virtue, and it destroys the freedom of other people along with their communities and livelihoods. There are such things as economic weapons of massive destruction. We have allowed them to be used against us, not just by public submission and regulatory malfeasance, but also by public subsidies, incentives, and sufferances impossible to justify.

We have failed to acknowledge this threat and to act in our own defense. As a result, our once-beautiful and bountiful countryside has long been a colony of the coal, timber, and agribusiness corporations, yielding an immense wealth of energy and raw materials at an immense cost to our land and our land's people. Because of that failure also, our towns and cities have been gutted by the likes of Wal-Mart, which have had the permitted luxury of destroying locally owned small businesses by means of volume discounts.

Because as individuals or even as communities we cannot protect ourselves against these aggressions, we need our state and national governments to protect us. As the poor deserve as much justice from our courts as the rich, so the small farmer and the small merchant deserve the same economic justice, the same freedom in the market, as big farmers and chain stores. They should not suffer ruin merely because their rich competitors can afford (for a while) to undersell them.

Furthermore, to permit the smaller enterprises always to be ruined by false advantages, either at home or in the global economy, is ultimately to destroy local, regional, and even national capabilities of producing vital supplies such as food and textiles. It is impossible to understand, let alone justify, a government's willingness to allow the human sources of necessary goods to be destroyed by the "freedom" of this corporate anarchy. It is equally impossible to understand how a government can permit, and even subsidize, the destruction of the land or of the land's productivity. Somehow we have lost or discarded any controlling sense of the interdependence of the Earth and the human capacity to use it well. The governmental obligation to protect these economic resources, inseparably human and natural, is the same as the obligation to protect us from hunger or from foreign invaders. In result, there is no difference between a domestic threat to the sources of our life and a foreign one.

It appears that we have fallen into the habit of compromising on issues that should not, and in fact cannot, be compromised. I have an idea that a large number of us, including even a large number of politicians, believe that it is wrong to destroy the Earth. But we have powerful political opponents who insist that an Earth-destroying economy is justified by freedom and profit. And so we compromise by agreeing to permit the destruction only of parts of the Earth, or to permit the Earth to be destroyed a little at a time—like the famous three-legged pig that was too well loved to be eaten all at once.

The logic of this sort of compromising is clear, and it is clearly fatal. If we continue to be economically dependent on destroying parts of the Earth, then eventually we will destroy it all.

So long a complaint accumulates a debt to hope, and I would like to end with hope. To do so I need only repeat something I said at the beginning: Our destructiveness has not been, and it is not, inevitable. People who use that excuse are morally incompetent, they are cowardly, and they are lazy. Humans don't have to live by destroying the sources of their life. People can change; they can learn to do better. All of us, regardless of party, can be moved by love of our land to rise above the greed and contempt of our land's exploiters. This of course leads to practical problems, and I will offer a short list of practical suggestions.

We have got to learn better to respect ourselves and our dwelling places. We need to quit thinking of rural America as a colony. Too much of the economic history of our land has been that of the export of fuel, food, and raw materials that have been destructively and too cheaply produced. We must reaffirm the economic value of good stewardship and good work. For that we will need better accounting than we have had so far.

We need to reconsider the idea of solving our economic problems by "bringing in industry." Every state government appears to be scheming to lure in a large corporation from somewhere else by "tax incentives" and other squanderings of the people's money. We ought to suspend that practice until we are sure that in every state we have made the most and the best of what is already there. We need to build the local economies of our communities and regions by adding value to local products and marketing them locally before we seek markets elsewhere.

We need to confront honestly the issue of scale. Bigness has a charm and a drama that are seductive, especially to politicians and financiers; but bigness promotes greed, indifference, and damage, and often bigness is not necessary. You may need a large corporation to run an airline or to manufacture cars, but you don't need a large corporation to raise a chicken or a hog. You don't need a large corporation to process local food or local timber and market it locally.

And, finally, we need to give an absolute priority to caring well for our land—for every bit of it. There should be no compromise with the destruction of the land or of anything else that we cannot replace. We have been too tolerant of politicians who, entrusted with our country's defense, become the agents of our country's destroyers, compromising on its ruin.

And so I will end this by quoting my fellow Kentuckian, a great patriot and an indomitable foe of strip mining, the late Joe Begley of Blackey: "Compromise, hell!"

Charlie Fisher

I DON'T IMAGINE Charlie Fisher told me everything he has done, but in the day and a half I spent with him I did find out that he was raised on a truck farm, that for a while he rode bulls and exhibited a trick horse on the rodeo circuit, that as a young man he worked for a dairyman, and that later he had a dairy farm of his own. His interest in logging and in working horses began while he was a hired hand in the dairy. In the winter, between milkings, he and his elderly employer spent their time in the woods at opposite ends of a crosscut saw— which, Charlie says, made him tireder than it made the old dairyman. They cut some big timber and dragged out the logs with horses. The old dairyman saw that Charlie liked working horses and was good at it. And so it was that he became both a teamster and a logger.

Though he tried other employment, those two early interests stayed with him, and he has spent many years in logging with horses. There were times when he worked alone, cutting and skidding out the logs by himself. Later, his son, David, began to work with him, skidding out the logs while Charlie cut. David, who is now twenty-two, virtually grew up in the woods. He started skidding logs with a team when he was nine, and he is still working with his father, as both teamster and log cutter.

Nine years ago, near Andover in northeast Ohio, Charlie Fisher and Jeff Green formed a company, Valley Veneer, which involves both a logging operation and a sawmill. Charlie buys the standing timber, marks the trees that are to be cut, and supervises the logging crews, while Jeff keeps things going at the mill and markets the lumber.

The mill employs eight or nine hands, and it saws three million board feet a year. It provides a local market for local timber. This obviously is good for the economy of the Andover neighborhood, but it also is good for the forest. By establishing the mill, Charlie and Jeff have invested in the neighborhood and formed a permanent connection to it, and so they have an inescapable interest in preserving the productivity of the local forest. Thus a local forest economy, if it is complex enough, will tend almost naturally to act as a conserver of the local forest ecosystem. Valley Veneer, according to Charlie and Jeff, has been warmly received into the neighborhood. The company deals with the only locally-owned bank in the area. The bankers have been not only cooperative but also friendly, at times offering more help than Charlie and Jeff asked for.

The mill yard is the neatest I have ever seen. The logs are sorted and ricked according to species. Veneer logs are laid down separately with one end resting on a pole, so that they can be readily examined by buyers. The mill crew is skillful in salvaging good lumber from damaged or inferior trees. This is extremely important, as is Jeff's marketing of lumber from inferior species such as soft maple, for it means that the cutting in the woods is never limited to the best trees. Charlie marks the trees, knowing that whatever the woodland can properly yield—soft maple or fine furniture-quality cherry or trees damaged by disease or wind—can be sawed into boards and sold. The mill seemed to me an extraordinarily efficient place, where nothing of value is wasted. Twenty percent of the slabs are sold for firewood; the rest go to the chipper and are used for pulp. The sawdust is sold to farmers, who use it as bedding for animals.

The woods operation—Charlie's end of the business— consists of three logging crews, each made up of one log cutter and two teamsters. Each of the teamsters works two horses on a logging cart or "logging arch." And so Charlie routinely employs nine men and twelve horses. At times, the cutter also will do some skidding, and this increases the number of teams in use. The three crews will usually be at work at three different sites.

Mostly they log small, privately-owned woodlots within a radius of forty or fifty miles. Charlie recently counted up and found that he had logged 366 different tracts of timber in the last three years. And there are certain advantages to working on this scale. In a horse logging operation, it is best to limit the skidding distance to five or six hundred feet, though Charlie says they sometimes increase it to a thousand, and they can go somewhat farther in winter when snow or freezing weather reduces the friction. Big tracts, however, involve longer distances, and eventually it becomes necessary either to build a road for the truck or to use a bulldozer to move the logs from where the teamsters yard them in the woods to a second yarding place accessible from the highway. For this purpose, in addition to a log truck equipped with a hydraulic loading boom, Valley Veneer owns two bulldozers, one equipped with a fork, one with a blade, and both with winches. Even so, about 98 percent of the logs are moved with horses.

The logging crews work the year round and in all weather except pouring rain. The teamsters, who furnish their own teams and equipment, receive forty dollars per thousand board feet. Two of his teamsters, Charlie says, make more than thirty thousand dollars a year each.

The logging arch, in comparison to a mechanical skidder, is a very forthright piece of equipment. Like the forecart that is widely used for fieldwork, it is simply a way to provide a drawbar for a team of horses. There are a number of differences in design, but the major one is that the logging arch's drawbar is welded on edge-up and has slots instead of holes. The slots are made so as to catch and hold the links of a log chain. Each cart carries an eighteen-foot chain with a grab hook at each end. Four metal hooks (which Charlie calls "log grabs," but which are also called "J-hooks" or "logging dogs") are linked to rings and strung on the chain, thus permitting the cart to draw as many as four logs at a time. The chain can also be used at full-length if necessary to reach a hard-to-get-to-log. Larger logs require the use of tongs, which the teamsters also carry with them, or two grabs driven into the log on opposite sides. The carts are equipped also with a cant hook and a "skipper" with which to drive the grabs into the log and knock them out again.

The slotted drawbar permits the chain to be handily readjusted as the horses work a log into position for skidding. When the log is ready to go, it is chained as closely as possible to the drawbar, so that when the horses tighten, the fore end of the log is raised off the ground. This is the major efficiency of the logging arch: By thus raising the log, the arch both keeps it from digging and reduces its friction against the ground by more than half.

We watched a team drag out a twelve-foot log containing about 330 board feet. They were well-loaded but were not straining. Charlie says that a team can handle up to five or six hundred board feet. For bigger logs, they use an additional team or a bulldozer. A good teamster can skid 3000 to 3500 board feet a day in small logs. The trick, Charlie says, is to know what your horses can do, and then see that they do that much on every pull. Overload, and you're resting too much. Underload, and you're wasting energy and time. The important thing is to keep loaded and keep moving.

Charlie Fisher is a man of long experience in the woods and extensive knowledge of the timber business and of logging technology. He has no prejudice against mechanical equipment as such, but uses it readily according to need; for a time, during his thirties, he used mechanical skidders. That this man greatly prefers horses for use in the woods is therefore of considerable interest. I asked him to explain.

His first reason, and the most important, is one I'd heard before from draft horsemen: "I've always liked horses." Charlie and David are clearly the sort of men who can't quite live without horses. Between them, they own six excellent, very large Belgian geldings and two Belgian mares. Charlie, as he explained, owns three and a half horses, and David four and a half. The two halves, fortunately, belong to the same horse, which Charlie and David own in partnership. Charlie has long been an enthusiastic participant in pulling contests, and David has followed in his father's footsteps in the arena as in the woods. Last season, David participated in twenty-three contests and Charlie in five, which for him was many fewer than usual. Charlie and his wife, Becky, showed us several shelves crowded with trophies, many of which were David's. It looked to me like they are going to need more shelves. Charlie and Becky are very proud of David, who is an accomplished logger and horseman. David, Charlie says, is an exceptionally quiet hand with a team—unlike Charlie, who confessed, "I holler." Since they would have the horses anyhow, Charlie said, they might as well put them to work in the woods, which keeps them fit and allows them to earn their keep.

Charlie's second reason for using horses in the woods, almost as important as the first, is that he likes the woods, and horses leave the woods in better condition than a skidder. A team and a logging arch require a much narrower roadway than a skidder; unlike a skidder, they don't bark trees; and they leave their skidding trails far less deeply rutted. "The horse," Charlie says, "will always be the answer to good logging in a woods."

A third attractive feature of the horse economy in the woods is that the horse logger both earns and spends his money in the local community, whereas the mechanical skidder siphons money away from the community and into the hands of large corporate suppliers. Moreover, the horse logger's kinder treatment of the woods will, in the long run, yield an economic benefit.

And, finally, horses work far more cheaply and cost far less than a skidder, thus requiring fewer trees to be cut per acre, and so permitting the horse logger to be more selective and conservative.

(Another issue involved in the use of horses for work is that of energy efficiency. Legs are more efficient than wheels over rough ground—something that will quickly be apparent to you if you try riding a bicycle over a plowed field.)

Well ahead of the logging crews, Charlie goes into the woods to mark the trees that are to be cut. Except when he is working for a "developer" who is going to clear the land, Charlie never buys or marks trees with the idea of taking every one that is marketable. His purpose is to select a number of trees, often those that need cutting because they are diseased or damaged or otherwise inferior, which will provide a reasonable income to landowner and logger alike, without destroying the wood-making capacity of the forest. The point can best be understood by considering the difference between a year's growth added to a tree fourteen inches in diameter and that added to a tree four inches in diameter. Clear-cutting or any other kind of cutting that removes all the trees of any appreciable size radically reduces the wood-making capacity of the forest. After such a cutting, in Charlie's part of the country, it will be sixty to a hundred years before another cutting can be made. Of a clear-cut woodland that adjoined one of his own tracts, Charlie said, "In fifty years there still won't be a decent log in it."

Charlie does not believe that such practices are good for the forest or the people—or, ultimately, for the timber business. He stated his interest forthrightly in economic terms, but his is the right kind of economics: "I hope maybe there'll be trees here for my son to cut in ten or twenty years." If you don't overdo the cutting, he says, a woodland can yield a cash crop every ten to fifteen years. We looked at one tract of twenty acres on which Charlie had marked about 160 trees and written the owner a check for $23,000. Charlie described this as "a young piece of timber," and he said that it "definitely" could be logged again in ten years—at which time he could both take more and leave more good trees than he will take and leave at this cutting.

Owners of wooded land should consider carefully the economics of this twenty-acre tract. If it is selectively and carefully logged every ten years, as Charlie says it can be, then every acre will earn $1,150 every ten years, or $115 per year. And this comes to the landowner without expense or effort. (These particular figures, of course, apply only to this particular woodlot. Some tracts might be more productive, others less.)

We looked at marked woodlands, at woodlands presently being logged, and finally, at the end of the second day of our visit, at a woodland that one of Charlie's crews had logged three years ago. The last, a stand predominantly of hard and soft maples, provided convincing evidence of the good sense of Charlie's kind of forestry. Very few of the remaining trees had been damaged by trees felled during the logging. I saw not a single tree that had been barked by a skidded log. The skid trails had completely healed over; there was no sign of erosion. And, most striking, the woodland was still ecologically intact. It was still a diverse, uneven-aged stand of trees, many of which were over sixteen inches in diameter. We made a photograph of three trees, standing fairly close together, which varied in diameter from seventeen to twenty-one inches. After logging, the forest is still a forest, and it will go on making wood virtually without interruption or diminishment. It seems perfectly reasonable to think that, if several generations of owners were so inclined, this sort of forestry could eventually result in an "old growth" forest that would have produced a steady income for two hundred years.

I was impressed by a good many things during my visit with Charlie Fisher, but what impressed me most is the way that Charlie's kind of logging achieves a complex fairness or justice to the several interests that are involved: the woods, the land-owner, the timber company, the woods crews and their horses.

Charlie buys standing trees, and he marks every tree he buys. Within a fairly narrow margin of error, Charlie knows what he is buying, and the landowner knows what he is getting paid for. When Charlie goes in to mark the trees, he is thinking not just about what he will take, but also about what he will leave. He sees the forest as it is, and he sees the forest as it will be when the logging job is finished. I think he sees it too as it will be in ten or fifteen or twenty years, when David or another logger will return to it. By this long-term care, he serves the forest and the landowner as well as himself. As he marks the trees he is thinking also of the logging crew that will soon be there. He marks each tree that is to be cut with a slash of red paint. Sometimes, where he has seen a leaning deadfall or a dead limb or a flaw in the trunk, he paints an arrow above the slash, and this means "Look up!" The horses, like the men, are carefully borne in mind. Everywhere, the aim is to do the work in the best and the safest way.

Moreover, these are not competing interests, but seem rather to merge into one another. Thus one of Charlie's economic standards—"I hope maybe there'll be trees here for my son to cut in ten or twenty years"—becomes, in application, an eco-logical standard. And the ecological standard becomes, again, an economic standard as it proves to be good for business.

Most landowners, Charlie says, care how their woodlands are logged. Though they may need the income from their trees, they don't want to sacrifice the health or beauty of their woods in order to get it. Charlie's way of logging recom-mends itself to such people; he does not need to advertise. As we were driving away from his house on the morning of our second day, one of the neighbors waved us to a stop. This man makes his living selling firewood, and he had learned of two people who wanted their woodlands logged by a horse logger. That is the way business comes to him, Charlie said. Like other horse loggers, he has all the work he can do, and more. It has been ten years since he has had to hunt for wood-lots to log. He said, "Everybody else has buyers out running the roads, looking for timber." But he can't buy all that he is offered.

I don't know that I have ever met a man with more enthusiasms than Charlie Fisher. I have mentioned already his abounding interest in his family, in forestry, and in working and pulling horses, but I have neglected to say that he is also a coon hunter. This seems to me a most revealing detail. Here is a man who makes his living by walking the woods all day, and who then entertains himself by walking the woods at night.

He told me that he had a list of several things he had planned to do when he retired, but that now, at sixty-six, he is busier than ever.

"Well," I said, "you seem to be enjoying it."

"Oh," he said, "I *love* it!"

The Way of Ignorance

> In order to arrive at what you do not know
> You must go by a way which is the way of ignorance.
> T. S. Eliot, "East Coker"

OUR PURPOSE HERE is to worry about the predominance of the supposition, in a time of great technological power, that humans either know enough already, or can learn enough soon enough, to foresee and forestall any bad consequences of their use of that power. This supposition is typified by Richard Dawkins's assertion, in an open letter to the Prince of Wales, that "our brains . . . are big enough to see into the future and plot long-term consequences."

When we consider how often and how recently our most advanced experts have been wrong about the future, and how often the future has shown up sooner than expected with bad news about our past, Mr. Dawkins's assessment of our ability to know is revealed as a superstition of the most primitive sort. We recognize it also as our old friend hubris, ungodly ignorance disguised as godly arrogance. Ignorance plus arrogance plus greed sponsors "better living with chemistry," and produces the ozone hole and the dead zone in the Gulf of Mexico. A modern science (chemistry or nuclear physics or molecular biology) "applied" by ignorant arrogance resembles much too closely an automobile being driven by a six-year-old or a loaded pistol in the hands of a monkey. Arrogant ignorance promotes a global economy while ignoring the global exchange of pests and diseases that must inevitably accompany it. Arrogant ignorance makes war without a thought of peace.

We identify arrogant ignorance by its willingness to work on too big a scale, and thus to put too much at risk. It fails to foresee bad consequences not only because some of the consequences of all acts are inherently unforeseeable, but also because the arrogantly ignorant often are blinded by money invested; they cannot afford to foresee bad consequences.

*

377

Except to the arrogantly ignorant, ignorance is not a simple subject. It is perhaps as difficult for ignorance to be aware of itself as it is for awareness to be aware of itself. One can hardly begin to think about ignorance without seeing that it is available in several varieties, and so I will offer a brief taxonomy.

There is, to begin with, the kind of ignorance we may consider to be inherent. This is ignorance of all that we cannot know because of the kind of mind we have—which, I will note in passing, is neither a computer nor exclusively a brain, and which certainly is not omniscient. We cannot, for example, know the whole of which we and our minds are parts. The English poet and critic Kathleen Raine wrote that "we cannot imagine how the world might appear if we did not possess the groundwork of knowledge which we do possess; nor can we in the nature of things imagine how reality would appear in the light of knowledge which we do not possess."

A part of our inherent ignorance, and surely a most formidable encumbrance to those who presume to know the future, is our ignorance of the past. We know almost nothing of our history as it was actually lived. We know little of the lives even of our parents. We have forgotten almost everything that has happened to ourselves. The easy assumption that we have remembered the most important people and events and have preserved the most valuable evidence is immediately trumped by our inability to know what we have forgotten.

There are several other kinds of ignorance that are not inherent in our nature but come instead from weaknesses of character. Paramount among these is the willful ignorance that refuses to honor as knowledge anything not subject to empirical proof. We could just as well call it materialist ignorance. This ignorance rejects useful knowledge such as traditions of imagination and religion, and so it comes across as narrowmindedness. We have the materialist culture that afflicts us now because a world exclusively material is the kind of world most readily used and abused by the kind of mind the materialists think they have. To this kind of mind, there is no longer a legitimate wonder. Wonder has been replaced by a research agenda, which is still a world away from demonstrating the impropriety of wonder. The materialist conservationists need to tell us how a materialist culture can justify its contempt and destructiveness of material goods.

A related kind of ignorance, also self-induced, is moral ignorance, the invariable excuse of which is objectivity. One of the purposes of objectivity, in practice, is to avoid coming to a moral conclusion. Objectivity, considered a mark of great learning and the highest enlightenment, loves to identify itself by such pronouncements as the following: "You may be right, but on the other hand so may your opponent," or "Everything is relative," or "Whatever is happening is inevitable," or "Let me be the devil's advocate." (The part of devil's advocate is surely one of the most sought after in all the precincts of the modern intellect. Anywhere you go to speak in defense of something worthwhile, you are apt to encounter a smiling savant writhing in the estrus of objectivity: "Let me play the devil's advocate for a moment." As if the devil's point of view will not otherwise be adequately represented.)

There is also ignorance as false confidence, or polymathic ignorance. This is the ignorance of people who know "all about" history or its "long-term consequences" in the future. And this is closely akin to self-righteous ignorance, which is the failure to know oneself. Ignorance of one's self and confident knowledge of the past and future often are the same thing.

Fearful ignorance is the opposite of confident ignorance. People keep themselves ignorant for fear of the strange or the different or the unknown, for fear of disproof or of unpleasant or tragic knowledge, for fear of stirring up suspicion and opposition, or for fear of fear itself. A good example is the United States Department of Agriculture's panic-stricken monopoly of inadequate meat inspections. And there is the related ignorance that comes from laziness, which is the fear of effort and difficulty. Learning often is not fun, and this is well-known to all the ignorant except for a few "educators."

And finally there are for-profit ignorance, which is maintained by withholding knowledge, as in advertising, and for-power ignorance, which is maintained by government secrecy and public lies. ·

Kinds of ignorance (and there must be more than I have

named) may thus be sorted out. But having sorted them out, one must scramble them back together again by acknowledging that all of them can be at work in the same mind at the same time, and in my opinion they frequently are.

I may be talking too much at large here, but I am going to say that a list of kinds of ignorance comprises half a description of a human mind. The other half, then, would be supplied by a list of kinds of knowledge.

At the head of that list let us put the empirical or provable knowledge of the materialists. This is the knowledge of dead certainty or dead facts, some of which at least are undoubtedly valuable, undoubtedly useful, but at best this is static, smallish knowledge that always is what it always was, and it is rather dull. A fact may thrill us once, but not twice. Once available, it is easy game; we might call it sitting-duck knowledge. This knowledge becomes interesting again when it enters experience by way of use.

And so, as second, let us put knowledge as experience. This is useful knowledge, but it involves uncertainty and risk. How do you know if it is going to rain, or when an animal is going to bolt or attack? Because the event has not yet happened, there is no empirical answer; you may not have time to calculate the statistical probability even on the fastest computer. You will have to rely on experience, which will increase your chance of being right. But then you also may be wrong.

The experience of many people over a long time is traditional knowledge. This is the common knowledge of a culture, which it seems that few of us any longer have. To have a culture, mostly the same people have to live mostly in the same place for a long time. Traditional knowledge is knowledge that has been remembered or recorded, handed down, pondered, corrected, practiced, and refined over a long time.

A related kind of knowledge is made available by the religious traditions and is not otherwise available. If you premise the falsehood of such knowledge, as the materialists do, then of course you don't have it and your opinion of it is worthless.

There also are kinds of knowledge that seem to be more strictly inward. Instinct is inborn knowledge: how to suck, bite, and swallow; how to run away from danger instead of toward it. And perhaps the prepositions refer to knowledge that is more or less instinctive: up, down, in, out, etc.

Intuition is knowledge as recognition, a way of knowing without proof. We know the truth of the Book of Job by intuition.

What we call conscience is knowledge of the difference between right and wrong. Whether or not this is learned, most people have it, and they appear to get it early. Some of the worst malefactors and hypocrites have it in full; how else could they fake it so well? But we should remember that some worthy people have believed conscience to be innate, an "inner light."

Inspiration, I believe, is another kind of knowledge or way of knowing, though I don't know how this could be proved. One can say in support only that poets such as Homer, Dante, and Milton seriously believed in it, and that people do at times surpass themselves, performing better than all you know of them has led you to expect. Imagination, in the highest sense, is inspiration. Gifts arrive from sources that cannot be empirically located.

Sympathy gives us an intimate knowledge of other people and other creatures that can come in no other way. So does affection. The knowledge that comes by sympathy and affection is little noticed—the materialists, I assume, are unable to notice it—but in my opinion it cannot be overvalued.

Everybody who has done physical work or danced or played a game of skill is aware of the difference between knowing how and being able. This difference I would call bodily knowledge.

And finally, to be safe, we had better recognize that there is such a thing as counterfeit knowledge or plausible falsehood.

I would say that these taxonomies of mine are more or less reasonable; I certainly would not claim that they are scientific. My only assured claim is that any consideration of ignorance and knowledge ought to be at least as complex as this attempt of mine. We are a complex species—organisms surely, but also living souls—who are involved in a life-or-death negotiation, even more complex, with our earthly circumstances, which are complex beyond our ability to guess, let alone know. In dealing with those circumstances, in trying "to see into the future and plot long-term consequences," the human mind is

neither capacious enough nor exact nor dependable. We are
encumbered by an inherent ignorance perhaps not significantly
reducible, as well as by proclivities to ignorance of other kinds,
and our ways of knowing, though impressive within human
limits, have the power to lead us beyond our limits, beyond
foresight and precaution, and out of control.

What I have said so far characterizes the personal minds of
individual humans. But because of a certain kind of arrogant
ignorance, and because of the gigantic scale of work permitted
and even required by powerful technologies, we are not safe in
dealing merely with personal or human minds. We are obliged
to deal also with a kind of mind that I will call corporate, although it is also political and institutional. This is a mind that
is compound and abstract, materialist, reductionist, greedy,
and radically utilitarian. Assuming as some of us sometimes do
that two heads are better than one, it ought to be axiomatic
that the corporate mind is better than any personal mind, but
it can in fact be much worse—not least in its apparently limitless ability to cause problems that it cannot solve, and that
may be unsolvable. The corporate mind is remarkably narrow.
It claims to utilize only empirical knowledge—the preferred
term is "sound science," reducible ultimately to the "bottom
line" of profit or power—and because this rules out any explicit recourse to experience or tradition or any kind of inward
knowledge such as conscience, this mind is readily susceptible
to every kind of ignorance and is perhaps naturally predisposed
to counterfeit knowledge. It comes to its work equipped with
factual knowledge and perhaps also with knowledge skillfully
counterfeited, but without recourse to any of those knowl-
edges that enable us to deal appropriately with mystery or with
human limits. It has no humbling knowledge. The corporate
mind is arrogantly ignorant by definition.

Ignorance, arrogance, narrowness of mind, incomplete knowledge, and counterfeit knowledge are of concern to us because
they are dangerous; they cause destruction. When united with
great power, they cause great destruction. They have caused
far too much destruction already, too often of irreplaceable
things. Now, reasonably enough, we are asking if it is possible, if it is even thinkable, that the destruction can be stopped.
To some people's surprise, we are again backed up against the

fact that knowledge is not in any simple way good. We have often been a destructive species, we are more destructive now than we have ever been, and this, in perfect accordance with ancient warnings, is because of our ignorant and arrogant use of knowledge.

Before going further, we had better ask what it is that we humans need to know. We need to know many things, of course, and many kinds of things. But let us be merely practical for the time being and say that we need to know who we are, where we are, and what we must do to live. These questions do not refer to discrete categories of knowledge. We are not likely to be able to answer one of them without answering the other two. And all three must be well answered before we can answer well a further practical question that is now pressing urgently upon us: How can we work without doing irreparable damage to the world and its creatures, including ourselves? Or: How can we live without destroying the sources of our life?

These questions are perfectly honorable, we may even say that they are perfectly obvious, and yet we have much cause to believe that the corporate mind never asks any of them. It does not care who it is, for it is not anybody; it is a mind perfectly disembodied. It does not care where it is so long as its present location yields a greater advantage than any other. It will do anything at all that is necessary, not merely to live, but to aggrandize itself. And it charges its damages indifferently to the public, to nature, and to the future.

The corporate mind at work overthrows all the virtues of the personal mind, or it throws them out of account. The corporate mind knows no affection, no desire that is not greedy, no local or personal loyalty, no sympathy or reverence or gratitude, no temperance or thrift or self-restraint. It does not observe the first responsibility of intelligence, which is to know when you don't know or when you are being unintelligent. Try to imagine an official standing up in the high councils of a global corporation or a great public institution to say, "We have grown too big," or "We now have more power than we can responsibly use," or "We must treat our employees as our neighbors," or "We must count ourselves as members of this community," or "We must preserve the ecological integrity of our work places," or "Let us do unto others as we would have them do unto us"—and you will see what I mean.

The corporate mind, on the contrary, justifies and encourages the personal mind in its worst faults and weaknesses, such as greed and servility, and frees it of any need to worry about long-term consequences. For these reliefs, nowadays, the corporate mind is apt to express noisily its gratitude to God.

But now I must hasten to acknowledge that there are some corporations that do not simply incorporate what I am calling the corporate mind. Whether the number of these is increasing or not, I don't know. These organizations, I believe, tend to have hometowns and to count themselves as participants in the local economy and as members of the local community.

I would not apply to science any stricture that I would not apply to the arts, but science now calls for special attention because it has contributed so largely to modern abuses of the natural world, and because of its enormous prestige. Our concern here has to do immediately with the complacency of many scientists. It cannot be denied that science, in its inevitable applications, has given unprecedented extremes of scale to the technologies of land use, manufacturing, and war, and to their bad effects. One response to the manifest implication of science in certain kinds of destruction is to say that we need more science, or more and better science. I am inclined to honor this proposition, if I am allowed to add that we also need more than science.

But I am not at all inclined to honor the proposition that "science is self-correcting" when it implies that science is thus made somehow "safe." Science is no more safe than any other kind of knowledge. And especially it is not safe in the context of its gigantic applications by the corporate mind. Nor is it safe in the context of its own progressivist optimism. The idea, common enough among the universities and their ideological progeny, that one's work, whatever it is, will be beneficently disposed by the market or the hidden hand or evolution or some other obscure force is an example of counterfeit knowledge.

The obvious immediate question is, How *soon* can science correct itself? Can it correct itself soon enough to prevent or correct the real damage of its errors? The answer is that it cannot correct itself soon enough. Scientists who have made a plausible "breakthrough" hasten to tell the world, including of course the corporations. And while science may have corrected itself, it is not necessarily able to correct its results or its influence.

We must grant of course that science in its laboratories may be well under control. Scientists in laboratories did not cause the ozone hole or the hypoxic zones or acid rain or Chernobyl or Bhopal or Love Canal. It is when knowledge is corporatized, commercialized, and applied that it goes out of control. Can science, then, make itself responsible by issuing appropriate warnings with its knowledge? No, because the users are under no obligation to heed or respect the warning. If the knowledge is conformable to the needs of profit or power, the warning will be ignored, as we know. We are not excused by the doctrine of scientific self-correction from worrying about the influence of science on the corporate mind, and about the influence of the corporate mind on the minds of consumers and users. Humans in general have got to worry about the origins of the permission we have given ourselves to do large-scale damage. That permission is our problem, for by it we have made our ignorance arrogant and given it immeasurable power to do harm. We are killing our world on the theory that it was never alive but is only an accidental concatenation of materials and mechanical processes. We are killing one another and ourselves on the same theory. If life has no standing as mystery or miracle or gift, then what signifies the difference between it and death?

To state the problem more practically, we can say that the ignorant use of knowledge allows power to override the question of scale, because it overrides respect for the integrity of local ecosystems, which respect alone can determine the appropriate scale of human work. Without propriety of scale, and the acceptance of limits which that implies, there can be no form —and here we reunite science and art. We live and prosper by form, which is the power of creatures and artifacts to be made whole within their proper limits. Without formal restraints, power necessarily becomes inordinate and destructive. This is why the poet David Jones wrote in the midst of World War II that "man as artist hungers and thirsts after form." Inordinate size has of itself the power to exclude much knowledge.

*

What can we do? Anybody who goes on so long about a problem is rightly expected to have something to say about a solution. One is expected to "end on a positive note," and I mean to do that. But I also mean to be careful. The question, What can we do? especially when the problem is large, implies the expectation of a large solution.

I have no large solution to offer. There is, as maybe we all have noticed, a conspicuous shortage of large-scale corrections for problems that have large-scale causes. Our damages to watersheds and ecosystems will have to be corrected one farm, one forest, one acre at a time. The aftermath of a bombing has to be dealt with one corpse, one wound at a time. And so the first temptation to avoid is the call for some sort of revolution. To imagine that destructive power might be made harmless by gathering enough power to destroy it is of course perfectly futile. William Butler Yeats said as much in his poem "The Great Day":

> Hurrah for revolution and more cannon shot!
> A beggar upon horseback lashes a beggar on foot.
> Hurrah for revolution and cannon come again!
> The beggars have changed places, but the lash goes on.

Arrogance cannot be cured by greater arrogance, or ignorance by greater ignorance. To counter the ignorant use of knowledge and power we have, I am afraid, only a proper humility, and this is laughable. But it is only partly laughable. In his political pastoral "Build Soil," as if responding to Yeats, Robert Frost has one of his rustics say,

> I bid you to a one-man revolution—
> The only revolution that is coming.

If we find the consequences of our arrogant ignorance to be humbling, and we are humbled, then we have at hand the first fact of hope: We can change ourselves. We, each of us severally, can remove our minds from the corporate ignorance and arrogance that is leading the world to destruction; we can honestly confront our ignorance and our need; we can take guidance from the knowledge we most authentically possess, from experience, from tradition, and from the inward promptings of affection, conscience, decency, compassion, even inspiration.

This change can be called by several names—change of heart,

rebirth, metanoia, enlightenment—and it belongs, I think, to all the religions, but I like the practical way it is defined in the Confucian *Great Digest*. This is from Ezra Pound's translation:

> The men of old wanting to clarify and diffuse throughout the empire that light which comes from looking straight into the heart and then acting, first set up good government in their own states; wanting good government in their states, they first established order in their own families; wanting order in the home, they first disciplined themselves; desiring self-discipline, they rectified their own hearts; and wanting to rectify their hearts, they sought precise verbal definitions of their inarticulate thoughts [the tones given off by the heart]; wishing to attain precise verbal definitions, they set to extend their knowledge to the utmost.

This curriculum does not rule out science—it does not rule out knowledge of any kind—but it begins with the recognition of ignorance and of need, of being in a bad situation.

If the ability to change oneself is the first fact of hope, then the second surely must be an honest assessment of the badness of our situation. Our situation is extremely bad, as I have said, and optimism cannot either improve it or make it look better. But there is hope in seeing it as it is. And here I need to quote Kathleen Raine again. This is a passage written in the aftermath of World War II, and she is thinking of T. S. Eliot's poem *The Waste Land*, written in the aftermath of World War I. In *The Waste Land*, Eliot bears unflinching witness to the disease of our time: We are living the death of our culture and our world. The poem's ruling metaphor is that of a waterless land perishing for rain, an image that becomes more poignant as we pump down the aquifers and dry up or pollute the rivers.

> But Eliot [Kathleen Raine said] has shown us what the world is very apt to forget, that the statement of a terrible truth has a kind of healing power. In his stern vision of the hell that lies about us . . . , there is a quality of grave consolation. In his statement of the worst, Eliot has always implied the whole extent of the reality of which that worst is only one part.

Honesty is good, then, not just because it is a virtue, but for a practical reason: It can give us an accurate description of our problem, and it can set the problem precisely in its context.

Honesty, of course, is not a solution. As I have already said, I don't think there are solutions commensurate with our problems. I think the great problems call for many small solutions. But for that possibility to attain sufficient standing among us, we need not only to put the problems in context but also to learn to put our work in context. And here is where we turn back from our ambitions to consult both the local ecosystem and the cultural instructions conveyed to us by religion and the arts. All the arts and sciences need to be made answerable to standards higher than those of any art or science. Scientists and artists must understand that they can honor their gifts and fulfill their obligations only by living and working as human beings and community members rather than as specialists. What this may involve may not be predictable even by scientists. But the best advice may have been given by Hippocrates: "As to diseases make a habit of two things—to help, or at least, to do no harm."

The wish to help, especially if it is profitable to do so, may be in human nature, and everybody wants to be a hero. To help, or to try to help, requires only knowledge; one needs to know promising remedies and how to apply them. But to do no harm involves a whole culture, and a culture very different from industrialism. It involves, at the minimum, compassion and humility and caution. The person who will undertake to help without doing harm is going to be a person of some complexity, not easily pleased, probably not a hero, probably not a billionaire.

The corporate approach to agriculture or manufacturing or medicine or war increasingly undertakes to help at the risk of harm, sometimes of great harm. And once the risk of harm is appraised as "acceptable," the result often is absurdity: We destroy a village in order to save it; we destroy freedom in order to save it; we destroy the world in order to live in it.

The apostles of the corporate mind say, with a large implicit compliment to themselves, that you cannot succeed without risking failure. And they allude to such examples as that of the Wright brothers. They don't see that the issue of risk raises directly the issue of scale. Risk, like everything else, has an appropriate scale. By propriety of scale we limit the possible damages of the risks we take. If we cannot control scale so as to limit the effects, then we should not take the risk. From this, it is clear that some risks simply should not be taken. Some experiments should not be made. If a Wright brother wishes to risk failure, then he observes a fundamental decency in risking it alone. If the Wright airplane had crashed into a house and killed a child, the corporate mind, considering the future profitability of aviation, would count that an "acceptable" risk and loss. One can only reply that the corporate mind does not have the householder's or the parent's point of view.

I am aware that invoking personal decency, personal humility, as the solution to a vast risk taken on our behalf by corporate industrialism is not going to suit everybody. Some will find it an insult to their sense of proportion, others to their sense of drama. I am offended by it myself, and I wish I could do better. But having looked about, I have been unable to convince myself that there is a better solution or one that has a better chance of working.

I am trying to follow what T. S. Eliot called "the way of ignorance," for I think that is the way that is appropriate for the ignorant. I think Eliot meant us to understand that the way of ignorance is the way recommended by all the great teachers. It was certainly the way recommended by Confucius, for who but the ignorant would set out to extend their knowledge to the utmost? Who but the knowingly ignorant would know there is an "utmost" to knowledge?

But we take the way of ignorance also as a courtesy toward reality. Eliot wrote in "East Coker":

> The knowledge imposes a pattern, and falsifies,
> For the pattern is new in every moment
> And every moment is a new and shocking
> Valuation of all we have been.

This certainly describes the ignorance inherent in the human condition, an ignorance we justly feel as tragic. But it also is a way of acknowledging the uniqueness of every individual creature, deserving respect, and the uniqueness of every moment, deserving wonder. Life in time involves a great freshness that is falsified by what we already know.

And of course the way of ignorance is the way of faith. If enough of us will accept "the wisdom of humility," giving due honor to the ever-renewing pattern, accepting each moment's "new and shocking/Valuation of all we have been," then the corporate mind as we now have it will be shaken, and it will cease to exist as its members dissent and withdraw from it.

Quantity vs. Form

MY FAMILY AND I had a good friend I will call Lily. Lily was industrious and generous, a good neighbor. She was especially well-loved by her neighbors' children and grandchildren, though she had no children of her own. She lived a long time, surviving her husband by many years. At last, permanently ill and debilitated, she had to leave the small house that she and her husband had bought in their latter years and go to the nursing home. My brother, who was her lawyer, never until then much needed, arranged for the sale of her house and all her worldly goods.

I went to visit her a day or two after the sale. She was bedfast, sick and in some pain, but perfectly clear in her mind. We talked of the past and of several of our old neighbors, long gone. And then, speaking of the sale of her possessions, she said, "I'm all finished now. Everything is done."

She said this so cheerfully that I asked her, "Lily, is it a load off your mind?"

She said, "Well, Wendell, it hurt me. I laid here the night when I knew it was all gone, and I could *see* it all, all the things I'd cared for so long. But, yes, it is a load off my mind."

I was so moved and impressed by what she said that I wrote it down. She had lived her life and met her hardships bravely and cheerfully, and now she faced her death fully aware and responsible and with what seemed to me a completed grace. I didn't then and I don't now see how she could have been more admirable.

The last time I saw Lily she was in the hospital, where the inevitable course of her illness had taken her. By then, in addition to a seriously afflicted heart, she had not recovered from a bout of pneumonia, and because of osteoporosis she had several broken bones. She was as ill probably as a living creature can be and in great pain. She was dying. But in talking with the resident physician, I discovered that he had taken her off her pain medication to increase her appetite in the hope, he said, of "getting her back on her feet."

And so a life in every sense complete had to suffer at its end

391

this addendum of useless and meaningless pain. I don't think this episode is unusual or anomalous at the present time. The doctor's stupidity and cowardice are in fact much mitigated by being perfectly conventional. The medical industry now instructs us all that longevity is a good in itself. Plain facts and simple mercy, moreover, are readily obscured by the supposed altruism of the intent to "heal."

I am obliged now to say that I am by no means an advocate of euthanasia or "assisted suicide." My purpose here is only to notice that the ideal of a whole or a complete life, as expressed in Psalm 128 or in Tiresias' foretelling of the death of Odysseus, now appears to have been replaced by the ideal merely of a *long* life. And I do not believe that these two ideals can be reconciled.

As a man growing old, I have not been able to free my mind of the story of Lily's last days or of other stories like it that I know, and I have not been able to think of them without fear. This fear is only somewhat personal. It is also a cultural fear, the fear that something valuable and necessary to our life is being lost.

To clarify my thoughts I have been in need of some further example, and recently the associations of reading led me to Robert Southey's account of the Battle of Trafalgar in his biography of Lord Nelson. I am by conviction a pacifist, but that does not prevent me from being moved and instructed by the story of a military hero. What impresses me in Southey's account is the substantial evidence that Nelson went into the battle both expecting and fully prepared to die.

He expected to die because he had refused any suggestion that he should enter the battle in disguise in order to save himself. Instead, he would wear, Southey wrote, "as usual, his Admiral's frockcoat, bearing on the left breast four stars of the different orders with which he was invested." He thus made himself the prime target of the engagement; he would live and fight as himself, though it meant that he would die unmistakably as himself. As for his decorations: "In honor I gained them, and in honor I will die with them." And before the battle he wrote out a prayer, asking for a British victory but also for humanity afterward toward the enemy. "For myself individually," he wrote, "I commit my life to Him who made me . . ."

At the end of his account, published in 1813, eight years after the battle, Southey wrote of the admiral's death a verdict undoubtedly not so remarkable then as the succeeding two centuries have made it: "There was reason to suppose, from the appearance upon opening the body, that in the course of nature he might have attained, like his father, to a good old age. Yet he cannot be said to have fallen prematurely whose work was done . . ."

Nelson was killed at the age of forty-seven, which would seem to us in our time to be a life cut "tragically short." But Southey credited to that life a formal completeness that had little to do with its extent and much to do with its accomplishments and with Nelson's own sense of its completeness: "Thank God, I have done my duty."

The issue of the form of a lived life is difficult, for the form as opposed to the measurable extent of a life has as much to do with inward consciousness as with verifiable marks left on the world. But we are already in the thick of the problem when we have noticed that there does seem to be such a thing as a good life; that a good life consists, in part at least, of doing well; and that this possibility is an ancient one, having apparently little to do with the progress of science or how much a person knows. And so we must ask how it is that one does not have to know everything in order to do well.

The answer, apparently, is that one does so by accepting formal constraints. We are excused from the necessity of creating the universe, and most of us will not have even to command a fleet in a great battle. We come to form, we in-form our lives, by accepting the obvious limits imposed by our talents and circumstances, by nature and mortality, and thus by getting the scale right. Form permits us to live and work gracefully within our limits.

In *The Soil and Health*, his light-giving book of 1947, Sir Albert Howard wrote:

It needs a more refined perception to recognize throughout this stupendous wealth of varying shapes and forms the principle of stability. Yet this principle dominates. It dominates by means of an ever-recurring cycle, a cycle which, repeating itself silently and ceaselessly, ensures the continuation of living matter. This cycle is constituted of the successive and repeated processes of birth, growth, maturity, death, and decay.

Following, as he said, "an eastern religion," Howard speaks of this cycle as "The Wheel of Life." The life of nature depends upon the uninterrupted turning of this wheel. Howard's work rested upon his conviction, obviously correct, that a farm needed to incorporate within its own working the entire revolution of the wheel of life, so that it too might remain endlessly alive and productive by obeying "Nature's law of return." When, thirty years ago, I wrote in a poem, "The farm is an infinite form," this is what I meant.

The wheel of life is a form. It is a natural form, and it can become an artistic form insofar as the art of farming and the work of a farm can be made to conform to it. It can be made a form also of the art of living, but that, I think, requires an additional step. The wheel of human—that is, of *fully* human—life would consist over the generations of birth, growth, maturity, *ripeness*, death, and decay.

"Ripeness" is implicit in the examples of Lord Nelson and my friend Lily, but the term itself comes from act V, scene 2, of *King Lear*, in which Edgar says to his father:

> Men must endure
> Their going hence, even as their coming hither.
> Ripeness is all.

By "ripeness" Edgar means a perfect readiness for death, and his sentence echoes "The readiness is all" in act V, scene 2, of *Hamlet*. In the wheel of human life, "ripeness" adds to the idea of biological growth the growth in a living soul of the knowledge of time and eternity in preparation for death. And after the addition of "ripeness," "decay" acquires the further sense of the "plowing in" of experience and memory, building up the cultural humus. The art of living thus is practiced not only by individuals, but by whole communities or societies. It

is the work of the long-term education of a people. Its purpose,
we may say, is to make life conform gracefully both to its natural course and to its worldly limits. And this is in fulfillment
of what Vermont Chief Justice Jeffrey L. Amestoy says is "our
common responsibility . . . to imagine humanity the heart
can recognize."

What is or what should be the goal of our life and work? This is
a fearful question and it ought to be fearfully answered. Probably it should not be answered for anybody in particular by
anybody else in particular. But the ancient norm or ideal seems
to have been a life in which you perceived your calling, faithfully followed it, and did your work with satisfaction; married,
made a home, and raised a family; associated generously with
neighbors; ate and drank with pleasure the produce of your
local landscape; grew old seeing yourself replaced by your children or younger neighbors, but continuing in old age to be
useful; and finally died a good or a holy death surrounded by
loved ones.

   Now we seem to have lost any such thought of a completed
life. We no longer imagine death as an appropriate end or as
a welcome deliverance from pain or grief or weariness. Death
now apparently is understood, and especially by those who
have placed themselves in charge of it, as a punishment for
growing old, to be delayed at any cost.

   We seem to be living now with the single expectation that
there should and will always be more of everything, including
"life expectancy." This insatiable desire for more is the result of
an overwhelming sense of incompleteness, which is the result
of the insatiable desire for more. This is the wheel of death. It is
the revolving of this wheel that now drives technological progress. The more superficial and unsatisfying our lives become,
the faster we need to progress. When you are skating on thin
ice, speed up.

   The medical industry's invariable unction about life-saving,
healing, and the extended life expectancy badly needs a meeting on open ground with tragedy, absurdity, and moral horror.
To wish for a longer life is to wish implicitly for an extension
of the possibility that one's life may become a burden or even
a curse. And what are we to think when a criminal becomes a

medical emergency by the beneficence of nature, is accorded the full panoply of technological mercy, and is soon back in practice? The moral horror comes when the suffering or dementia of an overly extended life is reduced to another statistical verification of the "miracle" of modern medicine; or when a mental disease, such as the inability to face death or an ungovernable greed for more of everything, is exploited for profit.

Perhaps there is nobody now who has not benefited in some urgently personal way from the technology of the modern medical industry. To disregard the benefit is a falsehood, and to be ungrateful is inexcusable. But even gratitude does not free us of the obligation to be critical when criticism is needed. And there can be no doubt that the rapid development of industrial technology in medicine—and, as I am about to show, in agriculture—is much in need of criticism. We need to study with great concern the effects of introducing the mechanical and chemical procedures and quantitative standards of industry into the organic world and into the care of creatures. If this has given us benefits, it has also charged us and our world with costs that, typically, have been ignored by the accountants of progress. There is never, at best, an exact fit between the organic world and industrial technology. At worst, there is contradiction, opposition, and serious damage.

Industrial technology tends to obscure or destroy the sense of appropriate scale and of propriety of application. The standard of performance tends to be set by the capacity of the technology rather than the individual nature of places and creatures. Industrial technology, instead of adapting itself to life, attempts to adapt life to itself by treating life as merely a mechanical or chemical process, and thus it inhibits the operation of love, imagination, familiarity, compassion, fear, and awe. It reduces responsibility to routine, and work to "processing." It destroys the worker's knowledge of what is being worked upon.

II

The opposition of quantity and form in agriculture is not so immediately painful as in medicine, but it is more obvious. The medical industry has lifted the "norm" of life expectancy out of reach by proposing to extend longevity ad infinitum. Likewise, agricultural science, agribusiness, and the food industry propose to increase production ad infinitum, and this is their only aim. They will increase production by any means and at any cost, even at the cost of future productivity, for they have no functioning idea of ecological or agricultural or human limits. And since the agricultural economy is controlled by agribusiness and the food industry, their fixation on quantity is too easily communicated to farmers.

The art of farming, as I said earlier, fashions the farm's cycle of productivity so that it conforms to the wheel of life. That is Sir Albert Howard's language. In Wes Jackson's language, the art of farming is to mimic on the farm the self-renewing processes of the local ecosystem. But that is not all. The art of farming is also the art of living on a farm. The form of a farm is partly in its embodied consciousness of ecological obligation, and thus in its annual cycle of work, but it is also in the arrangement of fields and buildings in relation to the life of the farm's human family whose focus is the household. There is thus a convergence or even a coincidence between the form of a farm and the form of a farming life. The art of sustaining fertility and the art of living on a farm are mutually enhancing and mutually reinforcing.

A long view of an old agricultural landscape, in America and even more in Europe, would show how fencerows and fields have conformed over time both to natural topography and to human use, and how the location of dwellings, barns, and outbuildings reveals the established daily and seasonal patterns of work. In talking now about such farms in such landscapes, we are talking mostly about the past. Such farms were highly diversified and formally complex, and sometimes they were impressively sensitive (though perhaps no farm can ever be sensitive enough) both to the requirements of the place and to human need. The pursuit of higher and higher productivity has replaced those complex forms with the form (if it can be called that) of a straight line. The minimal formality of the straight line is even further attenuated because the line really is an arrow pointing toward nothing at all that is present, but toward the goal of even more production in the future. The line, it is proposed, will go on and on from one record yield to another.

And the line of this determination is marked on the ground by longer and longer rows, which is to say larger and larger farms.

The exclusive standard of productivity destroys the formal integrity of a farm just as the exclusive standard of longevity destroys the formal integrity of a life. The quest for higher and higher production on farms leads almost inevitably to specialization, ignoring the natural impulsion toward diversity; specialization in turn obliterates local proprieties of scale and proportion and obscures any sense of human connection. Driven by fashion, debt, and bad science, the desire for more overrides completely the idea of a home or a home place or a home economy or a home community. The desire for quantity replaces the desire for wholeness or holiness or health. The sense of right proportion and scale cannot survive the loss of the sense of relationship, of the parts to one another and to the whole. The result, inevitably, is ugliness, violence, and waste.

Those of us who have watched, and have cared, have seen the old diverse and complex farm homesteads dissolving into an oversimplified, overcapitalized, market-determined agriculture that destroys farms and farmers. The fences, the fencerow plants and animals, the woodlots, the ponds and wetlands, the pastures and hayfields, the grassed waterways all disappear. The farm buildings go from disuse to neglect to decay and finally to fire and the bulldozer. The farmhouse is rented, dishonored, neglected until it too goes down and disappears. A neighborhood of home places, a diverse and comely farmed landscape, is thus replaced by a mechanical and chemical, entirely-patented agricultural desert. And this is a typical reductionist blunder, the success story of a sort of materialist fundamentalism.

By indulging a limitless desire for a supposedly limitless quantity, one gives up all the things that are most desirable. One abandons any hope of the formal completeness, grace, and beauty that come only by subordinating one's life to the whole of which it is a part, and thus one is condemned to the life of a fragment, forever unfinished and incomplete, forever greedy. One loses, that is, the sense of human life as an artifact, a part made imaginatively whole.

Renewing Husbandry

I REMEMBER WELL a summer morning in about 1950 when my father sent a hired man with a McCormick High Gear No. 9 mowing machine and a team of mules to the field I was mowing with our nearly new Farmall A. That memory is a landmark in my mind and my history. I had been born into the way of farming represented by the mule team, and I loved it. I knew irresistibly that the mules were good ones. They were stepping along beautifully at a rate of speed in fact only a little slower than mine. But now I saw them suddenly from the vantage point of the tractor, and I remember how fiercely I resented their slowness. I saw them as "in my way." For those who have had no similar experience, I was feeling exactly the outrage and the low-grade superiority of a hot-rodder caught behind an aged dawdler in urban traffic. It is undoubtedly significant that in the summer of 1950 I passed my sixteenth birthday and became eligible to solve all my problems by driving an automobile.

This is not an exceptional or a remarkably dramatic bit of history. I recite it here to confirm that the industrialization of agriculture is a part of my familiar experience. I don't have the privilege of looking at it as an outsider. It is not incomprehensible to me. The burden of this essay, on the contrary, is that the industrialization of agriculture is a grand oversimplification, too readily comprehensible, to me and to everybody else.

We were mowing that morning, the teamster with his mules and I with the tractor, in the field behind the barn on my father's home place, where he and before him his father had been born, and where his father had died in February of 1946. The old way of farming was intact in my grandfather's mind until the day he died at eighty-two. He had worked mules all his life, understood them thoroughly, and loved the good ones passionately. He knew tractors only from a distance, he had seen only a few of them, and he rejected them out of hand because he thought, correctly, that they compacted the soil.

Even so, four years after his death his grandson's sudden resentment of the "slow" mule team foretold what history would bear out: The tractor would stay and the mules would go. Year after year, agriculture would be adapted more and more to the technology and the processes of industry and to the rule of industrial economics. This transformation occurred with astonishing speed because, by the measures it set for itself, it was wonderfully successful. It "saved labor," it conferred the prestige of modernity, and it was highly productive.

Though I never entirely departed from farming or at least from thoughts of farming, and my affection for my homeland remained strong, during the fourteen years after 1950 I was much away from home and was not giving to farming the close and continuous attention I have given to it in the forty years since.

In 1964 my family and I returned to Kentucky, and in a year were settled on a hillside farm in my native community, where we have continued to live. Perhaps because I was a returned traveler intending to stay, I now saw the place more clearly than before. I saw it critically, too, for it was evident at once that the human life of the place, the life of the farms and the farming community, was in decline. The old self-sufficient way of farming was passing away. The economic prosperity that had visited the farmers briefly during World War II and for a few years afterward had ended. The little towns that once had been social and economic centers, thronged with country people on Saturdays and Saturday nights, were losing out to the bigger towns and the cities. The rural neighborhoods, once held together by common memories, common work, and the sharing of help, had begun to dissolve. There were no longer local markets for chickens or eggs or cream. The spring lamb industry, once a staple of the region, was gone. The tractors and other mechanical devices certainly were saving the labor of the farmers and farm hands who had moved away, but those who had stayed were working harder and longer than ever.

Because I remembered with affection and respect my grandparents and other country people of their generation, and because I had admirable friends and neighbors with whom I was again farming, I began to ask what was happening, and why. I began to ask what would be the effects on the land, on the community, on the natural world, and on the art of farming. And these questions have occupied me steadily ever since.

The effects of this process of industrialization have become so apparent, so numerous, so favorable to the agribusiness corporations, and so unfavorable to everything else, that by now the questions troubling me and a few others in the 1960s and 1970s are being asked everywhere.

There are no doubt many ways of accounting for this change, but for convenience and brevity I am going to attribute it to the emergence of context as an issue. It has become increasingly clear that the way we farm affects the local community, and that the economy of the local community affects the way we farm; that the way we farm affects the health and integrity of the local ecosystem, and that the farm is intricately dependent, even economically, upon the health of the local ecosystem. We can no longer pretend that agriculture is a sort of economic machine with interchangeable parts, the same everywhere, determined by "market forces" and independent of everything else. We are not farming in a specialist capsule or a professionalist department; we are farming in the world, in a webwork of dependences and influences more intricate than we will ever understand. It has become clear, in short, that we have been running our fundamental economic enterprise by the wrong rules. We were wrong to assume that agriculture could be adequately defined by reductionist science and determinist economics.

If you can keep the context narrow enough (and the accounting period short enough), then the industrial criteria of labor saving and high productivity seem to work well. But the old rules of ecological coherence and of community life have remained in effect. The costs of ignoring them have accumulated, until now the boundaries of our reductive and mechanical explanations have collapsed. Their collapse reveals, plainly for all to see, the ecological and social damages that they were meant to conceal. It will seem paradoxical to some that the national and global corporate economies have narrowed the context for thinking about agriculture, but it is merely the truth. Those large economies, in their understanding and in their accounting, have excluded any concern for the land and the people. Now, in the midst of much unnecessary human and ecological damage, we are facing the necessity of a new start in agriculture.

*

And so it is not possible to look back at the tableau of team and tractor on that morning in 1950 and see it as I saw it then. That is not because I have changed, though obviously I have; it is because, in the fifty-four years since then, history and the law of consequence have widened the context of the scene as circles widen on water around a thrown stone.

My impatience at the slowness of the mules, I think, was a fairly representative emotion. I thought I was witnessing a contest of machine against organism, which the machine was bound to win. I did not see that the team arrived at the field that morning from the history of farming and from the farm itself, whereas the tractor arrived from almost an opposite history, and by means of a process reaching a long way beyond that farm or any farm. It took me a long time to understand that the team belonged to the farm and was directly supportable by it, whereas the tractor belonged to an economy that would remain alien to agriculture, functioning entirely by means of distant supplies and long supply lines. The tractor's arrival had signaled, among other things, agriculture's shift from an almost exclusive dependence on free solar energy to a total dependence on costly fossil fuel. But in 1950, like most people at that time, I was years away from the first inkling of the limits of the supply of cheap fuel.

We had entered an era of limitlessness, or the illusion thereof, and this in itself is a sort of wonder. My grandfather lived a life of limits, both suffered and strictly observed, in a world of limits. I learned much of that world from him and others, and then I changed; I entered the world of labor-saving machines and of limitless cheap fossil fuel. It would take me years of reading, thought, and experience to learn again that in this world limits are not only inescapable but indispensable.

My purpose here is not to disturb the question of the use of draft animals in agriculture—though I doubt that it will sleep indefinitely. I want instead to talk about the tractor as an influence. The means we use to do our work almost certainly affect the way we look at the world. If the fragment of autobiography I began with means anything, it means that my transformation from a boy who had so far grown up driving a team to a boy driving a tractor was a sight-changing experience.

Brought up as a teamster but now driving a tractor, a boy almost suddenly, almost perforce, sees the farm in a different way: as ground to be got over by a means entirely different, at an entirely different cost. The team, like the boy, would grow weary, but that weariness has all at once been subtracted, and the boy is now divided from the ground by the absence of a living connection that enforced sympathy as a practical good. The tractor can work at maximum speed hour after hour without tiring. There is no longer a reason to remember the shady spots where it was good to stop and rest. Tirelessness and speed enforce a second, more perilous change in the way the boy sees the farm: Seeing it as ground to be got over as fast as possible and, ideally, without stopping, he has taken on the psychology of a traveler by interstate highway or by air. The focus of his attention has shifted from the place to the technology.

I now suspect that if we work with machines the world will seem to us to be a machine, but if we work with living creatures the world will appear to us as a living creature. Be that as it may, mechanical farming certainly makes it easy to think mechanically about the land and its creatures. It makes it easy to think mechanically even about oneself, and the tirelessness of tractors brought a new depth of weariness into human experience, at a cost to health and family life that has not been fully accounted.

Once one's farm and one's thoughts have been sufficiently mechanized, industrial agriculture's focus on production, as opposed to maintenance or stewardship, becomes merely logical. And here the trouble completes itself. The almost exclusive emphasis on production permits the way of working to be determined, not by the nature and character of the farm in its ecosystem and in its human community, but rather by the national or the global economy and the available or affordable technology. The farm and all concerns not immediately associated with production have in effect disappeared from sight. The farmer too in effect has vanished. He is no longer working as an independent and loyal agent of his place, his family, and his community, but instead as the agent of an economy that is fundamentally adverse to him and to all that he ought to stand for.

After mechanization it is certainly possible for a farmer to maintain a proper creaturely and stewardly awareness of the

lives in her keeping. If you look, you can still find farmers who
are farming well on mechanized farms. After mechanization,
however, to maintain this kind of awareness requires a distinct
effort of will. And if we ask what are the cultural resources that
can inform and sustain such an effort of will, I believe that we
will find them gathered under the heading of *husbandry*, and
here my essay arrives finally at its subject.

The word *husbandry* is the name of a connection. In its original sense, it is the name of the work of a domestic man, a
man who has accepted a bondage to the household. We have
no cause here, I think, to raise the issue of "sexual roles." We
need only to say that our earthly life requires both husbandry
and housewifery, and that nobody, certainly no household, is
excused from a proper attendance to both.

Husbandry pertains first to the household; it connects the
farm to the household. It is an art wedded to the art of housewifery. To husband is to use with care, to keep, to save, to
make last, to conserve. Old usage tells us that there is a husbandry also of the land, of the soil, of the domestic plants and
animals—obviously because of the importance of these things
to the household. And there have been times, one of which is
now, when some people have tried to practice a proper human
husbandry of the nondomestic creatures in recognition of the
dependence of our households and domestic life upon the wild
world. Husbandry is the name of all the practices that sustain
life by connecting us conservingly to our places and our world;
it is the art of keeping tied all the strands in the living network
that sustains us.

And so it appears that most and perhaps all of industrial agriculture's manifest failures are the result of an attempt to make
the land produce without husbandry. The attempt to remake
agriculture as a science and an industry has excluded from it
the age-old husbandry that was central and essential to it, and
that denoted always the fundamental domestic connections
and demanded a restorative care in the use of the land and its
creatures.

This effort had its initial and probably its most radical success in separating farming from the economy of subsistence.
Through World War II, farm life in my region (and, I think,

nearly everywhere) rested solidly upon the garden, dairy, poultry flock, and meat animals that fed the farm's family. Especially in hard times these farm families, and their farms too, survived by means of their subsistence economy. This was the husbandry and the housewifery by which the farm lived. The industrial program, on the contrary, suggested that it was "uneconomic" for a farm family to produce its own food; the effort and the land would be better applied to commercial production. The result is utterly strange in human experience: farm families who buy everything they eat at the store.

An intention to replace husbandry with science was made explicit in the renaming of disciplines in the colleges of agriculture. "Soil husbandry" became "soil science," and "animal husbandry" became "animal science." This change is worth lingering over because of what it tells us about our susceptibility to poppycock. When any discipline is made or is called a science, it is thought by some to be much increased in preciseness, complexity, and prestige. When "husbandry" becomes "science," the lowly has been exalted and the rustic has become urbane. Purporting to increase the sophistication of the humble art of farming, this change in fact brutally oversimplifies it.

"Soil science," as practiced by soil scientists, and even more as it has been handed down to farmers, has tended to treat the soil as a lifeless matrix in which "soil chemistry" takes place and "nutrients" are "made available." And this, in turn, has made farming increasingly shallow—literally so—in its understanding of the soil. The modern farm is understood as a surface on which various mechanical operations are performed, and to which various chemicals are applied. The under-surface reality of organisms and roots is mostly ignored.

"Soil husbandry" is a different kind of study, involving a different kind of mind. Soil husbandry leads, in the words of Sir Albert Howard, to understanding "health in soil, plant, animal, and man as one great subject." We apply the word "health" only to living creatures, and to soil husbandry a healthy soil is a wilderness, mostly unstudied and unknown, but teemingly alive. The soil is at once a living community of creatures and their habitat. The farm's husband, its family, its crops and animals, all are members of the soil community; all belong to the character and identity of the place. To rate the farm family merely as "labor" and its domestic plants and animals merely as "production" is thus an oversimplification, both radical and destructive.

"Science" is too simple a word to name the complex of relationships and connections that compose a healthy farm—a farm that is a full membership of the soil community. If we propose, not the reductive science we generally have, but a science of complexity, that too will be inadequate, for any complexity that science can comprehend is going to be necessarily a human construct, and therefore too simple.

The husbandry of mere humans of course cannot be complex enough either. But husbandry always has understood that what is husbanded is ultimately a mystery. A farmer, as one of his farmer correspondents once wrote to Liberty Hyde Bailey, is "a dispenser of the 'Mysteries of God.'" The mothering instinct of animals, for example, is a mystery that husbandry must use and trust mostly without understanding. The husband, unlike the "manager" or the would-be objective scientist, belongs inherently to the complexity and the mystery that is to be husbanded, and so the husbanding mind is both careful and humble. Husbandry originates precautionary sayings like "Don't put all your eggs into one basket" and "Don't count your chickens before they hatch." It does not boast of technological feats that will "feed the world."

Husbandry, which is not replaceable by science, nevertheless uses science, and corrects it too. It is the more comprehensive discipline. To reduce husbandry to science, in practice, is to transform agricultural "wastes" into pollutants, and to subtract perennials and grazing animals from the rotation of crops. Without husbandry, the agriculture of science and industry has served too well the purpose of the industrial economy in reducing the number of landowners and the self-employed. It has transformed the United States from a country of many owners to a country of many employees.

Without husbandry, "soil science" too easily ignores the community of creatures that live in and from, that make and are made by, the soil. Similarly, "animal science" without husbandry forgets, almost as a requirement, the sympathy by which we recognize ourselves as fellow creatures of the animals. It forgets that animals are so called because we once believed them to be endowed with souls. Animal science has led us away from that belief or any such belief in the sanctity of animals. It has led us instead to the animal factory, which, like the concentration camp, is a vision of Hell. Animal husbandry, on the contrary, comes from and again leads to the psalmist's vision of good grass, good water, and the husbandry of God.

(It is only a little off my subject to notice also that the high and essential art of housewifery, later known as "home economics," has now become "family and consumer science." This presumably elevates the intellectual standing of the faculty by removing family life and consumption from the context—and the economy—of a home or household.)

Agriculture must mediate between nature and the human community, with ties and obligations in both directions. To farm well requires an elaborate courtesy toward all creatures, animate and inanimate. It is sympathy that most appropriately enlarges the context of human work. Contexts become wrong by being too small—too small, that is, to contain the scientist or the farmer or the farm family or the local ecosystem or the local community—and this is crucial. "Out of context," as Wes Jackson has said, "the best minds do the worst damage."

Looking for a way to give an exact sense of this necessary sympathy, the *feeling* of husbandry at work, I found it in a book entitled *Feed My Sheep* by Terry Cummins. Mr. Cummins is a man of about my age, who grew up farming with his grandfather in Pendleton County, Kentucky, in the 1940s and early 50s. In the following sentences he is remembering himself at the age of thirteen, in about 1947:

When you see that you're making the other things feel good, it gives you a good feeling, too.

The feeling inside sort of just happens, and you can't say this did it or that did it. It's the many little things. It doesn't seem that taking sweat-soaked harnesses off tired, hot horses would be something that would make you notice. Opening a barn door for the sheep standing out in a cold rain, or throwing a few grains of corn to the chickens are small things, but these little things begin

to add up in you, and you can begin to understand that
you're important. You may not be real important like
people who do great things that you read about in the
newspaper, but you begin to feel that you're important
to all the life around you. Nobody else knows or cares
too much about what you do, but if you get a good feeling inside about what you do, then it doesn't matter if
nobody else knows. I do think about myself a lot when
I'm alone way back on the place bringing in the cows or
sitting on a mowing machine all day. But when I start
thinking about how our animals and crops and fields and
woods and gardens sort of all fit together, then I get that
good feeling inside and don't worry much about what
will happen to me.

This passage goes to the heart of what I am trying to say, because it goes to the heart of farming as I have known it. Mr.
Cummins's sentences describe an experience regrettably and
perhaps dangerously missing now from the childhood of most
children. They also describe the communion between the
farmer as husband and the well-husbanded farm. This communion is a cultural force that can exist only by becoming
personal. To see it so described is to understand at once how
necessary and how threatened it now is.

I have tried to say what husbandry is, how it works, and why
it is necessary. Now I want to speak of two paramount accomplishments of husbandry to which I think we will have to pay
more deliberate attention, in our present circumstances, than
we ever have before. These are local adaptation and local coherence of form. It is strange that a science of agriculture founded
on evolutionary biology, with its practical emphasis on survival,
would exempt the human species from these concerns.

    True husbandry, as its first strategy of survival, has always
striven to fit the farming to the farm and to the field, to the
needs and abilities of the farm's family, and to the local economy. Every wild creature is the product of such an adaptive
process. The same process once was a dominant influence on
agriculture, for the cost of ignoring it was hunger. One striking and well-known example of local adaptation in agriculture

is the number and diversity of British sheep breeds, most of which are named for the localities in which they were developed. But local adaptation must be even more refined than this example suggests, for it involves consideration of the individuality of every farm and every field.

Our recent focus upon productivity, genetic and technological uniformity, and global trade—all supported by supposedly limitless supplies of fuel, water, and soil—has obscured the necessity for local adaptation. But our circumstances are changing rapidly now, and this requirement will be forced upon us again by terrorism and other kinds of political violence, by chemical pollution, by increasing energy costs, by depleted soils, aquifers, and streams, and by the spread of exotic weeds, pests, and diseases. We are going to have to return to the old questions about local nature, local carrying capacities, and local needs. And we are going to have to resume the breeding of plants and animals to fit the region and the farm.

The same obsessions and extravagances that have caused us to ignore the issue of local adaptation have at the same time caused us to ignore the issue of form. These two issues are so closely related that it is difficult to talk about one without talking about the other. During the half century and more of our neglect of local adaptation, we have subjected our farms to a radical oversimplification of form. The diversified and reasonably self-sufficient farms of my region and of many other regions have been conglomerated into larger farms with larger fields, increasingly specialized, and subjected increasingly to the strict, unnatural linearity of the production line.

But the first requirement of a form is that it must be comprehensive; it must not leave out something that essentially belongs within it. The farm that Terry Cummins remembers was remarkably comprehensive, and it was not any one of its several enterprises alone that made him feel good, but rather "how our animals and crops and fields and woods and gardens sort of all fit together."

The form of the farm must answer to the farmer's feeling for the place, its creatures, and its work. It is a never-ending effort of fitting together many diverse things. It must incorporate the life cycle and the fertility cycles of animals. It must bring crops and livestock into balance and mutual support. It must be a pattern on the ground and in the mind. It must be at once ecological, agricultural, economic, familial, and neighborly. It must be inclusive enough, complex enough, coherent, intelligible, and durable. It must have within its limits the completeness of an organism or an ecosystem, or of any other good work of art.

The making of a form begins in the recognition and acceptance of limits. The farm is limited by its topography, its climate, its ecosystem, its human neighborhood and local economy, and of course by the larger economies, and by the preferences and abilities of the farmer. The true husbandman shapes the farm within an assured sense of what it cannot be and what it should not be. And thus the problem of form returns us to that of local adaptation.

The task before us, now as always before, is to renew and husband the means, both natural and human, of agriculture. But to talk now about renewing husbandry is to talk about unsimplifying what is in reality an extremely complex subject. This will require us to accept again, and more competently than before, the health of the ecosystem, the farm, and the human community as the ultimate standard of agricultural performance.

Unsimplification is difficult, I imagine, in any circumstances; our present circumstances will make it especially so. Soon the majority of the world's people will be living in cities. We are now obliged to think of so many people demanding the means of life from the land, to which they will no longer have a practical connection, and of which they will have little knowledge. We are obliged also to think of the consequences of any attempt to meet this demand by large-scale, expensive, petroleum-dependent technological schemes that will ignore local conditions and local needs. The problem of renewing husbandry, and the need to promote a general awareness of everybody's agricultural responsibilities, thus becomes urgent.

How are we to do this? How can we restore a competent husbandry to the minds of the world's producers and consumers?

For a start of course we must recognize that this effort is already in progress on many farms and in many urban consumer groups scattered across our country and the world. But we must recognize too that this effort needs an authorizing focus and force that would grant it a new legitimacy, intellectual rigor, scientific respectability, and responsible teaching. There are many reasons to hope that this might be supplied by our colleges of agriculture, and there are some reasons to think that this hope is not fantastical.

With that hope in mind, I want to return to the precaution that I mentioned earlier. The effort of husbandry is partly scientific, but it is entirely cultural, and a cultural initiative can exist only by becoming personal. It will become increasingly clear, I believe, that agricultural scientists, and the rest of us as well, are going to have to be less specialized, or less isolated by our specialization. Agricultural scientists will need to work as indwelling members of agricultural communities or of consumer communities. Their scientific work will need to accept the limits and the influence of that membership. It is not irrational to propose that a significant number of these scientists should be farmers, and so subject their scientific work, and that of their colleagues, to the influence of a farmer's practical circumstances. Along with the rest of us, they will need to accept all the imperatives of husbandry as the context of their work. We cannot keep things from falling apart in our society if they do not cohere in our minds and in our lives.

The Burden of the Gospels

ANYBODY HALF AWAKE these days will be aware that there are many Christians who are exceedingly confident in their understanding of the Gospels, and who are exceedingly self-confident in their understanding of themselves in their faith. They appear to know precisely the purposes of God, and they appear to be perfectly assured that they are now doing, and in every circumstance will continue to do, precisely God's will as it applies specifically to themselves. They are confident, more-over, that God hates people whose faith differs from their own, and they are happy to concur in that hatred.

Having been invited to speak to a convocation of Christian seminarians, I at first felt that I should say nothing until I con-fessed that I do not have any such confidence. And then I un-derstood that this would have to be my subject. I would have to speak of the meaning, as I understand it, of my lack of con-fidence, which I think is not at all the same as a lack of faith.

It is a fact that I have spent my life, for the most part will-ingly, under the influence of the Bible, particularly the Gospels, and of the Christian tradition in literature and the other arts. As a child, sometimes unwillingly, I learned many of the Bible's stories and teachings, and was affected more than I knew by the language of the King James Version, which is the trans-lation I still prefer. For most of my adult life I have been an urgently interested and frequently uneasy reader of the Bible, particularly of the Gospels. At the same time I have tried to be a worthy reader of Dante, Milton, Herbert, Blake, Eliot, and other poets of the Christian tradition. As a result of this reading and of my experience, I am by principle and often spontaneously, as if by nature, a man of faith. But my read-ing of the Gospels, comforting and clarifying and instructive as they frequently are, deeply moving or exhilarating as they frequently are, has caused me to understand them also as a burden, sometimes raising the hardest of personal questions, sometimes bewildering, sometimes contradictory, sometimes apparently outrageous in their demands. This is the confession of an unconfident reader.

*

I will begin by dealing with the embarrassing questions that the Gospels impose, I imagine, upon any serious reader. There are two of these, and the first is this: If you had been living in Jesus's time and had heard Him teaching, would you have been one of His followers? To be an honest taker of this test, I think you have to try to forget that you have read the Gospels and that Jesus has been a "big name" for two thousand years. You have to imagine instead that you are walking past the local courthouse and you come upon a crowd listening to a man named Joe Green or Green Joe, depending on judgments whispered among the listeners on the fringe. You too stop to listen, and you soon realize that Joe Green is saying something utterly scandalous, utterly unexpectable from the premises of modern society. He is saying: "Don't resist evil. If somebody slaps your right cheek, let him slap your left cheek too. Love your enemies. When people curse you, you must bless them. When people hate you, you must treat them kindly. When people mistrust you, you must pray for them. This is the way you must act if you want to be children of God." Well, you know how happily *that* would be received, not only in the White House and the Capitol, but among most of your neighbors. And then suppose this Joe Green looks at you over the heads of the crowd, calls you by name, and says, "I want to come to dinner at your house."

I suppose that you, like me, hope very much that you would say, "Come ahead." But I suppose also that you, like me, had better not be too sure. You will remember that in Jesus's lifetime even His most intimate friends could hardly be described as overconfident.

The second question is this—it comes right after the verse in which Jesus says, "If ye love me, keep my commandments." Can you be sure that you would keep His commandments if it became excruciatingly painful to do so? And here I need to tell another story, this time one that actually happened.

In 1569, in Holland, a Mennonite named Dirk Willems, under threat of capital sentence as a heretic, was fleeing from arrest, pursued by a "thief-catcher." As they ran across a frozen body of water, the thief-catcher broke through the ice. Without help, he would have drowned. What did Dirk Willems do then?

Was the thief-catcher an enemy merely to be hated, or was he a neighbor to be loved as one loves oneself? Was he an enemy whom one must love in order to be a child of God? Was he "one of the least of these my brethren"?

What Dirk Willems did was turn back, put out his hands to his pursuer, and save his life. The thief-catcher, who then of course wanted to let his rescuer go, was forced to arrest him. Dirk Willems was brought to trial, sentenced, and burned to death by a "lingering fire."

I, and I suppose you, would like to be a child of God even at the cost of so much pain. But would we, in similar circumstances, turn back to offer the charity of Christ to an enemy? Again, I don't think we ought to be too sure. We should remember that "Christian" generals and heads of state have routinely thanked God for the deaths of their enemies, and that the persecutors of 1569 undoubtedly thanked God for the capture and death of the "heretic" Dirk Willems.

Those are peculiar questions. I don't think we can escape them, if we are honest. And if we are honest, I don't think we can answer them. We humans, as we well know, have repeatedly been surprised by what we will or won't do under pressure. A person may come to be, as many have been, heroically faithful in great adversity, but as long as that person is alive we can only say that he or she did well but remains under the requirement to *do* well. As long as we are alive, there is always a next time, and so the questions remain. These are questions we must live with, regarding them as unanswerable and yet profoundly influential.

The other burdening problems of the Gospels that I want to talk about are like those questions in that they are not solvable but can only be lived with as a sort of continuing education. These problems, however, are not so personal or dramatic but are merely issues of reading and making sense.

As a reader, I am unavoidably a writer. Many years of trying to write what I have perceived to be true have taught me that there are limits to what a human mind can know, and limits to what a human language can say. One may believe, as I do, in inspiration, but one must believe knowing that even the most inspired are limited in what they can tell of what they know.

We humans write and read, teach and learn, at the inevitable cost of falling short. The language that reveals also obscures. And these qualifications that bear on any writing must bear of course on the Gospels.

I need to say also that, as a reader, I am first of all a literalist, as I think every reader should be. This does not mean that I don't appreciate Jesus's occasional irony or sarcasm ("They have their reward"), or that I am against interpretation, or that I don't believe in "higher levels of meaning." It certainly does not mean that I think every word of the Bible is equally true, or that "literalist" is a synonym for "fundamentalist." I mean simply that I expect any writing to make literal sense before making sense of any other kind. Interpretation should not contradict or otherwise violate the literal meaning. To read the Gospels as a literalist is, to me, the way to take them as seriously as possible.

But to take the Gospels seriously, to assume that they say what they mean and mean what they say, is the beginning of troubles. Those would-be literalists who yet argue that the Bible is unerring and unquestionable have not dealt with its contradictions, which of course it does contain, and the Gospels are not exempt. Some of Jesus's instructions are burdensome, not because they involve contradiction, but merely because they are so demanding. The proposition that love, forgiveness, and peaceableness are the only neighborly relationships that are acceptable to God is difficult for us weak and violent humans, but it is plain enough for any literalist. We must either accept it as an absolute or absolutely reject it. The same for the proposition that we are not permitted to choose our neighbors ahead of time or to limit neighborhood, as is plain from the parable of the Samaritan. The same for the requirement that we must be perfect, like God, which seems as outrageous as the Buddhist vow to "save all sentient beings," and perhaps is meant to measure and instruct us in the same way. It is, to say the least, unambiguous.

But what, for example, are we to make of Luke 14:26: "If any man come to me, and hate not his father, and mother, and wife, and children, and brethren, and sisters, yea, and his own life also, he cannot be my disciple." This contradicts not only the fifth commandment but Jesus's own instruction to "Love

thy neighbor as thyself." It contradicts His obedience to his
mother at the marriage in Cana of Galilee. It contradicts the
concern He shows for the relatives of his friends and followers.
But the word in the King James Version is "hate." If you go to
the New English Bible or the New Revised Standard Version,
looking for relief, the word still is "hate." This clearly is the sort
of thing that leads to "biblical exegesis." My own temptation
is to become a literary critic, wag my head learnedly, and say,
"Well, this obviously is a bit of hyperbole—the sort of exagger-
ation a teacher would use to shock his students awake." Maybe
so, but it is not obviously so, and it comes perilously close to
"He didn't really mean it"—always a risky assumption when
reading, and especially dangerous when reading the Gospels.
Another possibility, and I think a better one, is to accept our
failure to understand, not as a misstatement or a textual flaw or
as a problem to be solved, but as a question to live with and a
burden to be borne.

We may say with some reason that such apparent difficulties
might be resolved if we knew more, a further difficulty being
that we *don't* know more. The Gospels, like all other written
works, impose on their readers the burden of their incompleteness. However partial we may be to the doctrine of the
true account or "realism," we must concede at last that reality
is inconceivably great and any representation of it necessarily
incomplete.

St. John at the end of his Gospel, remembering perhaps
the third verse of his first chapter, makes a charming acknowledgement of this necessary incompleteness: "And there
are also many other things which Jesus did, the which, if they
should be written every one, I suppose that even the world
itself could not contain the books that should be written." Our
darkness, then, is not going to be completely lighted. Our ignorance finally is irremediable. We humans are never going to
know everything, even assuming we have the capacity, because
for reasons of the most insistent practicality we can't be told
everything. We need to remember here Jesus's repeated admonitions to his disciples: You don't know; you don't understand;
you've got it wrong.

The Gospels, then, stand at the opening of a mystery in which
our lives are deeply, dangerously, and inescapably involved.

This is a mystery that the Gospels can only partially reveal, for it could be fully revealed only by more books than the world could contain. It is a mystery that we are condemned but also are highly privileged to live our way into, trusting properly that to our little knowledge greater knowledge may be revealed. It is this privilege that should make us wary of any attempt to reduce faith to a rigmarole of judgments and explanations, or to any sort of familiar talk about God. Reductive religion is just as objectionable as reductive science, and for the same reason: Reality is large, and our minds are small.

And so the issue of reality—What is the *scope* of reality? What is real?—emerges as the crisis of this discussion. Right at the heart of the religious impulse there seems to be a certain solicitude for reality: the fear of foreclosing it or of reducing it to some merely human estimate. Many of us are still refusing to trust Caesar, in any of his modern incarnations, with the power to define reality. Many of us are still refusing to entrust that power to science. As inhabitants of the modern world, we are religious now perhaps to the extent of our desire to crack open the coffin of materialism, and to give to reality a larger, freer definition than is allowed by the militant materialists of the corporate economy and their political servants, or by the mechanical paradigm of reductive science. Or perhaps I can make most plain what I'm trying to get at if I say that many of us are still withholding credence, just as properly and for the same reasons, from any person or institution claiming to have the definitive word on the purposes and the mind of God.

It seems to me that all the religions I know anything about emerge from an instinct to push against any merely human constraints on reality. In the Bible such constraints are conventionally attributed to "the world" in the pejorative sense of that term, which we may define as the world of the creation *reduced* by any of the purposes of selfishness. The contrary purpose, the purpose of freedom, is stated by Jesus in the fourth Gospel: "I am come that they might have life, and that they might have it more abundantly."

This astonishing statement can be thought about and understood endlessly, for it is endlessly meaningful, but I don't think

THE WAY OF IGNORANCE

it calls for much in the way of interpretation. It does call for a very strict and careful reading of the word "life."

To talk about or to desire more abundance of anything has probably always been dangerous, but it seems particularly dangerous now. In an age of materialist science, economics, art, and politics, we ought not to be much shocked by the appearance of materialist religion. We know we don't have to look far to find people who equate more abundant life with a bigger car, a bigger house, a bigger bank account, and a bigger church. They are wrong, of course. If Jesus meant only that we should have more possessions or even more "life expectancy," then John 10:10 is no more remarkable than an advertisement for any commodity whatever. Abundance, in this verse, cannot refer to an abundance of material possessions, for life does not require a material abundance; it requires only a material sufficiency. That sufficiency granted, life itself, which is a membership in the living world, is already an abundance.

But even life in this generous sense of membership in creation does not protect us, as we know, from the dangers of avarice, of selfishness, of the wrong kind of abundance. Those dangers can be overcome only by the realization that in speaking of more abundant life, Jesus is not proposing to free *us* by making us richer; he is proposing to set *life* free from precisely that sort of error. He is talking about life, which is only incidentally our life, as a limitless reality.

Now that I have come out against materialism, I fear that I will be expected to say something in favor of spirituality. But if I am going to go on in the direction of what Jesus meant by "life" and "more abundantly," then I have got to avoid that duality of matter and spirit at all costs.

As every reader knows, the Gospels are overwhelmingly concerned with the conduct of human life, of life in the human commonwealth. In the Sermon on the Mount and in other places Jesus is asking his followers to see that the way to more abundant life is the way of love. We are to love one another, and this love is to be more comprehensive than our love for family and friends and tribe and nation. We are to love our neighbors though they may be strangers to us. We are to love our enemies. And this is to be a practical love; it is to be practiced, here and now. Love evidently is not just a feeling but is indistinguishable from the willingness to help, to be useful to one another. The way of love is indistinguishable, moreover, from the way of freedom. We don't need much imagination to imagine that to be free of hatred, of enmity, of the endless and hopeless effort to oppose violence with violence, would be to have life more abundantly. To be free of indifference would be to have life more abundantly. To be free of the insane rationalizations for our desire to kill one another—that surely would be to have life more abundantly.

And where more than in the Gospels' teaching about love do we see that famously estranged pair, matter and spirit, melt and flow together? There was a Samaritan who came upon one of his enemies, a Jew, lying wounded beside the road. And the Samaritan had compassion on the Jew and bound up his wounds and took care of him. Was this help spiritual or material? Was the Samaritan's compassion earthly or heavenly? If those questions confuse us, that is only because we have for so long allowed ourselves to believe, as if to divide reality impartially between science and religion, that material life and spiritual life, earthly life and heavenly life, are two different things.

To get unconfused, let us go to a further and even more interesting question about the parable of the Samaritan: Why? Why did the Samaritan reach out in love to his enemy, a Jew, who happened also to be his neighbor? Why was the unbounding of this love so important to Jesus?

We might reasonably answer, remembering Genesis 1:27, that all humans, friends and enemies alike, have the same dignity, deserve the same respect, and are worthy of the same compassion because they are, all alike, made in God's image. That is enough of a mystery, and it implies enough obligation, to waylay us a while. It is certainly something we need to bear anxiously in mind. But it is also too human-centered, too potentially egotistical, to leave alone.

I think Jesus recommended the Samaritan's loving-kindness, what certain older writers called "holy living," simply as a matter of propriety, for the Samaritan was living in what Jesus understood to be a holy world. The foreground of the Gospels is occupied by human beings and the issues of their connection to one another and to God. But there is a background, and the background more often than not is the world in the best sense of the word, the world as made, approved, loved, sustained, and finally to be redeemed by God. Much of the action and the talk of the Gospels takes place outdoors: on mountainsides, lake shores, river banks, in fields and pastures, places populated not only by humans but by animals and plants, both domestic and wild. And these non-human creatures, sheep and lilies and birds, are always represented as worthy of, or as flourishing within, the love and the care of God.

To know what to make of this, we need to look back to the Old Testament, to Genesis, to the Psalms, to the preoccupation with the relation of the Israelites to their land that runs through the whole lineage of the prophets. Through all this, much is implied or taken for granted. In only two places that I remember is the always implicit relation—the practical or working relation—of God to the creation plainly stated. Psalm 104:30, addressing God and speaking of the creatures, says, "Thou sendest forth thy spirit, they are created . . ." And, as if in response, Elihu says to Job (34:14–15) that if God "gather unto himself his spirit and his breath; All flesh shall perish together . . ." I have cut Elihu's sentence a little short so as to leave the emphasis on the phrase "all flesh."

Those also are verses that don't require interpretation, but I want to stretch them out in paraphrase just to make as plain as possible my reason for quoting them. They are saying that not just humans but *all* creatures live by participating in the life of God, by partaking of His spirit and breathing His breath.* And so the Samaritan reaches out in love to help his enemy, breaking all the customary boundaries, because he has clearly seen in his enemy not only a neighbor, not only a fellow human or a fellow creature, but a fellow sharer in the life of God.

When Jesus speaks of having life more abundantly, this, I think, is the life He means: a life that is not reducible by division, category, or degree, but is one thing, heavenly and earthly, spiritual and material, divided only insofar as it is embodied

---

*We now know that this relationship is even more complex, more utterly inclusive and whole, than the biblical writers suspected. Some scientists would insist that the conventional priority given to living creatures over the nonliving is misleading. Try, for example, to separate life from the lifeless minerals on which it depends.

in distinct creatures. He is talking about a finite world that is infinitely holy, a world of time that is filled with life that is eternal. His offer of more abundant life, then, is not an invitation to declare ourselves as certified "Christians," but rather to become conscious, consenting, and responsible participants in the one great life, a fulfillment hardly institutional at all.

To be convinced of the sanctity of the world, and to be mindful of a human vocation to responsible membership in such a world, must always have been a burden. But it is a burden that falls with greatest weight on us humans of the industrial age who have been and are, by any measure, the humans most guilty of desecrating the world and of destroying creation. And we ought to be a little terrified to realize that, for the most part and at least for the time being, we are helplessly guilty. It seems as though industrial humanity has brought about phase two of original sin. We all are now complicit in the murder of creation. We certainly do know how to apply better measures to our conduct and our work. We know how to do far better than we are doing. But we don't know how to extricate ourselves from our complicity very surely or very soon. How could we live without degrading our soils, slaughtering our forests, polluting our streams, poisoning the air and the rain? How could we live without the ozone hole and the hypoxic zones? How could we live without endangering species, including our own? How could we live without the war economy and the holocaust of the fossil fuels? To the offer of more abundant life, we have chosen to respond with the economics of extinction.

If we take the Gospels seriously, we are left, in our dire predicament, facing an utterly humbling question: How must we live and work so as not to be estranged from God's presence in His work and in all His creatures? The answer, we may say, is given in Jesus's teaching about love. But that answer raises another question that plunges us into the abyss of our ignorance, which is both human and peculiarly modern: How are we to make of that love an economic practice?

That question calls for many answers, and we don't know most of them. It is a question that those humans who want to answer will be living and working with for a long time—if they are allowed a long time. Meanwhile, may Heaven guard us from those who think they already have the answers.

WHAT MATTERS?
(2010)

Money Versus Goods

MY ECONOMIC point of view is from ground-level. It is a point of view sometimes described as "agrarian." That means that in ordering the economy of a household or community or nation, I would put nature first, the economies of land use second, the manufacturing economy third, and the consumer economy fourth. The basis of such an economy would be broad, the successive layers narrowing in the order of their diminishing importance.

The first law of such an economy would be what the agriculturalist Sir Albert Howard called "the law of return." This law requires that what is taken from nature must be given back: The fertility cycle must be maintained in continuous rotation. The primary value in this economy would be the capacity of the natural and cultural systems to renew themselves. An authentic economy would be based upon renewable resources: land, water, ecological health. These resources, if they are to stay renewable in human use, will depend upon resources of culture that also must be kept renewable: accurate local memory, truthful accounting, continuous maintenance, un-wastefulness, and a democratic distribution of now-rare practical arts and skills. The economic virtues thus would be honesty, thrift, care, good work, generosity, and (since this is a creaturely and human, not a mechanical, economy) imagination, from which we have compassion. That primary value and these virtues are essential to what we have been calling "sustainability."

A properly ordered economy, putting nature first and consumption last, would start with the subsistence or household economy and proceed from that to the economy of markets. It would be the means by which people provide to themselves and to others the things necessary to support life: goods coming from nature and human work. It would distinguish between needs and mere wants, and it would grant a firm precedence to needs.

A proper economy, moreover, would designate certain things as priceless. This would not be, as now, the "pricelessness" of things that are extremely rare or expensive, but would refer to things of absolute value, beyond and above any price that could be set upon them by any market. The things of absolute value would be fertile land, clean water and air, ecological health, and the capacity of nature to renew herself in the economic landscapes. Our nearest cultural precedent for this assignment of absolute value is biblical, as in Psalm 24 ("The earth is the Lord's, and the fulness thereof . . .") and Leviticus 25:23 ("The land shall not be sold forever . . ."). But there are precedents in all societies and traditions that have understood the land or the world as sacred—or, speaking practically, as possessing a suprahuman value. The rule of pricelessness clearly imposes certain limits upon the idea of land ownership. Owners would enjoy certain customary privileges, necessarily, as the land would be entrusted to their intelligence and responsibility. But they would be expected to use the land as its servants and on behalf of all the living.

The present and now-failing economy is just about exactly opposite to the economy I have just described. Over a long time, and by means of a set of handy prevarications, our economy has become an anti-economy, a financial system without a sound economic basis and without economic virtues.

It has inverted the economic order that puts nature first. This economy is based upon consumption, which ultimately serves, not the ordinary consumers, but a tiny class of excessively wealthy people for whose further enrichment the economy is understood (by them) to exist. For the purpose of their further enrichment, these plutocrats and the great corporations that serve them have controlled the economy by the purchase of political power. The purchased governments do not act in the interest of the governed and their country; they act instead as agents for the corporations.

That this economy is, or was, consumption-based is revealed by the remedies now being proposed for its failure: stimulate, spend, create jobs. What is to be stimulated is spending. The government injects into the failing economy money to be spent, or to be loaned to be spent. If people have money to spend and are eager to spend it, demand for products will increase, creating jobs; industry will meet the demand with more products, which will be bought, thus increasing the amount of money in circulation; the greater amount of money in circulation will increase demand, which will increase spending, which will increase production—and so on until the old fantastical economy of limitless economic growth will have "recovered."

But spending is not an economic virtue. Miserliness is not an economic virtue either. Saving is. Not-wasting is. To encourage spending with no regard at all to what is being purchased may be pro-finance, but it is anti-economic. Finance, as opposed to economy, is always ready and eager to confuse wants with needs. From a financial point of view, it is good, even patriotic, to buy a new car whether you need one or not. From an economic point of view, however, it is wrong to buy anything you do not need. It is unpatriotic too: If you love your country, you don't want to burden or waste it by frivolous wants. Only in a financial system, an anti-economy, can it seem to make sense to talk about "what the economy needs." In an authentic economy, we would ask what the land and the people need. People do need jobs, obviously. But they need jobs that serve natural and human communities, not arbitrarily "created" jobs that serve only the economy.

From an economic point of view, a society in which every schoolchild "needs" a computer, and every sixteen-year-old "needs" an automobile, and every eighteen-year-old "needs" to go to college is already delusional and is well on its way to being broke.

In a so-called economy that is dependent on indiscriminate spending, "job creation" often implies an ability to "create" new "needs." Until lately this economy has been able to create jobs by creating needs. But this has involved much confusion and a kind of fraud, because it gives no priority to the meeting of needs, and cannot distinguish needs from wants. Our economy, having confused necessities with products or commodities that are merely marketable, deliberately reduces the indispensable service of providing needed goods to "selling" or "marketing" products, some of which have never been and will never be needed by anybody. The gullibility of the public thus becomes an economic resource.

The category of things sold that are not needed now includes even legally marketed foods and drugs. This involves the

art (taught and learned in universities) of lying about products.
A friend of mine remembers a teacher who said that advertising
is "the manufacture of discontent." And so we have come to
live in a world in which every brand of painkiller is better than
every other brand, in which we have a "service economy" that
does not serve and an "information economy" that does not
distinguish good from bad or true from false.

The manufacturing sector of a financial system, which does
not or cannot distinguish between needs and induced wants,
will come willy-nilly into the service of wants, not needs. So
it has happened with us. If in some state of emergency our
manufacturers were suddenly called upon to supply us with
certain necessities—shoes, for example—we would be out of
luck. "Outsourcing" the manufacture of frivolities is at least
partly frivolous; outsourcing the manufacture of necessities is
entirely foolish.

As for the land economies, the academic and political economists seem mainly to ignore them. For years, as I have read
articles on the economy, I have waited in vain for the author
to "factor in" farming or ranching or forestry. The expert assumption appears to be that the products of the soil are not
included in the economy until after they have been taken at
the lowest possible cost from those who did the actual work of
production, at which time they enter the economy as raw materials for the food, fiber, timber, and lately the fuel industries.
The result is inevitable: The industrial system is disconnected
from, is unconcerned about, and takes no responsibility for,
its natural and human sources. The further result is that these
sources are not maintained but merely used and thus are made
as exhaustible as the fossil fuels.

As for nature herself, virtually nobody—not the "environmentalist," let alone the economist—regards nature as an
economic resource. Nature, especially where she has troubled
herself to be scenic, is understood to have a recreational and
perhaps an aesthetic value that is to some extent economic. But
for her accommodation of our needs to eat, drink, breathe, and
be clothed and sheltered, our industrial and financial systems
grant her no recognition, honor, or care.

Far from assigning an absolute value to those things we absolutely need, the financial system puts a price, though a highly

variable price, on everything. We know from much experience that everything that is priced will sooner or later be sold. And from the accumulating statistics of soil loss, land loss, deforestation, overuse of water, various sorts of pollution, etc., we have reason to fear that everything that is sold will be ruined. When everything has a price, and the price is made endlessly variable by an economy without a stable relation to necessity or to real goods, then everything is disconnected from history, knowledge, respect, and affection—from anything at all that might preserve it—and so is implicitly eligible to be ruined.

What we have been pleased to call our economy does not acknowledge and apparently does not even recognize its continuing absolute dependence on the natural world, on the land economies, and on the work of farmers, ranchers, and foresters—all of which, given the use of available knowledge and precautions, would be self-renewing. At the same time, with a remarkable lack of foresight or even the sight to see what is presently obvious, this economy has made itself absolutely dependent on resources that are either exhaustible by nature or have been made exhaustible by our wastefulness and our refusal to husband and reuse: fossil fuels, metals, and other mined materials. By standards that are utterly absurd, it has long been "too expensive" to salvage perfectly good and usable materials from old buildings, which we knock down or blow up and haul to landfills, and so make even bricks and stones valueless and irrecoverable. Because of falsely cheap materials and energy, we have a "bubble" of houses too big to be heated efficiently or cheaply, or even to be paid for.

To use our agricultural land for the production of "biofuel," as some are now doing, is immediately to raise the question whether it can ever be right to replace food production by the production of a fuel to be burned. If this fuel is produced, like most of our food at present, without the close and loving care that the land requires, then the land becomes an exhaustible resource. Biofuel may be a product of the land and our world-changing technology, but it is just as much a product of ignorance and moral carelessness.

As commodities, the fossil fuels are in a category strictly their own. Unlike other minerals that (in a sensible economy) can be reused, and unlike waterpower that uses water and releases it to be used again, the fossil fuels can be made useful only by being destroyed. They are useful and therefore valuable only in the instant in which they are burning.

To be available for their brief usefulness, these fuels must be dug or pumped from the ground. Their extraction has nearly always damaged, often irreparably, the places and the human communities from which they are taken. For coal to feed the fires by which we live, whole landscapes are destroyed, forests and their soils and creatures are obliterated, streams are covered over, watersheds are degraded and polluted, poisonous residues are left behind, communities are degraded or flooded by toxic wastes or runoff from denuded watersheds, the people are exploited and endangered, their houses damaged, their drinking water poisoned, their complaints and needs ignored. When the fossil fuels, extracted at such a cost to people and nature, are burned, they pollute the atmosphere of all the world, with consequences that are fearful, infamous, and continuing.

In a consciously responsible economy, such abuses would be inconceivable. They could not happen. To damage or destroy an otherwise permanent resource for the sake of a temporary advantage would be readily perceived as senseless by every practical measure and, by the measure of human wholeness, as insane. To value human wants above all the natural and human resources that supply human needs, as the now-failing economy has done, is to run risks and defy paradoxes by which it was and is bound to fail. If we pursue limitless "growth" now, we impose ever-narrower limits on the future. If we put spending first, we put solvency last. If we put wants first, we put needs last. If we put consumption first, we put health last. If we put money first, we put food last. If for some spurious reason such as "economic growth" or "economic recovery," we put people and their comfort first, before nature and the land-based economies, then nature sooner or later will put people last.

But the fossil fuels, which involve destruction for the sake of production and again destruction as a consequence of production, are not the only typical products of our anti-economy. Also typical are products that replace, at high cost, goods that once were cheap or free. The genius of marketing and selling has given us, for example, bottled tapwater, for which we pay more than we pay for gasoline, because of our perfectly rational fear that our unbottled tapwater is polluted. The system of industry, finance, and "marketing" thus makes capital of its own viciousness and of the ignorance and gullibility of a supposedly educated public. By the influence of marketers and sellers, citizens and members are transformed into suckers. And so we have an alleged economy that is not only fire-dependent and consumption-dependent but also sucker-dependent.

For another example, consider the money-drenched entertainment industry. The human species, which has apparently outlived the name Homo sapiens, is said to be something like 200,000 years old. Except for the last seventy-five or so years of their life so far, and except for their decadent ruling classes, most humans have entertained themselves by remembering and telling stories, singing, dancing, playing games, and even by their work of providing themselves with necessities and things of beauty, which usually were the same things. All of this entertainment came free of charge, as a sort of overflow of human nature, local culture, and daily life. Even the beauty of good work and well-made things was a value added at no charge. The entertainment industry has improved upon this great freedom by providing at a high cost, in money but also in health and sanity, an egregiously overpaid corps of entertainers and athletes who tell or perform stories, sing, dance, and play games for us or sell games to us as we passively consume their often degrading productions. The wrong here may be at root only that of an inane and expensive redundancy. If you can read and have more imagination than a doorknob, what need do you have for a "movie version" of a novel?

This strange economy produces, typically and in the ordinary course of business, products that are destructive or fraudulent or unnecessary or useless, or all four at once. Another of its typical enterprises is remarkable for the production of what I suppose we will have to call no-product, or no product but money (to the extent that this works). The best-known or longest infamous example of a no-product financial industry is the practice of usury, which is to say the lending of money at exorbitant interest or (some have said) at any interest. In our

cultural tradition, as opposed to financial precedent, the condemnation of usury seems to be unanimous.

The Hebrew Bible speaks emphatically against usury in ten
of its chapters (by my count), calling it by name, but without
much explanation, assuming apparently that its wrongfulness is
obvious. From the context it is clear that usury is understood
as an injustice and an offense against charity. It is a way for
people of wealth to exploit the poor, whom they have been instructed to care for. Only the wealthy have a surplus of money
to lend, and they should not use it to take advantage of the
needs of others. Usury, moreover, cannot be consistent with
the command (Leviticus 19:18) that "thou shalt love thy neighbor as thyself."

Aristotle in *The Nicomachean Ethics* also condemns usury
and in language that is remarkably consistent with my description of our own economic malpractice. He classes usurers with
pimps, as people who take "anything from any source" or who
"take more than they ought and from wrong sources" (the
Oxford edition, translated by Sir David Ross).

Dante is perfectly consistent with the Bible and Aristotle
when he places the usurers in Hell (*Inferno* XI) with others
who are guilty of violence against God. Virgil, explaining this
fault to Dante, makes the case clearly and usefully. Usury is a
violence against God because it is a violence against nature.
Nature is the art of God, just as productive work, the making
of useful things, is the art of humans. Humans prosper rightly
when their goods come from nature by their good work.
Usurers prosper, on the contrary, by making money grow from
itself (by "making their money work for them," as we say), thus
holding in contempt both nature and work, both divine art
and human art.

Ezra Pound, a poet of our own time, was in Dante's tradition
when he wrote the two versions of his eloquent poem against
usury (*Cantos* XLV and LI). Pound who was (I hope) insane
when at his worst, was perfectly sane when he wrote this:

> With usury has no man a good house
> made of stone, no paradise on his church wall
> With usury the stone cutter is kept from his stone
> the weaver is kept from his loom by usura

Wool does not come into market
the peasant does not eat his own grain
the girl's needle goes blunt in her hand
The looms are hushed one after another
.   .   .   .

Usury kills the child in the womb
And breaks short the young man's courting
Usury brings age into youth; it lies between the bride
and the bridegroom
Usury is against Nature's increase.

The point—as I understand it, though I understand also that
this poem offers far more than a point—is that when money is
misused to grow from itself into heaps in the possession inevitably of fewer and fewer people, it cannot be rightly used for
the production of goods or even to maintain the subsistence of
the people. Workers will not be well paid for good work. The
arts will not flourish, and neither will nature.

I need to say here that this issue of usury is far from simple,
and that I am not capable even of giving usury a proper definition. The issue is simple only if usury is defined as the taking
of *any* interest. It is so defined by Jesus in the Gospel of Luke
(6:34–35):

> And if ye lend to them of whom ye hope to receive, what
> thank have ye? for sinners also lend to sinners, to receive
> as much again.
>
>   But love ye your enemies, and do good, and lend, hop-
> ing for nothing again . . .

Such free lending would be possible among neighbors or in a
small local economy, but in general we appear to be far from
that, the churches along with the rest. In an extensive economy using money, banks appear to be necessary. If the poor,
for instance, are to rise above their poverty, or if the young are
to acquire houses or farms or small businesses, there probably
needs to be an established means of lending them money, and
that would be banking. If we are to have banks and banking,
then we have to build and equip and maintain the necessary
buildings, return a fair dividend to the necessary investors,
and pay fair wages or salaries to the necessary employees. The

needed funds would have to come to a considerable extent from interest on loans. If the money were to be loaned at no charge, there soon would be no lending institution and nobody to make loans.

And so we come to the uneasy question of what rate of interest would be neither too little nor too much. If too little, loans cannot be made. If too much, then lending becomes, not a service, but the exploitation and even the ruin of borrowers. I don't think a fair rate can be determined according to standards that are only financial. It would have to be determined by responsible bankers, acting also as community members, in the context of their community, local nature, and the local economy.

Such a determination, I believe, can take place only in a bank that is locally owned, conforming in scale to the size and needs of the local community, and by bankers who are aware that the prosperity of the bank is not and can never be separated from the prosperity of the community.

I know from my own experience and observation that a bank of community scale, owned principally by local investors, understanding its dependence on responsible service first of all to local customers—even in a fevered and delirious economy—can function usefully and considerately as a part of the community. Such a bank does not, because if it is to survive it cannot, adopt the lending practices that resulted in our recent housing bubble. In such a bank the loan officers understand necessarily that their responsibility is to the borrowers as much as to the bank. In a locally owned community bank, the lender is a neighbor of the borrower. You don't put your neighbors into trouble or into ruin by misleading them to assume debts they cannot pay—which ultimately, of course, would ruin the lender.

It is clear that if interest rates are not limited by a reasonable, workable concept of fairness, enforceable by law, then they will become exorbitant. Moreover, they are apt to become highly variable according to the whim of lenders inclined to "take more than they ought."

Among its other wrongs, usury destabilizes the relation of money to goods. So does inflation. So does the speculative trading in mortgages, "futures," and "commercial paper" that gives a monetary value to commodities having no present existence or no existence at all. To inflate or obscure the value of money in relation to goods is in effect to steal both from those who spend and from those who save. It is to subordinate real value to a value that is false.

By destabilizing the relation of money to goods, a financial system usurps an economy. Then, instead of the exchange of money for goods or goods for money, we have the conversion of goods into money, in the process often destroying the goods. Money, instead of a token signifying the value of goods, becomes a good in itself, which the wealthy can easily manipulate in their own favor. This is sometimes justified (by the favored) as freedom, as in "free trade" or "the free market," but such a freedom is calculated to reduce substantially the number of the free. The tendency of this freedom necessarily is toward monopoly—toward the one economic entity that will own or control everything. The undisguised aim of Monsanto, for example, is to control absolutely the economy of food. It would do so by setting its own price on its products sold to dependent purchasers who can set a price neither on what they buy nor on what they sell.

To permit so much wealth, power, influence, and ambition to one corporation is an egregious error in a polity supposedly democratic. From the point of view of nature and agriculture, it is an error even larger and more dangerous. For by this error agriculture is forced to subserve the rule of industrialism, which is in most respects antithetical to the healthful practice of agriculture and to the laws of nature, by which, and only by which, agriculture can be made sustainable.

The dominance of agriculture by agribusiness is made possible by the dominance of the economy by interests that are industrial or purely financial. Agribusiness is immensely more profitable monetarily than agriculture, which customarily for the last fifty or sixty years has been either barely profitable or unprofitable. Hence the drastic decline in the agricultural population. One cost of this error is economic injustice, characteristic of industrialism, to the people who do the work: ranchers, farmers, and farm workers. Another cost is first agricultural and then ecological: Under the rule of industrialism the land is forced to produce but is not maintained; the fertility cycle is broken; soil nutrients become water pollutants; toxic chemicals and fossil energy replace human work.

We have allowed, and even justified as "progress," a fundamental disconnection between money and food. And so we are led to the assumption, by ignorant leaders who apparently believe it, that if we have money we will have food, an assumption that is destructive of agriculture and food. It is a superstition just as wicked, and hardly different from, the notion that the world is conformable to our wants and we can be whatever we want to be.

Apparently it takes a lot of money, a lot of power, and even a lot of education to obscure the knowledge that food comes from the land and from the human ability to cause the land to produce and to remain productive. Under the rule of an economy perverted by industrial and financial presumptions, we are destroying both the land and the human means of using the land and caring for it.

We are destroying the land by exposing it to erosion, by infusing it year after year with toxic chemicals (which incidentally poison the water), by surface mining, and by so-called development. We are destroying the cultures and the communities of land use and land husbandry by deliberately slanting the economy of the food system against the primary producers.

We are losing and degrading our agricultural soils because we no longer have enough competent people available to take proper care of them. And we will not produce capable and stewardly farmers, ranchers, and foresters by what we are calling "job creation." The fate of the land is finally not separable from the fate of the people of the land (and the fate of country people is finally not different from the fate of city people). Industrial technology does not and cannot adequately replace human affection and care. Industrial and financial procedures cannot replace stable rural communities and their cultures of husbandry. One farmer, if that name applies, cannot farm thousands of acres of corn and soybeans in the Midwest without production costs that include erosion and toxicity, which is to say damages that are either long-term or permanent.

The farm population has now declined almost to nonexistence

because, since the middle of the last century, we have deliberately depressed farm income while allowing production costs to rise, for the sake of "cheap food" and to favor agribusiness. No wonder that farm-raised young people have been moving into the cities and suburbs by millions for two generations, leaving the farms without heirs or successors. The young people decide against too much investment and too much work for too little return. Even if they love farming or ranching enough to want to stay, paying the inevitable economic and personal penalties, they are more than likely to find that they cannot buy land and pay for it by using it. The one reason for this is the disequilibrium between the economy of money and the land economies. Professional people in the cities, who have done well financially, have been "investing" in farmland and rangeland and so lifting the market value of the land above the reach of farmers and ranchers who are *not* doing well economically. The result is that we have an enormous population of dependent people with the subservient mentality of industrial employees, helpless to feed themselves, who are being fed by the tiniest minority of exploited people and from land that is more cruelly exploited than the people.

If we are destroying both the productive land and the rural communities and cultures, how can we assume that money will somehow attract food to us whenever we need it? If, on the contrary, we should decide to right the economic balance by paying a just price to producers, then money could revert to its proper function of encouraging and supporting both food production and the proper husbanding of the land. This, if it could happen, would solve a number of problems. The right answer to urban sprawl, for example, is to make agriculture pay well enough that farmers and ranchers would want to keep the land in use, and their children would want to inherit it to use.

To a ground-level observer, it is obvious that the economic failures I have described involve moral issues of the gravest sort. An essentially immoral system of economy-as-finance, or an economy run by the sole standard of monetary profit, has been allowed to flourish to the point of catastrophe by a fairly general consent to the proposition that economy and morality are two professional specialties that either do not converge, or that can be made to converge by a simple moral manipulation, as follows.

In 1986 the "conservative" columnist William Safire wrote that "Greed is finally being recognized as a virtue . . . the best engine of betterment known to man." This was not, I think, the news that Mr. Safire thought it was, but was merely a repetition of a time-worn rationalization. What may have been new was the "professional" falsehood that greed is the exclusive motive in every choice—that, for example, the *only* way to have good teachers or good doctors is to pay them a lot of money.

Mr. Safire's error, and that of the people he spoke for, is in the idea that everybody can be greedy up to some limit—that, once you have made greed a virtue, it will not crowd out other virtues such as temperance or justice or charity. The virtuously greedy perhaps would agree to let one another be greedy, so long as one person's greed did not interfere with the greed of another person. This would be the Golden Rule of greedy persons, who no doubt would thank God for it.

But that rule appears to be honored entirely in the breach. There are still a good many people who choose or accept a vocation that will not make them rich—many teachers, for instance, and most writers. But for the greedy there appears to be no such concept as *greedy enough*. The greedy consume the poor, the moderately prosperous, and each other with the same relish and with an ever-growing appetite.

Part two of Mr. Safire's error is his assumption that we can restrict the honor of virtuehood to greed alone, leaving the other sins to pine away in customary disfavor: "I hold no brief," he said, "for Anger, Envy, Lust, Gluttony, Pride . . . or Sloth." But he was already too late. A glance at magazine advertising in 1986 would have suggested that these sins had been virtues of commerce long enough already to be taken for granted. As we have sometimes been told, the sins, like the virtues, are inclined to enjoy one another's company.

Mr. Safire's announcement was not a moral innovation, but rather a confession of the depravity of what in 1986 we were calling, and are still calling, "the economy"—a ramshackle, propped-up, greed-enforced anti-economy that is delusional, vicious, wasteful, destructive, hard-hearted, and so fundamentally dishonest as to have resorted finally to "trad-ing" in various pure-nothings. Might it not have been better and safer to have assumed that there is no partition between economy and morality, that the test of both is practicality, and that morality is long-term practicality?

The problem with "the economy" is not only that it is anti-economic, destructive of the natural and human bases of any authentic economy, but that it has been out of control for a long time. At the root of our problem, we now need to sup-pose, is industrialism and the Industrial Revolution itself. As the original Luddites saw clearly and rightly, the purpose of industrialism from the first has been to replace human work-ers with machines. This has been justified and made unques-tionable by the axiom that machines, according to standards strictly mechanical, work more efficiently and cheaply than people. They answer directly the perpetual need of the greedy to get more for less. This is yet another of our limitless "pro-gressive" ideas: The industrial academics or academic indus-trialists who subserve the technological cutting edge are now nominating robots as substitutes for parents, nurses, and sur-geons. Soon, surely, we will have robots that can worship and make love faster and cheaper than we mere humans, who have been encumbered in those activities by flesh and blood and our old-fashioned ways.

But to replace people by machines is to raise a difficult, and I would say an urgent, question: What are the replaced people to do? Or, since this is a question not all replaced people have been able to answer satisfactorily for themselves, What is to be done with or for them? This question has never received an honest answer from either liberals or conservatives, commu-nists or capitalists. Replaced people have entered into a condi-tion officially euphemized as "mobility." If you have left your farm or your country town and found a well-paying city job or entered a profession, then you are said to have been "up-wardly mobile." If you have left the country for the city with visions of bright lights and more money, or if you have gone to the city because you have been replaced as a farm worker by machines and you have no other place to go and you end up homeless or living in a slum without a job, then I suppose you are downwardly mobile—but this is still "progress," for at least you have been relieved of "the idiocy of rural life" or the "mind-numbing work" of agriculture.

When replacement leads to "mobility" or displacement, and displacement leads to joblessness or homelessness, then we have a problem as characteristic of the industrialized world as land waste and pollution. To this problem the two political sides have produced nonsolutions that are hopeless and more cynical (I hope) than many of their advocates realize: versions of "Get a job," job training, job retraining, "better" education, job creation, and "safety nets" such as welfare, Social Security, varieties of insurance, retirement funds, etc. All of these "solutions," along with joblessness itself, serve the purposes of an economy of bubbling money. And every one of them fails to address the problem of "mobility," which is to say a whole society that is socially and economically unstable. In this state of perpetual mobility, even the most lucratively employed are likely to be homeless, if "home" means anything at all, for they are endlessly moving at the dictates of their careers or at the whims of their employers.

To escape the cynicism, heartlessness, and damage implicit in all this mobility, it is necessary to ask another question: Might it not have been that these replaced and displaced people were needed in the places from which they were displaced? I don't mean to suggest that this is a question easily answered, or that anybody should be required to stay put. I do mean that the question ought to be asked. It ought to be asked if only because it calls up another question that might lead to actual thinking: By what standard, or from what point of view, are we permitted to suppose that the displaced people were not needed in their original places? According to the industrial standard and point of view, persons are needed only when they perform a service valuable to an employer. When a machine can perform the same service, a person then is not needed.

Not-needed persons must graduate into mobility, which will take them elsewhere to a job newly vacated or "created," or to job training, or to some safety net, or to netlessness, joblessness, and homelessness. But this version of "not-needed" fits uncomfortably into the cultural pattern by which we define ourselves as civilized or humane or human. It grates achingly against the political and religious traditions that have affirmed for us the inherent worth and even preciousness of individual people. Our mobility, whether enforced or fashionable, has dismembered and scattered families and communities. Politicians and opinion dealers from far left to far right predictably and loudly regret these disintegrations, prescribing for them (in addition to the "solutions" already mentioned) year-round schools, day care, expert counseling, drugs, and prisons.

And so: Might it not be that the displaced persons were needed by their families and their neighbors, not only for their economic assistance to the home place and household, but for their love and understanding, for their help and comfort in times of trouble? Of the Americans known to me, only the Amish have dealt with such questions openly and conscientiously as families, neighbors, and communities. The Amish are Amish by choice. There is no requirement either to subscribe to the religion or to stay in the community. The Amish have their losses and their failures, as one would expect. Lately some of their communities have become involved in the failure of the larger economy. But their families and communities nevertheless have been held together by principle and by the deliberate rejection of economic and technological innovations that threaten them. With the Amish—as once with the rest of us—a family member or a neighbor is by definition needed, and is needed not according to any standard of usefulness or any ratio of cost and price, but according to the absolute standards of kindness, mutuality, and affection. Unlike the rest of us, the Amish have remembered that the best, most dependable, most kind safety net or social security or insurance is a coherent, neighborly, economically sound, local community.

To speak of the need for affection and loyalty and social stability is not at all to slight the need for life-supporting work. Of course people need to work. Everybody does. And in a money-using economy, people need to earn money by their work. Even so, to speak of "a job" as if it were the only economic need a person has, as if it doesn't matter what the job is or where a person must go in order to have it, is brutally reductive. To speak so is to leave out virtually everything that is humanly important: family and community ties, connection to a home place, the questions of vocation and good work. If you have "a job," presumably, you won't mind being a stranger among strangers in a strange place, doing work that is demeaning or unethical or work for which you are unsuited by talent or calling.

When people accept mobility as a condition of work, it means that they have accepted a kind of homelessness. It used to be a part of good manners to ask a person you had just met, "Where are you from?" That question has now become a social embarrassment, for it is too likely to be answered, "I'm not from anywhere." But to be not from anywhere is part of the definition of helplessness. Mobility is a condition in which you can do little or nothing to help yourself, and in which you live apart from family and old neighbors who would be the people most likely to help you.

Usury, for example, is "a job." But it happens to be a job that nobody ought to do. It is a violence against fellow humans who happen to be in need, a violence against work, or against good work, a violence against nature, and therefore (for those to whom it matters) a violence against God. It is a job also that estranges and isolates one from other people, who are perceived by the usurer, not as neighbors, but as potential victims.

To be mobile is not only to be in a new sense homeless. It is also to be in an old sense landless. If you have plenty of money to buy the necessities of life, and the stores are well-stocked with those necessities, then you may not see landlessness as a threat. But suppose you are a poor migrant, black or white, from the cotton or cane fields of the South or the Appalachian coal fields, and you wind up jobless in some "inner city." You have come from the country, and now, cooped up in a strange and unyielding place, without the mutual usefulness of a functioning neighborhood, you experience a helplessness that is new to you: the practical difficulty or impossibility of helping or being helped by somebody you know. A most significant part of that helplessness is the impossibility of helping yourself, and this is the condition of landlessness. I am not talking here about owning land, but merely of having access to it or the use of it. In your new circumstances of displacement, you have no place to grow a garden or keep a few chickens or gather firewood or hunt or fish. Maybe you were, by the official definition, poor where you came from, but there your abilities to do for yourself and others were given scope and efficacy by the landscape. You have come, in short, to the difference, defined by Paul Goodman a long time ago, between competent poverty and abject poverty. A home landscape enables personal subsistence but also generosity. It enables a community to exist and function.

When country people leave home to find work, even when "jobs" are available, they incur liabilities that cannot easily be discounted. The liabilities of homelessness and landlessness may not be noticeable in times of easy money and lots of stuff to buy. But in a time of economic failure and rising unemployment, as now, the liabilities once again rise undeniably into view.

Now the following sentences by Lowell H. Harrison and James C. Klotter, in *A New History of Kentucky*, make a different sense than they would have made to most readers a year ago:

> Yet [in the 1930s] the commonwealth weathered the drought and the floods and survived the depression better than many places. . . . [The] general absence of industry meant relatively little damage there, the overall lack of wealth left people only a little way to fall, and the rural nature of the commonwealth allowed families to live off the land. In fact, people returned to Kentucky, and the decade of the 1930s saw the state's population increase faster than the national average. . . . From distant places, those who had migrated in search of jobs that were now gone came home to crowd in with their families . . .

They "came home" because at home they still had families who were growing a garden, keeping a milk cow, raising chickens, fattening hogs, and gathering their cooking and heating fuel from the woods. Now, eighty years and much "progress" later, where will the jobless go? Not home, for there are no 1930s homes to go to.

Since the end of the Great Depression, and even more since the end of World War II, country people have crowded into the cities. They have come because they have attended colleges and been "overeducated" for country life. They have come for

available jobs. They have come because television and the movies have taught them to be unhappy in their "provincial" or
"backward" or "nowhere" circumstances. They have come because machines have displaced them from their work and their
homes. Many who have come were already poor, and were entirely unprepared for a life away from home. Immense numbers
of them have ended up in slums. Some live from some variety
of "safety net." Some, the homeless or insane or addicted poor,
sleep in doorways or under bridges. Some beg or steal.

In the long run, these surplus people, the not-needed, have
overfilled the "labor pool" and so have made labor relatively
cheap. If we run short of exploitable poor people in the United
States, then we "outsource" our work to the exploitable poor of
other countries. Industrial workers and labor unions are having
a hard time, and so are farmers, ranchers, and farm workers.
People who do the actual work of producing actual products
must expect to work cheap, for they are not of the quality of
the professionals who "deserve" to charge too much for their
services or the financial nobility who sell worthless mortgages.
As an exploitable underclass, those who perform actual work
have raised a vexing question for their superiors, and they seem
to have fallen somewhat short of the right answer: How could
they get the cheapest work out of their workers and still pay
them enough to afford the products they have made? Though
mere workers may be crippled by debt for their houses or farms
or their children's education, they must still be able with some
frequency to buy a new car or pickup truck or television set
or motorboat or tractor or combine. If they have such things
along with an occasional stunt in Outer Space, then maybe
they won't covet a financial noble's private jet and three or four
"homes."

Decades of cheap labor, cheap energy, and cheap food (all
more expensive than has been imagined) have allowed our
society to incorporate itself in a material structure that will
have to be seen as top-heavy. We have flooded the country,
the roadsides and landfills, with shoddy "consumer goods."
We have too many houses that are too big, too many public
buildings that are gigantic, too much useless space enclosed
in walls that are too high and under roofs that are too wide.
We replaced an until-then-adequate system of railroads with

an interstate highway system, expensive to build, disruptive of neighborhoods and local travel, increasingly expensive to maintain and use. We replaced an until-then-adequate system of local schools with consolidated schools, letting the old buildings tumble down, replacing them with bigger ones, breaking the old ties between neighborhoods and schools, and making education entirely dependent on the fossil fuels. Every rural school now runs a fleet of buses for the underaged and provides a large parking lot for those over sixteen who "need" a car to go to school. Education has been oversold, overbuilt, overelectrified, and overpriced. Colleges have grown into universities. Universities have become "research institutions" full of undertaught students and highly accredited "professionals" who are overpaid by the public to job-train the young and to invent cures and solutions for corporations to "market" for too much money to the public. And we have balanced this immense superstructure, immensely expensive to use and maintain, upon the frail stem of the land economy that we conventionally abuse and ignore.

There is no good reason, economic or otherwise, to wish for the "recovery" and continuation of the economy we have had. There is no reason, really, to expect it to recover and continue, for it has depended too much on fantasy. An economy cannot "grow" forever on limited resources. Energy and food cannot stay cheap forever. We cannot continue forever as a tax-dependent people who do not wish to pay taxes. Delusion and the future cannot serve forever as collateral. An untrustworthy economy dependent on trust cannot beguile the people's trust forever. The old props have been kicked away. The days when we could be safely crazy are over. Our airborne economy has turned into a deadfall, and we have got to jack it down. The problem is that all of us are under it, and so we have got to jack it down with the least possible suffering to our land and people. I don't know how this is to be done, and I am inclined to doubt that anybody does. You can't very confidently jack something down if you didn't know what you were doing when you jacked it up.

I do know that the human economy as a whole depends, as it always has, on nature and the land economy. The economy

of land use is our link with nature. Though economic failure
has not yet called any official attention to the land economy
and its problems, those problems will have to be rightly solved
if we are to solve rightly our other economic problems. Before we can make authentic solutions to the problems of credit
and spending, we have got to begin by treating our land with
the practical and effective love that alone deserves the name of
patriotism. From now on, if we would like to continue here,
our use of our land will have to be ruled by the principles of
stewardship and thrift, using as the one indispensable measure,
not monetary profit or industrial efficiency or professional success, but ecological health. And so I will venture to propose
the following agenda of changes that would amount to a new,
long-term agricultural policy:

1. There should be no further price supports or subsidies
   without production controls. This is because surplus
   production is an economic weapon, allowing corporations to reduce income to farmers while increasing their
   own income.
2. Return to 100 percent parity between agriculture and
   industry. Parity (fair) prices for agricultural products
   would make proposed payments for "ecological services" unnecessary, and would solve other problems as
   well.
3. Enforce anti-trust and anti-monopoly laws. Don't let
   any corporation get big, rich, or powerful enough to
   hold the nation for ransom. This applies with exceptional force to agribusiness and food corporations.
4. Help young farmers to own farms. In a sane economy
   such help would be unnecessary, but the departure of
   farm-raised young people from farming is now an agricultural crisis of the greatest urgency. And we don't
   have enough farm-raised young people. Others need to
   be drawn in. Here are some measures we should consider. We should set appropriate and reasonable acreage
   limits, according to region, for family-scale farms and
   ranches. Taxes should be heavy on holdings above those
   limits. Holdings within the prescribed limits should be

taxed at their agricultural value. There should be inheritance taxes on large holdings; none on small holdings. No-interest loans should be made available to young farmers and ranchers buying acreages under the limits. (These suggestions raise a lot of problems, and I flinch in making them. Acreage limits are hard to set appropriately, as we learned from the homestead laws. Also some of these measures would be unnecessary if land prices were not inflated above agricultural value, and if food prices were not deflated below their actual economic and ecological cost.)

5. Phase out toxic chemicals, which are inconsistent with the principles of good agriculture, and which are polluting the rivers and the oceans.

6. Phase out biofuels as quickly as possible. We have got to observe a strict distinction between fire and food, driving and eating. We can't "feed the hungry" and feed automobiles from the same land, using the same land-destroying technologies and methods, forever.

7. Phase in perennial plants—for pasture, winter forage, and grain crops—to replace annual crops requiring annual soil disturbance or annual applications of "no-till" chemicals. This would bring a substantial reduction of soil erosion and toxicity.

8. Set and enforce high standards of water quality.

9. High water quality standards (enforced) and a program to replace annual crops with perennials would tend strongly toward the elimination of animal factories. But let us be forthright on this issue. We should get rid of animal factories, those abominations, as quickly as we can. Get the farm animals, including hogs and chickens, back on grass. Put the animals where they belong, and their manure where it belongs.

10. Animal production should be returned to the scale of localities and communities. Do away with subsidies, incentives, and legislation favorable to gigantism in dairy, meat, and egg production.

11. Encourage the development of local food economies, which make more sense agriculturally and economically

than our present overspecialized, too-concentrated, long-distance food economy. Local food economies are desirable also from the standpoints of public health, "homeland security," and the energy economy. Provide economic incentives and supportive legislation for the establishment of local, small-scale food-processing plants, canneries, year-round farmers' markets, etc.

12. Local food economies, to be genuine, require local adaptation of domestic species and varieties of plants and animals. The universal evolutionary requirement of local adaptation has unaccountably been waived with respect to humans. But this waiver is potentially disastrous. We need ways of agriculture that are preservingly adapted to the ecological mosaic and even to individual farms and ranches. For the sake of local adaptation, and the genetic diversity that is necessary to it, we need to put an end to the U.S. Department of Agriculture's proposed National Animal Identification System, to the patenting of species, and to genetic engineering—all of which aim at a general agricultural uniformity and corporate control of agriculture and food. Central planning and its inevitable goal of uniformity cannot work in agriculture because of the requirement of local adaptation and the consequent need for local intelligence. Central planning and uniformity are effective only for the diminishment of genetic and biological diversity and the destruction of small producers.

13. Help and encourage small-scale forestry and owners of small woodlands. See that current market prices for sawlogs and other forest products are readily available everywhere. Tax fairly.

14. Study and teach sustainable forestry, using examples such as the Menominee Forest in Wisconsin and the Pioneer Forest in Missouri.

15. Promote the good use and care of farm woodlands as assets integral to the economy of farms.

16. Encourage the development, in forested regions, of local forest economies, providing economic incentives for local processing and value-adding, as for food.

Would such measures increase significantly the number of people at work in the land economy? Of course they would. This would be an authentic version, for a change, of "job creation." This work would help our economy, our people, and our country all at the same time. And that is the authentic test of practicality, for it makes complete economic sense.

Faustian Economics

THE GENERAL REACTION to the apparent end of the era of cheap fossil fuel, as to other readily foreseeable curtailments, has been to delay any sort of reckoning. The strategies of delay, so far, have been a sort of willed oblivion, or visions of large profits to the manufacturers of such "biofuels" as ethanol from corn or switchgrass, or the familiar unscientific faith that "science will find an answer." The dominant response, in short, is a dogged belief that what we call "the American way of life" will prove somehow indestructible. We will keep on consuming, spending, wasting, and driving, as before, at any cost to anything and everybody but ourselves.

This belief was always indefensible—the real names of global warming are "waste" and "greed"—and by now it is manifestly foolish. But foolishness on this scale looks disturbingly like a sort of national insanity. We seem to have come to a collective delusion of grandeur, insisting that all of us are "free" to be as conspicuously greedy and wasteful as the most corrupt of kings and queens. (Perhaps by devoting more and more of our already abused cropland to fuel production, we will at last cure ourselves of obesity and become fashionably skeletal, hungry but—Thank God!—still driving.)

The problem with us is not only prodigal extravagance, but also an assumed godly limitlessness. We have obscured the issue by refusing to see that limitlessness is a godly trait. We have insistently, and with relief, defined ourselves as animals or as "higher animals." But to define ourselves as animals, given our specifically human powers and desires, is to define ourselves as *limitless* animals—which of course is a contradiction in terms. Any definition is a limit, which is why the God of Exodus refuses to define Himself: "I am that I am."

Even so, that we have founded our present society upon delusional assumptions of limitlessness is easy enough to demonstrate. A recent "summit" in Louisville, Kentucky, was entitled "Unbridled Energy: The Industrialization of Kentucky's Energy Resources." Its subjects were "clean-coal generation, biofuels, and other cutting-edge applications," the conversion of coal to "liquid fuels," and the likelihood that all this will be "environmentally friendly." These hopes, which "can create jobs and boost the nation's security," are to be supported by government "loan guarantees . . . investment tax credits and other tax breaks." Such talk we recognize as completely conventional. It is, in fact, a tissue of clichés that is now the common tongue of promoters, politicians, and journalists. This language does not allow for any question about the *net* good of anything proposed. The entire contraption of "Unbridled Energy" is supported only by a rote optimism: "The United States has 250 billion tons of recoverable coal reserves —enough to last 100 years even at double the current rate of consumption." We humans have inhabited the earth for many thousands of years, and now we can look forward to surviving for another hundred by doubling our consumption of coal? *This* is national security? The world-ending fire of industrial fundamentalism may already be burning in our furnaces and engines, but if it will burn for a hundred more years, that will be fine. Surely it would be better to intend straightforwardly to contain the fire and eventually put it out? But once greed has been made an honorable motive, then you have an economy without limits, a contradiction in terms. This supposed economy has no plan for temperance or thrift or the ecological law of return. It will do anything. It is monstrous by definition.*

In keeping with our unrestrained consumptiveness, the commonly accepted basis of our present economy is the fantastical possibility of limitless growth, limitless wants, limitless wealth, limitless natural resources, limitless energy, and limitless debt. The idea of a limitless economy implies and requires a doctrine of general human limitlessness: *All* are entitled to pursue without limit whatever they conceive as desirable—a license that classifies the most exalted Christian capitalist with the lowliest pornographer.

This fantasy of limitlessness perhaps arose from the coincidence of the industrial revolution with the suddenly exploitable resources of the "new world." Or perhaps it comes from

---

*This is abundantly demonstrated by the suddenly ubiquitous rationalizations of "clean" and "safe" nuclear energy, ignoring the continuing problem of undisposable waste.

the contrary apprehension of the world's "smallness," made possible by modern astronomy and high-speed transportation. Fear of the smallness of our world and its life may lead to a kind of claustrophobia and thence, with apparent reasonableness, to a desire for the "freedom" of limitlessness. But this desire paradoxically reduces everything. The life of this world *is* small to those who think it is, and the desire to enlarge it makes it smaller, and can reduce it finally to nothing.

However it came about, this credo of limitlessness clearly implies a principled wish, not only for limitless possessions, but also for limitless knowledge, limitless science, limitless technology, and limitless progress. And necessarily it must lead to limitless violence, waste, war, and destruction. That it should finally produce a crowning cult of political limitlessness is only a matter of mad logic.

The normalization of the doctrine of limitlessness has produced a sort of moral minimalism: the desire to be "efficient" at any cost, to be unencumbered by complexity. The minimization of neighborliness, respect, reverence, responsibility, accountability, and self-subordination—this is the "culture" of which our present leaders and heroes are the spoiled children.

Our national faith so far has been "There's always more." Our true religion is a sort of autistic industrialism. People of intelligence and ability seem now to be genuinely embarrassed by any solution to any problem that does not involve high technology, a great expenditure of energy, or a big machine. Thus an X marked on a paper ballot no longer fulfills our idea of voting. One problem with this state of affairs is that the work now most needing to be done—that of neighborliness and caretaking—cannot be done by remote control with the greatest power on the largest scale. A second problem is that the economic fantasy of limitlessness in a limited world calls fearfully into question the value of our monetary wealth, which does not reliably stand for the real wealth of land, resources, and workmanship, but instead wastes and depletes it.

That human limitlessness is a fantasy means, obviously, that its life expectancy is limited. There is now a growing perception, and not just among a few experts, that we are entering a time of inescapable limits. We are not likely to be granted another world to plunder in compensation for our pillage of this one. Nor are we likely to believe much longer in our ability to outsmart, by means of science and technology, our economic stupidity. The hope that we can cure the ills of industrialism by the homeopathy of more technology seems at last to be losing status. We are, in short, coming under pressure to understand ourselves as limited creatures in a limited world.

This constraint, however, is not the condemnation it may seem. On the contrary, it returns us to our real condition and to our human heritage, from which our self-definition as limitless animals has for too long cut us off. Every cultural and religious tradition that I know about, while fully acknowledging our animal nature, defines us specifically as *humans* —that is, as animals (if the word still applies) capable of living, not only within natural limits, but also within cultural limits, self-imposed. As earthly creatures we live, because we must, within natural limits, which we may describe by such names as "earth" or "ecosystem" or "watershed" or "place" or "neighborhood." But as humans we may elect to respond to this necessary placement by the self-restraints implied in neighborliness, stewardship, thrift, temperance, generosity, care, kindness, friendship, loyalty, and love.

In our limitless selfishness, we have tried to define "freedom," for example, as an escape from all restraint. But, as my friend Bert Hornback has explained in his book *The Wisdom in Words*, "free" is etymologically related to "friend." These words come from the same Germanic and Sanskrit roots, which carry the sense of "dear" or "beloved." We set our friends free by our love for them, with the implied restraints of faithfulness or loyalty. This suggests that our "identity" is located not in the impulse of selfhood but in deliberately maintained connections.

Thinking of our predicament has sent me back again to Christopher Marlowe's *Tragical History of Doctor Faustus.* This is a play of the Renaissance: Faustus, a man of learning, longs to possess "all nature's treasury," to "Ransack the ocean . . . / And search all corners of the new-found world . . ." To assuage his thirst for knowledge and power, he deeds his soul to Lucifer, receiving in compensation for twenty-four years the services of the subdevil Mephistophilis, nominally his slave but in fact his master. Having the subject of limitlessness in mind, I was astonished on this reading to come upon Mephistophilis' description of hell. When Faustus asks, "How comes it then that thou art out of hell?" Mephistophilis replies, "Why, this is hell, nor am I out of it." A few pages later he explains:

> Hell hath no limits, nor is circumscribed
> In one self place, but where we [the damned] are is hell,
> And where hell is must we ever be.

For those who reject heaven, hell is everywhere, and thus is limitless. For them, even the thought of heaven is hell.

It is only appropriate, then, that Mephistophilis rejects any conventional limit: "Tut, Faustus, marriage is but a ceremonial toy. If thou lovest me, think no more of it." Continuing this theme, for Faustus' pleasure the devils present a sort of pageant of the seven deadly sins, three of which—Pride, Wrath, and Gluttony—describe themselves as orphans, disdaining the restraints of parental or filial love.

Seventy or so years later, and with the issue of the human definition more than ever in doubt, John Milton in Book VII of *Paradise Lost* returns again to a consideration of our urge to know. To Adam's request to be told the story of creation, the "affable Archangel" Raphael agrees "to answer thy desire / Of knowledge *within bounds* [my emphasis] . . . ," explaining that

> Knowledge is as food, and needs no less
> Her temperance over appetite, to know
> In measure what the mind may well contain;
> Oppresses else with surfeit, and soon turns
> Wisdom to folly, as nourishment to wind.

Raphael is saying, with angelic circumlocution, that knowledge without wisdom, limitless knowledge, is not worth a fart; he is not a humorless archangel. But he also is saying that knowledge without measure, knowledge that the human mind cannot appropriately use, is mortally dangerous.

I am well aware of what I risk in bringing this language of religion into what is normally a scientific discussion—if economics is in fact a science. I do so because I doubt that we can define our present problems adequately, let alone solve them, without some recourse to our cultural heritage. We are, after all, trying now to deal with the failure of scientists, technicians, and politicians to "think up" a version of human continuance that is economically probable and ecologically responsible, or perhaps even imaginable. If we go back into our tradition, we are going to find a concern with religion, which at a minimum shatters the selfish context of the individual life and thus forces a consideration of what human beings are and ought to be.

This concern persists at least as late as our Declaration of Independence, which holds as "self-evident, that all men are created equal; that they are endowed by their Creator with certain inalienable rights . . ." Thus among our political roots we have still our old preoccupation with our definition as humans, which in the Declaration is wisely assigned to our Creator; our rights and the rights of all humans are not granted by any human government but are innate, belonging to us by birth. This insistence comes, not from the fear of death or even extinction, but from the ancient fear, readily documentable in our cultural tradition, that in order to survive we might become inhuman or monstrous.

Our cultural tradition is in large part the record of our continuing effort to understand ourselves as beings specifically human—to say that, as humans, we must do certain things and we must not do certain things. We must have limits or we will cease to exist as humans; perhaps we will cease to exist, period. At times, for example, some of us humans have thought that human beings, properly so-called, did not make war against civilian populations, or hold prisoners without a fair trial, or use torture for any reason.

Some of us would-be humans have thought too that we should not be free at anybody else's expense. And yet in the phrase "free market," the word "free" has come to mean unlimited economic power for some, with the necessary consequence of economic powerlessness for others. Several years ago, after I had spoken at a meeting, two earnest and obviously troubled young veterinarians approached me with a question: How could they practice veterinary medicine without serious economic damage to the farmers who were their clients? Underlying their question was the fact that for a long time veterinary help for a sheep or a pig has been likely to cost more than the animal is worth. I had to answer that, in my opinion, so long as their practice relied heavily on selling patented drugs, they had no choice, since the market for medicinal drugs was entirely controlled by the drug companies, whereas most farmers had no control at all over the market for agricultural products. My questioners were asking in effect if a predatory economy can have a beneficent result. The answer usually is No. And that is because there is an absolute discontinuity between the economy of the seller of medicines and the economy of the buyer, as there is in the health industry as a whole. The drug industry is interested in the survival of patients, we have to suppose, because surviving patients will continue to consume drugs.

Now let us consider a contrary example. Recently at another meeting I talked for some time with an elderly, some would say old-fashioned, farmer from Nebraska. Unable to farm any longer himself, he had rented his land to a younger farmer on the basis of what he called "crop share" instead of a price paid or owed in advance. Thus, as the old farmer said of his renter, "If he has a good year, I have a good year. If he has a bad year, I have a bad one." This is what I would call community economy. It is a sharing of fate. It assures an economic continuity and a common interest between the two partners to the trade. This is as far as possible from the economy in which the young veterinarians were caught, in which the economically powerful are limitlessly "free" to trade to the disadvantage, and ultimately the ruin, of the powerless.

It is this economy of community destruction that, wittingly or unwittingly, most scientists and technicians have served for the last two hundred years. These scientists and technicians have justified themselves by the proposition that they are the vanguard of progress, enlarging human knowledge and power. Thus have they romanticized both themselves and the predatory enterprises that they have served.

As a consequence, our great need now is for sciences and technologies of limits, of domesticity, of what Wes Jackson of the Land Institute in Salina, Kansas, has called "homecoming." These would be specifically human sciences and technologies, working, as the best humans always have worked, within self-imposed limits. The limits would be, as they always have been, the accepted contexts of places, communities, and neighborhoods, both natural and human.

I know that the idea of such limitations will horrify some people, maybe most people, for we have long encouraged ourselves to feel at home on "the cutting edges" of knowledge and power or on some "frontier" of human experience. But I know too that we are talking now in the presence of much evidence that improvement by outward expansion may no longer be a good idea, if it ever was. It was not a good idea for the farmers who "leveraged" secure acreage to buy more during the 1970s. It has proved tragically to be a bad idea in a number of recent wars. If it is a good idea in the form of corporate gigantism, then we must ask, For whom? Faustus, who wants all knowledge and all the world for himself, is a man supremely lonely and finally doomed. I don't think Marlowe was kidding. I don't think Satan is kidding when he says in *Paradise Lost*, "myself am Hell."

If the idea of appropriate limitation seems unacceptable to us, that may be because, like Marlowe's Faustus and Milton's Satan, we confuse limits with confinement. But that, as I think Marlowe and Milton and others were trying to tell us, is a great and potentially a fatal mistake. Satan's fault, as Milton understood it and perhaps with some sympathy, was precisely that he could not tolerate his proper limitation; he could not subordinate himself to anything whatsoever. Faustus' error was his unwillingness to remain "Faustus, and a man." In our age of the world it is not rare to find writers, critics, and teachers of literature, as well as scientists and technicians, who regard Satan's and Faustus' defiance as salutary and heroic.

On the contrary, our human and earthly limits, properly understood, are not confinements, but rather are inducements to formal elaboration and elegance, to *fullness* of relationship and meaning. Perhaps our most serious cultural loss in recent centuries is the knowledge that some things, though limited, can be inexhaustible. For example, an ecosystem, even that of a working forest or farm, so long as it remains ecologically intact, is inexhaustible. A small place, as I know from my own experience, can provide opportunities of work and learning, and a fund of beauty, solace, and pleasure—in addition to its difficulties—that cannot be exhausted in a lifetime or in generations.

To recover from our disease of limitlessness, we will have to give up the idea that we have a right to be godlike animals, that we are at least potentially omniscient and omnipotent, ready to discover "the secret of the universe." We will have to start over, with a different and much older premise: the naturalness and, for creatures of limited intelligence, the necessity of limits. We must learn again to ask how we can make the most of what we are, what we have, what we have been given. If we always have a theoretically better substitute available from somebody or someplace else, we will never make the most of anything. It is hard enough to make the most of one life. If we each had two lives, we would not make much of either. One of my best teachers said of people in general: "They'll never be worth a damn as long as they've got two choices."

To deal with the problems, which after all are inescapable, of living with limited intelligence in a limited world, I suggest that we may have to remove some of the emphasis we have lately placed on science and technology and have a new look at the arts. For an art does not propose to enlarge itself by limitless extension, but rather to enrich itself within bounds that are accepted prior to the work.

It is the artists, not the scientists, who have dealt unremittingly with the problem of limits. A painting, however large, must finally be bounded by a frame or a wall. A composer or playwright must reckon, at a minimum, with the capacity of an audience to sit still and pay attention. A story, once begun, must end somewhere within the limits of the writer's and the reader's memory. And of course the arts characteristically impose limits that are artificial: the five acts of a play, or the fourteen lines of a sonnet. Within these limits artists achieve elaborations of pattern, of sustaining relationships of parts with one another and with the whole, that may be astonishingly complex. Probably most of us can name a painting, a piece of music, a poem or play or story that still grows in meaning and remains fresh after many years of familiarity.

We know by now that a natural ecosystem survives by the same sort of formal intricacy, ever-changing, inexhaustible, and perhaps finally unknowable. We know further that if we want to make our economic landscapes sustainably and abundantly productive, we must do so by maintaining in them a living formal complexity something like that of natural ecosystems. We can do this only by raising to the highest level our mastery of the arts of agriculture, animal husbandry, forestry, and, ultimately, the art of living.

It is true that insofar as scientific experiments must be conducted within carefully observed limits, scientists also are artists. But in science one experiment, whether it succeeds or fails, is logically followed by another in a theoretically infinite progression. According to the underlying myth of modern science, this progression is always replacing the smaller knowledge of the past with the larger knowledge of the present, which will be replaced by the yet larger knowledge of the future.

In the arts, by contrast, no limitless sequence of works is ever implied or looked for. No work of art is necessarily followed by a second work that is necessarily better. Given the methodologies of science, the law of gravity and the genome were bound to be discovered by somebody; the identity of the discoverer is incidental. But in the arts there are no second chances. We must assume that we had one chance each for *The Divine Comedy* and *King Lear*. If Dante and Shakespeare had died before they wrote those poems, nobody ever would have written them.

The same is true of our arts of land use, our economic arts, which are our arts of living. With these it is once-for-all. We will have no chance to redo our experiments with bad agriculture leading to soil loss. The Appalachian mountains and forests we have destroyed for coal are gone forever. It is now and forevermore too late to use thriftily the first half of the world's supply of petroleum. In the art of living we can only start again with what remains.

As we confront the phenomenon of "peak oil," we are really confronting the end of our customary delusion of "more." Whichever way we turn, from now on, we are going to find a limit beyond which there will be no more. To hit these limits at top speed is not a rational choice. To start slowing down, with the idea of avoiding catastrophe, *is* a rational choice, and a viable one if we can recover the necessary political sanity. Of course it makes sense to consider alternative energy sources, provided *they* make sense. But also we will have to reexamine the economic structures of our life, and conform them to the tolerances and limits of our earthly places. Where there is no more, our one choice is to make the most and the best of what we have.

IMAGINATION IN PLACE
(2010)

Imagination in Place

B<small>Y AN</small> interworking of chance and choice, I have happened to live nearly all my life in a place I don't remember not knowing. Most of my forebears for the last two hundred years could have said the same thing. I was born to people who knew this place intimately, and I grew up knowing it intimately. For a long time the intimacy was not very conscious, but I certainly did not grow up here thinking of the place as "subject matter," and I have never thought of it in that way. I have not lived here, or worked with my neighbors and my family, or listened to the storytellers and the rememberers, in order to be a writer. The place is precedent to my work, especially my fiction, and is, as I shall try to show, inevitably different from it.

By the same interworking of chance and choice, though somewhat expectably, I have lived here as a farmer. Except for one great-grandfather, all of my family that I know about have been farming people, and I grew up under instruction, principally from my father but also from others, to learn farming, to know the difference between good farming and bad, to regard the land as of ultimate value, and to admire and respect those who farmed well. I never heard a farmer spoken of as "just a farmer" or a farm woman as "just a housewife." To my father and his father especially, the knowledge of land and of farming was paramount. They thought the difference between a good farmer and a bad one was just as critical as the difference between a good politician and a bad one.

In 1964, after several years of wandering about, my wife Tanya and I returned to Kentucky with our two children and bought the property known as Lanes Landing, on the Kentucky River, about a mile from the house where my mother was born and raised and about five miles from my father's home place. The next summer we fixed up the house and moved in. We have been here ever since. Or Tanya and I have; our children are farming nearby.

Before we moved here, I had known this place for thirty-one years, and we have now lived here for thirty-nine. We raised our children here. We have taken from this place most of our food, much of our fuel, and always, despite the difficulties and frustrations of a farming life, a sustaining pleasure. Also, nearly everything I have written has been written here. When I am asked how all this fits together, I have to say, "Awkwardly." Even so, this has been the place of my work and of my life.

This essay is most immediately obstructed by the difficulty of separating my work from my life, and the place from either. The place included in some of my work is also the place that has included me as a farmer and as a writer.

In the course of my life and of my work as a farmer, I have come to know familiarly two small country towns and about a dozen farms. That is, I have come to know them well enough at one time or another that I can shut my eyes and see them as they were, just as I can see them now as they are. The most intimate "world" of my life is thus a small one. The most intimate "world" of my fiction is even smaller: a town of about a hundred people, "Port William," and a few farms in its neighborhood. Between these two worlds, the experienced and the imagined, there is certainly a relationship. But it is a relationship obscure enough as it is, and easy to obscure further by oversimplification. Another difficulty of this essay is the temptation to oversimplify.

As a lot of writers must know, it is easy for one's family or neighbors to identify fictional characters with actual people. A lot of writers must know too that these identifications are sometimes astonishingly wrong, and are always at least a little wrong. The inevitability of this sort of error is explainable, and it is significant.

Some of my own fiction has seemed to me to be almost entirely imagined. Some of it has drawn maybe as close as possible to actual experience. The writing has sometimes grown out of a long effort to come to terms with an actual experience. But one must not be misled by the claims of "realism." There is, true enough, a kind of writing that has an obligation to tell the truth about actual experience, and therefore it is obliged to accept the limits of what is actually or provably known. But works of imagination come of an impulse to transcend the limits of experience or provable knowledge in order to make a thing that is whole. No human work can become whole by including everything, but it can become whole in another way: by accepting its formal limits and then answering within those limits all the questions it raises. Any reasonably literate reader can understand Homer without the benefit of archaeology, or Shakespeare without resort to his literary sources.

It seems to me that my effort to come to terms in writing with an actual experience has been, every time, an effort to imagine the experience, to see it clear and whole in the mind's eye. One might suppose, reasonably enough, that this could be accomplished by describing accurately what one actually knows from records of some sort or from memory. But this, I believe, is wrong. What one actually or provably knows about an actual experience is never complete; it cannot, within the limits of memory or factual records, be made whole. Imagination "completes the picture" by transcending the actual memories and provable facts. For this reason, I have often begun with an actual experience and in the end produced what I have had to call a fiction. In the effort to tell a whole story, to see it whole and clear, I have had to imagine more than I have known. "There's no use in telling a pretty good story when you can tell a really good one," my mother's father told me once. In saying so, he acknowledged both a human limit and a human power, as well as his considerable amusement at both.

I believe I can say properly that my fiction originates in part in actual experience of an actual place: its topography, weather, plants, and animals; its language, voices, and stories. The fiction I have written here, I suppose, must somehow belong here and must be different from any fiction I might have written in any other place. I am pleased to suppose so, but the issue of influence is complex and obscure, and the influence of this place alone cannot account for the fiction and the other work I have written here.

Both my writing and my involvement with this place have been in every way affected by my reading. My work would not exist as it is if the influence of this place were somehow subtracted from it. Just as certainly it would not exist as it is, if at all, without my literary mentors, exemplars, teachers, and guides. Lists are dangerous, but as a placed writer I have depended on the examples of Andrew Marvell at Appleton

House, Jane Austen in Hampshire, Thomas Hardy in Dorset, Mark Twain in Hannibal, Thoreau in Concord, Sarah Orne Jewett on the Maine coast, Yeats in the west of Ireland, Frost in New England, William Carlos Williams in Rutherford, William Faulkner and Eudora Welty in Mississippi, Wallace Stegner in the American West, and in Kentucky, James Still, Harlan Hubbard, and Harry Caudill—to name only some of the dead and no contemporaries. I have kept fairly constantly in my mind the Bible, Homer, Dante, Shakespeare, Herbert, Milton, and Blake. I have taken much consolation and encouragement from Paul Cézanne's devotion to his home landscapes in Provence and from Samuel Palmer's work at Shoreham. I have remembered often the man of Psalm 128 who shall eat the labor of his hands, and Virgil's (and Ronsard's) old Cilician of *Georgics* IV. Over the last twenty years or so, I have contracted a large debt to certain writers about religious and cultural tradition, principally Ananda Coomaraswamy, Titus Burckhardt, Kathleen Raine, and Philip Sherrard —again, to name only the dead. Now that I have listed these names, I am more aware than before how incomplete any such list necessarily must be, and how necessarily confusing must be the issue of influence.

I will allow the list to stand, not as an adequate explanation, but as a hint at the difficulty of locating the origins of a work of fiction by me (or, I assume, by anybody else). And I must add further to the difficulty by saying that I don't believe I am conscious of all the sources of my work. I dislike learned talk about "the unconscious," which always seems to imply that the very intelligent are able somehow to know what they don't know, but I mean only to acknowledge that much of what I have written has taken me by surprise. What I know does not yield a full or adequate accounting for what I have imagined. It seems to have been "given." My experience has taught me to believe in inspiration, about which I think nobody can speak with much authority.

My fiction, anyhow, has come into being within the contexts of local geography and local culture, of the personal culture of reading, listening, and looking, and also within the contexts of what is not known and of the originating power we call inspiration. But there is another context, that of agriculture, which I will need to deal with at more length.

I was brought up, as I have said, by agrarians and was conscientiously instructed in a set of assumptions and values that could be described only as agrarian. But I never saw that word in print or heard it pronounced until I was a sophomore at the University of Kentucky. At that time I was in a composition class whose instructor, Robert D. Jacobs, asked us to write an argument. I wrote, as I recall, a dialogue between two farmers on the condemnation of land for the construction of a highway or an airport. The gist of my argument was that the land was worth more than anything for which it might be destroyed. Dr. Jacobs didn't think much of my argument, but he did me a valuable service by identifying it as "agrarian" and referring me to a group of writers, the Southern Agrarians, who had written a book called *I'll Take My Stand*. I bought the book and read at least part of it about three years later, in 1956. It is a valuable book, in some ways a wonder, and I have returned to it many times since. My debt to it has increased.

I must have become a good deal interested in the Southern Agrarians during my last years at the university, for with my friend and fellow student Mac Coffman (Edward M. Coffman, the historian) I drove up to Kenyon College to talk with John Crowe Ransom on that subject. But it is hard now for me to tell how much I may have been influenced by the Southern Agrarians and their book at that time. (Ransom by then was disaffected from *I'll Take My Stand*, though his elegant introduction, "A Statement of Principles," is still the best summary of agrarian principles versus the principles of industrialism). And I think I encountered not much at the University of Kentucky that would have confirmed my native agrarianism. It seems to me now that my agrarian upbringing and my deepest loyalties were obscured by my formal education. Only after I returned to Kentucky in 1964 did I begin to reclaim what I had been taught at home as a growing boy. Once I was home again, the purpose and point of that teaching became clear to me as it had not before, and I became purposefully and eagerly an agrarian. Moreover, because I had settled here as a farmer, I knew that I was not a literary agrarian merely but also a practical one.

*

In 1970 I published in *The Southern Review* a small essay, "The Regional Motive," that I suppose was descended from, or at least a cousin to, the essays of *I'll Take My Stand*. But in my essay I said that "the withdrawal of the most gifted of [the Southern Agrarians] into . . . Northern colleges and universities invalidated their thinking, and reduced their effort to the level of an academic exercise." Whatever the amount of truth in that statement, and there is some, it is also a piece of smartassery.

I received in response a letter from Allen Tate. As I knew, Tate could be a combative man, and so I was moved, as I still am, by the kindness of his letter. He simply pointed out to me that I did not know the pressing reasons why he and his friends had moved to the North. And so when I reprinted my essay I added a footnote apologizing for my callowness and ignorance, but saying, even so—and, as I remember, with Tate's approval —that I might appropriately "warn that their departure should not be taken either as disproof of the validity of their [agrarian] principles, or as justification of absentee regionalism (agrarianism without agriculture)."

The parentheses around that concluding phrase suggest to me now that I was making a point I had not quite got. The phrase, which appears to have been only an afterthought thirty-two years ago, indicates what to me now seems the major fault of *I'll Take My Stand*: The agrarianism of most of the essays, like the regionalism of most of them, is abstract, too purely mental. The book is not impractical—none of its principles, I believe, is in conflict with practicality—but it is too often remote from the issues of practice. The legitimate aim (because it is the professed aim) of agrarianism is not some version of culture but good farming, though a culture complete enough may be implied in that aim. By 1970 I had begun to see the flaws and dangers of absentee regionalism, and especially of Southern absentee regionalism. Identifying with "The South," as if it were somehow all one and the same place, would not help you to write any more than it would help you to farm. As a regional book, *I'll Take My Stand* mostly ignores the difficulty and the discipline of locality. As an agrarian book, it mostly ignores also the difficulty and the discipline of farming, but this problem is more complicated, and dealing with it took me longer.

Of the twelve essayists, only Andrew Lytle and John Donald Wade appear to speak directly from actual knowledge of actual farming in an actual place. And a passage of Andrew Lytle's essay, "The Hind Tit," points the direction I now must take with this essay of mine. He has begun to write about "a type" of farmer who has two hundred acres of land, but he does so with a necessary precaution:

> This example is taken, of course, with the knowledge that the problem on any two hundred acres is never the same: the richness of the soil, its qualities, the neighborhood, the distance from market, the climate, water, and a thousand such things make the life on every farm distinctly individual.

Thus he sets forth the fundamental challenge, not only to all forms of industrial land use, but to all other approaches to land use, including agrarianism, that are abstract.

The most insistent and formidable concern of agriculture, wherever it is taken seriously, is the distinct individuality of every farm, every field on every farm, every farm family, and every creature on every farm. Farming becomes a high art when farmers know and respect in their work the distinct individuality of their place and the neighborhood of creatures that lives there. This has nothing to do with the set of personal excuses we call "individualism" but is akin to the holy charity of the Gospels and the political courtesy of the Declaration of Independence and the Bill of Rights. Such practical respect is the true discipline of farming, and the farmer must maintain it through the muddles, mistakes, disappointments, and frustrations, as well as the satisfactions and exultations, of every actual year on an actual farm.

And so it has mattered, undoubtedly it has mattered to my fiction, that I have lived in this place both as a farmer and as a writer. I am not going to pretend here to a judgment or criticism of the writing I have done. I mean only to say something about the pressures and conditions that have been imposed on my writing by my life here as a farmer. Rather than attempt to say what I have done, I will attempt to speak of farming as an influence.

Having settled even in so marginal a place as this, undertaking to live in it even by such marginal farming as I have done, one is abruptly and forcibly removed from easy access to the abstractions of regionalism, politics, economics, and the academic life. To farm is to be placed absolutely. To do the actual work of an actual farm, one must shed the clichés that constitute "The South" or "My Old Kentucky Home" and come to the ground.

One may begin as an agrarian, as some of us to our good fortune have done, but for a farmer agrarianism is not enough. Southern agrarianism is not enough, and neither is Kentucky agrarianism or Henry County agrarianism. None of those can be local enough or particular enough. To live as a farmer, one has to come into the local watershed and the local ecosystem, and deal well or poorly with them. One must encounter directly and feelingly the topography and the soils of one's particular farm, and treat them well or poorly.

If one wishes to farm well, and agrarianism inclines to that wish above all, then one must submit to the unending effort to change one's mind and ways to fit one's farm. This is a hard education, which lasts all one's life, never to be completed, and it almost certainly will involve mistakes. But one does not have to do this alone, or only with one's own small intelligence. Help is available, as one had better hope.

In my farming I have relied most directly on my family and my neighbors, who have helped me much and taught me much. And my thoughts about farming have been founded on a few wonderful books: *Farmers of Forty Centuries* by F. H. King, *An Agricultural Testament* and *The Soil and Health* by Sir Albert Howard, *Tree Crops* by J. Russell Smith, and *A Sand County Almanac* by Aldo Leopold. These writers bring the human economy face to face with ecology, the local landscape, and the farm itself. They teach us to think of the ecological problems and obligations of agriculture, and they do this by seeing in nature the inescapable standard and in natural processes the necessary pattern for any human use of the land. Their thinking has had its finest scientific result thus far in the Natural Systems Agriculture of the Land Institute in Salina,

Kansas. Natural Systems Agriculture returns to the classical conception of art as an imitation of nature. But whereas Hamlet saw art as holding a mirror up to nature, and thus in a sense taking its measure, these agricultural thinkers have developed the balancing concept of nature as the inevitable mirror and measure of art.

In addition to books specifically about agriculture and ecology, I have been steadily mindful, as a farmer, of the writers mentioned earlier as literary influences. And I have depended for many years on the writing and the conversation of my friends Gene Logsdon, Maurice Telleen, Wes Jackson, and David Kline. I have been helped immeasurably also by the examples of Amish agriculture, of the traditional farming of Tuscany as I saw it more than forty years ago, of the ancient agricultures of the Peruvian Andes and the deserts of the American Southwest, of the also ancient pastoral landscapes of Devonshire, and of the best farming here at home as I knew it in the 1940s and early '50s, before industrialization broke up the old pattern.

What I have learned as a farmer I have learned also as a writer, and vice versa. I have farmed as a writer and written as a farmer. For the sake of clarity, I wish that this were more divisible or analyzable or subject to generalization than it is. But I am talking about an experience that is resistant to any kind of simplification. It is an experience of what I will go ahead and call complexification. When I am called, as to my astonishment I sometimes am, a devotee of "simplicity" (since I live supposedly as a "simple farmer"), I am obliged to reply that I gave up the simple life when I left New York City in 1964 and came here. In New York, I lived as a passive consumer, supplying nearly all my needs by purchase, whereas here I supply many of my needs from this place by my work (and pleasure) and am responsible besides for the care of the place.

My point is that when one passes from any abstract order, whether that of the consumer economy or Ransom's "Statement of Principles" or a brochure from the Extension Service, to the daily life and work of one's own farm, one passes from a relative simplicity into a complexity that is irreducible except by disaster and ultimately is incomprehensible. It is the complexity of the life of a place uncompromisingly itself, which is at the same time the life of the world, of all Creation. One meets not only the weather and the wildness of the world, but also the limitations of one's knowledge, intelligence, character, and bodily strength. To do this, of course, is to accept the place as an influence.

My further point is that to do this, if one is a writer, is to accept the place and the farming of it as a literary influence. One accepts the place, that is, not just as a circumstance, but as a part of the informing ambience of one's mind and imagination. I don't dare to claim that I know how this "works," but I have no doubt at all that it is true. And I don't mind attempting some speculations on what might be the results.

To begin with, the work of a farmer, or of the sort of farmer I have been, is particularizing work. As farmers themselves never tire of repeating, you can't learn to farm by reading a book. You can't lay out a fence line or shape a plowland or fell a tree or break a colt merely by observing general principles. You can't deal with things merely according to category; you are continually required to consider the distinct individuality of an animal or a tree, or the uniqueness of a place or a situation, and to do so you draw upon a long accumulation of experience, your own and other people's. Moreover, you are always under pressure to explain to somebody (often yourself) exactly what needs to be done. All this calls for an exactly particularizing language. This is the right kind of language for a writer, a language developing, so to speak, from the ground up. It is the right kind of language for anybody, but a lot of our public language now seems to develop downward from a purpose. Usually, the purpose is to mislead, the particulars being selected or invented to suit the purpose; or the particulars dangle loosely and unregarded from the dislocated intellectuality of the universities. This is contrary to honesty and also to practicality.

The ability to speak exactly is intimately related to the ability to know exactly. In any practical work such as farming the penalties for error are sometimes promptly paid, and this is valuable instruction for a writer. A farmer who is a writer will at least call farming tools and creatures by their right names, will be right about the details of work, and may extend the same courtesy to other subjects.

A writer who is a farmer will in addition be apt actually to know some actual country people, and this is a significant advantage. Reading some fiction, and this applies especially to some Southern fiction, one cannot avoid the impression that the writers don't know any country people in particular and are in general afraid of them. They fill the blank, not with anybody they have imagined, but with the rhetorically conjured stereotype of the hick or hillbilly or redneck who is the utter opposite of the young woman with six arms in the picture by the late ("Alas") Emmeline Grangerford, and perhaps is her son. He comes slouching into the universe with his pistol in one hand, his penis in another, his Bible in another, his bottle in another, his grandpappy's cavalry sword in another, his plug of chewing tobacco in another. This does harm. If you wish to steal farm products or coal or timber from a rural region, you will find it much less troubling to do so if you can believe that the people are too stupid and violent to deserve the things you wish to steal from them. And so purveyors of rural stereotypes have served a predatory economy. Two of the Southern Agrarians, I should add, countered this sort of thing with knowledge. I am thinking of John Donald Wade's essay "The Life and Death of Cousin Lucius" in *I'll Take My Stand*, and *A Wake for the Living* by Andrew Lytle.

If you understand that what you do as a farmer will be measured inescapably by its effect on the place, and of course on the place's neighborhood of humans and other creatures, then if you are also a writer, you will have to wonder too what will be the effect of your writing on that place. Obviously this is going to be hard for anybody to know, and you yourself may not live long enough to know it, but in your own mind you are going to be using the health of the place as one of the indispensable standards of what you write, thus dissolving the university and "the literary world" as adequate contexts for literature. It also is going to skew your work away from the standard of realism. "How things really are" is one of your concerns, but by no means the only one. You have begun to ask also how things will be, how you want things to be, how things ought to be. You want to know what are the meanings, both temporal and eternal, of the condition of things in this world. "Realism," as Kathleen Raine said, "cannot show us what we are, but only our failure to become that to which the common man and the common woman inadequately, but continually, aspire and strive." If, in other words, you want to write a whole story about whole people—living souls, not "higher animals"—you must reach for a reality which is inaccessible merely to observation or perception but which in addition requires imagination, for imagination knows more than the eye sees, and also inspiration, which you can only hope and pray for. You will find, I think, that this effort involves even a sort of advocacy. Advocacy, as a lot of people will affirm, is dangerous to art, and you must beware the danger, but if you accept the health of the place as a standard, I think the advocacy is going to be present in your work. Hovering over nearly everything I have written is the question of how a human economy might be conducted with reverence, and therefore with due respect and kindness toward everything involved. This, if it ever happens, will be the maturation of American culture.

I have tried (clumsily, I see) to define the places, real and imagined, where I have taken my stand and done my work. I have made the imagined town of Port William, its neighborhood and membership, in an attempt to honor the actual place where I have lived. By means of the imagined place, over the last fifty years, I have learned to see my native landscape and neighborhood as a place unique in the world, a work of God, possessed of an inherent sanctity that mocks any human valuation that can be put upon it. If anything I have written in this place can be taken to countenance the misuse of it, or to excuse anybody for rating the land as "capital" or its human members as "labor" or "resources," my writing would have been better unwritten. And then to hell with any value anybody may find in it "as literature."

American Imagination and the Civil War

S OME SENTENCES of the Irish poet Patrick Kavanagh have been prominent in my thoughts for many years:

> Parochialism and provincialism are opposites. The provincial has no mind of his own; he does not trust what his eyes see until he has heard what the metropolis . . . has to say . . . The parochial mentality on the other hand is never in any doubt about the social and artistic validity of his parish.

In spite of necessary qualifications, which I will get to in a minute, Kavanagh's distinction has become indispensable to me in thinking about my native place and history. In Kentucky, a state famously characterized as barefooted, we might oversimplify Kavanagh by saying that those of us who are always admiring our shoes are provincial, whereas the unself-consciously bare or shod are parochial. Or we could more legitimately paraphrase him by saying that people who fear they are provincial are provincial.

I believe I can say truthfully that my particular part of Kentucky, at the time of my growing up in it, was in Kavanagh's terms more parochial than provincial. The parochial in any locality probably always is subject to qualification and inexact in geographical extent. I grew up in a county at that time almost exclusively preoccupied with farming: the county of Henry, a few miles south of the Ohio River. But the country truly native to my family and my experience is in the watersheds of Town Branch of Drennon Creek, Emily's Run, and Cane Run.

During my first twenty or so years, the "social validity" of that place at that time certainly was impaired by racial segregation. That phrase now has the currency of an abstraction, but segregation itself could be experienced only in particular. We were living in the history of segregation, but we were living in it in our place, with our neighbors, and as ourselves. In our small communities segregation involved the wicked prejudice on which it was based, but it also involved much familiarity and many exceptions. Racial inequality was a theory that performed its customary disservices and sometimes justified horrors, but that theory was inevitably qualified by the daily life in which the two races were separate only to an extent. In those places the history of segregation was lived out familiarly by black and white people who knew one another, told stories about one another to one another, helped or harmed one another, liked or disliked one another, and often worked together. Separate and different as the races were, it is impossible to imagine a white person of that place and time whose knowledge did not include the stories, songs, sayings, teachings, and characters of black persons. An honest accounting of the ancestry of my own mind would have to include prominently several black people. Despite segregation, the communities of my young life were, in function and in their consciousness of themselves, more intact than they are now.

As for the "artistic validity" of our place at that time, I have to be both careful and modest. We did have a local music that came to the fore at square dances, though not everybody granted much value to it, and it had begun to be supplanted by music from the radio and jukebox. Most of us were familiar with the Protestant hymns and the King James Bible. But we were not greatly concerned with issues of art, local or otherwise.

The arts that we took for granted, and that did gather us all together, were the arts of farming, gardening, cooking, and talking. Our economy was either agricultural or in service to agriculture. Vegetable gardens, grape arbors, and fruit trees were still commonplace. It was still ordinary to see poultry flocks, fattening hogs, or milkcows in the back yards or back lots of the towns. The grocery stores still bought surplus produce from the farms. Most of the food was homegrown, and excellent cooking was customary. Most of the cooking was done by women, but everybody talked about it.

Everybody, in fact, talked about everything. It seems to me that I grew up totally immersed in talk. Talk was a fifth element: talk in hayfields and tobacco patches, in tobacco barns and stripping rooms, in kitchens and living rooms, on porches and out in the yards. Sometimes, as we sat out in the yard or on the porch after a hot day, the dark would gradually disembody us, and we would become just voices going on until weariness re-embodied us and we would go into the house to bed. My best gift as a writer was that circumstance of talk. We had no cultivated art of conversation. Our talk was practical, local in reference, but was carried on also for pleasure or comfort. It was sometimes crude enough, but it was also articulate enough, humorous, precise, expressive, and sometimes beautifully so.

Though by then most of us had listened to the radio and seen at least a few movies, our talk as yet bore no hint of apology for the way we talked, or for our status as country or small-town people. We knew we were not Yankees, for we had heard Yankees talk and we knew we did not talk like them. We also had listened to people from "down South," and we knew we did not have what we called a "Southern accent." Maybe I can be excused for concluding, when I got old enough to read a map, that I spoke a perfectly average language, Standard American, since I could see that I lived at about the middle of the north–south axis. And maybe I can elicit a little sympathy for my surprise when, having clung to this notion all the way to some literary party in California, I delivered an undoubtedly sophisticated opinion to a literary young lady, whose eyes thereupon grew round with recognition. "Wayull!" she said in Yankee-Southern. "Wheah *you*'all frum, honey chile?"

And so I turned out to be a Southerner—legitimately so, as that term is used. I was born on the south side of the Ohio, was descended from slave-owners, and certainly did not talk like a Yankee.

The problem, as I am hardly the first to know, is that being a Southerner is less a condition than a job. The job, unendingly, is to distinguish between local life and the abstractions that we have allowed to obscure it. There is a huge difference between knowledge and classification. "South" and "Southerner" are not terms that are invariably useful. They belong sometimes to a taxonomy of clichés, stereotypes, and prejudices that have intruded between ourselves and our actual country. These shallow, powerful abstractions have worked to depreciate local knowledge and provincialize local life, and so have denied us the imaginative realizations that alone could have saved our country from the damage that has befallen it.

These old habits of mind and speech have continued in the babbled-to-nonsense polarity of "conservative" and "liberal," and of "red states" and "blue states." This oversimplified language of the media and politics is as far as possible from the best of the local speech I heard as a child, which was like no other in the world because it was of and about our place, which was like no other in the world. In it we were at least beginning to imagine ourselves somewhat as we actually were, and even somewhat as we should have been. Now, under the influence of media speech, we can only pretend and try to be like everybody else.

The problem is that there can be no general or official or sectional or national imagination. The chief instrument of economic and political power now is a commodified speech, wholly compatible with the old clichés, that can distinguish neither general from particular nor false from true. Local life is now a wren's egg brooded by an eagle or a buzzard. As Guy Davenport saw, nothing now exists that is so valuable as whatever theoretically might replace it. Every place must anticipate the approach of the bulldozer. No place is free of the threat implied in such phrases as "economic growth," "job creation," "natural resources," "human capital," "bringing in industry," even "bringing in culture"—as if every place is adequately identified as "the environment" and its people as readily replaceable parts of a machine. Devotion to any particular place now carries always the implication of heartbreak.

I suppose that human minds have always been threatened by the slur and blur of general bias, but it seems to me that this curse fell upon us Americans with a great fatefulness in the circumstances leading to the Civil War, and that the curse has persisted.

The Civil War and the rhetoric associated with it become penetrable by actual thought only when one asks why. Why could people of good sense on both sides not have treated slavery as a problem with a practical solution short of war? The answers, I suppose, are foolishness, fanaticism, sectional loyalty and pride, the wish to protect one's faults from correction by others, moral outrage, self-righteousness, the desire to punish sinners, and sectional hatred.

The Civil War was caused undoubtedly by disagreements over slavery and secession. It was contested so fiercely and so long by the Confederacy undoubtedly because of a truth that our federal government has never learned: People generally don't like to be invaded. But why was there no lenity?

Shakespeare's Henry V, incongruously in the midst of his invasion of France, gives "lenity" a pertinent definition:

> . . . we give express charge that in our marches through the country there be nothing compelled from the villages, nothing taken but paid for, none of the French upbraided or abused in disdainful language; for when lenity and cruelty play for a kingdom, the gentler gamester is the soonest winner.

The word occurs more credibly in Edmund Burke's *Speech On American Taxation*, in which he pleads desperately against the impositions that brought on the American Revolution:

> Yet now, even now, I should confide in the prevailing virtue and efficacious operation of lenity, though working in darkness and in chaos, in the midst of all this unnatural and turbid combination: I should hope it might produce order and beauty in the end.

The American Revolution may have been another "irrepressible conflict," but Burke, who saw it as a civil war, seems never to have doubted that there were two other possibilities: reconciliation on terms of justice or amicable separation.

Lenity can be understood as lenience or gentleness or mercy, and there was too little of it in Burke's England in 1774. There was too little in our North and South from 1861 to 1865, and before, and after. Failing lenity in any conceivable form, relishing its differences, savoring its animosities and divergent patriotisms, the nation divided and went to war. The two sides met in a series of great battles, and at last the strongest won in the name of emancipation and union.

That is the official version, and it is right enough as far as it goes. But to grant a just complexity to this history let us add a third side: that of the dead. Armies, by the necessity and purpose of military organization, are abstractions. We think of battles as convergences not of individuals but of "units."

Survivors, in their memoirs, speak as participants. Only in the aftermath of battle, on the nighttime battlefields horribly littered with the dead and the dying, do the individual soldiers begin to enter our imagination in their mere humanity. Imagination gives status in our consciousness and our hearts to a suffering that the statisticians would undoubtedly render in gallons of blood and gallons of tears. Maybe I am speaking only for myself, though I doubt it, when I say that to me the dead in Mathew Brady's photographs don't look like Unionists or Confederates; they look like dead boys, once uniquely themselves, undiminished by whichever half of the national quarrel they died for. In those photographs we meet war as a great maker of personal tragedies, not as a great enterprise of objectives.

Mathew Brady was by no means the first to show us this, nor was Shakespeare, but Shakespeare did show us, with a poignance unsurpassed in my reading, the tragedy specifically of civil war. In *Henry VI, Part III*, there is a battle scene in which first "a Son" and then "a Father," not identified as to side, enter separately, each bearing the body of a dead man whom he has killed and whom he now looks at. The Son says:

Who's this? O God! it is my father's face . . .

And the Father says:

But let me see: is this our foeman's face?
Ah, no, no, no, it is mine only son!

Of our own civil war Walt Whitman saw clearly the pageantry and glamour and "all the old mad joy" of battle that Robert E. Lee acknowledged. But he saw also the personal tragedy much as Shakespeare saw it. With the same anonymity as to side, he speaks of coming at dawn upon three of the dead lying covered near a hospital tent:

Curious I halt and silent stand,
Then with light fingers I from the face of the nearest the
    first just lift the blanket;
Who are you elderly man so gaunt and grim, with well-gray'd
    hair, and flesh all sunken about the eyes?
Who are you my dear comrade?

Then to the second I step—and who are you my child and
     darling?
Who are you sweet boy with cheeks yet blooming?

Then to the third—a face nor child nor old, very calm,
     as of beautiful yellow-white ivory;
Young man I think I know you—I think this face is the
     face of the Christ himself,
Dead and divine and brother of all, and here again he lies.

Once dead, the dead in war are conscripted again into abstraction by political leaders and governments, and this is a
great moral ugliness. The dead are made hostages of policy to
sanctify the acts and intentions of their side: These have died
in a holy cause; that they may not have died in vain, more
must be killed. And to benefit the victors, there is always the
calculation, frequently alluded to but never openly performed:
At the cost of so many deaths, so much suffering, so much destruction, so much money or so much debt, we have got what
we wanted, and at a fair price.

There is no doubt that wars may have moral purposes. Union
and emancipation were moral purposes. So were secession and
independence, however muddied by the immoral purpose of
slavery. But battles don't have the same purposes as wars. The
only purpose of a battle, once joined, is victory. And any price
for victory is acceptable to the generals and politicians of the
victorious side, who are under great pressure to say that it is acceptable. But the accounting is conventionally not attempted.
Victors do not wish to evaluate their victory as a net gain, for
fear that it will prove a net loss.

I doubt that such a calculation is possible, even if somebody
were willing to try it. But that should not stop us from asking,
if only to keep the question open, what we gained, as a people,
by the North's expensive victory. My own impression is that
the net gain was more modest and more questionable than is
customarily said.

The Northern victory did preserve the Union. But despite
our nationalist "mystique," our federation of states is a practical condition maintained only by the willing consent of the

states and the people. And secession, today, is still not a dead issue. There is now, for instance, a vigorous and strictly principled secession movement in Vermont.

The other large Northern objective, the emancipation of the slaves, also was achieved. But this too appears in retrospect to be an achievement painfully limited. It does not seem unreasonable to say that emancipation was achieved and, almost by the same stroke, botched. The slaves were set free only to remain an exploited people for another hundred years. My guess is that, after the decision was taken to make slavery an issue of war, emancipation was inevitably botched. The North in effect abandoned the ex-slaves to the mercy of its embittered and still dissident former enemy, to whom they would be ever-present reminders, symbols virtually, of defeat.

Furthermore, we have remained a people in need of a racially designated underclass of menial laborers to do the work that the privileged (of whatever race) are too good, too well educated, and too ignorant to do for themselves. Our Stepanfetchits at present are Mexican immigrants, whom we fear for the familiar reasons that we exploit them and that we depend on them.

And so our Civil War raised questions that have been raised a number of times since: Can you force people to change their hearts and minds? Can you make them good by violence? Again and again human nature has replied no. Again and again, ignoring human nature and history, politicians have answered yes. And yet it seems true that Martin Luther King Jr. and his followers, by refusing to answer violence with violence, did more to alter racial attitudes in the South than was done by all the death and damage of the Civil War.

Is this reading of history too idealistic and unforgiving? Probably. Must we not say, pragmatically, that a botched emancipation is better than legal slavery? Well, I am a farmer, therefore a pragmatist: Half a crop beats none; a botched emancipation is better than none. But, as I am a farmer, I am also a critic, and I know the difference between a bad result and a good one. Of our history, though we cannot change it, we must still try for a true accounting. And to me it seems that the resort to violence is the death of imagination. Once the killing has started, lenity and the hope for order and beauty vanish along with causes and aims. Edmund Wilson's logic of the two sea slugs, the larger eating the smaller, then goes into effect: "not virtue but . . . the irrational instinct of an active power organism in the presence of another such organism . . ."

Once opponents become enemies, then the rhetoric of violence prevents them from imagining each other. Or it reduces imagination to powerlessness. Men such as Lincoln and Lee, from what I have read of them, seem not to have been destitute of imagination; of this I take as a sign their grief, their regret for the war even while they fought it. I see them as figures of tragedy, each an instrument of an immense violence which, once begun, was beyond their power to mitigate or stop, and which made of their imagination only a feckless suffering of the suffering of others. Once the violence has started, the outcome must be victory for one side, defeat for the other—with perhaps unending psychological and historical consequences.

When my thoughts circle about, trying to give my disturbance a location that is specific and familiar enough, they light sooner or later on "The Battle Hymn of the Republic." This song has a splendid tune, but the words are perfectly insane. Suppose, if you doubt me, that an adult member of your family said to you, without the music but with the same triumphal conviction, "Mine eyes have seen the glory of the coming of the Lord"—would you not, out of fear and compassion, try to find help? And yet this sectional hymn, by an alchemy obscure to me, seems finally to have given us all—North and South, East and West—a sort of official judgment of our history. It renders our ordeal of civil war into a truly terrifying simplemindedness, in which we can still identify Christ with military power and conflate "the American way of life" with the will of God.

I have made clear, I hope, my failure to perceive the glory of the coming of the Lord in the Civil War and its effects. The North was not uniformly abolitionist; the South was not uniformly proslavery or even prosecession. Theirs was not a conflict of pure good and pure evil. The Civil War was our first great industrial war, which was good for business, like every war since. The Civil War established violence against noncombatants as acceptable military policy. The Army of the United States, no longer the Northern army, proceeded from the liberation of the slaves to racist warfare against the native tribespeople of the West. Moreover, as the historian Donald Worster has said, the Civil War supplanted the "slave power" of the South with the "money power" of the North: "The fact of the matter is we have not even today figured out how to come to terms with the money power that replaced the slave power . . ." The great advantage of the aftermath went, certainly not to the ex-slaves or to the farmers and small tradesmen of either side, not to the people Wallace Stegner called "stickers," but rather to those he called "boomers": the speculators and exploiters, the main-chancers, the Manifest Destinarians, the railroads, the timber and mineral companies.

My purpose in reciting these problems is not to suggest that a Southern victory would have been better—which I doubt —but only to point out that the Northern victory set the tone of overconfidence, of self-righteousness and assumed privilege, that became the political tone of the whole nation.

The Civil War was followed, perhaps as a matter of course —and would have been followed, no matter who won—by the industrial exploitation of our land and people that still continues. While we have stood at our school desks or in our church pews asserting the divine prerogative of "The Battle Hymn," we have been destroying our country. This is not an impression. By measures empirical enough, we have wasted perhaps half of our country's topsoil; we are destroying by "development" thousands of acres every day; we have polluted the atmosphere and the water cycle; we have destroyed or damaged or brought under threat all of our natural ecosystems; in our agriculture and forestry we are treating renewable resources as carelessly as we have burned the fossil fuels; we have severely damaged all of our human communities. We have established unregarding violence as our means of choice in everything from international relations to land use to entertainment.

What are we to conclude? Only, I fear, that violence is its own way, which is entirely unlike the ways of thought or dialogue or work or art or any manner of caretaking. Once you have committed yourself to the way of violence, you can only suffer it through to exhaustion and accept the always unforeseen results.

I have been describing an enormous failure, and to me this appears to be a failure of imagination. Though we are now

far advanced in the destruction of our country, we have only
begun to imagine it. We are destroying it *because* of our failure
to imagine it.

By "imagination" I do not mean the ability to make things
up or to make a realistic copy. I mean the ability to make real
to oneself the life of one's place or the life of one's enemy—and
therein, I believe, is implied imagination in the highest sense.
When I use this word I never forget its definitions by Coleridge
and Blake, but for present purposes I am going to refer to the
writings of William Carlos Williams, whose understanding of
imagination, though compatible with that of his English predecessors, is peculiarly American in its urgency.

Three generations and more ago, Williams was fretting
about the inclination of Americans to debase their land and,
with it, themselves,

> as if the earth under our feet
> were
> an excrement of some sky
>
> and we degraded prisoners
> destined
> to hunger until we eat filth . . .

We were, he thought, "like a chicken with a broken neck, that
aims where it cannot peck and pecks where it cannot aim,
which a hog-plenty everywhere prevents from starving to
death . . ."

Williams seems to have been one of the few so far who could
see the vulnerability of a highly centralized economy. In a letter
to James Laughlin on November 28, 1950, he tells of the dis-
ruptions of a recent storm, and then he writes:

> But witnessing what one small storm can do to a com-
> munity in these parts I am awonder over the thought of
> what a single small atom bomb might not accomplish.
> Disruption of every service, now become more and more
> centralized, would starve us out in 3 days . . .

Against such craziness he set the "single force" of imagination:
"To refine, to clarify, to intensify that eternal moment in which
we alone live . . ." And imagination, in this sense, is not passively holding up a mirror to nature; it is a changing force. It

does not produce illusions, or copies of reality, or "plagiarism
after nature." And yet it does not produce artificiality. It does
not lead away from reality but toward it. It can be used to
show relationships. By it "the old facts of history" are "re-
united in present passion." Thus I have pieced together Williams' thoughts from the prose fragments of *Spring and All*.
Thirty or so years later, in "The Host," one of the devotional
poems in *The Desert Music*, he lays it out more plainly, giving
it, like Coleridge and Blake, a religious significance:

> There is nothing to eat,
>   seek it where you will,
>   but of the body of the Lord.
>
> The blessed plants
>   and the sea, yield it
>   to the imagination
> intact. And by that force
>   it becomes real . . .

If what we see and experience, if our country, does not become
real in imagination, then it never can become real to us, and
we are forever divided from it. And for Williams, as for Blake,
imagination is a particularizing and a local force, native to the
ground underfoot. If that ground is not in a great cultural
center, but only in a New Jersey suburb, so be it. Imagination
is as urgently necessary in Rutherford, New Jersey, or in Knott
County, Kentucky, or in Point Coupée Parish, Louisiana, as
it is in San Francisco or New York. As I am understanding it,
imagination in this high sense shatters the frameworks of realism in the arts and empiricism in the sciences. It does so by
placing the world and its creatures within a context of sanctity
in which their worth is absolute and incalculable.

The particularizing force of imagination is a force of justice
with obvious crucial correspondences in biology and in our
legal system. Robert Ulanowicz says that "in ecosystems com-
prised of hundreds or thousands of distinguishable organisms,
one must reckon not just with the occasional unique event,
but with *legions* of them. Unique, singular events are occurring
all the time, everywhere!" And, except for identical twins, every creature that comes into being by way of sexual reproduction is genetically unique. Recognition of the uniqueness of

creatures and events is the reason for the standing we humans grant (when we do grant it) to one another before the law, and it is the reason we "return thanks" (when we do so) for food and other gifts that come to us from the living world. Without imagination there is no right appreciation of these rarities—no lenity, amity, or mercy. And, I think, there is no satisfaction either. Imagination, amply living in a place, brings what we want and what we have ever closer to being the same. It is the power that can save us from the prevailing insinuation that our place, our house, our spouse, and our automobile are not good enough.

Historians and scientists work toward generalizations from their knowledge, just as all of us do. We must do this, for generalization is a part of our means of making sense. But generalization alone, without the countervailing, particularizing power of imagination, is dehumanizing and destructive.

"The South," for example, as the name of an historical side, can have a reckonable and useful meaning. But as the name merely of a part of the country, it means less. If "region" means anything at all, then the South, like the North or the West, is a region of many regions. But so is Kentucky. My county has several distinct regions. My neighbors don't look like Southerners or Kentuckians to me. The better I know them, the more they look like *themselves*. The better I know my place, the less it looks like other places and the more it looks like itself. It is imagination, and only imagination, that can give standing to these distinctions.

If imagination is to have a real worth to us, it needs to have a practical, an *economic*, effect. It needs to establish us in our places with a practical respect for what is there besides ourselves. I think the highest earthly result of imagination is probably local adaptation. If we could learn to belong fully and truly where we live, then we would all finally be native Americans, and we would have an authentic multiculturalism.

And yet the problem I began with has never been resolved: How do we equilibrate or even negotiate between local identity and the abstractions of regional or national identity with the attendant clichés of "economic growth"? Obviously there can be no general answer to this question. If we see the need for an answer, then we must attempt it for ourselves in our

communities. I believe that there is hope in the increasing uneasiness of people who see themselves as dispossessed or displaced and therefore as economically powerless. Growing out
of this uneasiness, there is now a widespread effort toward local
economy, local self-determination, and local adaptation. In this
there is the potential of a new growth of imagination, and at
last an authentic settlement of our country.

But we must not fool ourselves. This movement toward local adaptation necessarily is being led from the bottom. And
it confronts a leadership from the top—in government, in the
corporate economy, in the universities—that is utterly lacking
in imagination, local loyalty, and local knowledge. Both conservatives and liberals, having accepted the ecological and social damages of industrialism as inevitable, even normal, have
conceived the individual as subject alone either to the economy or to the government. In this official numbness, though
it is clearly self-doomed, there is for the moment an almost
overwhelming power.

Let me give you an example of the way a failure of imagination works against people and land. At present, in the eastern
mountains of my state, the coal companies are blasting the tops
off the mountains and pushing them into the valleys, covering
the streams. They are doing this without concern for the land,
the topsoil, the forest, the waterways and the water, or for the
homes and lives of the people. This total, permanent destruction is not anomalous in our economy or without causes in
our history. I need not delay you here by retelling the history
of the corporate pillage of eastern Kentucky, which you will
find well told in *Night Comes to the Cumberlands* and other
books by Harry M. Caudill, or by describing the culmination
of that malignant history in "mountaintop removal," which
Erik Reece has accomplished fully in *Lost Mountain*. But if you
want to know how this hardly credible or bearable waste could
have happened, consider the chapter "Mountain Passes of the
Cumberland" in *The Bluegrass Region of Kentucky and Other
Kentucky Articles* by James Lane Allen, published in 1892.

Allen was a "genteel" writer of the Bluegrass and an outsider to the mountains, about which he had a curiously divided
mind. On the one hand he regarded the then mostly unspoiled
forests of that region with a sort of rapture. But he was equally

rapturous about the economics and machinery of industrialization. Between these enthusiasms he saw no possible contradiction. The beautiful forest, of course, was invaluable in itself,
and it would be nearly completely cut down by the middle
of the next century. Beneath the forest lay enormously valuable deposits of coal. Allen accurately foresaw the industrial
exploitation of the region, but he thought only good could
come of it.

As for the local people, he characterized them in the person
of "a faded, pinched, and meager mountain boy" driving a
team of oxen and eating from a sack of candy:

> In one dirty claw-like hand he grasped a small paper bag,
> into the open mouth of which he had thrust the other
> hand . . . He had just bought . . . some sweetmeat of
> civilization which he was about for the first time to taste.

These people, according to Allen, needed to be civilized and
Christianized—which was the official rationale of the federal
campaigns against the Indians in the same era.

I don't think Allen was an evil man. Probably, like us, he
was pretty good. But he also was ignorant and naive, as we too
have been about our continuing bonfire of coal and oil. Being
only a prophet, Allen had no doubt about the beneficence of
industrialization; he thought it was the coming of the Lord:
"You begin with coke and end with Christianity." Compare
this bit of prophecy with James Still's *River of Earth*, written
half a century later, and you will see what I mean by failure of
imagination.

I would like to end by turning to the work of a Southern writer
who, for several reasons, is exemplary and dear to me. Ernest
J. Gaines inherited the harder side of the racial history that I
inherited, and I believe I know some of the questions he faced
as he made his way into his work. Could he imagine sympa-
thetically a Southern white person? Could he imagine, so as
to require us to imagine, an uneducated black farmhand as a
person of dignity, wisdom, and eloquence? Yes, as we know, he
could. He has imagined also the community of his people as a
part of the life of their place and the hardships of that community. He has imagined the community's belonging to its place,
the houses that had the names of people, the flower-planted

dooryards, the church, the graveyard, the shared history and experience, the shared stories, the talk of old people on the galleries in the summer evenings and the young people listening. He has imagined also the loss of those things.

In a time when the provincial fear of provinciality has brought the local into suspicion, Ernest Gaines has been true to his place, his people, and their story. He has shown that the local, fully imagined, becomes universal. He has brought his place and his people to such a pitch of realization that again and again as I read him he seems to speak also for me and mine. He has done this in a language like no other, belonging to a place like no other.

His novel *In My Father's House* contains a passage that alarms and consoles me every time I read it. The reason for my alarm is obvious, for the passage is about somebody's damage that becomes everybody's danger. But I consent to the author's understanding of the damage, and I am consoled by the companionship of that.

The main character in the novel, Phillip Martin, gives a ride to a terribly angry young black man named Billy. Billy has been to Vietnam. He is now a would-be revolutionary. He would like to burn the whole country by setting fire to the gasoline in every filling station. In his fury, this Billy is thoroughly frightening. I am frightened of him and for him. But then Billy says:

> You see all them empty fields round here, mister . . .
> Go all over this place—empty fields, empty houses, empty
> roads. Where the people used to be—nothing. Machines.
> Every time they build another machine that takes work
> from the people, they hire another hundred cops to keep
> the people quiet.

And I am caught. I see that Billy and I are joined by a mutual sense of calamity and loss. From my well-wishing in the safety of my chair, I have been carried into the trouble itself that has so nearly consumed Billy. Suddenly, in the midst of his rant, he has spoken from the grief felt by many rural Americans, of whatever race, and certainly by me. I know well that it is possible for me, like Billy, to respond with anger and despair. But I know also that it is possible for me, as for Ernest Gaines, to respond with work, hope, and love.

Sweetness Preserved

WHAT I AM GOING TO DO is talk about some poems—lyric poems—as the products of a story. Most poems, whether or not they tell or contain stories, come out of stories, and often they bear reference to the stories they come out of. For a couple of generations now, critics and teachers have not thought it wise to approach poems by way of stories. They have thought that poems should be read "as poems" or "as texts," as words written or printed on a page, ignoring the story the poems come out of.

I am willing to suppose (for the sake of supposing) that such poems as I am going to talk about must finally shrug off their stories, and all else they do not explicitly contain, and stand before us on their own. I am aware that some poems stand before us on their own because they must; we do not know the stories they came from. But in reading poems that are perfectly anonymous, we still know that there is more to them than the "text." We think of a poem and in the same thought think of what it is about, if it is about anything. Literature involves more than literature, or we would not be grateful for it.

Suppose we know not only a body of poetry but also the story that the poetry came out of. How then are we to help knowing what we know? How are we to help knowing, for instance, that some poets are pilgrims, and that their poems are not just objets d'art, but records, reports, road signs, or trail markers? By what curious privilege are we allowed to ignore what we know?

But I would like to go a little further still, and honor the possibility that the stories that poems come out of are valuable in themselves, so far as they are known. Those who are living and writing at a given time are not isolated poetry dispensers more or less equivalent to soft-drink machines, awaiting the small change of critical approval. We are, figuratively at least, members of a community, joined together by our stories. We are inevitably collaborators. We are never in any simple sense the authors of our own work. The body of work we make for ourselves in our time is only remotely a matter of literary history. The work we make is the work we are living by, and not in the hope of making literary history, but in the hope of using, correcting so far as we are able, and passing on the art of human life, of human flourishing, which includes the arts of reading and writing poetry.

There is a danger of presumption or imposition in what I am about to do, for I am not an authority on the story I am going to talk about. The story of a person's life cannot be entirely known or told. However, I do know unavoidably and unforgettably that Jane Kenyon's poems came out of a story. I know as well that Donald Hall's did. And I know that the poems of both poets came to a considerable extent out of the same story—or perhaps out of the intersection or overlapping of two stories; I want my language to be accurate and courteous, and am not confident of my ability to make it so.

The story of Donald Hall and Jane Kenyon is not my story. And yet their story is not absolutely distinct from mine, for their story is one I have depended on, and have spent a good deal of time telling over to myself and thinking about. And I have been attentive to the poems that came out of it. Their story and their poems have affected, instructed, troubled, consoled, and clarified my understanding of my own story.

Because I am a storyteller and was from childhood a hearer and reader (and believer) of stories, I have always known that people live in stories. And so it has been a little shocking to me to realize also that it is possible for people to wander outside their stories. When Donald Hall and I first met, at a literary party in Manhattan in the winter of 1963, both of us were living outside our stories. I find it readily supposable that Don didn't know what to make of me, even if it is supposable that he tried to make anything of me at all. I certainly didn't know what to make of him, and the reason was that I didn't know what to make of anything. Later, when both of us were again living inside our stories, we would recognize each other and become friends. This happened, I think, because we both loved our grandparents and we both derived from childhood homeplaces that we did not like to forsake. We have corresponded in two ways.

At the time of the party in 1963, the two of us were in "exile." I give that word an emphasis because it was so important and applied so peculiarly to young writers in our generation. We came to our calling in the shadow (and the glamour) of eminent literary exiles: James, Pound, Eliot, Joyce, Stein, Hemingway, and others. Moreover, those in charge of our education tended to think that they were preparing us for careers, not for settling down someplace. The question before us seemed to be, not how we might fit ourselves and our book knowledge into our home landscapes, but how we would fit into our careers, which is to say our exile.

This is confirmed by Donald Hall's early poem entitled "Exile." Looking back at that poem now, I find nothing in it that surprises me. It is a good poem. It is also an inevitable product of the poet's era and education; it *had* to be written by somebody. It states the case beautifully:

> Imagining, by exile kept from fact,
> We build of distance mental rock and tree,
> And make of memory creative act . . .

This is an exact enough description of the poet's job of work in Don's "Elegy for Wesley Wells," also an early poem. It too is a good poem—I don't mean at all to be denigrating this work of "exile." In the elegy, the poet mourns and celebrates his grandfather Wesley Wells, a New Hampshire farmer, a good one apparently, but one belonging to an age that, at the time of the poem, is "bygone." The poet (in the way of a young elegist, which I was once myself) is hoping to grant a measure of immortality to his grandfather by means of his poem. It is, to me, an extraordinarily moving poem. I have never read it without being moved by it, though by now I have lived beyond the notion that immortality can be conferred by a poem, and though by now my reading of the poem is influenced by my knowledge of a story that the poem, so to speak, does not know. When I read "Elegy for Wesley Wells" now, I feel a humorousness and a sadness that the poem did not anticipate.

In immortalizing his grandfather Wells, Donald Hall the young elegist is also immortalizing a part of his own life which he now considers to be finished. That life, if it is to have a present life, must have the immortal life of art. Maybe you are outside your life when you think your past has ended. Maybe you are outside your life when you think you are outside it. I don't know what Donald Hall in later life would say. I know only what I in later life would say. I would say, partly from knowing the story I am talking about, that though you may get a new life, you can't get a new past. You don't get to leave your story. If you leave your story, then how you left your story *is* your story, and you had better not forget it.

Now I want to speak of another poem that is a landmark both in the story I am dealing with and in my own consciousness of poetry and of the world. This is the poem called "Maple Syrup," written about twenty-five years after "Elegy for Wesley Wells." The poem tells about an experience shared by a couple designated merely as "we." Since I am observing no critical conventions here, I will say that this "we" refers to the poet Donald Hall and the poet Jane Kenyon, who have returned to the house of Donald Hall's grandparents Kate and Wesley Wells. The two poets, married to each other, will live their life together in this house on this farm, relinquished and immortalized in the "Elegy" so many years before. In the poem "Maple Syrup" they go through the house together, to "the back chamber" full of artifacts and relics, and then down into "the root cellar," where they find a quart of maple syrup left there by Wesley Wells. And here I must let the poem speak:

> Today
> we take my grandfather's last
> quart of syrup
> upstairs, holding it gingerly,
> and we wash off twenty-five years
> of dirt, and we pull
> and pry the lid up, cutting the stiff
> dried rubber gasket, and dip our fingers
> in, you and I both, and taste
> the sweetness, you for the first time,
> the sweetness preserved, of a dead man . . .

We (that is, now, the poet and his readers) have come a long way from "Elegy for Wesley Wells." We have left the immortality of art and have come, by way of a sort of mortalization, to a communion of lovers with one another and with the dead, which is to say that we have come to a marriage rite, joining two mortals to one another, to a place, and to other mortals, bringing them perhaps within imaginable reach of a more authentic idea of immortality. Donald Hall, who in the "Elegy" is maybe a sort of bard, has now become, in the full and mysterious sense, a love poet. Or we might say that, having started out to be a professional, he has become an amateur, working (like the best kind of professional) for love. The sign of this is that the memory of Wesley Wells, once elegized into a mental landscape of the finished past, has become a living faculty of the poet's mind and imagination. The sweetness of the dead man, now, is not preserved in an artifact but in the lives of those who taste it.

One more thing. Because this rite of marriage occurs in this story, it does not give new life just to the couple, who now enter into its "one flesh"; it gives new life also to the dead and to an old house. It matters that this is an old house that is familiar to the bridegroom. If the house had been sold to strangers, according to the common destiny of old houses in our day, Wesley Wells' quart of syrup, if found, would have been thrown away. It would have seemed fearfully old and fearfully anonymous. To Don, and to Jane, trusting Don, it was mortal and everlasting, old and new, and sweet.

Having set up (so to speak) these two landmarks, an elegy and a celebration of marriage, I am much more moved than I would be by either of the poems alone, for I know the story that joins them. The two poems are joined by this story because the story of Donald Hall had become also (to the degree that separate stories do converge) the story of Jane Kenyon.

What had happened was that these two stories had converged in one of the stages of Donald Hall's exile, teaching at the University of Michigan, and their convergence had made him free to return to the family house in New Hampshire. The agent of this freedom was Jane Kenyon, who said, according to her husband, "Why are we thinking of *here*, when there is New Hampshire?"

Not long after Don and Jane were settled at Eagle Pond, Don wrote to me, telling what they had done, and I wrote back some advice: Don't take on too much farming too quickly. Don has pointed out that the advice was wasted, since he did not intend to take on any farming at all—leaving me with the consolation that, anyhow, if he had needed it, it would have been good advice.

I am not sure when I met Jane, except that it was a good while ago, when she and Don were still heating their house with a very handsome wood-burning stove. I was on a speaking and reading trip in New England, and was able to stop by just for a short visit and lunch at a local eating place. I remember a tour of the house, but not much that was said. I remember being impressed by Jane's self-possession and dignity and quietness. These qualities continued to impress me after I knew her better. She was a writer, but she appeared to be watching "the literary world" without anxiety or great excitement.

Now the requirement of honesty is going to embarrass me, for I have to confess that I didn't read anything by Jane for a long time after I met her. For one reason, I felt a certain complicated sympathy for her—a poet who had set up shop smack in the middle of another poet's subject. The other poet's claim to this subject was well established; the other poet was her husband. It was easy to wish that she might have been, say, a painter. Another reason was that I liked her, and if she was a bad poet I did not want to know.

And then Bert Hornback invited Don and Galway Kinnell and Seamus Heaney and me to give a reading at the University of Michigan in January of 1986. For this there was a reason and a real reason. The reason was the public reading on Friday night. The real reason was that Bert wanted his students to have a late breakfast and conversation with the visiting poets on Saturday. In this age of careerist "research professors," Bert is a real teacher who thinks nothing of the trouble it takes to capture poets alive to talk with his students.

The visitors gathered at Bert's house for supper before the reading. When I came into the kitchen as the mingling and the talking began, Jane was standing by the refrigerator, watching the situation develop with the composure that I mentioned before. For the sake of political correctness I have been trying to avoid saying that Jane was beautiful, but of course she was, and of course I could see that she was. When we greeted each other, she said, "Wendell, I can't give you a hug. I have a bad cold." Baffled utterly by this generosity, I remember thinking that I had nothing better to do than catch a bad cold.

I have to go ahead and confess also that I do not greatly love literary occasions. The reading on Friday night was as readings are. The occasion beginning at breakfast on Saturday, however, was a literary occasion that surpassed itself. It was a *friendly* occasion, one of the loveliest that I have known. What I so liked about it was that everybody was talking for pleasure. There was no contention. Nobody defended a "position." There was much laughter. The students were hesitant to take part, but after a while they too entered into the conversation, and we had that additional pleasure.

Finally, late in the day, somebody—I don't remember who; it wasn't me—said, "Jane, why don't you read us a poem?"

Jane, who had been sitting almost outside the room, saying little, perhaps nothing at all, during the conversation, fished up from somewhere a page that she had brought with her and spread it open to read. For me, this was the only uncomfortable moment of that day. I don't remember what I thought, but it would have been like me to have started trying to think of some ambiguous compliment to make in case I thought the poem was bad—something like "Well, Jane, you certainly do write poetry." And then that quiet woman read beautifully her poem "Twilight: After Haying":

> Yes, long shadows go out
> from the bales; and yes, the soul
> must part from the body:
> what else could it do?
>
> The men sprawl near the baler,
> reluctant to leave the field.
> They talk and smoke,
> and the tips of their cigarettes
> blaze like small roses
> in the night air. (It arrived
> and settled among them
> before they were aware.)
>
> The moon comes
> to count the bales,
> and the dispossessed—
> *Whip-poor-will, Whip-poor-will*
> —sings from the dusty stubble.

> These things happen . . . the soul's bliss
> and suffering are bound together
> like the grasses. . . .
>
> The last, sweet exhalations
> of timothy and vetch
> go out with the song of the bird;
> the ravaged field
> grows wet with dew.

I hope I have adequately prepared you to imagine my relief. Now I must ask you to imagine something else. However many poets there may be who know from experience the subject of this poem by Jane Kenyon, I surely am one of them. I have lived countless times through that moment at the end of a day's work when its difficulty and heat and weariness take on a kind of sublimity and you know that you are alive both in the world and in something greater, when it is time to go and yet you stay on, charmed. I had never tried to write a poem about that moment, and on that day, when I had heard Jane read her poem, I knew that I would not need to write one; Jane had written better about it than I could. Sometimes I feel competitive or jealous when I *suspect* that somebody has written better than I can about something I know. When I am *certain* that somebody has done so, then I am relieved and I feel happy. "Twilight: After Haying" made me happy that day in 1986, and it has made me happy every time I have read it since.

Wittgenstein said that "In art [and, I assume, in writing about art] it is hard to say anything as good as: saying nothing." I believe and honor that, and I keep it in mind. But also we obviously need to speak from time to time of the things that have moved us. We need to wonder, for instance, why we remember some things and forget others. I have remembered Jane's reading of her poem that day, I think, because it was impossible to mistake the revelation of the event: Here was a poet present in her work with an authority virtually absolute. I don't mean that she is in the poem personally, but that all her gifts are in it: her quietness, gentleness, compassion, elegance, and clarity, her awareness of mystery, her almost severe good sense. This poem, like just about every one of her poems, is unconditional; it is poetry without qualification. It has no irony, no cynicism, no self-conscious reference to literary history, no anxiety about its place in literary history, no glance at the reader, no anticipation of the critic, no sensationalism, no self-apology or self-indulgence. How many poets of our time have been so unarmed as to say, "The moon comes / to count the bales . . ."? As she herself said (in the next poem of *The Boat of Quiet Hours*):

> These lines are written
> by an animal, an angel,
> a stranger sitting in my chair;
> by someone who already knows
> how to live without trouble
> among books, and pots and pans. . . .

—which is to say that she was authentically a poet of inspiration.

And this, to return to the story, seems to have settled pretty quickly the artistic problem of a shared life and subject. My wife, Tanya, has pointed out to me, from her knowledge of her own story, that Jane Kenyon had become, in fact, an exile in the very place that her husband had once felt himself to be exiled from. For a while after coming to Eagle Pond, she seems to have remembered "Ruth amid the alien corn":

> I'm the one who worries
> if I fit in with the furniture
> and the landscape.

And:

> Maybe
> I don't belong here.
> Nothing tells me that I don't.

But such lines as these testify to a radically different approach to the problem of exile. The modern American version of exile is a rootless and wandering life in foreign lands or (amounting to about the same thing) in American universities. Jane Kenyon, like Ruth of old, understood her exile as resettlement. Very few American exiles, and not many American settlers, have asked "if I fit in with . . . the landscape" or worried about belonging to a place. And already one is aware of her originality, as one continues always to be aware of it. I mean "originality" in what I take to be the best sense: not the enactment of a certain kind of literary intention or ambition, but the grace to submit to influence—the influence of places, passages of scripture, works of art; the influence of all her subjects—and the grace and patience to find within herself the means to respond. Her contribution to this story is hers distinctly.

When I read a disparagement of the book *Otherwise* in *The Hudson Review*, I was offended, but also puzzled. How could anybody able to read fail to see the quality of that book? But after a while, I believe, I figured it out. Jane Kenyon's work, in fact, makes an unnegotiable demand upon a reader. It doesn't demand great intellect or learning or even sympathy; it demands quiet. It demands that in this age of political, economic, educational, and recreational pandemonium, and a concomitant rattling in the literary world, one must somehow become quiet enough to listen. Her poems raise unequivocally the issue of the quality of the poet's ear.

A true poem, we know, forms itself within hearing. It must live in the ear before it can live in the mind or the heart. The ear tells the poet when and how to break the silence, and when enough has been said. If one has no ear, then one has no art and is no poet. There is no appeal from this. If one has no ear, it does not matter what or how one writes. Without an ear, the traditional forms will not produce Andrew Marvell, nor will "free verse" give us William Carlos Williams.

Jane Kenyon had a virtually faultless ear. She was an exquisite master of the art of poetry. Her voice always carries the tremor of feeling disciplined by art. This is what over and over again enabled her to take the risk of plainness, or of apparent plainness. Her ear controls rhythm and sound, and also tone. It is tone as much as anything that makes one able to say what is unusual or unexpected. It is because of her perfection of tone that Jane Kenyon is able to say, "The moon comes / to count the bales . . ."

It is her perfection of tone that makes her poems able to accommodate sudden declarations of spiritual knowledge or religious faith, and that gives to so many of her poems the quality of prayer. It rules in her poems and passages of humor. It is the enabling principle in the political stroke of the poem entitled "Fat," and of the affirmation always present in her poems of sorrow. I am suggesting what I suppose cannot be demonstrated: that there is a practical affinity between the life of her soul and the technique of her poems.

The poems assemble themselves with a seeming arbitrariness, which is perhaps a comment. The poet looks at her subjects and experiences as they come to her and sees that they are ordinary; they are the stuff of life in this world; they could have come to anybody, at any time, in any order. They are revelations of ordinary satisfactions, joys, sufferings, deliverances which, in being revealed, become somehow numinous and resonant—extraordinary. In seeing that the poems are revelations, you see that they are not arbitrary but inevitable; in the course of the poem, form has occurred.

Sometimes the poems are poems of suspense; everything waits for the final line, as in the poem called "Things":

> The hen flings a single pebble aside
> with her yellow, reptilian foot.
> Never in eternity the same sound—
> a small stone falling on a red leaf.
>
> The juncture of twig and branch,
> scarred with lichen, is a gate
> we might enter, singing.
>
> The mouse pulls batting
> from a hundred-year-old quilt.
> She chewed a hole in a blue star
> to get it, and now she thrives. . . .
> Now is her time to thrive.
>
> Things: simply lasting, then
> failing to last: water, a blue heron's
> eye, and the light passing
> between them: into light all things
> must fall, glad at last to have fallen.

The poem gathers itself as quietly as a snowy night, and then by the end a kind of dawn has come and everything is shining. That seems to be about all there is to say. This poem confirms for me as well as any I have read what I think is the fundamental fact of poetry: If you can explain it, it is something else.

Nor am I able to say much more about this story that I have undertaken to talk about. It is, I think, a good and valuable story. Two poets entered into it together, consenting to its foretold cost, lived it out, met its occasions, and made, separately and together, a life and a body of work that, for some of us, the world is now unimaginable without. They tasted a sweetness stored up by others; they stored up a sweetness to be tasted by others. And what are we friends and beneficiaries to say? Well, finally, maybe no more than "Thank you."

The Uses of Adversity

IT HAS BEEN useful to me to think of *As You Like It* and *King Lear* as versions of the same archetypal story, belonging to human experience both before and after the plays. This is the story: In the instituted life of a society "things fall apart" because the people of power have grown selfish, cruel, and dishonest. The effect of this is centrifugal; the powerless and the disempowered are sent flying from their settled domestic life into the wilderness or the world's wildness—the state of nature. Thus deprived of civil society and exposed to the harshness of the natural world and its weather, they suffer correction, and their suffering eventually leads to a restoration of civility and order.

The outline of this story is clearly apparent in *As You Like It*. In *King Lear* the story is subjected to nearly intolerable stresses, and yet the outline remains unbroken; it is the major source of the play's coherence and meaning. What I believe is the proper understanding of both plays depends on our ability to take seriously the assumptions of the archetypal story—on how we answer the following questions: Do all human societies have in them the seeds of their failure? Are those seeds likely to be the selfishness and dishonesty of the dominant people? Does failure typically reduce the society, or persons in it, to some version of the state of nature? And is there something possibly instructive and restorative in this reduction?

For most readers nowadays these questions will be an unwelcome dose. We have read some history, and we do not doubt that other societies have failed, but we are not much inclined to credit the possible failure of our own, even though we are less and less able to deny the implications of our propensity to waste or to mechanical violence, or of our entire dependence on cheap petroleum. We have pretty much made a virtue of selfishness as the mainstay of our economy, and we have provided an abundance of good excuses to dishonesty. Most of us give no thought to the state of nature as the context of our lives, because we conventionally disbelieve in natural limits.

Another problem is that there is a considerable overlap

between this archetypal story and the pastoral tradition. In
the pastoral tradition, as Shakespeare was fully aware, there is
a prominent strain of frivolity. What is frivolous is the sentimentalization of rural life, which is supposedly always pretty,
pleasant, and free of care. The famous example is Christopher
Marlowe's:

> Come live with me and be my love,
> And we will all the pleasures prove
> That valleys, groves, hills, and fields,
> Woods, or steepy mountains yields.

To this Sir Walter Raleigh justly and just as famously replied:

> The flowers do fade, and wanton fields
> To wayward winter reckoning yields . . .

What neither poet acknowledged is the possibility of a real
need, as Robert Frost put it, "of being versed in country
things."

Shakespeare knew of course the pastoral conventions represented by Marlowe's poem. But he was a countryman, and
he knew the truth of Raleigh's admonition; he knew also the
need of being versed in country things. He knew that "a
true laborer" might have something to say to a courtier that
the courtier might need to hear—because, for one reason, the
courtier lives by eating country things.

Another obstacle between modern readers and the archetypal story underlying these plays is our popular, and uncritical,
egalitarianism. To us, the order of the natural world is horizontal, and so, we would like to think, is the order of human
society: Any creature is as important as any other; any citizen is
as important as any other.

But to Shakespeare the order of the world, as of human society, is also vertical and hierarchical. The order of created things
descends in a Chain of Being from God down to the simplest
organisms. In human society, order descends downward from
the monarch. Every creature and every human has a place in
this hierarchy according to "degree." Ulysses' discourse on degree in the first act of *Troilus and Cressida* can serve as a clarifying prologue to a reading of *As You Like It* and *King Lear*:

O, when degree is shaked,
Which is the ladder of all high designs,
The enterprise is sick. How could communities,
Degrees in schools, and brotherhoods in cities,
Peaceful commerce from dividable shores,
The primogenity and due of birth,
Prerogative of age, crowns, sceptres, laurels,
But by degree, stand in authentic place?
· · · · · · · · · · · · · · · · · · · ·
                              right and wrong,
Between whose endless jar justice resides,
Should lose their names, and so should justice too;
Then everything include itself in power,
Power into will, will into appetite.
And appetite, an universal wolf,
So doubly seconded with will and power,
Must make perforce an universal prey
And last eat up himself.

This speech, by which Ulysses calls the "tortive and errant" Greeks to order, tells us precisely how to understand Orlando's complaint at the beginning of *As You Like It*. Orlando's oldest brother, Oliver, charged by their father, now dead, with Orlando's education, has forsaken his duty. Orlando's "keeping," he says to his old servant Adam, "differs not from the stalling of an ox." As the younger brother, lacking the "primogenity . . . of birth," Orlando is a man of lower degree than Oliver. But he is, even so, a man, his father's son, and Oliver's brother; Oliver's mistreatment of him, as if he were no more than a beast, is an affront to order, both human and natural; it is a symptom of a sick enterprise.

The trouble, for Oliver as for the villains of *King Lear* and other Shakespearean villains, is that the human place in the order of things, between the angels and the animals, is precisely and narrowly delimited, and it is precarious. To fall from one's rightful place, to become less than human, is not to become an animal; it is to become monstrous. And so Oliver's mere dislike and neglect of Orlando decline fairly predictably to a plot to kill him, which forces Orlando into exile.

In scene iii of Act I, a parallel estrangement occurs. The scene is in the palace of Duke Frederick, who has usurped the place of his brother, the carelessly named Duke Senior. Duke Senior, as we have already learned, is in exile in the Forest of Arden where he and some "merry men," his followers, "live like the old Robin Hood of England." Duke Senior's daughter, Rosalind, has been permitted to remain in the palace as the companion of Celia, Duke Frederick's daughter. The two young women not only are cousins and companions but are dearest friends. The two, Celia says, are "coupled and inseparable." She says to Rosalind, "Thou and I am one." And it is Celia also who, in attempting to console Rosalind, states one of the main themes common to this play and *King Lear*, that of affection or gentleness or generosity versus force: "what he [her father] hath taken away from thy father perforce, I will render thee again in affection." This affection is soon tested by Duke Frederick's determination to send Rosalind into exile:

> Within these ten days if that thou beest found
> So near our public court as twenty miles,
> Thou diest for it.

His reason is that he does not trust her. His distrust originates of course in his knowledge that he himself is not trustworthy. His daughter and niece, by contrast, possess in full the trust and the trustworthiness now lacking in the court, and so they must leave. In proof both of their friendship and of Duke Frederick's failure to know them, they decide to disguise themselves as "Ganymede" and "Aliena" and run away to join Duke Senior in Arden.

Act II, I think, is the paramount act of the play and is one of the greatest acts in all of Shakespeare. Both its poetry and its drama are exceedingly fine. It is also the crisis of the play for its readers, who have to decide here whether or not to take the play seriously. From what I have read and seen, some readers and directors have found it easy to understand the play as a pastoral diversion, merely sentimental and "comic," which I think is an insult to the play and its poet.

The test comes immediately with Duke Senior's speech that begins the first scene of Act II. The speech develops a standard pastoral theme: the honesty of the pastoral or rural life in contrast to life at court; it is the same theme expounded by Meliboe in *The Faerie Queene*, VI. The duke asks rhetorically, "Are not these woods / More free from peril than the envious court?" And we know the answer as well as his fellow exiles: These woods are free from the envy, jealousy, hypocrisy, power-hunger, and fraud that imperil the court or any other center of power.

There is an editorial crux in line five that we have to settle before reading further. The duke says, "Here feel we not [or: but] the penalty of Adam . . ." I am quoting the new Pelican edition, in which the editor chooses "not." But that usage, if it stands, reduces the speech to nonsense, and the duke to a fool. The problem with this reading is that the duke is *not* a fool, and the exiles, according to the play, are still subject to the penalty of Adam—that is, to mortality, discord, and the need to earn their living. And so the line necessarily is "Here feel we *but* the penalty of Adam . . ." The intended contrast is not between Eden and the fallen world, but between the unadorned life of the forest and the "painted pomp" of the court.

That "the icy fang / And churlish chiding of the winter's wind" are not flatterers but "feelingly persuade me what I am" we may take without argument to be merely true. "Sweet are the uses of adversity" may oversweeten the point, and yet we know that adversity can be corrective, and is sometimes indispensably so.

For modern readers, the largest difficulty in this speech may come in the last three lines, in which the duke proclaims that

> this our life, exempt from public haunt,
> Finds tongues in trees, books in the running brooks,
> Sermons in stones, and good in everything.

To the modern ear, this is likely to sound naive—an instance of the "pathetic fallacy," an almost cartoonish sentimentalization of nature. And yet this is a play solidly biblical and Christian in its moral basis, and this is one of its passages that most insistently depends on our knowledge of scripture. The overarching concept is that of the "good in everything," and the authority for this is Genesis 1:31: "And God saw every thing that he had made, and, behold, it was very good." As for "tongues in trees,

books in the running brooks, / Sermons in stones," Shakespeare may be paraphrasing Job 12:7–9:

> Aske now the beastes, and they shal teache thee, and the
>   foules of the heaven, and they shal tel thee:
> Or speake to the earth, and it shal shewe thee: or the fishes
>   of the sea, and they shal declare unto thee.
> Who is ignorant of all these, but that the hand of the Lord
>   hathe made these?

And he could as well be alluding to the long tradition in which
nature is seen as a second or supplementary revelation.

The third scene of Act II parallels thematically the third
scene of Act I. In the earlier scene Rosalind is confronted by
Duke Frederick, sentenced to exile, and she and Celia make
their plan to escape together in disguise. In II, iii, Adam, a
servant loyal to Orlando, warns his young master that he, like
Rosalind, must go into exile, for his envious and vengeful
brother is plotting to kill him.

This play is not an allegory, but some of its characters have
a semiallegorical or representative function; they represent
human qualities or kinds. Adam, for one, is "the old Adam,"
father of us all, the fallen humanity which we all share, but
he is furthermore the old Adam redeemed by good and faithful service to his master. He was first the servant of Orlando's
father, the good Sir Rowland de Boys. In this scene, out of
loyalty to the father and love for the son, he makes an absolute
gift of his service and his fortune to Orlando, trusting that in
his old age he will be comforted by him "that doth the ravens
feed" and "caters for the sparrow." Thus, as a true servant to
good men, he understands himself as a true servant of God.
Orlando reciprocates by saying, like Celia in Act I, that the two
of them will join their fates: "we'll go along together" in the
belief that, before they have spent all of Adam's savings, they
will "light upon some settled low content." The idea of a "settled low content" is the moral baseline of the play. It is what
human beings most authentically have a right to expect and
to achieve. It is the possibility that adversity most usefully and
sweetly reveals. A settled low content is what Thomas Jefferson
wished for America's small farmers; it is what Henry Thoreau
was seeking at Walden Pond.

In scene iv, having arrived in the Forest of Arden, Rosalind and Celia encounter two other representative figures: Silvius, the young shepherd, and the old shepherd Corin. Silvius, classically named, represents what is most artificial in literary pastorals. He is an "uncouth swain" stricken by love into utter silliness and uselessness; wherever his sheep are, he is not going to think of them during this play.

Corin, by contrast as Englishly named as Spenser's Colin Clout or Hardy's Hodge, is strongly drawn as an individual and at the same time as a representative countryman. He is an "ideal character" of the same honest family as Chaucer's Plowman, who was "A trewe swinkere and a good . . ." Another critical question that this play imposes on its readers and directors is what to make of Corin. Here I have to depart from the sequence of the action to quote Corin's characterization of himself to Touchstone in III, ii:

> Sir, I am a true laborer; I earn that I eat, get that I wear, owe no man hate, envy no man's happiness, glad of other men's good, content with my harm; and the greatest of my pride is to see my ewes graze and my lambs suck.

To many readers that last clause would seem fatally countrified; from them the best rating it could hope for would be "quaint." Many Americans now would see this speech unhesitatingly as the utterance of a "hick" or a "redneck," hopelessly "retro." Nevertheless, any husbander of livestock would recognize Corin as a good shepherd, and Thomas Jefferson would have appraised him highly. In his independence he is democratic, and in his charity, fortitude, and humility he is Christian. Shakespeare knew that the human world survives by the work and responsibility of such people, and Corin's character is one of the standards by which we are to measure the other persons of the play.

In II, iv, Touchstone, assuming the role of sophisticated urbanite, sees Corin on their first encounter as a hick and addresses him accordingly: "Holla, you clown!" But Rosalind, as Ganymede, displaying her extraordinary good sense, recognizes him immediately for what he is: "Good even to you, friend." And Corin replies with perfect courtesy: "And to you, gentle sir, and to you all." Corin, offering hospitality to the strangers, is obliged to reveal that he is poorly paid:

> I am shepherd to another man
> And do not shear the fleeces that I graze.
> My master is of a churlish disposition
> And little recks to find the way to heaven
> By doing deeds of hospitality.

This ungenerous master, moreover, is preparing to sell his flock and land. Rosalind and Celia arrange with Corin to buy "the cottage, pasture, and the flock," Celia promising, "we will mend thy wages." Receiving gratefully this offer of economic justice, Corin sounds again the play's theme of the good servant: "I will your very faithful feeder be . . ."

A fourth representative character is Jaques, whose dominant trait is self-indulgence. "The melancholy Jaques," as he is called in Act II, scene i, manages to be both sentimental and cynical. He is uselessly sensitive and intellectual, a dilettante of his own moods, a boastfully free-speaking critic who corrects nothing. In II, i, Duke Senior speaks of his proper regret at having to kill the deer of the forest for food. But Jaques, as his fellows report, sentimentalizes this regret, making the same equation between human beings and animals as some animal rights advocates of our own day. And, like them, he offers no practicable alternative.

In II, v, after Amiens has sung a song that closely paraphrases Duke Senior's speech in II, i, Jaques responds by supplying a verse of his own which suggests that the forest company are asses and fools. So far he has been a peripheral character, looking on and commenting from the margin as a sort of fecklessly disapproving "chorus." Presently he will serve the play much more vitally, though still passively.

The next scene is brief, containing only two speeches, but to fail to take it seriously enough is again to be seriously in error about the play. Old Adam, weakened by hunger, cannot go on: "Here lie I down and measure out my grave. Farewell, kind master." As a "fallen" man, Adam cannot save himself. Nor can he survive as the servant of Orlando. But *As You Like It* is a play of transformations, and this scene presents the first one.

Adam has completed his servanthood. As a servant, he knows, he is as good as dead. His life now depends upon a change in Orlando. And Orlando changes; he becomes his servant's servant—as Edgar in *King Lear*, his father being reduced to helplessness, becomes his father's parent. Shakespeare is relying again on our knowledge of scripture, and the reference here is to Matthew 20:25–27:

> Ye knowe that the lords of the Gentiles have domination over them, and they that are great, exercise autoritie over them.
> But it shall not be so among you: but whosoever wil be great among you, let him be your servant . . .

The apparent lightheartedness of Orlando's reply must be understood as tenderness: as his attempt to lighten the heart of old Adam and as his pledge of service. His words also recall the measure of a "settled low content":

> Live a little, comfort a little, cheer thyself a little . . .
> For my sake be comfortable; hold death awhile at the arm's end. I will here be with thee presently, and if I bring thee not something to eat, I will give thee leave to die . . .

Here as elsewhere, and despite his allegiance to "degree," there is a strong democratic impulse in Shakespeare. But he is a democrat, not in the fashion of Jefferson, but in the fashion of Christ. "The least of these my brethren" also have their place in the order of things and their entitlement to be loved and served.

What is the relevance of this to the archetypal story that is my interest? Let us remember, to start with, that this play begins after the old state of things, the old "power structure," has fallen. We don't know what the error or fault of Duke Senior might have been; we know only that he became so weakened —perhaps so misled by flattery—that he was driven into exile by his power-hungry brother. Also the good Sir Rowland de Boys has died, and has been replaced by his selfish eldest son, Oliver. There is nothing more disorderly and disordering in civilized life than the selfishness of people of power—that is, their failure to be servants either to God or to their subjects. ("Public servants," as they and we too often forget, are meant not to rule but to serve the people.) The corrective to this is begun in the exiles by their recognition of the need to serve. And, in exile, this need is insistently practical. Outcasts in the forest—or on the stormy heath—cannot survive by selfishness.

In the long seventh and final scene of Act II, the theme of the forest (or adversity) as the corrective of selfishness and misrule, the theme of the necessity of servanthood, and the theme of affection or gentleness versus force are all joined in the play's moral climax. In my opinion, this scene threatens also to be the play's dramatic climax—to be both more dramatic and more moving than anything in the three acts that follow. Shakespeare's problem (and I assume a director's also) is to make the rest of the play worthy in moral interest and drama of what he has done in the first two acts.

The seventh scene begins with a leisurely, bantering conversation at first about Jaques and then between Jaques and Duke Senior. Jaques, having encountered Touchstone in the forest, wishes that he too could be a fool, apparently without in the least suspecting that he already is one. If, he says, he were given the liberty of a fool—that is, if the duke should grant him an official tolerance, permitting him to speak the truth as he sees it—then he would prove himself so purgative a critic as to "Cleanse the foul body of th' infected world . . ." The duke says that as such a critic Jaques would necessarily be a hypocrite, "For thou thyself hast been a libertine . . ." Jaques thereupon discourses on the universality of sin and hypocrisy in a speech that prefigures a much better one by the maddened King Lear.

Jaques' speech is interrupted by the entrance of Orlando with his sword drawn, and the scene then gets serious. Dinner has been laid out in the camp of the exiles, and Orlando is desperately in need of food for Adam and for himself. His sword is drawn because, like Touchstone in his encounter with Corin, he is mistaken about the circumstances. He assumes, as he will presently say, that he is in a "savage" place, and therefore will have to take the food by force. In his own savagery, then, he finds himself comically and wonderfully reproved by the duke in the name of "good manners" and "civility." Having fled from the failed civility of civilization, he has come into the presence of a civility reconstituted in the "savage" forest. Instead of drawing his own sword to defend his dinner, the duke welcomes Orlando as a guest:

> What would you have? Your gentleness shall force
> More than your force move us to gentleness.
> . . . .
> Sit down and feed, and welcome to our table.

Orlando, surprised, acknowledges his error and apologizes. He and Duke Senior then speak an antiphonal celebration of their common tradition of charity. Orlando says:

> If ever you have looked on better days,
> If ever been where bells have knolled to church,
> If ever sat at any good man's feast,
> If ever from your eyelids wiped a tear
> And know what 'tis to pity and be pitied,
> Let gentleness my strong enforcement be;
> In the which hope I blush, and hide my sword.

And the duke replies:

> True is it that we have seen better days,
> And have with holy bell been knolled to church,
> And sat at good men's feasts, and wiped our eyes
> Of drops that sacred pity hath engendered;
> And therefore sit you down in gentleness,
> And take upon command what help we have
> That to your wanting may be ministered.

But Orlando is not yet ready to sit down. He remains true to his promise to Adam, and he asks the company to "forbear your food a little while . . ." When he speaks of Adam now his kindness is forthright: "like a doe, I go to find my fawn / And give it food." There could be no more tender expression of loving servanthood, and no more apt a simile.

While Orlando is away, Jaques, in response to no encouragement, delivers his famous speech on the seven ages of man. This is a dandy set piece, but it is also utterly cynical. It is the life history of a lone specimen, such as one might find in a

modern zoology manual. The last age, which is described most
heartlessly,

> Is second childishness and mere oblivion,
> Sans teeth, sans eyes, sans taste, sans everything.

What Shakespeare thought of this may be inferred from the
stage direction that immediately follows: "Enter Orlando, with
Adam." That Orlando enters carrying Adam in his arms we
know from Duke Senior's next speech, which also seems a rebuke to Jaques: "Welcome. Set down your venerable burden /
and let him feed." Far from "sans everything," old Adam has a
young friend who is his faithful servant—and who moreover,
seeing that Adam is in his "second childishness," treats him
with a mother's tenderness.

The scene ends with Duke Senior's recognition of Orlando,
in which he implicitly affirms love as the right bond between
generations and the members of a community: "Be truly welcome hither. I am the duke / That loved your father."

After Act II, *As You Like It* becomes a play of lovers, and the
comedy of it, I think, is brilliant enough to follow worthily the
eminent scene I have just described. The theme of transformation is worked out in greatest detail and most delightfully
in the courtship of Rosalind and Orlando. In this courtship,
which is both farcical and serious, Rosalind in the guise of
Ganymede assumes the role of "Rosalind," so that Orlando, in
the guise of his love-maddened self, may practice as a lover and
so be "cured." The premise of this masquerade is set forth by
Rosalind in III, ii: "Love is merely a madness . . ." There is
good sense in this. She and Orlando fell in love "at first sight"
in Act I. Rosalind, who is as smart and resourceful as she needs
to be, realizes that such a love requires testing. Lovers in the
madness of new love are, as Albany says of Goneril in *King
Lear*, "self-covered." Rosalind's "cure," as it turns out, is a trial
for herself as well as for Orlando. It removes the "cover" of
selfhood; it tests them and proves them worthy of each other
and ready for marriage. It is important to notice that these lovers do not turn seriously toward each other and toward their
marriage until each of them has explicitly rejected the company
of the cynical and sentimental Jaques.

The issue, for Rosalind and for the play, is how to make a

civil thing of the wildness of sexual love. The forest is the right place for courtship, which puts lovers in the state of nature. By the same token it is the right place to transform "mad" lovers, if they wish, into grown-up lovers fully prepared for the marriage rite and the "blessed bond of board and bed" with which the play ends.

By the end of the play its "self-covered" villains also have been transformed: Oliver by becoming the conscious and grateful beneficiary of his brother's courage and forgiveness, and Duke Frederick by his encounter with "an old religious man" in "the skirts of this wild wood." Jaques even has resolved to go and learn from "these convertites." Duke Senior and his fellow exiles, as we know from Act II, will return from the forest to a domestic world far better than the one they fled, for they too have been changed, renewed in their specifically human nature, their civility and charity, by this time of adversity in the natural world.

Thus by the play's end all of its principal characters have been changed, and for the better, by their time in the forest. Shakespeare saw, and wants us to see, that the forest can be corrective and restorative to disordered human life. But he goes further. At once explicitly and indirectly he invests the forest with a mysterious and even a mystical transformative power. Partly this is accomplished by Touchstone, speaking with implication apparently beyond his intention. In III, ii, he says to Rosalind, "You have said; but whether wisely or no, let the forest judge." And, twitting Audrey in III, iii, he says, "here we have no temple but the wood . . ." Also Rosalind, in her masquerade with Orlando, alludes to "an old religious uncle of mine" and to "a magician" she has "conversed with." Orlando, in V, iv, conflates the two when he speaks of Ganymede's uncle,

> Whom he reports to be a great magician,
> Obscurèd in the circle of this forest.

Is there, then, a great magician in the forest? Is the forest a holy place of judgment and magical or miraculous transformation? We must ask, but we must not answer. The play must not answer. *As You Like It* is not the voice out of the whirlwind. Once upon a time several people fled from a disordered and murderous society into a forest, and there they were profoundly changed. That is all we know.

II.

In *King Lear*, both the Lear story and the Gloucester story
grow out of corruption at the center of wealth and power, just
as does the action of *As You Like It*. Initially, in *King Lear*, this
is the corruption merely of selfishness: self-complacency, self-
indulgence, self-ignorance, the lack of critical self-knowledge.
From this selfishness grows, in turn, an infection of monstrous
proportions that is described, though unwittingly, by Gloucester's bastard son, Edmund, in Act I, scene ii, as he works his
deception upon Edgar, his legitimate elder brother:

> . . . unnaturalness between the child and the parent;
> death, dearth, dissolutions of ancient amities; divisions in
> state, menaces and maledictions against king and nobles;
> needless diffidences, banishment of friends, dissipation of
> cohorts, nuptial breaches . . .

Because Lear is king, his self-absorption becomes in effect state
policy. Like any head of state he is able, temporarily, to invest
his fantasies with power. His fantasy is a primitive instance of
"early retirement." He believes that by dividing his kingdom
among his daughters he can free himself of care and responsibility while retaining the initiative and the privileges of his
kingship. This will prove to be almost limitlessly foolish. He is
an "idle old man," as his daughter Goneril calls him, "That still
would manage those authorities / That he hath given away."
His daughter Regan also is right when she says of him that
"he hath ever but slenderly known himself." Because he does
not know himself, he cannot know others. He has failed disas-
trously to know Goneril and Regan. He fails to learn of them
in the play's first scene what is obvious to everybody else: that
they are eloquent, clever, heartless, false, and greedy.

Having apparently determined already the portions of land
that he will give to his daughters, Lear involves them point-
lessly and cruelly in a contest in which they are to compete for
his "largest bounty" by declaring their love for him. Goneril
and Regan, good poets and good actors, give him precisely
the groveling flattery he has asked for. Only his third daughter,
Cordelia, who lacks neither sense nor eloquence, and who in
fact truly loves him, refuses to tell him more than the plain
truth: She loves him as she ought. She loves him *completely* as

she ought, and the play will reveal this, but her refusal to participate in the love contest is entirely proper. It is a refusal to
falsify her love by indulging her father's frivolous abuse of his
power, which she both disdains and fears.

Predictably infuriated, Lear disinherits Cordelia. In doing so,
Martin Lings argues in *The Secret of Shakespeare*, Lear banishes
"the Spirit," by which Lings means "the Holy Spirit" or "the
pearl of great price." I am unwilling so to allegorize the play,
but I think nevertheless that Lings has pointed us in the right
direction. In disinheriting Cordelia, in making her "a stranger
to my heart and me . . . forever," Lear has, in the face of
great evil, estranged himself from goodness. He then deepens
and ratifies this estrangement by exiling the Earl of Kent, who
has dared to call folly by its right name.

Thus in *King Lear*, exactly as in *As You Like It*, corruption
at the center of power sets loose a centrifugal force that ultimately will send the powerless and the defeated into the wildness of the natural world. But in the six or so years between the
two plays Shakespeare evidently saw a need to raise the stakes.
The archetypal story is, after all, not necessarily a comedy. Selfishness does not necessarily involve one in a limited evil. Evil
people are not necessarily relenting or easily converted. The
state of nature is not necessarily the relatively hospitable Forest
of Arden. The uses of adversity are not necessarily "sweet."
The terms and the affirmation of *As You Like It* now required
a harsher test.

Shakespeare brought the earlier play to trial by imagining a
set of villains who in the course of the play will reveal—and discover for themselves, to their cost—that they have limitlessly
consigned themselves to evil. Lacking self-knowledge and too
"self-covered" even to suspect that he does, Lear rids his court
of love, goodness, and honesty, and thus in effect abandons
himself to the purposes of Goneril, Regan, and Regan's husband, the Duke of Cornwall. These three, like Lear, are selfish,
but there is a difference. Lear, in his selfishness, is self-deluded:
He thinks he is a loving and generous father, as no doubt he
wishes to be. Goneril, Regan, and Cornwall, by contrast, are
selfish by policy; there is no inconsistency between what they
are doing and what they think they are doing. By dividing his
kingdom, by isolating himself from Cordelia and Kent, Lear
places himself in a deadly trap. Escape will cost him everything

he has, or everything he thinks he has in the opening scene. In his quarrel with Kent, he unknowingly foretells his fate: "So be my grave my peace . . ."

In outline, through II, i, the Gloucester story exactly parallels that of Lear. Gloucester has two sons, Edgar and Edmund. Edgar loves his father as Cordelia loves hers. Edmund, the illegitimate younger son, is as contemptuous of Gloucester as Goneril and Regan are of Lear. Edmund, like those daughters, is a good actor and flatterer. He wants to cheat Edgar out of their father's estate, and he succeeds in convincing Gloucester that Edgar is planning to kill him. Edgar is then forced to flee to save his own life.

By the beginning of scene iii of Act II, when the fugitive Edgar transforms himself into Tom o'Bedlam, the villains are successful and in control; their schemes are working and they have what they want. Love, goodness, honesty, and fidelity have been directly confronted by evil, and evil so far has won. But this working of evil, by its very successes, has instigated a countermovement, and in parallel to *As You Like It* this movement is the work of good and faithful servants. The proscriptions against Cordelia, Kent, and Edgar have set them free to serve Lear and Gloucester.

Disguise in this play is just as important as in *As You Like It*, and more portentously so. Goneril, Regan, and Cornwall are "self-covered." Their better selves have been utterly and finally renounced. Edmund is only barely, but significantly, less self-obscured than they. Gloucester, like Lear, is a naively selfish old man. Neither is sinful or evil beyond the measure of ordinary human behavior, but both are obscured, obscured most consequentially to themselves, by foolishness and complacency. They are deluded and self-deluded. They are deludable *because* they are self-deluded. Goneril's husband, the Duke of Albany, for the time being is disguised to himself because of his hesitancy in recognizing and denouncing the evil character of his wife. In order to serve Lear and Gloucester in their time of greatest need, Kent and Edgar must serve in disguise. Of all the major characters, Cordelia alone always appears, to us and to herself, only as she is. She is good, and her understanding of her goodness is constant, profound, and absolutely assured. Much of the drama and meaning of the play comes from the actions of the characters in relation to their disguises, and we understand those actions by the measure of Cordelia's transparency, clarity, and candor.

In II, i, the villains of the Lear plot, recognizing their own kind in Edmund, claim him as an ally. Cornwall tells him, with terrible import, "you shall be ours . . . You we first seize on." And so, early in the play, the party of evil, of power-lust and greed, recognizes superficially the usefulness of cooperation, and for a while they are a coherent force. By contrast the party of goodness—the party of Cordelia, Kent, the Fool, Edgar, and, finally, Albany—in its early defeat is widely scattered. As the play proceeds, however, the party of evil, because of the nature of evil, disintegrates while the members of the other party recognize one another and draw together.

Edmund's soliloquy in I, ii, introduces another set of contraries into the play as he subordinates his specifically human nature to nature:

> Thou, Nature, art my goddess; to thy law
> My services are bound. Wherefore should I
> Stand in the plague of custom, and permit
> The curiosity of nations to deprive me,
> For that I am some twelve or fourteen moonshines
> Lag of a brother? Why bastard? Wherefore base . . . ?
> . . . .
> Legitimate Edgar, I must have your land.

If we are fair-minded, we must see the justice of Edmund's indictment of the prejudice against bastards as "base born," just as we see the justice of Goneril and Regan's perception that their father is foolish and intemperate. But evil characteristically supports and disguises itself by such partial claims of justice. In his dire intention to deceive his father and his brother, putting both their lives at risk, Edmund offends against moral law, specifically the fifth, sixth, eighth, ninth, and tenth of the Ten Commandments, and against the order of "degrees" as set forth in Ulysses' speech in *Troilus and Cressida*. Edmund understands "nature" as exclusive self-interest, which he by implication ascribes to all "natural" creatures. As a person self-consciously "enlightened," later in the same scene he rejects his father's astrological determinism, and so accepts full moral responsibility for what he is doing—and again we are tempted to sympathize. But in rejecting his father's superstition, he defines himself as self-determined. By thus subordinating human nature to nature, he means that he accepts *no* subordination. By putting himself at the service of nature's law, he means, perhaps more absolutely than he intends, that he rejects all service to anybody but himself, and will honor *no* law. This speech of Edmund's is answered in IV, vi, by the "Gentleman" who says of Cordelia in an apostrophe to Lear, "Thou hast one daughter / Who redeems nature . . ."

And so the thrusting-out of Lear and Gloucester into the wild world is as profoundly and purposefully thematic in this play as is the forest exile of the sufferers in *As You Like It*. When Lear speaks of Goneril and Regan as "unnatural," he means that they have, like Edmund, subscribed to nature's supposed law of entire selfishness, as opposed to human nature's laws of filiality and love. These virtually opposite uses of the word "nature" may be confusing, but the word in fact has this duplicity in our language, and Shakespeare exploits it fully, to serious purpose, in both plays.

By the unnaturalness of his bad daughters Lear is driven out into nature. Nature now is not the Forest of Arden, but the open heath in the midst of a "pitiless storm." The pitilessness of the storm, which is set before us in its full extremity in the dialogue, is the measure of the pitilessness of Goneril, Regan, and Cornwall—though the pitilessness of the storm, unlike that of these familial villains, is not unkind, as Lear understands and says in III, ii. The heath and the storm belong to the moral landscape of the tragedy, just as the forest belongs to the moral landscape of the comedy. And Lear's dreadful exile upon the heath in the storm and the darkness forces almost immediately a change upon his character. Even as he announces to the Fool that he is going mad—"My wits begin to turn"—he speaks for the first time unselfishly, in compassion and concern for the Fool's suffering: "How dost, my boy? Art cold?" And so his wits are turning, we may say, not just to madness, but through his madness, which is the utter frustration and destruction of his sanity as of Act I, to a better sanity.

Lear's adversity is not "sweet" but it *is* useful: It has made him tender; it has feelingly persuaded him what he is; it has reduced him from a king to a mere human, sharing the lot of other humans. And in III, iv, he speaks in compassion, confession, and repentance, his words recalling both Duke Senior's speech on the uses of adversity and Rosalind and Celia's act of justice toward Corin. These two themes of *As You Like It* recur, with heightened urgency and purpose, in *King Lear*. The disguised Kent, the faithful servant, has led the old king and the Fool to no welcome in Arden, but to a "hovel" that will provide them some meager shelter from the storm. At the doorway Lear says:

> Poor naked wretches, wheresoe'er you are,
> That bide the pelting of this pitiless storm,
> How shall your houseless heads and unfed sides,
> Your looped and windowed raggedness, defend you
> From seasons such as these? O, I have ta'en
> Too little care of this! Take physic, pomp;
> Expose thyself to feel what wretches feel,
> That thou mayst shake the superflux to them
> And show the heavens more just.

Lear's admission, "O, I have ta'en / Too little care of this!" is the turning point of his story. He has heretofore "ta'en care" mainly of himself; that has now become his calamity, and he knows it. His reproof, "Take physic, pomp," recalls Duke Senior's denunciation of "painted pomp" in *As You Like It*, at the same time that it takes up with greater force the earlier play's concern for economic justice. Recognition of the suffering of "Poor naked wretches" leads directly here to the biblical imperative of charity to the poor, for as long as people are painfully in want there is an implicit cruelty in anybody's "superflux" of wealth. This theme is repeated in full by Gloucester in IV, i, after he has given his purse to "Poor Tom":

> Heavens, deal so still!
> Let the superfluous and lust-dieted man,
> That slaves your ordinance, that will not see
> Because he does not feel, feel your pow'r quickly;
> So distribution should undo excess,
> And each man have enough.

*

In revenge for his kindness and service to the king during the storm, Gloucester has been captured by Cornwall and Regan, who bind him and put out his eyes. He too is then thrust out, blind and (as his tormentors believe) alone, into the world and the weather—to "smell / His way to Dover," as Regan says in as cruel a speech as was ever written.

But immediately after the terrible scene of his blinding, we find that Gloucester is not after all alone. He is helped first by an elderly servant who, in the little he tells of himself, answers exactly to the description of the old Adam of *As You Like It*:

> O my good lord,
> I have been your tenant, and your father's tenant,
> These fourscore years.

And then he is helped by Edgar in the guise of Tom o'Bedlam or Poor Tom, who is in fact his father's faithful servant, guide, and teacher, and who at last "save[s] him from despair."

Cast out into the storm and the darkness, Lear too is accompanied—first by the Fool and then by Kent and then by Gloucester, at what cost we know, and then by Cordelia and Albany. These are the good and faithful servants of this play, and they continue here *As You Like It*'s theme of service and, with it, the earlier play's theme of affection and generosity as opposed to force. But good service in *King Lear* is more costly than in *As You Like It*, also less effective, and thus it emerges in what Shakespeare must have concluded is its true character. Kent is exiled because he understands faithful service, not only as loyalty, as faithful help, but also as truth-telling, requiring even opposition. In this Kent is contrasted with Oswald, who is a bad servant because he connives in the evil of his masters and does as he is told; and Oswald is again contrasted with Cornwall's "First Servant," who opposes his master's cruelty to Gloucester and dies for his insubordination.

By the time he wrote *King Lear*, Shakespeare clearly had begun to doubt that it is possible to consent to use by policy a little or a limited evil to serve some perceived "good," and then to stop before the evil has enlarged to some unforeseen perfection of itself. This perfection, as Shakespeare saw it, was the destruction of the evildoers along with whatever else might be destroyed by them. (This is the tragedy itself of *Macbeth*, which was written in the next year.)

Self-destruction, after selfishness has been accepted as a policy, is merely a matter of logic, as explained in Ulysses' speech. The key insight is given by Cornwall in III, vii, when, before the blinding of Gloucester, he speaks of "our wrath, which men / May blame, but not control"; and again by Goneril in V, iii: "the laws are mine . . ." If the laws belong to individual persons—if Goneril, as queen, does not rule "under the law" —then those persons are in effect lawless. The party of evil is by definition out of control from the start. Its members are out of control as individuals dedicated to self-interest. People who are united by the principle of unrestrained self-interest have inevitably a short-lived union. However large and however costly to their victims their successes may be, their failure is assured. But their espousal of evil as a deliberate policy assures also that they will be unrelenting while they last.

To this great force of relentless if self-doomed evil Shakespeare opposes the counterforce of good and faithful service. As Lear and Gloucester are made powerless, poor, and helpless, the theme of help manifests itself in the presence and the acts of people entirely dedicated to serving them. But Lear and Gloucester in their selfishness are too vulnerable, and the wickedness of their adversaries is too great, to permit to the good servants any considerable practical success. They can give no victory and achieve no restoration, as the world understands such things. Their virtues do not lead certainly or even probably to worldly success, as some bad teachers would have us believe. They stand by, suffering what they cannot help, as parents stand by a dying or disappointing child. This assures only the survival in this world of faithfulness, compassion, and love —which is no small thing.

But this play refuses to stop at what the world understands of service or success. For Lear and Gloucester worldly failure is fully assured; it is too late for worldly vindication. What the good servants can do, and this they succeed in doing, is restore those defeated old men to their true nature as human beings. They can waken them to love and save them from despair.

*

It is obvious by now that I have begun to argue against what we might call the "dark interpretation" of *King Lear*. The dark interpretation is well represented by Stephen Orgel, editor of the new Pelican edition of "the traditional conflated text," who sums the meaning of the play as follows:

> The world is an instrument of torture, and the only comfort is in the nothing, the never, of death. The heroic vision is of suffering, unredeemed and unmitigated.

It is impossible to see this nihilistic reading of the play as valid, and hard to see it as "heroic." There is a kind of modern mind that finds Hell more imaginable and believable than Heaven and nihilism more palatable than redemption. What is "heroic" to this mind is the "courage" to face the immitigable pointlessness of human experience. This is the same mind that, in default of any structure of meaning, finds all bad outcomes, political or economic or ecological, to be "inevitable."

Before even approaching the issues of this play's ending, one ought at least to consider the biblical context within which Shakespeare oriented his work. *King Lear* was written in reference to three passages in the Gospel of Matthew. Like *As You Like It*, it alludes repeatedly, and more insistently and sternly, to the call to service in Matthew 20 that I quoted earlier. And more forcibly than in the earlier play it extends the obligation of service to "the least of these my brethren" (Matthew 25:40), for who more than Lear and Gloucester in their injury and helplessness could be counted as "least"? And unlike the comedy, the tragic play broods constantly on the idea, in Matthew 10:39 and also in the other three Gospels, of losing one's life in order to find it. This theme is stated plainly in II, i, when the King of France says to Cordelia, "Thou losest here, a better where to find," and this strikes so nearly to the heart of the play as to be virtually its subject.

In I, iv, Kent, newly exiled and in disguise, his old life thus lost, says to himself:

> If thou canst serve where thou dost stand condemned,
> So may it come thy master whom thou lov'st
> Shall find thee full of labors.

This is a literal description of Kent's predicament in the play, if we read "thy master" as King Lear, and "condemned" as Kent's exile. But it is also, and just as literally, a description of the human predicament and consequent obligation, if we read "thy master" as Christ, and "where thou dost stand condemned" as the fallen world.

In III, iv, Edgar as Poor Tom, in his feigning madness, recites a number of pertinent biblical laws, four of them from the Ten Commandments.

In IV, ii, when Albany says to Goneril, "Wisdom and goodness to the vile seem vile . . . ," Shakespeare could be recalling Isaiah 5:20: "Wo unto them that speake good of evil, and evil of good . . ."

Cordelia's sentence, lines 23–24 in IV, iv, has the same ambiguity as Kent's speech cited earlier: "O dear father, / It is thy business I go about." This either is an apostrophe to Lear or it is a prayer recalling Luke 2:49, in which Jesus says to his parents, "knewe ye not that I must go about my fathers busines?"

In IV, vi, when Edgar, seeing Lear in his madness "bedecked with weeds," exclaims, "O thou side-piercing sight!" he is recalling John 19:34: "But one of the soldiers with a spear pierced his side . . ."

When Lear in his mad sermon in IV, vi, says, "See how yond justice rails upon yond simple thief. Hark in thine ear: change places and, handy-dandy, which is the justice, which is the thief?" his words paraphrase Romans 2:1: "Therefore thou art inexcusable, o man, whosoever thou art that judgest: for in that thou judgest another, thou condemnest thy self: for thou that judgest, doest the same things."

Gloucester's prayer in the same scene—

> You ever-gentle gods, take my breath from me;
> Let not my worser spirit tempt me again
> To die before you please

—is the daunting submission of the Lord's Prayer and of Christ's agony in Gethsemane: "Thy wil be done . . ."

Cordelia's lamentation over her father in IV, vii—". . . and wast thou fain, poor father, / To hovel thee with swine and rogues forlorn . . ."—recalls the story of the Prodigal Son. She wakens Lear a few lines later in what he perceives as a

resurrection. And her forgiveness of his offenses against her—
"No cause, no cause"—is Christ's: not a mere human excusing
or overlooking of an error, but the cancellation of its cause in
Lear's fallen nature; his wrong no longer exists to be forgiven.

The trumpets that sound in V, iii, must have sounded in
Shakespeare's imagination, and to the ears of many in his audi-
ences, like the trumpets of Revelation, for they are a summon-
ing to judgment.

When Albany in V, iii, offers to "friends . . . / The wages of
their virtue," the words evoke Romans 6:23: "For the wages
of sinne is death; but the gifte of God is eternal life . . ."

And when Kent, in his final speech, says, "My master calls
me; I must not say no," he is confirming earlier suggestions of
his impending death. But here we have again that term "master," which in I, iv, we could take to be ambiguous, but which
here we are bound to understand as referring to Christ. To assume that "master" refers to Lear is to assume that Kent thinks
Lear will require his services in the hereafter, a sentimentality
that puts Kent far out of character.

The play, furthermore, contains three references to the miraculous, always in circumstances of great misery, reminding
us that Christ's miracles are almost always performed on behalf
of those who are seemingly beyond help. Kent, in the stocks in
II, ii, says, "Nothing almost sees miracles / But misery." And
Edgar, after Gloucester's "suicide" in IV, vi, says to him in the
playacting of his "cure," but in fact urging the realization on
him: "Thy life's a miracle." A few lines further on, he says, "the
clearest gods who make them honors / Of men's impossibilities have preserved thee."

The foregoing list of biblical references may be incomplete
or otherwise at fault. But it is at least sufficient to show that
Shakespeare thought of the action of this play as occurring in
a context far larger than that of what we have come to mean
by "realism."

Anybody looking for meaninglessness or nihilism in *King Lear*
can find it in abundance, but nearly all of it is in the deeds
and the implicit principles of the villains. There are, however,
three statements in the play that are explicitly and pointedly
nihilistic.

The first is in one of Lear's speeches in III, iv, when in the storm he is reduced, as he thinks, virtually to nothing, and in his madness he adopts a fierce reductionism of his own. "Is man no more than this?" he asks of the nearly naked Edgar. And then, addressing Edgar, he says, "Thou art the thing itself; unaccommodated man is no more but such a poor, bare, forked animal as thou art." This, now under the weight of tragedy, is Jaques' conclusion in his "seven ages" speech; "unaccommodated man" is a lone specimen "sans everything."

A considerable part of the purpose of this play is to answer such statements, and this one is answered simply by the circumstances in which it is uttered. Lear's despair at this point is over the failure of a mere man to be successfully selfish. He cannot secure for himself his own wishes, and he cannot, alone, save himself even from the weather. But as bad as his predicament is, as nearly hopeless as it is, he is not "unaccommodated." Like the old Adam of *As You Like It*, he is not alone. Kent, the Fool, and Edgar are with him. A little later Gloucester enters with a torch to offer what help he can. With one most consequential exception, the good people of the play are going to be with him, doing all they can for him, to the end. What they can do is not enough, but they stand nonetheless for all that is opposite to his trouble and his suffering, his rage and his despair. They stand for the faithfulness that is opposed to treachery and the gentleness that is opposed to force.

The second expression of utter despair needing some comment is blinded Gloucester's accusation against "th' gods" in IV, i. This phrase "th' gods" is in keeping with a parliamentary proscription of the use of the word "God" on the stage, which the Puritans thought to be blasphemous. And so Gloucester was reduced to blaming the Greek and Roman deities: "As flies to wanton boys are we to th' gods; / They kill us for their sport."

Gloucester says this just after he has met Edgar-as-Poor-Tom. From now until Gloucester's death, Edgar's ruling purpose is to save his father from despair. Gloucester's sentence, while avoiding the appearance of blasphemy so fearful to the Puritan politicians, is authentically blasphemous, as Edgar understands. It is blasphemous, desperate, and perfectly self-centered. It is self-pity in extremis, driving him to say what he can hardly bear to say and cannot know. To save him from despair is to save him from the death of "a poor, bare, forked animal" reduced to the self-indulgence of self-pity. And by the end of the Gloucester story Edgar has led his father to a proper care for his life ("Thy life's a miracle") and to the proper submission to divine will that I quoted earlier. Edgar's service to Gloucester is clearly to be understood as redemptive, and he is not being frivolous when he says that his father died "smilingly" between the two extremes of "joy and grief."

My final exhibit in this line of nihilism is from a speech of Edmund's. Near the end of the play he sends a "captain" to follow Lear and Cordelia to prison, with instructions to kill them. Here is his justification: "Know thou this, that men / Are as the time is." This is a crude, self-serving determinism, the counterpart of "It is inevitable." All the energy and passion of *King Lear* gathers to refute this speech. *Some* men are as the time is, some always are, and they have always said so in self-justification. But Cordelia is not as the time is, Kent is not, Edgar is not, Albany is not, even the Fool is not.

And so these three assertions of hopelessness and meaninglessness are answered with three resounding nos that are passionately affirmative: No, "unaccommodated man" is not the type specimen of humanity. No, you are not eligible to conclude that the gods kill us for their sport. No, all men are not "as the time is."

By the end of the play we can have no doubt that we have watched a deadly skirmish in the battle between good and evil. We have watched the passage of tormented souls and a human community through a profound disorder, in which they have been driven away from their comforts and customary assurances into the world's unaccommodating wildness. The consequences of this casting out are surely tragic and horrifying. The death of Cordelia, as Dr. Johnson and others have testified, is shocking; it is nearly unbearable. The survivors are clearly in shock themselves, barely able to speak. And so we now must ask if in fact *King Lear* conforms to the archetypal story I outlined at the beginning. Are Lear and Gloucester in any sense reformed or redeemed by their great suffering? Is there any promise of a return to civil order? Does the play (to quote J. A. Bryant Jr.) "satisfy the society's impulse to renewal"?

Well, before concluding with the proponents of darkness that the play merely demonstrates the meaninglessness of suffering, we need to deal patiently with certain facts. The first of these is that by the play's end every one of the villains is dead —and not one of them is dead by chance. The death of each has come as a logical consequence of the assumption that human nature can be satisfactorily subordinated to nature. This assumption has proved to be as uncontrollable as the storm on the heath. There is a right relation between nature and human nature, and to get it wrong is eventually to perish. Shakespeare does not present this as an issue of justice, for such wrongs may destroy the innocent as well as the guilty; he presents it as the natural result of unnatural (that is, inhuman) behavior. The conflict of the two natures is revealed in Edmund's dying effort to redeem himself: "Some good I mean to do, / Despite of mine own nature."

According to this view, it would be too much to expect a "comic" outcome, for great evil will victimize the good. But the good people, unlike the evil ones, are not inevitably destroyed. Another fact is that, as the play ends, Kent, Edgar, and Albany are still alive. Kent's life is in doubt, but Edgar and Albany are young and will live on. In them is a reasonable hope for the restoration of civil order. And not only have those three survived, but in the course of the play, together with Cordelia and the Fool, they have grown ever greater in our respect and love, as has Shakespeare himself for imagining such people.

Another fact hard to ignore is the work of forgiveness. Both Cordelia and Edgar freely forgive their erring fathers, and by this forgiveness those fathers are made more truly and fully human.

Now we must deal with the reconciliation of the two plots. This is not really a problem, except that it has been made so by bad reading. The "problem" for the dark interpreter is that the Lear and the Gloucester stories are parallel, each enlarging our understanding of the other by resonating with it. The "problem" is that the Gloucester story is explicitly redemptive, for Edgar intends, as he says, to save his father from despair, and he succeeds, whereas the Lear story, according to the dark interpretation, ends in despair: "suffering, unredeemed and unmitigated." How can a play thus have two plots and two meanings that absolutely contradict each other and still deserve our respect? Stephen Orgel solves this problem by asserting that "Gloucester is effectively abandoned by the play." But this only raises a worse problem: Why would Shakespeare have given so much of his play, and so much magnificent poetry, to a secondary plot that he later "abandons"? And why should we indulge or forgive his doing so?

Another, more sensible way to deal with the supposed problem is to ask if Lear's story actually ends in despair, and thus in contradiction of Gloucester's story. To answer, we must look with the greatest care at Lear's final speech. It is possible, I suppose, to read or speak those lines as an unpausing scramble of outrage, grief, and despair. But the speech in fact has five parts, involving four profound changes of mind and mood. It begins with a complaint:

> And my poor fool is hanged: no, no, no life?

This is tenderness, heartbreaking enough, but it bears still a taint of the old selfishness. Cordelia is "*my* poor fool." Her death here is perceived as Lear's loss, not hers. Then comes a natural outrage:

> Why should a dog, a horse, a rat, have life,
> And thou no breath at all?

All grief over the death of the young, especially the death of one's own child, must bear the burden of such a question. And only after that ineluctable and futile question, which also comes from his own loss, can Lear turn his thoughts fully to the dead girl in his arms, and, forgetting himself, speak to her of her death:

> Thou'lt come no more,
> Never, never, never, never, never.

And then, turning to one of the bystanders, he says:

> Pray you undo this button. Thank you, sir.

This is literally meant. His clothes are somehow binding; he asks and receives help with a button. But the button is symbolic as well; it, or this small discomfort, is the last thing holding him to the world. This is not the renunciation of Gloucester's "suicide," but rather a profound submission and relinquishment of his will. At this point all the emotions of the preceding lines, and of his tragedy, pass from him, with the result that at last he *sees* Cordelia:

> Do you see this? Look on her! Look, her lips,
> Look there, look there—

Martin Lings, in *The Secret of Shakespeare*, understands Lear's story as at once a descent into Hell and a Purgatory, and he thinks that when Lear speaks these final two lines he is seeing Cordelia again, and this time in truth, as "a soul in bliss." I have no doubt that the play can be read or presented as Dr. Lings suggests. And yet I hesitate. The difficulty is that Shakespeare, as it seems to me, was not a visionary Christian; he was not Dante. The redemption he saw as possible for Gloucester and Lear did not come by way of an intercession from Heaven. It was earned, or lived out, or suffered out, in an unrelenting confrontation both with the unregenerate self, the self-covered self, and with the deliberate evil of others. The straight way was lost to Gloucester and Lear as it was to Dante, but it was recoverable to them by a self-loss more painful, and nearer too late, than Dante's.

I am content to rest with the more literal understanding that Cordelia, the play's only wholly undisguised character, has been disguised to Lear until the end by his self-preoccupying pride, anger, outrage, guilt, grief, and despair; and that, when his vision clears at last and he can see her as she was and is, he is entirely filled with love and wonder. And so the play may be said to show us at last a miracle: that Lear, dying, is more alive than he has ever been until this moment.

God, Science, and Imagination

Among the oddest bedevilers of our time are the eminent scientists who use their heaped-up credentials, achievements, and awards as pedestals from which to foretell the future and pronounce upon the ultimate questions of life and religion. One of the most recent of these is Steven Weinberg's essay "Without God" in *The New York Review of Books* for September 25, 2008.

The oddity of these ventures, of which Professor Weinberg's is fairly typical, is in their ready—and, it seems to me, their thoughtless—abandonment of scientific rigor and methodology. For example, despite his protest that he does not want "to try to talk anyone out of their [*sic*] religion," Prof. Weinberg sets forth an elaborate argument for the nonexistence of God, an argument obviously meant to be persuasive but one that is based entirely on opinion.

As a fundamentalist of science, like the fundamentalists of religion, he is clearly evangelizing, hoping to convert or at least to disturb those who disagree. And like the religious fundamentalists, he uses a language that presents belief as knowledge. But more troubling than the authority he grants to his own opinions is his claim to know what cannot be known. "As religious belief weakens," he writes, "more and more of us know that after death there is nothing." The only fact available here is that Prof. Weinberg and more and more of us do not, and will never, know any such thing. There is no proof of this "nothing," and there is no scientific or other procedure by which to attempt such a proof.

Prof. Weinberg is a physicist, and he says that he is "professionally concerned with finding out what is true." But as a mere person he evidently is concerned, like too many others, merely with investing his opinions with power. This is the concern of fundamentalists of all kinds: religious, atheistic, scientific, technological, economic, and political. They all seek power—they seek victory, in fact—by abandoning the proprieties that permit us to seek and to honor what is true while acknowledging the limits of our ability to know.

Not far into his essay, Prof. Weinberg says, with proper humility, "Of course, not everything has been explained, nor will it ever be." But, two paragraphs later, speaking of "religious conservatives," he abandons the careful and exacting speech of humility, and prognosticates with the absolute confidence and gleeful vengefulness of a religious conservative: "I can imagine how disturbed they will feel in the future, when at last scientists learn how to understand human behavior in terms of the chemistry and physics of the brain, and nothing is left that needs to be explained by our having an immaterial soul."

This is something else that he does not know. Nor does he hesitate over the apparent difficulty of a material proof of the nonexistence of something immaterial.

The argument about the existence of God necessarily must be conducted in the absence of evidence that would stand as proof in either a laboratory or a court of law. There is no objective or empirical or experimental evidence on either side. The argument, as such, is by definition hopeless—a piece of foolishness and a waste of time. Even so, it has long existed and no doubt it will long continue, but only for the paltry reason that it cannot be won. Chaucer defined the problem about six hundred years ago, and I doubt that it can be more clearly defined:

> A thousand tymes have I herd men telle
> That ther ys joy in hevene and peyne in helle,
> And I acorde wel that it ys so;
> But, natheles, yet wot I wel also
> That ther nis noon dwellyng in this contree,
> That eyther hath in hevene or helle ybe,
> Ne may of hit noon other weyes witen,
> But as he hath herd seyd, or founde it writen;
> For by assay ther may no man it preve.
> —*The Prologue to the Legend of Good Women*

People of religion, and not just fundamentalists, can speak with tiresome confidence of knowing what in fact they don't know but instead believe. None of us is immune to the temptation to do this. Modern science itself, ignoring its famous devotion to empirical proof and factuality, has pampered and marketed itself by beliefs that have proved to be empirically flimsy and unimaginably damaging. Chemistry, while helping us to "live better," has poisoned the whole world; the "elegant" science of nuclear physics, while making us "safe" from our enemies and offering us "cheap" and "peaceful" power, has littered the world with lethal messes that apparently are irremediable; and so on to genetic engineering and other giddy "miracles." These developments, at least in their origin, are scientific. But the science involved has not been comprehensive or humble or self-critical or neighborly or publicly responsible. Mere self-interest obliges us to doubt the scientific faith that facts alone can assure the proper or safe use of facts. Modern science, as we have known it and as it has represented itself to us, has encouraged a healthy skepticism of everything but itself. But surely it implies no disrespect for science if we regard it with the skepticism upon which it prides itself.

We human beings, because we are short-lived creatures of limited intelligence, are going to remain under the necessity of talking about things that we don't provably know. But respect for "what is true," for what we don't know, and for our neighbors and fellow creatures, requires us to know and to say when we *don't* know.

Prof. Weinberg understands religious belief as "the belief in facts about God or the afterlife." This is a mistake. If "fact" means what we have agreed it means, and if we respect the word, then we have to say with Chaucer that none of us knows any facts about God or the afterlife. If we did, there would be no issue of "belief." We know for sure that it is possible to speak of beliefs and opinions as facts, but that does not make the beliefs and opinions factual; it only makes them lies.

Most writers about religion, however, have not been scientists, or consciously subject to the methodological strictures of science. If they speak of knowledge, they may mean the things one knows from tradition or from unreplicable experience or "from the heart." Even so, in the Bible the language of belief often falls far short of the confidence of factual knowledge. It is most moving—and, to me, it seems most authentic—when it is honestly confronting its own imperfection, or the inadequacy or failure of knowledge. Far from the cocksureness of fundamentalism, the starting place of authentic belief or faith is not-knowing.

One of the primary characteristics of the biblical God is his irreducibility; he cannot be confined in any structure of human comprehension. And so in 1 Kings, having completed his temple, Solomon cries out, not with confident religiosity, but with despair; his mighty work has contradicted what apparently was its purpose:

> But will God indeed dwell on the earth? behold, the
> heaven and heaven of heavens cannot contain thee;
> how much less this house that I have builded? (8:27)

The supplication of Mark 9:24 is likewise authenticated by its honest unknowing, unconfidence, and sense of struggle: "Lord, I believe; help thou my unbelief." And Paul's letter to the Romans is precise and unrelenting in his definition of hope:

> For we are saved by hope: but hope that is seen is not
> hope: for what a man seeth, why doth he yet hope for?
> But if we hope for that we see not, then do we with
> patience wait for it. (8:24–25)

"Faith," at root, is related to "bide" and "abide." It has certainly the sense of belief, but also the sense of difficult belief —of waiting, of patience, of endurance, of hanging on and holding together.

And so Prof. Weinberg's definition of "belief" involves not only a misuse of the word "facts," but also an implicit misunderstanding of the word "faith."

I dislike very much the disciplinary provincialisms of the universities; therefore, as a literary person, I ought to be delighted that Prof. Weinberg finds literature as irresistible as religion. But I am obliged instead to regret that he speaks of it with complacent oversimplification and ineptitude. When he says, for example, that "nothing prevents those of us who have no religious belief from enjoying religious poetry," does he mean that such enjoyment is the same for believers and unbelievers? If so, how does he know this? Does he have a way of comparing objectively the degrees and kinds of enjoyment? I would

gladly agree that enjoyment is a desirable, maybe even a necessary, result of any art; but is enjoyment the only or the highest
effect of religious art? What is it about religious art that unbelievers enjoy? The underlying question here, and an important
one, is this: How do you authenticate, and make credible to
somebody else, your response to a work of art? Prof. Weinberg
seems not to have suspected that this question exists, or that it
implies a careful, difficult job of work.

He likewise suspects no danger in his assertion that "we see
already that little English language poetry written in the past
few decades owes anything to belief in God." But who is "we"?
How many decades does he have in mind? What is signified
by "little"? The existence of God is not a statistical issue to
be proved or disproved by quantities of belief or numbers of
believers. If God exists, then, like Prof. Weinberg, he exists
independently of anybody's knowledge or anybody's belief or
disbelief in his existence. If *nobody* believed in God, Prof. Weinberg would still have his case to make, and evidence would be
required of him that he cannot produce.

He says further that "very great poetry can be written without religion," using Shakespeare as his example, apparently
unaware that Shakespeare's religion is still a controversial issue among Shakespearean scholars. But you don't have to be
an expert—you have only to read the Bible and the plays—to
know of Shakespeare's frequent allusions to scripture and his
concern with scriptural themes such as mercy and forgiveness
and with scriptural characters such as the good and faithful servant. But if great poetry can be written without religion, what
does that prove about religion? It proves only that great poetry
can be written without it.

In the boring, pointless, and destructive quarrel between fundamentalist science and fundamentalist religion, it seems to me
that both sides are wrong. The religious fundamentalists are
wrong because their disrespect for the materiality of the world
involves, as a matter of course, disrespect for material evidence.
They are like a jury that sees no significance in a "smoking
gun" because its members don't believe in guns or smoke. The
fundamentalist scientists are wrong because they counter one

absolutism with another. Against their own history and tradition, they assume the posture of absolute certainty and unquestionability. Both sides assume that they are right now and forever. Neither can say "I don't know" or "I wonder." Both are bigoted, unforgiving, and humorless.

In his troubled and consoling last book of poems, *Second Space*, Czeslaw Milosz includes a poem of exemplary generosity:

> If there is no God,
> Not everything is permitted to man.
> He is still his brother's keeper
> And he is not permitted to sadden his brother,
> By saying there is no God.

This instruction, as Milosz undoubtedly knew, is perfectly reversible: If there is a God, that does not justify condescension or insult to your atheist neighbor. Such differences, so far as I can see, become issues of justice only when one side attempts to abridge or deny the freedom of the other.

If in fact the fundamentalist scientists were as smart as they think they are, and if the religious fundamentalists were as secure in their belief as they claim to be, then they would (except for issues of justice) leave one another in peace. They keep pestering each other because they need each other. The sort of mind that is inclined to fundamentalism is not content within itself or within its own convictions or principles. It needs to humiliate its opponents. It needs the sustenance of converts. It is fundamentally insecure and ungenerous.

Worst of all, the fundamentalists of both science and religion do not adequately understand or respect imagination. Is imagination merely a talent, such as a good singing voice, the ability to "make things up" or "think things up" or "get ideas"? Or is it, like science, a way of knowing things that can be known in no other way? We have much reason to think that it is a way of knowing things not otherwise knowable. As the word itself suggests, it is the power to make us *see*, and to see, moreover, things that without it would be unseeable. In one of its aspects it is the power by which we sympathize. By its means we may see what it was to be Odysseus or Penelope, or David or Ruth, or what it is to be one's neighbor or one's enemy. By it, we may "see ourselves as others see us." It is also the power by which we see the place, the predicament, or the story we are in.

To use what is by now the most notorious example, the creation story in Genesis is neither science nor pseudo-science, neither history nor pseudo-history. Like other traditional creation stories, it welled up out of the oldest, deepest human imagination to help us, even now, to see what it is to have a wondrous world that had a beginning in time. It is not true by the corroboration of contemporary documents or physical evidence. It is the imagination, in the high sense given to it by the greatest poets, that assents to its truth, just as it assents to the story of King Lear or Blake's rendering of Jacob's vision. The following lines, from Hayden Carruth's *Toward the Distant Islands*, rightly ignoring the unwinnable contest of science versus religion, were written with a proper deference to mystery and a proper respect for imagination:

> The *Iliad*, the *Odyssey*, the *Book of Genesis*,
> These were acts of love, I mean deeply felt gestures, which
>   continuously bestow upon us
> What we are.

As for the afterlife, it has been imagined by Homer, Virgil, the biblical writers, Dante, and others, with the result that at least some of us, their willing heirs, have imagined it also.

I don't see that scientists would suffer the loss of any skin from their noses by acknowledging the validity and the power of imaginative truths, which are harmless to the truths of science, even though imagination in the highest sense seems allied less to science than to religion. The first chapters of Genesis are imagined and imaginable, whereas the big bang theory is the result of calculation. If you have read Dante, you can imagine Hell, Purgatory, and Heaven, but reading Prof. Weinberg cannot help you to imagine "nothing."

Perhaps the most interesting thing that Prof. Weinberg says in his essay is this: "There are plenty of people without religious faith who live exemplary moral lives (as for example, me)." This of course is a joke, modeled on the shameless self-commendation of politicians, but it is a joke without a context sufficient to reveal how large and sad a joke it is. The large sad fact that gives the joke its magnitude and its cutting edge is that there is probably not one person now living in the United States who, by a strict accounting, could be said to be living an exemplary moral life.

We are still somewhere in the course of the most destructive centuries of human history. And, though I believe I know some pretty good people whom I love and admire, I don't know one who is not implicated, by direct participation and by proxies given to suppliers, in an economy, recently national and now global, that is the most destructive, predatory, and wasteful the world has ever seen. Our own country in only a few hundred years has suffered the loss of maybe half its arable topsoil, most of its original forest and prairie, much too much of its mineral wealth and underground water. Most of its surface water and all of its air are polluted. Its rural cultures—the cultures, at their best, of husbandry—have been almost annihilated. Many of its plants and animals, both wild and domestic, are extinct or in danger. It is littered with wastelands, landfills, and, most shameful and fearful of all, dumps, industrial sites, and whole landscapes made dangerous virtually forever by radioactive waste. An immense part of this damage has been done in the years after World War II, when the machinery and chemicals of industrial warfare were turned upon the land—to make production "efficient" by the most doubtful standards and to replace the people of the land economies. I have no doubt that the dualisms of body and soul, heaven and earth, too prominent among the religious, have been damaging both to people and to the world, for that division has made it easy to withhold the necessary protections from material things. But the materialists of the science-technology-industry complex, whose minds are not so divided, and who might have been expected to value highly the material world, have instead held it in contempt and damaged it more than anybody.

Scientists and scholars in the knowledge industry—corporate or academic, if there is a difference—are probably in the greatest moral jeopardy of anybody except political, military, and corporate leaders. All knowledge now is potentially a commodity, and there is no way for its originators to control, or even foresee, the uses to which it may be put.

*

I would like, in conclusion, to bring up a question of religion and politics that I think needs more attention from everybody, but maybe especially from atheists. Before going on, I had better say that I adhere absolutely to the First Amendment. Any form of religious coercion by religious organizations or by governments would be intolerable to me. The idea of the separation of church and state seems to me fairly clear when it is a matter simply of limiting the powers of institutions. But the world is not so simple as to allow a neat, clear separation of politics and religion or of politics and irreligion.

My question is about the origin and existence of human rights. How did we get them? How are they authenticated? In ancient traditional cultures such as those of surviving peasant or hunter-gatherer communities, the people may be said to possess certain rights by tradition and inheritance: Because they have possessed them immemorially, they possess them still. No modern government, younger and of shallower origin, could rightfully revoke or ignore them. A younger nation of recent immigrants such as the United States possesses no rights by immemorial tradition and inheritance. The founders understood this, and so they stated in our Declaration of Independence the principle that "all men" (which we now construe as "all people") "are endowed by their Creator with certain inalienable Rights."

I don't think it is adequately appreciated how essential, and what a stroke of political brilliance, that statement is. The purport of it is that, as humans, we have rights that precede the existence of any government. We therefore were not on our knees to the government of England then, and we are not kneeling to our own government now, beseeching a grant of rights. As a would-be free people we were, and we are, requiring any government whatever to recognize and honor the rights we have always possessed by divine gift. The difference between rights granted by a government and rights given by "our Creator" is critical, for it is the difference between rights that are absolute and rights that are contingent upon the will or the whim of those in power.

The possession of rights by divine endowment obviously is an article of faith, for it has no objective or empirical standing.

It would have no standing at all with a government in principle atheistic. (We had better hope, I think, that the separation of church and state implies the separation of institutional atheism and the state.) But vulnerable as this principle may be as an article merely of faith, I know of no other authorization of human rights that can adequately replace it.

It is easy to anticipate that some who will not allow any validity to divine rights will bring forward "natural rights" as an alternative. But this too would be an article of faith, and it forces upon us the probably unanswerable question of what, in the nature of nature, might bring forth and confer upon us rights specifically human.

I am not able to settle such questions, even to my own satisfaction. And so I am obliged to conclude by offering the possibility that we humans are by definition, and perhaps by nature, creatures of faith (which we are as likely to place in luck or science or the free market or our own intelligence as in some version of God); and that we are further defined by principles and cultural properties, not objectively verifiable, that we inherit.

FROM
IT ALL TURNS ON AFFECTION
(2012)

It All Turns on Affection

"Because a thing is going strong now, it need not go strong for ever," [Margaret] said. "This craze for motion has only set in during the last hundred years. It may be followed by a civilization that won't be a movement, because it will rest upon the earth."

E. M. Forster, *Howards End* (1910)

O NE NIGHT in the winter of 1907, at what we have always called "the home place" in Henry County, Kentucky, my father, then six years old, sat with his older brother and listened as their parents spoke of the uses they would have for the money from their 1906 tobacco crop. The crop was to be sold at auction in Louisville on the next day. They would have been sitting in the light of a kerosene lamp, close to the stove, warming themselves before bedtime. They were not wealthy people. I believe that the debt on their farm was not fully paid, there would have been interest to pay, there would have been other debts. The depression of the 1890s would have left them burdened. Perhaps, after the income from the crop had paid their obligations, there would be some money that they could spend as they chose. At around two o'clock the next morning, my father was wakened by a horse's shod hooves on the stones of the driveway. His father was leaving to catch the train to see the crop sold.

He came home that evening, as my father later would put it, "without a dime." After the crop had paid its transportation to market and the commission on its sale, there was nothing left. Thus began my father's lifelong advocacy, later my brother's and my own, and now my daughter's and my son's, for small farmers and for land-conserving economies.

The economic hardship of my family and of many others, a century ago, was caused by a monopoly, the American Tobacco Company, which had eliminated all competitors and thus was able to reduce as it pleased the prices it paid to farmers. The American Tobacco Company was the work of James B. Duke of Durham, North Carolina, and New York City, who, disregarding any other consideration, followed a capitalist logic to absolute control of his industry and, incidentally, of the economic fate of thousands of families such as my own.

My effort to make sense of this memory and its encompassing history has depended on a pair of terms used by my teacher, Wallace Stegner. He thought rightly that we Americans, by inclination at least, have been divided into two kinds: "boomers" and "stickers." Boomers, he said, are "those who pillage and run," who want "to make a killing and end up on Easy Street," whereas stickers are "those who settle, and love the life they have made and the place they have made it in." "Boomer" names a kind of person and ambition that is the major theme, so far, of the history of the European races in our country. "Sticker" names a kind of person and desire that is, so far, a minor theme of that history, but a theme persistent enough to remain significant and to offer, still, a significant hope.

The boomer is motivated by greed, the desire for money, property, and therefore power. James B. Duke was a boomer, if we can extend the definition to include pillage in absentia. He went, or sent, wherever the getting was good, and he got as much as he could take.

Stickers on the contrary are motivated by affection, by such love for a place and its life that they want to preserve it and remain in it. Of my grandfather I need to say only that he shared in the virtues and the faults of his kind and time, one of his virtues being that he was a sticker. He belonged to a family who had come to Kentucky from Virginia, and who intended to go no farther. He was the third in his paternal line to live in the neighborhood of our little town of Port Royal, and he was the second to own the farm where he was born in 1864 and where he died in 1946.

We have one memory of him that seems, more than any other, to identify him as a sticker. He owned his farm, having bought out the other heirs, for more than fifty years. About forty of those years were in hard times, and he lived almost continuously in the distress of debt. Whatever has happened in what economists call "the economy," it is generally true that the land economy has been discounted or ignored. My grandfather lived his life in an economic shadow. In an urbanizing and industrializing age, he was the wrong kind of man. In one of his difficult years he plowed a field on the lower part of a long slope and planted it in corn. While the soil was exposed, a heavy rain fell and the field was seriously eroded. This was heartbreak for my grandfather, and he devoted the rest of his life first to healing the scars and then to his obligation of care. In keeping with the sticker's commitment, he neither left behind the damage he had done nor forgot about it, but stayed to repair it, insofar as soil loss can be repaired. My father, I think, had his father's error in mind when he would speak of farmers attempting, always uselessly if not tragically, "to plow their way out of debt." From that time, my grandfather and my father were soil conservationists, a commitment that they handed on to my brother and to me.

It is not beside the point, or off my subject, to notice that these stories and their meanings have survived because of my family's continuing connection to its home place. Like my grandfather, my father grew up on that place and served as its caretaker. It has now belonged to my brother for many years, and he in turn has been its caretaker. He and I have lived as neighbors, allies, and friends. Our long conversation has often taken its themes from the two stories I have told, because we have been continually reminded of them by our home neighborhood and topography. If we had not lived there to be reminded and to remember, nobody would have remembered. If we, like most of our generation, had moved away, the place with its memories would have been lost to us and we to it.

Because I have never separated myself from my home neighborhood, I cannot identify myself to myself apart from it. I am fairly literally flesh of its flesh. It is present in me, and to me, wherever I go. This undoubtedly accounts for my sense of shock when, on my first visit to Duke University, and by surprise, I came face-to-face with James B. Duke in his dignity, his glory perhaps, as the founder of that university. He stands imperially in bronze in front of a Methodist chapel aspiring to be a cathedral. He holds between two fingers of his left hand a bronze cigar. On one side of his pedestal is the legend: INDUSTRIALIST. On the other side is another single word: PHILANTHROPIST. The man thus commemorated seemed to me terrifyingly ignorant, even terrifyingly innocent, of the connection between his industry and his philanthropy. But I did know the connection. I felt it instantly and physically. The connection was my grandparents and thousands of others more or less like them. If you can appropriate for little or nothing the work and hope of enough such farmers, then you may dispense the grand charity of "philanthropy."

After my encounter with the statue, the story of my grandfather's 1906 tobacco crop slowly took on a new dimension and clarity in my mind. I still remembered my grandfather as himself, of course, but I began to think of him also as a kind of man standing in thematic opposition to a man of an entirely different kind. And I could see finally that between these two kinds there was a failure of imagination that was ruinous, that belongs indelibly to our history, and that has continued, growing worse, into our own time.

The term "imagination" in what I take to be its truest sense refers to a mental faculty that some people have used and thought about with the utmost seriousness. The sense of the verb "to imagine" contains the full richness of the verb "to see." To imagine is to see most clearly, familiarly, and understandingly with the eyes, but also to see inwardly, with "the mind's eye." It is to see, not passively, but with a force of vision and even with visionary force. To take it seriously we must give up at once any notion that imagination is disconnected from reality or truth or knowledge. It has nothing to do either with clever imitation of appearances or with "dreaming up."

I will say, from my own belief and experience, that imagination thrives on contact, on tangible connection. For humans to have a responsible relationship to the world, they must imagine their places in it. To have a place, to live and belong in a place, to live from a place without destroying it, we must imagine it. By imagination we see it illuminated by its own unique character and by our love for it. By imagination we recognize with sympathy the fellow members, human and nonhuman, with whom we share our place. By that local experience we see the need to grant a sort of preemptive sympathy to all the fellow members, the neighbors, with whom we share the world. As imagination enables sympathy, sympathy enables affection.

And in affection we find the possibility of a neighborly, kind, and conserving economy.

Obviously there is some risk in making affection the pivot of an argument about economy. The charge will be made that affection is an emotion, merely "subjective," and therefore that all affections are more or less equal: People may have affection for their children and their automobiles, their neighbors and their weapons. But the risk, I think, is only that affection is personal. If it is not personal, it is nothing; we don't, at least, have to worry about governmental or corporate affection. And one of the endeavors of human cultures, from the beginning, has been to qualify and direct the influence of emotion. The word "affection" and the terms of value that cluster around it —love, care, sympathy, mercy, forbearance, respect, reverence —have histories and meanings that raise the issue of worth. We should, as our culture has warned us over and over again, give our affection to things that are true, just, and beautiful. When we give affection to things that are destructive, we are wrong. A large machine in a large, toxic, eroded cornfield is not, properly speaking, an object or a sign of affection.

My grandfather knew, urgently, the value of money, but only of such comparatively small sums as would have paid his debts and allowed to his farm and his family a decent prosperity. He certainly knew of the American Tobacco Company. He no doubt had read and heard of James B. Duke, but nothing in his experience could have enabled him to imagine the life of the man himself.

James B. Duke came from a rural family in the tobacco country of North Carolina. In his early life he would have known men such as my grandfather. But after he began his rise as an industrialist, the life of a small tobacco grower would have been to him a negligible detail incidental to an opportunity for large profits. In the minds of the "captains of industry," then and now, the people of the land economies have been reduced to statistical numerals. Power deals "efficiently" with quantities that affection cannot recognize.

It may seem plausible to suppose that the head of the American Tobacco Company would have imagined at least that a dependable supply of raw material to his industry would depend upon a stable, reasonably thriving population of farmers and upon the continuing fertility of their farms. But he imagined no such thing. In this he was like apparently all agribusiness executives. They don't imagine farms or farmers. They imagine perhaps nothing at all, their minds being filled to capacity by numbers leading to the bottom line. Though the corporations, by law, are counted as persons, they do not have personal minds, if they can be said to have minds. It is a great oddity that a corporation, which properly speaking has no self, is by definition selfish, responsible only to itself. This is an impersonal, abstract selfishness, limitlessly acquisitive, but unable to look so far ahead as to preserve its own sources and supplies. The selfishness of the fossil fuel industries by nature is self-annihilating; but so, always, has been the selfishness of the agribusiness corporations. Land, as Wes Jackson has said, has thus been made as exhaustible as oil or coal.

There is another difference between my grandfather and James B. Duke that may finally be more important than any other, and this was a difference of kinds of pleasure. We may assume that, as a boomer, moving from one chance of wealth to another, James B. Duke wanted only what he did not yet have. If it is true that he was in this way typical of his kind, then his great pleasure was only in prospect, which excludes affection as a motive.

My grandfather, on the contrary, and despite his life's persistent theme of hardship, took a great and present delight in the modest good that was at hand: in his place and his affection for it, in its pastures, animals, and crops.

He did not participate in the least in what we call "mobility." He died, after eighty-two years, in the same spot he was born in. He was probably in his sixties when he made the one longish trip of his life. He went with my father southward across Kentucky and into Tennessee. On their return, my father asked him what he thought of their journey. He replied: "Well, sir, I've looked with all the eyes I've got, and I wouldn't trade the field behind my barn for every inch I've seen."

In such modest joy in a modest holding is the promise of a stable, democratic society, a promise not to be found in "mobility": our forlorn modern progress toward something indefinitely, and often unrealizably, better. A principled dissatisfaction with whatever one has promises nothing or worse.

James B. Duke would not necessarily have thought so far of the small growers as even to hold them in contempt. The Duke trust exerted an oppression that was purely economic, involving a mechanical indifference, the indifference of a grinder to what it grinds. It was not, that is to say, a political oppression. It did not *intend* to victimize its victims. It simply followed its single purpose of the highest possible profit, and ignored the "side effects." Confronting that purpose, any small farmer is only one, and one lost, among a great multitude of others, whose work can be quickly transformed into a great multitude of dollars.

Corporate industrialism has tended to be, and as its technological and financial power has grown it has tended increasingly to be, indifferent to its sources in what Aldo Leopold called "the land-community": the land, all its features and "resources," and all its members, human and nonhuman, including of course the humans who do, for better or worse, the work of land use. Industrialists and industrial economists have assumed, with permission from the rest of us, that land and people can be divorced without harm. If farmers come under adversity from high costs and low prices, then they must either increase their demands upon the land and decrease their care for it, or they must sell out and move to town, and this is supposed to involve no ecological or economic or social cost. Or if there are such costs, then they are rated as "the price of progress" or "creative destruction."

But land abuse *cannot* brighten the human prospect. There is in fact no distinction between the fate of the land and the fate of the people. When one is abused, the other suffers. The penalties may come quickly to a farmer who destroys perennial cover on a sloping field. They *will* come sooner or later to a land-destroying civilization such as ours.

And so it has seemed to me less a choice than a necessity to oppose the boomer enterprise with its false standards and its incomplete accounting, and to espouse the cause of stable, restorative, locally adapted economies of mostly family-sized farms, ranches, shops, and trades. Naïve as it may sound now, within the context of our present faith in science, finance, and technology—the faith equally of "conservatives" and "liberals" —this cause nevertheless has an authentic source in the sticker's hope to abide in and to live from some chosen and cherished small place—which is the agrarian vision that Thomas Jefferson spoke for, a sometimes honored human theme, minor and even fugitive, but continuous from ancient times until now. Allegiance to it, however, is not a conclusion but the beginning of thought.

The problem that ought to concern us first is the fairly recent dismantling of our old understanding and acceptance of human limits. For a long time we knew that we were not, and could never be, "as gods." We knew, or retained the capacity to learn, that our intelligence could get us into trouble that it could not get us out of. We were intelligent enough to know that our intelligence, like our world, is limited. We seem to have known and feared the possibility of irreparable damage. But beginning in science and engineering, and continuing, by imitation, into other disciplines, we have progressed to the belief that humans are intelligent enough, or soon will be, to transcend all limits and to forestall or correct all bad results of the misuse of intelligence. Upon this belief rests the further belief that we can have "economic growth" without limit.

Economy in its original—and, I think, its proper—sense refers to household management. By extension, it refers to the husbanding of all the goods by which we live. An authentic economy, if we had one, would define and make, on the terms of thrift and affection, our connections to nature and to one another. Our present industrial system also makes those connections, but by pillage and indifference. Most economists think of this arrangement as "the economy." Their columns and articles rarely if ever mention the land-communities and land-use economies. They never ask, in their professional oblivion, why we are willing to do permanent ecological and cultural damage "to strengthen the economy."

In his essay, "Notes on Liberty and Property," Allen Tate gave us an indispensable anatomy of our problem. His essay begins by equating, not liberty and property, but liberty and *control* of one's property. He then makes the crucial distinction between ownership that is merely legal and what he calls "effective ownership." If a property, say a small farm, has one

owner, then the one owner has an effective and assured, if limited, control over it—as long as he or she can afford to own
it, and is free to sell it or use it, and (I will add) free to use it
poorly or well. It is clear also that effective ownership of a small
property is personal and therefore can, at least possibly, be intimate, familial, and affectionate. If, on the contrary, a person
owns a small property of stock in a large corporation, then
that person has surrendered control of the property to larger
shareholders. The drastic mistake our people made, as Tate believed and I agree, was to be convinced "that there is *one* kind
of property—just *property*, whether it be a thirty-acre farm
in Kentucky or a stock certificate in the United States Steel
Corporation." By means of this confusion, Tate said, "Small
ownership . . . has been worsted by big, dispersed ownership
—the giant corporation." (It is necessary to append to Tate's
argument the further fact that by now, owing largely to corporate influence, land ownership implies the right to destroy the
land-community entirely, as in surface mining, and to impose,
as a consequence, the dangers of flooding, water pollution, and
disease upon communities downstream.)

Tate's essay was written for the anthology *Who Owns America?*, the publication of which was utterly without effect. With
other agrarian writings before and since, it took its place on
the far margin of the national dialogue, dismissed as anachro-
nistic, retrogressive, nostalgic, or reactionary in the face of the
supposedly "inevitable" dominance of corporate industrialism.
Who Owns America? was published in the Depression year of
1936. It is at least ironic that talk of "effective property" could
have been lightly dismissed at a time when many rural people
who had migrated to industrial cities were returning to their
home farms to survive. In 1936, when to the dominant minds
a thirty-acre farm in Kentucky was becoming laughable, Tate's
essay would have seemed irrelevant as a matter of course. At
that time, despite the Depression, faith in the standards and
devices of industrial progress was nearly universal and could
not be shaken.

But now, three-quarters of a century later, we are no longer
talking about theoretical alternatives to corporate rule. We are
talking with practical urgency about an obvious need. Now
the two great aims of industrialism—replacement of people by

technology and concentration of wealth into the hands of a small plutocracy—seem close to fulfillment. At the same time the *failures* of industrialism have become too great and too dangerous to deny. Corporate industrialism itself has exposed the falsehood that it ever was inevitable or that it ever has given precedence to the common good. It has failed to sustain the health and stability of human society. Among its characteristic signs are destroyed communities, neighborhoods, families, small businesses, and small farms. It has failed just as conspicuously and more dangerously to conserve the wealth and health of nature. No amount of fiddling with capitalism to regulate and humanize it, no pointless rhetoric on the virtues of capitalism or socialism, no billions or trillions spent on "defense" of the "American dream," can for long disguise this failure. The evidences of it are everywhere: eroded, wasted, or degraded soils; damaged or destroyed ecosystems; extinction of species; whole landscapes defaced, gouged, flooded, or blown up; pollution of the whole atmosphere and of the water cycle; "dead zones" in the coastal waters; thoughtless squandering of fossil fuels and fossil waters, of mineable minerals and ores; natural health and beauty replaced by a heartless and sickening ugliness. Perhaps its greatest success is an astounding increase in the destructiveness, and therefore the profitability, of war.

In 1936, moreover, only a handful of people were thinking about sustainability. Now, reasonably, many of us are thinking about it. The problem of sustainability is simple enough to state. It requires that the fertility cycle of birth, growth, maturity, death, and decay—what Albert Howard called "the Wheel of Life"—must turn continuously in place, so that the law of return is kept and nothing is wasted. For this to happen in the stewardship of humans, there must be a cultural cycle, in harmony with the fertility cycle, also continuously turning in place. The cultural cycle is an unending conversation between old people and young people, assuring the survival of local memory, which has, as long as it remains local, the greatest practical urgency and value. This is what is meant, and is all that can be meant, by "sustainability." The fertility cycle turns by the law of nature. The cultural cycle turns on affection.

That we live now in an economy that is not sustainable is not the fault only of a few mongers of power and heavy equipment.

We all are implicated. We all, in the course of our daily economic life, consent to it, whether or not we approve of it. This is because of the increasing abstraction and unconsciousness of our connection to our economic sources in the land, the land-communities, and the land-use economies. In my region and within my memory, for example, human life has become less creaturely and more engineered, less familiar and more remote from local places, resources, pleasures, and associations. Our knowledge, in short, has become increasingly statistical.

Statistical knowledge once was rare. It was a property of the minds of great rulers, conquerors, and generals, people who succeeded or failed by the manipulation of large quantities that remained, to them, unimagined because unimaginable: merely accountable quantities of land, treasure, people, soldiers, and workers. This is the sort of knowledge we now call "data" or "facts" or "information." By means of such knowledge a category assumes dominion over its parts or members. With the coming of industrialism, the great industrialists, like kings and conquerors, became exploiters of statistical knowledge. And finally virtually all of us, in order to participate and survive in their system, have had to agree to their substitution of statistical knowledge for personal knowledge. Virtually all of us now share with the most powerful industrialists their remoteness from actual experience of the actual world. Like them, we participate in an absentee economy, which makes us effectively absent even from our own dwelling places. Though most of us have little wealth and perhaps no power, we consumer-citizens are more like James B. Duke than we are like my grandfather. By economic proxies thoughtlessly given, by thoughtless consumption of goods ignorantly purchased, now we all are boomers.

The failure of imagination that divided the Duke monopoly and such farmers as my grandfather seems by now to be taken for granted. James B. Duke controlled remotely the economies of thousands of farm families. A hundred years later, "remote control" is an unquestioned fact, the realization of a technological ideal, and we have remote entertainment and remote war. Statistical knowledge is remote, and it isolates us in our remoteness. It is the stuff itself of unimagined life. We may, as we say, "know" statistical sums, but we cannot imagine them.

It is by imagination that knowledge is "carried to the heart" (to borrow again from Allen Tate). The faculties of the mind —reason, memory, feeling, intuition, imagination, and the rest —are not distinct from one another. Though some may be favored over others and some ignored, none functions alone. But the human mind, even in its wholeness, even in instances of greatest genius, is irremediably limited. Its several faculties, when we try to use them separately or specialize them, are even more limited.

The fact is that we humans are not much to be trusted with what I am calling statistical knowledge, and the larger the statistical quantities the less we are to be trusted. We don't learn much from big numbers. We don't understand them very well, and we are not much affected by them. The reality that is responsibly manageable by human intelligence is much nearer in scale to a small rural community or urban neighborhood than to the "globe."

When people succeed in profiting on a large scale, they succeed for themselves. When they fail, they fail for many others, sometimes for us all. Propriety of scale in all human undertakings is paramount, and we ignore it. We are now betting our lives on quantities that far exceed all our powers of comprehension. We believe that we have built a perhaps limitless power of comprehension into computers and other machines, but our minds remain as limited as ever. Our trust that machines can manipulate to humane effect quantities that are unintelligible and unimaginable to humans is incorrigibly strange.

As there is a limit only within which property ownership is effective, so is there a limit only within which the human mind is effective and at least possibly beneficent. We must assume that the limit would vary somewhat, though not greatly, with the abilities of persons. Beyond that limit the mind loses its wholeness, and its faculties begin to be employed separately or fragmented according to the specialties or professions for which it has been trained.

In my reading of the historian John Lukacs, I have been most instructed by his understanding that there is no knowledge but human knowledge, that we are therefore inescapably central to our own consciousness, and that this is "a statement not of arrogance but of humility. It is yet another recognition of the inevitable limitations of mankind." We are thus isolated within our uniquely human boundaries, which we certainly cannot transcend or escape by means of technological devices.

But as I understand this dilemma, we are not *completely* isolated. Though we cannot by our own powers escape our limits, we are subject to correction from, so to speak, the outside. I can hardly expect everybody to believe, as I do (with caution), that inspiration can come from the outside. But inspiration is not the only way the human enclosure can be penetrated. Nature too may break in upon us, sometimes to our delight, sometimes to our dismay.

As many hunters, farmers, ecologists, and poets have understood, Nature (and here we capitalize her name) is the impartial mother of all creatures, unpredictable, never entirely revealed, not my mother or your mother, but nonetheless our mother. If we are observant and respectful of her, she gives good instruction. As Albert Howard, Wes Jackson, and others have carefully understood, she can give us the right patterns and standards for agriculture. If we ignore or offend her, she enforces her will with punishment. She is always trying to tell us that we are not so superior or independent or alone or autonomous as we may think. She tells us in the voice of Edmund Spenser that she is of all creatures "the equall mother, / And knittest each to each, as brother unto brother." Nearly three and a half centuries later, we hear her saying about the same thing in the voice of Aldo Leopold: "In short, a land ethic changes the role of *Homo sapiens* from conqueror of the land-community to plain member and citizen of it."

We cannot know the whole truth, which belongs to God alone, but our task nevertheless is to seek to know what is true. And if we offend gravely enough against what we know to be true, as by failing badly enough to deal affectionately and responsibly with our land and our neighbors, truth will retaliate with ugliness, poverty, and disease. The crisis of this line of thought is the realization that we are at once limited and unendingly responsible for what we know and do.

The discrepancy between what modern humans presume to know and what they can imagine—given the background of pride and self-congratulation—is amusing and even funny. It becomes more serious as it raises issues of responsibility. It becomes fearfully serious when we start dealing with statistical measures of industrial destruction.

To hear of a thousand deaths in war is terrible, and we "know" that it is. But as it registers on our hearts, it is not more terrible than one death fully imagined. The economic hardship of one farm family, if they are our neighbors, affects us more painfully than pages of statistics on the decline of the farm population. I can be heartstruck by grief and a kind of compassion at the sight of one gulley (and by shame if I caused it myself), but, conservationist though I am, I am not nearly so upset by an accounting of the tons of plowland sediment borne by the Mississippi River. Wallace Stevens wrote that "Imagination applied to the whole world is vapid in comparison to imagination applied to a detail"—and that appears to have the force of truth.

It is a horrible fact that we can read in the daily paper, without interrupting our breakfast, numerical reckonings of death and destruction that ought to break our hearts or scare us out of our wits. This brings us to an entirely practical question: Can we—and, if we can, *how* can we—make actual in our minds the sometimes urgent things we say we know? This obviously cannot be accomplished by a technological breakthrough, nor can it be accomplished by a big thought. Perhaps it cannot be accomplished at all. Yet another not very stretchable human limit is in our ability to tolerate or adapt to change. Change of course is a constant of earthly life. You can't step twice into exactly the same river, nor can you live two successive moments in exactly the same place. And always in human history there have been costly or catastrophic sudden changes. But with relentless fanfare, at the cost of almost indescribable ecological and social disorder, and to the almost incalculable enrichment and empowerment of corporations, industrialists have substituted what they fairly accurately call "revolution" for the slower, kinder processes of adaptation or evolution. We have had in only about two centuries a steady and ever-quickening sequence of industrial revolutions in manufacturing, transportation, war, agriculture, education, entertainment, homemaking and family life, health care, and so-called communications.

Probably everything that can be said in favor of all this has been said, and it is true that these revolutions have brought some increase of convenience and comfort and some easing of pain. It is also true that the industrialization of everything has incurred liabilities and is running deficits that have not been adequately accounted. All of these changes have depended upon industrial technologies, processes, and products, which have depended upon the fossil fuels, the production and consumption of which have been, and are still, unimaginably damaging to land, water, air, plants, animals, and humans. And the cycle of obsolescence and innovation, goaded by crazes of fashion, has given the corporate economy a controlling share of everybody's income.

The cost of this has been paid also in a social condition that apologists call "mobility," implying it has been always "upward" to a "higher standard of living," but which in fact has been an ever-worsening unsettlement of our people, and the extinction or near-extinction of traditional and necessary communal structures.

For this also there is no technological or large-scale solution. Perhaps, as they believe, the most conscientiously up-to-date people can easily do without local workshops and stores, local journalism, a local newspaper, a local post office, all of which supposedly have been replaced by technologies. But what technology can replace personal privacy or the coherence of a family or a community? What technology can undo the collateral damages of an inhuman rate of technological change?

The losses and damages characteristic of our present economy certainly cannot be stopped, let alone restored, by "liberal" or "conservative" tweakings of corporate industrialism, against which the ancient imperatives of good care, homemaking, and frugality can have no standing. The possibility of authentic correction comes, I think, from two already-evident causes. The first is scarcity and other serious problems arising from industrial abuses of the land-community. The goods of nature so far have been taken for granted and, especially in America, assumed to be limitless, but their diminishment, sooner or later unignorable, will enforce change.

A positive cause, still little noticed by high officials and the media, is the by now well-established effort to build or re-build local economies, starting with economies of food. This effort to connect cities with their surrounding rural landscapes rests exactly upon the recognition of human limits and the necessity of human scale. Its purpose, to the extent possible, is to bring producers and consumers, causes and effects, back within the bounds of neighborhood, which is to say the effective reach of imagination, sympathy, affection, and all else—including enough food—that neighborhood implies. An economy genuinely local and neighborly offers to localities a measure of security they cannot derive from a national or a global economy controlled by people who, by principle, have no local commitment.

In this age so bewildered by technological magnifications of power, people who stray beyond the limits of their mental competence typically find no guide except for the supposed authority of market price. "The market" thus assumes the standing of ultimate reality. But market value is an illusion, as is proven by its frequent changes; it is determined solely by the buyer's ability and willingness to pay.

By now our immense destructiveness has made clear that the actual value of some things exceeds human ability to calculate, and therefore must be considered absolute. For the destruction of these things there is never, under any circumstances, any justification. Their absolute value is recognized by the mortal need of those who do not have them, and by affection. Land, to people who do not have it and who are thus without the means of life, is absolutely valuable. Ecological health, in a land dying of abuse, is not worth "something"; it is worth everything. And abused land relentlessly declines in value to its present and succeeding owners, whatever its market price.

But we need not wait, as we are doing, to be taught the absolute value of land and of land health by hunger and disease. Affection can teach us, and soon enough, if we grant appropriate standing to affection. For this we must look to the stickers, who "love the life they have made and the place they have made it in."

All thoughtful people have begun to feel our eligibility to be instructed by ecological disaster and mortal need. But we endangered ourselves first of all by dismissing affection as an honorable and necessary motive. Our decision in the middle of the last century to reduce the farm population, eliminating the allegedly "inefficient" small farmers, was enabled by the discounting of affection. As a result, we now have barely enough farmers to keep the land in production, with the help of increasingly expensive industrial technology and at an increasing ecological and social cost. Far from the plain citizens and members of the land-community, as Aldo Leopold wished them to be, farmers are now too likely to be merely the land's exploiters.

I don't hesitate to say that damage or destruction of the land-community is morally wrong, just as Leopold did not hesitate to say so when he was composing his essay "The Land Ethic" in 1947. But I do not believe, as I think Leopold did not, that morality, even religious morality, is an adequate motive for good care of the land-community. The *primary* motive for good care and good use is always going to be affection, because affection involves us entirely. And here Leopold himself set the example. In 1935 he bought an exhausted Wisconsin farm and, with his family, began its restoration. To do this was morally right, of course, but the motive was affection. Leopold was an ecologist. He felt, we may be sure, an informed sorrow for the place in its ruin. He imagined it as it had been, as it was, and as it might be. And a profound, delighted affection radiates from every sentence he wrote about it.

Without this informed, practical, and *practiced* affection, the nation and its economy will conquer and destroy the country.

In thinking about the importance of affection, and of its increasing importance, I have been guided most directly by E. M. Forster's novel *Howards End*, published in 1910. By then, Forster was aware of the implications of "rural decay," and in this novel he spoke, with some reason, of his fear that "the literature of the near future will probably ignore the country and seek inspiration from the town. . . . and those who care for the earth with sincerity may wait long ere the pendulum swings back to her again." Henry Wilcox, the novel's "plain man of business," speaks the customary rationalization, which has echoed through American bureaus and colleges of agriculture, almost without objection, for at least sixty years: "the days for small farms are over."

In *Howards End*, Forster saw the coming predominance of the machine and of mechanical thought, the consequent deracination and restlessness of populations, and the consequent ugliness. He saw an industrial ugliness, "a red rust," already creeping out from the cities into the countryside. He seems to have understood by then also that this ugliness was the result of the withdrawal of affection from places. To have beautiful buildings, for example, people obviously must want them to be beautiful and know how to make them beautiful, but evidently they also must love the places where the buildings are to be built. For a long time, in city and countryside, architecture has disregarded the nature and influence of places. Buildings have become as interchangeable from one place to another as automobiles. The outskirts of cities are virtually identical and as depressingly ugly as the corn-and-bean deserts of industrial agriculture.

What Forster could not have foreseen in 1910 was the *extent* of the ugliness to come. We still have not understood how far at fault has been the prevalent assumption that cities could be improved by pillage of the countryside. But urban life and rural life have now proved to be interdependent. As the countryside has become more toxic, more eroded, more ecologically degraded and more deserted, the cities have grown uglier, less sustainable, and less livable.

The argument of *Howards End* has its beginning in a manifesto against materialism:

> It is the vice of a vulgar mind to be thrilled by bigness, to think that a thousand square miles are a thousand times more wonderful than one square mile . . . That is not imagination. No, it kills it. . . . Your universities? Oh, yes, you have learned men who collect . . . facts, and facts, and empires of facts. But which of them will rekindle the light within?

"The light within," I think, means affection, affection as motive and guide. Knowledge without affection leads us astray. Affection leads, by way of good work, to authentic hope.

The climactic scene of Forster's novel is the confrontation between its heroine, Margaret Schlegel, and her husband, the self-described "plain man of business," Henry Wilcox. The issue is Henry's determination to deal, as he thinks, "realistically" with a situation that calls for imagination, for affection, and then forgiveness. Margaret feels at the start that she is "fighting for women against men." But she is not a feminist in the popular or political sense. What she opposes with all her might is Henry's hardness of mind and heart that is "realistic" only because it is expedient and because it subtracts from reality the life of imagination and affection, of living souls. She opposes his refusal to see the practicality of the life of the soul.

Margaret's premise, as she puts it to Henry, is the balance point of the book: "It all turns on affection now . . . Affection. Don't you see?"

In a speech delivered in 2006, "Revitalizing Rural Communities," Frederick Kirschenmann quoted his friend Constance Falk, an economist: "There is a new vision emerging demonstrating how we can solve problems and at the same time create a better world, and it all depends on collaboration, love, respect, beauty, and fairness."

Those two women, almost a century apart, speak for human wholeness against fragmentation, disorder, and heartbreak. The English philosopher and geometer Keith Critchlow brings his own light to the same point: "The human mind takes apart with its analytic habits of reasoning but the human heart puts things together because it loves them . . ."

The great reassurance of Forster's novel is the wholeheartedness of his language. It is to begin with a language not disturbed by mystery, by things unseen. But Forster's interest throughout is in soul-sustaining habitations: houses, households, earthly places where lives can be made and loved. In defense of such dwellings he uses, without irony or apology, the vocabulary that I have depended on in this talk: truth, nature, imagination, affection, love, hope, beauty, joy. Those words are hard to keep still within definitions; they make the dictionary hum like a beehive. But in such words, in their resonance within their histories and in their associations with one another, we find our indispensable humanity, without which we are lost and in danger.

No doubt there always will be some people willing to do anything at all that is financially or technologically possible, who look upon the world and its creatures without affection and therefore as exploitable without limit. Against that limit-lessness, in which we foresee assuredly our ruin, we have only our ancient effort to define ourselves as human and humane. But this ages-long, imperfect, unendable attempt, with its magnificent record, we have virtually disowned by assigning it to the ever more subordinate set of school subjects we call "arts and humanities" or, for short, "culture." Culture, so iso-lated, is seen either as a dead-end academic profession or as a mainly useless acquisition to be displayed and appreciated "for its own sake." This definition of culture as "high culture" ac-tually debases it, as it debases also the presumably low culture that is excluded: the arts, for example, of land use, life support, healing, housekeeping, homemaking.

I don't like to deal in categorical approvals, and certainly not of the arts. Even so, I do not concede that the "fine arts," in general, are useless or unnecessary or even impractical. I can testify that some works of art, by the usual classification "fine," have instructed, sustained, and comforted me for many years in my opposition to industrial pillage.

But I would insist that the economic arts are just as honor-ably and authentically refinable as the fine arts. And so I am nominating economy for an equal standing among the arts and humanities. I mean, not economics, but economy, the making of the human household upon the earth: the *arts* of adapting kindly the many human households to the earth's many eco-systems and human neighborhoods. This is the economy that the most public and influential economists never talk about, the economy that is the primary vocation and responsibility of every one of us.

My grandparents were fortunate. They survived their debts and kept their farm. But in the century and more since that hard year of 1907, millions of others have not been so fortu-nate. Owing largely to economic constraints, they have lost their hold on the land, and the land has lost its hold on them.

They have entered into the trial of displacement and scattering that we try to dignify as "mobility."

The land and the people nevertheless have suffered together, as invariably they must. Under the rule of industrial economics, the land, our country, has been pillaged for the enrichment, supposedly, of those humans who have claimed the right to own or exploit it without limit. Of the land-community much has been consumed, much has been wasted, almost nothing has flourished.

But this has not been inevitable. We do not have to live as if we are alone.

About Civil Disobedience

MATT ROTHSCHILD has asked me to write a short piece about civil disobedience. This is a request I would ignore, if I had my druthers. But Matt was representing Duty, perhaps, and several hours of hard thought have not afforded me an adequate excuse.

The reason for civil disobedience, as I understand it, is simple necessity. If you belong to a political side with a significant grievance, and if your side is not represented—is in effect not heard—in the halls of your nominally representative government, then you have two choices: You can passively submit, accepting your grievance as irremediable, or you can, perhaps uselessly, insist by some act of civil disobedience upon your right to have your grievance redressed by your government. I am speaking here of *nonviolent* civil disobedience, of course. I subscribe to the commandment to love our neighbors, even if they are our enemies. But I am also in favor of making sense. Answering violence with violence is understandable, but it is also nonsense.

Matt asked me to write this piece because he knows that last February, in protest against coal mining by "mountaintop removal," I committed myself to an act of civil disobedience in the office of Kentucky's governor. In fact, I have made that commitment three times. The first was on June 3, 1979, in opposition to a nuclear power plant then being built at Marble Hill on the Ohio River near Madison, Indiana. The second was in Washington, D.C., on March 2, 2009, in protest, with a host of others, generally against mountaintop removal and air pollution by the burning of fossil fuels, and immediately against the burning of coal by a power plant within a few blocks of the national capitol. The third was on the eleventh of last February: the aforementioned attempt to discover conscience in official Frankfort.

Only one of these adventures resulted in actual civil disobedience and arrest. After we crossed the fence at Marble Hill, we were arrested and booked—and turned loose. In Washington, the number of us offering to get arrested—two or three

566

thousand, maybe—overwhelmed the police, who, thinking perhaps of the hours it would take to write down our names and addresses, declined the opportunity to know us better. Or so we thought. We then had to choose between climbing the fence, potentially a felony, or, after far too many speeches, dispersing. We dispersed. In Frankfort, the governor, somewhat delightfully, outsmarted us. Instead of calling the police, he invited us to camp in his waiting room, which we did, from Friday until Monday morning.

And so my career in civil disobedience, so far, has been an exercise in anticlimax. Also it has been, by any practical reckoning, pretty useless. Owing probably not much, if anything, to our civil disobedience, the power plant at Marble Hill finally was stopped. But nothing that my side has done has come anywhere near to stopping mountaintop removal.

At a time when virtuous behavior tends to be measured in degrees of misery, I had better confess that all three of these episodes were mostly pleasant. The police and other officials in Indiana were nice to us, and we were nice to them and to one another. The march in Washington, in spite of cold weather, was a social success, better by far than any cocktail party I ever attended. And our weekend in the Governor's office was, I think, for all of us, an extraordinarily happy time, even a joyful time. We were warned only that if we left the building we could not return; we were to that extent confined. We stayed put, we worked hard at getting our message out to the media, we told stories, we laughed a lot, we ate the good food sent in by allies, and slept well on bedding likewise sent in. The security people, the office people, and the police were kind to us, and we reciprocated. I am proud to say that we were model guests. We damaged nothing, and we cleaned up after ourselves.

It may seem odd to speak of pleasure as a result of trouble, but there is nothing wrong with decent pleasure, however it comes. It is a gift, and we should be grateful. The pleasures I have mentioned certainly do not reduce the seriousness of civil disobedience. I am sure that all who have undertaken it have felt intensely and complexly the seriousness of it. There are a number of considerations that come in a hurry and are inescapable. I will list them, not in the order of their importance, but as I have thought of them in my own efforts to decide.

Civil disobedience will likely be considered, first of all, as an inconvenience. It will, and not for a predictable length of time, interrupt one's life and one's work. I have always been suspicious of people who seem to devote their entire lives to forms of protest. We all ought to have better things to do. Ken Kesey once said that the reason not to resist evil is that such resistance is dependent on evil; it makes *you* dependent on evil. He was right. And Edward Abbey said that saving the world is a good hobby—though he worked hard to save at least parts of it. As for me, the older I get, the less happy I am to leave home. All the places I go seem to be getting farther away. Frankfort, Kentucky, now appears as far off as the planet Saturn, and I wish it more remote. Reluctance, then, may be a dependable enforcer of thoughtfulness. Protest becomes properly a part of a citizen's life and work after political and legal processes have failed, and other recourse is exhausted. Civil disobedience is properly the last resort.

It is also an unhappiness of citizenship. By it, you make yourself, publicly, an exception. It involves a kind of loneliness. I, at least, have felt no pleasure in opposing constitutional authority, however corrupt and irresponsible I have found it to be.

Civil disobedience is also plenty scary. At least to me it is. I have never felt one bit brave even in thinking about it. It involves a strange and risky paradox: You and your friends will be exploiting your obvious powerlessness to recover to your cause, and to your own citizenship, a just measure of power. But your acknowledged condition is powerlessness. Your commitment to nonviolence makes you vulnerable to violence. You can get hurt, or worse. It is fearful also to make yourself available to be treated with contempt. And you are, in effect, volunteering to go to jail.

During the Washington protest, some genius with a microphone asked me, "Do you want to go to jail?"

I said, "*Hell no!*"

There is a world of difference between wanting to go to jail and being willing to go. I thank God for every minute I am not in jail.

A final consideration, and the one you may ponder longest, is the possibility that your act of civil disobedience may be useless. In Kentucky, where the state government is owned outright by the coal industry and the "flagship university" is the coal industry's trophy wife, opposition to mountaintop removal faces the unhappiest of odds.

This means, I believe, that it is a mistake to make your opposition conditional upon winning. If you do that, you won't last. I am in this struggle with the firm intention of winning, but I don't forget that I first wrote against strip mining in 1965. If I had required even a reasonable expectation of victory, I would have given up long ago. It is unlikely that anybody now opposing mountaintop removal thinks that victory is in sight. But the opposition is now astonishingly well-populated in comparison to what it was forty years ago. Why do so many people go to so much trouble, learning the hard things they need to know, organizing, arguing with politicians, making speeches, marching, risking arrest, getting arrested, keeping on, making the same try again and again?

I think they do it, above all else, to keep alive the possibility of decency, and to refuse to accept as normal the indecency of public officials.

Forty-some years ago, maybe in 1966, I attended in Frankfort a hearing on coal company abuses. In those days some of the companies were getting the coal out by opening contour gashes along the sides of the mountains, pushing the "overburden" regardlessly downslope. They were doing this, at that time, under the infamous "broad form deeds," which, as construed by the courts, gave all rights to owners of the coal and none to the surface owners whose properties were being damaged or destroyed.

Among the audience at this hearing was a group of maybe fifteen people whose homes and lands were damaged or threatened by "contour stripping." There was, on that occasion, no "demonstrating," but those people were there nonetheless in protest. Sitting with them was a man in appearance somewhat different from them. He was wearing a well-fitting summer suit and a tie. He held a rather elegant straw hat on his lap.

I became curious about him, and at one of the recesses of the hearing, I approached him. We introduced ourselves. He was Courtney Wells, he said, of the town of Hazard.

I asked if he owned land that was affected by strip mining.

He said no, he had no land at risk. He was a lawyer.

"Oh," I said. "You're here to represent these people."

Again he said no. He represented only himself.

I said, "Well, then, why are you here?"

And he replied: "I want to be on the side of the right."

I have never forgotten him, for he gave me the one reason that will always be enough.

OUR ONLY WORLD
(2015)

Paragraphs from a Notebook

WE NEED TO acknowledge the formlessness inherent in the analytic science that divides creatures into organs, cells, and ever smaller parts or particles according to its technological capacities.

I recognize the possibility and existence of this knowledge, even its usefulness, but I also recognize the narrowness of its usefulness and the damage it does. I can see that in a sense it is true, but also that its truth is small and far from complete.

In and by all my thoughts and acts, I am opposed to any claim that such knowledge is adequate to the sustenance of human life or the health of the ecosphere.

Do even the professionals and experts believe in it, in the sense of acting on it in their daily lives? I doubt that they do.

To this science, the body is an assembly of parts provisionally joined, a "basket case" sure enough. A mountain is a heap of "resources" unfortunately mixed with substances that are not marketable.

There is an always significant difference between knowing and believing. We may know that the earth turns, but we believe, as we say, that the sun rises. We know by evidence, or by trust in people who have examined the evidence in a way that we trust is trustworthy. We may sometimes be persuaded to believe by reason, but within the welter of our experience reason is limited and weak. We believe always by coming, in some sense, to see. We believe in what is apparent, in what we can imagine or "picture" in our minds, in what we feel to be true, in what our hearts tell us, in experience, in stories—above all, perhaps, in stories.

We can, to be sure, see parts and so believe in them. But there has always been a higher seeing that informs us that parts, in themselves, are of no worth. Genesis is right: "It is not good that the man should be alone." The phrase "be alone" is a contradiction in terms. A brain alone is a dead brain. A man alone is a dead man.

We are thus as likely to be wrong in what we know as in what we believe.

We may know, or think we know, and often say, that humans are "only" animals, but we teach our children specifically human virtues—evidently because we believe that they are not "only" animals.

Another question of knowledge and belief that keeps returning to my mind is this: Are there not some things that cannot be known apart from belief? This question refers not just to matters of religion—as in Job 19:25: "I know that my redeemer liveth, and that he shall stand at the latter day upon the earth . . ."—but also to ordinary motives of family and community life such as love, compassion, and forgiveness. Do people who believe that such motives are genetically determined have the same knowledge as people who believe that they are the results of choice, culture, cultivation, and discipline? Or: Do people who believe in the sanctity or intrinsic worth of the world and its creatures have the same knowledge as people who recognize only market value? If there is no way to measure or prove such differences of knowledge, that does at least prove one of my points: There is more to us than some of us suppose.

We may know the anatomy of the body to the extent of the anatomy of atoms, and yet we love and instruct our children as whole persons. And we accept an obligation to help them to preserve their wholeness, which is to say their health. This is not an obligation that we can safely transfer to the subdivided and anatomizing medical industry, not even for the sake of cures. Cures, to industrial medicine, are marketable products extractable from bodies. To cure in this sense is not to heal. To heal is to make whole, not so ideologically definable or so technologically possible or so handily billable.

*

This applies as well to the industries of landscapes: agriculture, forestry, and mining. Once they have been industrialized, these enterprises no longer recognize landscapes as *wholes*, let alone as the homes of people and other creatures. They regard landscapes as sources of extractable products. They have "efficiently" shed any other interest or concern.

We have come to this by way of the disembodiment of thought —a mentalization, almost a puritanization, of thought— depriving us of the physical basis of a sympathy that might join us kindly to landscapes and their creatures, including their human creatures. This purity or sublimity of thought is hard to account for, for it has come about under the sponsorship of materialism. Perhaps it happened because materialists, instead of assigning ultimate value to materiality as would have been reasonable, have abstracted "material" to "mechanical," and thus have removed from it all bodily or creaturely attributes. Or perhaps the abstracting impulse branched in either of two directions: one toward the mechanical, the other toward the financial, which is to say toward the so-called economy of money as opposed to the actual economy (oikonomia or "house-keeping") of goods. Either way the result is the same: the scientific-industrial culture, founded nominally upon materialism, arrives at a sort of fundamentalist disdain for material reality. The living world is then treated as dead matter, the worth of which is determined exclusively by the market.

This highly credentialed, highly politicized disdain, now allied with the similar disdain of highly spiritualized religions, is limitlessly destructive. We cannot say that its destructiveness has been unnoticed as it has been happening, or that the dissolutions, and the dissoluteness, of mechanical thought have not been, by some, well understood. The poet William Butler Yeats prayed: "God guard me from the thoughts men think / In the mind alone . . ." ("A Prayer for Old Age"). He wrote in 1916: "We only believe in those thoughts which have been conceived not in the brain but in the whole body" (Introduction to *Certain Noble Plays of Japan*, by Ezra Pound and Ernest Fenollosa). In the same essay he spoke with foreboding of "a mechanical sequence of ideas."

*

As another example, more explicit, here is the critic and translator Philip Sherrard on the Greek poet Anghelos Sikelianos: "He saw [the western world of his time] as increasingly alienated from those principles which give life significance and beauty and as approaching the condition of a machine out of control and hastening towards destruction. . . . The organic sense of life was being shattered into countless unconnected fragments. . . . A system of learning which made extreme demands on the purely mechanical and sterile processes of memory had the effect of absorbing all the spontaneous movements of body and soul of the younger generations."

Or here is a passage, by the poet and critic John Crowe Ransom, pointed more directly at our specialist system, which he identified as a phase of the Puritanism that began in religion: "You may dissociate the elements of experience and exploit them separately. But then at the best you go on a schedule of small experiences, taking them in turn, and trusting that when the rotation is complete you will have missed nothing. And at the worst you will become so absorbed in some one small experience that you will forget to go on and complete the schedule; in that case you will have missed something. The theory that excellence lies in the perfection of the single functions, and that society should demand that its members be hard specialists, assumes that there is no particular harm in missing something."

A proper attention to our language, moreover, informs us that the Greek root of "anatomy" means "dissection," and that of "analysis" means "to undo." The two words have essentially the same meaning. Neither suggests a respect for formal integrity. I suppose that the nearest antonym to both is a word we borrow directly from Greek: *poiesis*, making or creation, which suggests that the work of the poet, the composer or maker, is the necessary opposite to that of the analyst and the anatomist. Some scientists, I think, are in this sense poets.

But we appear to be deficient in learning or teaching a competent concern for the way that parts are joined. We certainly are not learning or teaching adequately the arts of forming parts into wholes, or the arts of preserving the formal integrity of the things we receive as wholes already formed.

Without this concern and these arts, our efforts of conservation are probably futile. Without some sense of necessary connections and a competent awareness of human and natural limits, the issues of scale and form are not only pointless, but cannot even enter our consciousness.

My premise is that there is a scale of work at which our minds are as effective and even as harmless as they ought to be, at which we can be fully responsible for consequences and there are no catastrophic surprises. But such a possibility does not excite us.

What excites us is some sort of technological revolution: the fossil fuel revolution, the automotive revolution, the assembly line revolution, the antibiotic revolution, the sexual revolution, the computer revolution, the "green revolution," the genomic revolution, and so on. But these revolutions—all with something to sell that people or their government "must" buy— are mere episodes of the one truly revolutionary revolution perhaps in the history of the human race: the Industrial Revolution, which has proceeded from the beginning with only two purposes: to replace human workers with machines, and to market its products, regardless of their usefulness or their effects, at the highest possible profit—and so to concentrate wealth into ever fewer hands.

This revolution has, so far, fulfilled its purposes, with remarkably few checks or thwarts. I say "so far" because its great weakness obviously is its dependence on what it calls "natural resources," which it has used ignorantly and foolishly, and which it has progressively destroyed. Its weakness, in short, is that its days are numbered. Having squandered nature's "resources," it will finally yield to nature's correction, which in prospect grows ever harsher.

We have formed our present life, including our economic and intellectual life, upon specialization, professionalism, and competition. Certified smart people expect to solve all problems

by analysis, dividing wholes into ever smaller parts. Science
and industry do give room to synthesis, but by that they do
not mean putting back together the things once together that
they have taken apart; they mean making something "syn-
thetic." They mean engineering the disassembled parts, by
some manner of violence, into profitable new commodities. In
such a state of things we don't see or, apparently, suspect the
complexity of connections among ecology, agriculture, food,
health, and medicine (if by "medicine" we mean healing). Nor
can we see how this complexity is necessarily contained within,
and at the mercy of, human culture, which in turn is necessarily contained within the not very expandable limits of human
knowledge and human intelligence.

We have accumulated a massive collection of "information" to
which we may have "access." But this information, by being accessible, does not become knowledge. We might find, if such a
computation were possible, that the amount of human knowledge over many millennia has remained more or less constant
—that is it has always filled the available mental capacity—and
therefore that learning invariably involves forgetting. To have
the Renaissance, we had to forget the Middle Ages. To the extent that we have learned about machines, we have forgotten
about plants and animals. Every nail we drive in, as I believe
C. S. Lewis said, drives another out.

The thing most overlooked by scientists, and by the enviers
and emulators of science in the humanities, is the complicity of
science in the Industrial Revolution, which science has served
not by supplying the "scientific" checks of skepticism, doubt,
criticism, and correction, but by developing of marketable
products, from refined fuels to nuclear bombs to computers
to poisons to pills.

It has been remarkable how often science has hired out to the
ready-made markets of depravity, as when it has served the
military-industrial complex, which is solidly founded upon
the hopeless logic of revenge, or the medical and pharmaceuti-
cal industries, which are based somewhat on the relief of suffering but also on greed, on the vicious circles of hypochondria,

and on the inducible fear of suffering yet to come. The commodification of genome-reading rides upon the same fears of the future that palmistry and phrenology rode upon.

We may say with some confidence that the most apparently beneficent products of science and industry should be held in suspicion if they are costly to consumers or bring power to governments or profits to corporations.

There are, we know, scientists who are properly scrupulous, responsible, and critical, who call attention to the dangers of oversold and under-tested products, and who are almost customarily ignored. They are often called "independent scientists," and the adjective is significant, for it implies not only certain moral virtues but also political weakness. The combination of expertise, prestige, wealth, and power, incapable of self-doubt or self-criticism, is hardly to be deterred by a few "independent scientists."

Scientists in general, like humanists and artists in general, have accepted the industrialists' habit, or principle, of ignoring the contexts of life, of place, of community, and even of economy.

The capitalization of fear, weakness, ignorance, bloodthirst, and disease is certainly financial, but it is not, properly speaking, economic.

Criticism of scientific-industrial "progress" need not be balked by the question of how we would like to do without anesthetics or immunizations or antibiotics. Of course there have been benefits. Of course there have been advantages—at least to the advantaged. But valid criticism does not deal in categorical approvals and condemnations. Valid criticism attempts a just description of our condition. It weighs advantages against disadvantages, gains against losses, using standards more general and reliable than corporate profit or economic growth. If criticism involves computation, then it aims at a full accounting and an honest net result, whether a net gain or a net loss. If we are to hope to live sensibly, correcting mistakes that need correcting, we need a valid general criticism.

*

Scared for health, afraid of death, bored, dissatisfied, vengeful, greedy, ignorant, and gullible—these are the qualities of the ideal consumer. Can we imagine a way of education that would turn passive consumers into active and informed critics, capable of using their own minds in their own defense? It will not be the purely technical education-for-employment now advocated by the most influential "educators" and "leaders."

We have good technical or specialized criticism: A given thing is either a good specimen of its kind or it is not. A valid *general* criticism would measure work against its context. The health of the context—the body, the community, the ecosystem—would reveal the health of the work.

The Commerce of Violence

O N THE DAY of the bombing in Boston, *The New York Times* printed an op-ed piece by a human being who had been imprisoned at Guantanamo for more than eleven years, uncharged and of course untried. The occurrence of these two events on the same day was a coincidence, but that does not mean that they are unrelated.

What connects them is our devaluation, and when convenient our disvaluation, of human life as well as the earthly life of which human life is a dependent part. This cheapening of life, and the violence that inevitably accompanies it, is surely the dominant theme of our time. The ease and quickness with which we resort to violence would be astounding if it were not conventional.

In the Appalachian coal fields we mine coal by destroying a mountain, its forest, its waterways, and its human community without counting the destruction as a cost. Our military technicians, our representatives, sit in armchairs and kill our enemies, and our enemies' children, by remote control. In the Guantanamo prison, guards force their fasting prisoners to live, and they do so as routinely as in other circumstances they would kill them.

And the Boston bombing? Like most people, I was not there and I don't know anybody who was, but I was grieved and frightened by the news. This fearful grief has grown familiar to me since I first felt it at the start of World War II, but at each of its returns it is worse. Each new resort to violence enlarges the argument against our species, and the task of hope becomes harder.

I am absolutely in sympathy with those who suffered the bombing in Boston and with their loved ones. They have been singled out by a violence that was general in its intent, not aimed particularly at anybody. The oddity, the mystery, of a particular hurt from a general violence—the necessity to ask, "Why me? Why my loved ones?"—must compound the suffering. What I am less and less in sympathy with is the rhetoric and the tone of official indignation. Public officials cry out for

581

justice against the perpetrators. I too wish them caught and
punished. But I am unwilling to have my wish spoken for me
in a tone of surprise and outraged innocence. The event in
Boston is not unique or rare or surprising or in any way new.
It is only another transaction in the commerce of violence: the
unending, the not foreseeably endable, exchange of an eye for
an eye, with customary justifications on every side, in which
we fully participate; and beyond that, it is our willingness to
destroy anything, any place, or anybody standing between us
and whatever we are "manifestly destined" to have.

We congratulate ourselves perpetually upon our Civil War
by which the slaves were, in a manner of speaking, "freed."
We forget, if we have ever learned, that the same army that
"freed the slaves" established for us the "right" of military violence against a civilian population, and then acted upon that
"right" by a war of extermination against the native people
of the West. Nobody who knows our history, from the "Indian wars" to our contemporary foreign wars of "homeland
defense," should find anything unusual in the massacre of civil-
ians and their children.

It is not possible for us to reduce the value of life, including
human life, to nothing *only* to suit our own convenience or our
own perceived need. By making this reduction for ourselves,
we make it for everybody and anybody, even for our enemies,
even for the maniacs whose enemies are schoolchildren or
spectators at a marathon.

We forget also that violence is so securely founded among
us—in war, in forms of land use, in various methods of economic "growth" and "development"—because it is immensely
profitable. People do not become wealthy by treating one another or the world kindly and with respect. Do we not need to
remember this? Do we have a single eminent leader who would
dare to remind us?

On the second day after the catastrophe in Boston, Thomas
L. Friedman announced in the *Times* that "the right reaction
is: Wash the sidewalk, wipe away the blood, and let whoever
did it know that . . . they have left no trace on our society or
way of life." We should, said Mr. Friedman, "let there be no
reminder whatsoever." And he asserted, with a shocking indifference to evidence and his own language, that "the benefits

—living in an open society—always outweigh the costs." He is speaking to (among others) people whose loved ones have been killed and people who will never again stand on their own legs. How can he think that all the traces of any violence can be easily wiped away? How would he wipe away the traces of a bombed village or a strip mine or a gullied field or a wrecked forest? The dead in Boston no longer live in an open society. How have *they* benefitted?

Mr. Friedman, like other journalists, asks us, as he wrote, to "notice how many people came *running toward the blast* within seconds to help." And that is very well. To know that people would run to help, perhaps at the risk of their lives, is consoling and reassuring. But we have got to acknowledge that the help that comes after the violence has been done, though it undeniably helps, is not a solution to violence.

The solution, many times more complex and difficult, would be to go beyond our ideas, obviously insane, of war as the way to peace and of permanent damage to the ecosphere as the way to wealth. Actually to help our suffering of one man-made horror after another, we would have to revise radically our understanding of economic life, of community life, of work, and of pleasure. We employ thousands of scientists and spend billions of dollars to reduce matter to its smallest particles and to search for farther stars. How many scientists and how many dollars are devoted to harmony between economy and ecology, or to amity and lenity in the face of hatred and killing?

To learn to meet our needs without continuous violence against one another and our only world would require an immense intellectual and practical effort, requiring the help of every human being perhaps to the end of human time.

This would be work worthy of the name "human." It would be fascinating and lovely.

A Forest Conversation

AN ECOSYSTEM, the web of relationships by which a place and its creatures sustain a mutual life, ultimately is mysterious, like life itself. We can know enough, and probably only enough, to tell us how little we know and to make us careful. At present, too ignorant to know how ignorant we are, we believe that we are free to impose our will upon the land with the utmost power and speed to gain the largest profit in the shortest time, and we believe that there are no penalties for this.

Industrial farming and forestry use extreme methodologies that barter the long-term health and fertility, which is to say the long-term productivity, of local ecosystems for a short-term monetary gain. Like mining, they tend, though not so suddenly, toward the same totality of ruin. The issue of land *use* is not on the agenda of most conservation organizations, which have been primarily concerned throughout their history with the preservation of wilderness and wildlife habitat, even though most land is being used, and used badly, and though no wilderness or wildlife can survive the prolonged abuse of the economic landscapes. Governments typically are preoccupied with the politics and commerce related to land use rather than the ever more pressing issues of land use itself: soil erosion, toxicity, ecological degradation, the destruction of rural communities, and the substitution of the global consumer culture for local cultures of husbandry. The considerable force of the colleges of agriculture has been applied mostly to promotion of industrial technologies and the economics of agribusiness, ignoring the great teachers and advocates of good land use: Liberty Hyde Bailey, J. Russell Smith, Hugh Hammond Bennett, Albert Howard, Aldo Leopold, and others. So utterly dominant has industrial agriculture become that the officially recommended soil conservation technology, "no-till agriculture," undertakes to "save" the soil by poisoning it. By this method, the noncommercial plant cover is killed with an herbicide, and then the "undisturbed" soil is planted in corn or soybeans genetically engineered for "herbicide resistance." Among the results of continuous no-till cropping on sloping land is severe soil erosion. Everything human and natural is sacrificed for the sake of annual production.

On the mostly wooded valley sides where I live, near the lower end of the Kentucky River, the same bad trade is under way. The woodlands here, which for the last three quarters of a century or so have enjoyed the flimsy beneficence of neglect, are now often as heartlessly cropped as the fields.

The story of a bad job of logging is easy to imagine. "We need money," a forest-owning family decides. "We'll sell our trees." "The environment" being in fashion, the family takes comfort in the thought that after the marketable trees are gone, many smaller trees will remain as the "next crop." And so they sell their "standing timber" to a logging company, whose representative comes in and marks every tree that can be sold.

And then, all too often, a sawyer and a driver of a mechanical skidder, employees of the logging company, arrive with, inevitably, the single purpose of cutting and removing the marked trees as quickly, which is to say as cheaply, as possible. The logging company is in every sense an absentee, and of all the short-term economies in forestry, the absentee logging company's is the shortest. Any concern for the "next crop" cherished by the owners, who also have in effect made themselves absentees, has become forceless. Standing trees, if they are unmarked, are now regarded merely as obstacles. There is little concern for directional felling. Many of the smaller trees are broken off or permanently bent or in other ways damaged by the marked trees as they fall. As the tree-length logs are dragged out of the woods, sometimes straight up or down a steep slope, the trees of the "next crop" are damaged by the skidder or by the dragged logs.

The woods is left a shambles, for nobody thought of the forest rather than the trees. All along the way, the economic interest was shifted from the forest to the marketable "standing timber," though the *source* of the timber is not the trees but the forest. Finally there is no significant difference between forest ecology and a long-term forest economy. To forget this profound kinship is to abandon the forest to bad work.

The example I have given is a worst case scenario, but my point is that when land abuse is normal and not a public issue, and when local land economies have disappeared into a global financial system, the worst case can happen easily. I asked William Martin, a forest ecologist living in Lexington, "What percentage of Kentucky's forest is sustainably managed?" He replied, "I would say less than ten." Jim Finley, a forester at Penn State, would apply the same fraction to the forest of Pennsylvania, but he suggests that "healthy" is a better term than "sustainably managed."

The damage extends to the human community when, as often happens, the cut timber is transported out of the neighborhood and even out of the state without so much as a stop at a sawmill, thus yielding the least possible local benefit from a local resource.

The U.S. Department of Agriculture and the land grant colleges of agriculture seem to regard forestry as a kind, or division, of farming. This risks confusion and requires thought. For example, the use of "harvesting" as a synonym for "logging" seems at first to be merely euphemistic; people of delicacy wish to have their woodlands "harvested" as they wish to have their meat "processed." But "harvest" (Middle English *hervest*, meaning "autumn") was associated originally with annual grain crops that ripen in the fall. When we harvest a crop of corn, we take all that is of economic worth—which, with some special exceptions, is a bad way to treat a forest.

Like the grasses and forbs of farm pastures and hayfields, trees are perennials. But *unlike* the grasses and forbs, which mature and can be cut once, or more than once, every year, trees keep growing bigger year after year, sometimes for hundreds of years. Forests also are more complex in structure and diversity of species. You can't learn to manage a forest by managing a pasture.

Farming, moreover, must always be to some extent a compromise with the local ecosystem. Whereas the farmer requires from the farm, specifically for human use, several plants and animals that the ecosystem in its natural state would not produce, the best foresters ask the local forest ecosystem only to continue to produce what, according to its nature, it produces best: primarily trees of the native species. A properly diversified forest economy produces more than trees, of course. And a healthy forest contributes to the health of the soil, water, air, and all other constituents of the natural world. The forest submits to compromise only by allowing the foresters to take timber and other forest products for human use, if they can do so without destroying the integrity and the continuous productivity of the ecosystem.

Forest owners, however, may have a good deal to learn from livestock farmers—little as such farmers may know about forestry. No sane farmer would sell all her brood cows, keep their heifer calves, and wait for another calf crop until the heifers have become old enough to breed and calve. But forest owners do substantially that when they sell off every marketable tree—except that the forest owners (or their descendants) may have to wait for generations, not years, for another marketable "crop."

Nor would a sane cattle farmer "highgrade" his herd by selling his best cows and keeping and breeding the worst. But people who highgrade their woodlands do exactly the same thing, selling the best and keeping the worst—which, if not thoughtless, as it usually is, would be insane.

The best logging, however, we may rightly call culling. The foresters' name for it is "worst-first single tree selection." By this method, the trees in the area to be logged are looked at individually, evaluated according to standards of worth and health, and the worst are carefully removed, leaving only small openings in the canopy, and doing the least possible damage to the trees, young and old, that remain, and to the forest floor. The "worst" are trees that are diseased or dying, leaning trees that are more likely to fall as they grow more top-heavy, trees that branch too low or are otherwise inferior in conformation.

But I need to interrupt myself here to notice that my need to be clear has led me into oversimplification. A stand consisting only of the "best" trees would be a tree farm, not a forest. A healthy forest would necessarily include some less valuable or unmarketable species and some diseased and dying trees, so as to sustain the finally unimaginable diversity of creatures that belong to the natural forest. Good foresters know this and are charitable toward members of the forest community that by the standards merely of economics would be considered worthless.

The essential point is that by applying the standard of ecological health, which is the standard of long-term economy, the good forester will leave standing numerous marketable trees because they are healthy and flourishing, and by doing so will maintain the forest at its highest productivity. The explanation for this comes from elementary geometry: A quarter inch of annual growth on a tree two feet in diameter is far more than a quarter inch of annual growth on a tree six inches in diameter. A flourishing large tree makes more marketable wood per year than many small ones. A woodland logged in this way can be logged fairly frequently, at intervals of ten or fifteen or twenty years. And at each successive logging the quality and market-ability of the trees that are cut will have increased.

Forestry of this kind, though it is still far too rare, is not new. That it is more than four hundred years old we know from Aldo Leopold's essay, "The Last Stand," in *The River of the Mother of God*:

> I know a hardwood forest called the Spessart, covering a mountain on the north flank of the Alps. Half of it has sustained cuttings since 1605, but was never slashed. The other half was slashed during the 1600s, but has been under intensive forestry during the last 150 years. Despite this rigid protection, the old slashing now produces only mediocre pine, while the unslashed portion grows the finest cabinet oak in the world; one of those oaks fetches a higher price than a whole acre of the old slashings. On the old slashings the litter accumulates without rotting, stumps and limbs disappear slowly, natural reproduction is slow. On the unslashed portion litter disappears as it falls, stumps and limbs rot at once, natural reproduction is automatic. Foresters attribute the inferior performance of the old slashing to its depleted microflora, meaning that underground community of bacteria, molds, fungi, insects, and burrowing mammals which constitute half the environment of a tree.

By "slashed" Leopold evidently meant something like what we now mean by "clear-cut." "Slashing" destroyed the integrity of the forest ecosystem.

The literature on what we are now calling "sustainable agri-culture," laying out the ecological principles of good land use, is fairly substantial, but not nearly so much has been written on sustainable forestry. To learn about that, it is necessary to become familiar with one or more of the few exemplary forests where good practice is well established and for which there is some published accounting, or one may study the ways and the work of exemplary loggers and forest managers. Bad logging, like bad farming, comes from estrangement between the industrial economy and nature's ecosystems. Most land users, as they come more and more to adopt industrial technologies and attitudes, know nothing or think nothing of ecology. And few ecologists or people oriented to ecology are involved with, or familiar with or even interested in, the economies of land use. When you find a farmer or a forester who has united the inescapable economic concern with an equally compelling interest in ecology, that is when you had better stop and take notice.

I first met Jason Rutledge a good many years ago when both of us were visiting Washington and Lee University in Lexington, Virginia, which is in Jason's home country. He was talking then, and he is talking still, on the closely related themes of sustainable forestry and logging with horses. Since that first encounter, our paths have crossed fairly often in the woods and at various field days and conferences. "Worst-first single tree selection" is language I first heard from Jason, a logger who uses horses (*Suffolk* horses, he would want me to say) to skid the logs out of the woods. He is also a teacher whose alumni are scattered about in the woods, horse logging on their own and earning a living by doing so.

It was Jason Rutledge who introduced me to Troy Firth. Like Jason, Troy is an advocate for sustainable forestry and is worth listening to because his advocacy, like Jason's, rests upon practice. Troy is a forest owner whose holdings are extensive. He is a forester and logger who operates his own sawmill. He is also owner and manager of Firth Maple Products, the chief product of which is maple syrup. This year Troy and his crew tapped 17,000 sugar maple trees and made 6,100 gallons of syrup. Next year they plan to tap 22,000 trees.

Troy has lived and worked all his life in the neighborhood near Spartansburg, Pennsylvania, long occupied by his family; his house is on Firth Road. Troy's involvement in the woodland economy of his region is elaborate and of long standing. He is not a waster of words, and I think this is because he is confident both of his knowledge and of the limits of his knowledge. He has a reliable sense of humor and is generous with his knowledge, but he generally keeps quiet until he has something worthwhile to say. The result is that when he talks you listen, and you are apt to be surprised at how much you remember of what he has said.

From his student days, Troy has been inclined toward the woods. While he was in college he took two years off to work as a logger. He accumulated some money and in 1972 bought woodland, which at that time was cheap. He counted his purchase as an investment, but it was more than that. His interest in forestry had become a passion, as it still is.

In college he majored in history, not forestry, which he thinks was a good thing, for a person's education should be broad rather than "vocational." As he went on with his work in the woods, he realized that the conventional (and recommended) brand of forestry was wrong. It went against nature and was not sustainable. I asked him how long it took him to see this. He said, "Not long. It was obvious."

He had, he says, "no role model or mentor." Nor was he acquainted with examples to show him what to try for. This part of his education, the postgraduate part, was up to him. He simply rejected the conventional practices and tolerances that over time reduced the productivity of the forest, and from there he was on his own. The change of mind from forestry as an extractive industry to the study of sustainability must always involve a radical shift in perspective. One ceases to think of the source of sawlogs as trees, which can be cut according to wishes or needs or standards that are merely economic, and begins the understanding—far more complex and difficult, but also far more interesting—that the source is the forest ecosystem.

In 1985 Troy bought more woodland, and he has continued to do so at the rate of "a property or two a year." He and his wife, Lynn, got married in 1988. The wedding took place on a beautiful day, out in the snowy woods, in January. Such a wedding certainly suggests a strong liking for the woods, on the part of both bride and groom. And I would say that it suggests, more than most weddings, a strong mutual interest in getting married. Lynn and Troy have a daughter who, like Troy, is an only child.

Marriage and parenthood, for responsible people, lead to considerations of futurity and mortality, duties and limits. Troy and Lynn now had their daughter's future to think of, along with the future of their woodlands. This led them to the deliberations and arrangements of "estate planning." Like in fact a good many people, Troy thinks there ought to be a limit to a parent's bequest to a child. Too much makes things too easy. Eventually he asked his mother to change her will, removing him as her heir, and leaving her entire estate to his daughter. His mother's estate, he says, amounted to "something, not too much."

Providing for the future of the woodlands was a more difficult problem, for it involved more time, more unknowns, and evidently there were no precedents. Good management for the length of a human life is certainly a gift to the forest, but it is pathetically brief. A single tree may live several times as long as a human. A forest, kindly used, may outlive unimaginably any of its trees. Troy doesn't know whether he or his lawyer thought of it first, but the idea for the Firth Family Foundation emerged in 2004 from the need to provide the Firths' forest properties an inheritor that might keep them intact and well managed beyond the limits of a human lifetime. This is the same impulse that has led to the formation of land trusts to hold development rights or "conservation easements" in order to preserve farmland or open space beyond the lifetimes of present owners. But Troy's vision for his foundation is more complex, and it addresses more problems.

One of the most daunting problems in forestry, especially in forested lands that are privately owned, is fragmentation. Small holdings in farmland make sense, or can be made to make sense. But small holdings in a forested landscape—and the average privately owned forest acreage in Kentucky is just under 20 acres; in Pennsylvania it is 16.7—are too small both economically and ecologically. A small woodlot on a farm can contribute significantly to the economy and the life of the farm, but a scattering of small forest properties, even if well used, cannot sustain a forest ecosystem.

Another problem is short-term management of large tracts of forest by financiers or investors who, for an acceptable return in fifteen years, may highgrade their properties, or cut everything, and then subdivide the land for "second homes." The problem, predictably, is that "short term" land management leads almost inevitably to long-term exhaustion.

As the land resource declines, in forestry as in farming, so necessarily does the local land economy and the local human community. As the human community declines, there are always fewer actual or potential good husbanders of the land. As this cycle turns, whatever the nation and the national economy may be doing, the country and its native health and wealth spiral downward.

Troy understood that the good health and productivity of his own woodlands depends on good use sustained over a long time. But good health cannot be bounded and isolated like a property. The good health of part of a forest depends on the good health of the whole forest. It would not be enough for the foundation to serve merely as heir and steward of the Firths' personal holdings; the foundation could serve as an heir also to other forest owners, who might bequeath or donate their properties to it. Beyond that, the need is to exemplify and promote sustainable forestry for the sake of the forest ecosystem as a whole. The goal of the Firth Family Foundation was "to promote and practice long-term sustainable forestry while conserving working timberlands throughout Pennsylvania, New York and the surrounding states."

By 2010 the Firth Family Foundation had transformed itself into the Foundation for Sustainable Forests: A Land Trust. The name had become more general, but the regional boundary had been retained. Troy hopes that eventually there might be "numerous regional associations that could work cooperatively." The goal "is an abundance of intact forest ecosystems that provide for the widest range of native biodiversity possible, sustainable forest products, the economic viability of rural communities, and recreational opportunities."

What Troy had forming in his mind—what apparently is still forming in his mind—is the idea of a sustainable forest economy. Like any sustainable land economy, it would have to form locally, taking shape in response to local conditions, limits, and opportunities. The influence of its example and working principles would then extend to anybody who might be interested.

Troy's ultimate wish is "to do whatever can be done to protect land." An example of sustainable forestry adapted to one place cannot be applied like a stamp to a different place. What *can* be applied are right principles and standards, and some understanding of what is involved in the effort of adaptation.

But a forest economy, however well adapted, however concordant with its ecosystem, cannot stand or continue alone. For a local economy to become truly sustainable, it must function as a belonging, a support and an artifact, of a local culture. The great need we are talking about is to hold the local forest ecosystem together, but that leads to a second need, just as great, which is to hold the local forest economy and therefore the local human community together. The two needs could be answered only by a thriving, confident, stable local culture in which the young would learn from their elders. Change and innovation would naturally occur, but would not be imposed from the outside or for the benefit primarily of outsiders. Nor would the changes and innovations arrive at the breakneck speed of industrial innovation. The community would understand that it exists, in part, to cherish the forest and to preserve its knowledge of the forest.

There is a good deal of supposing in that paragraph. In America, or in the history of Old World races in America, we have no example of an economic enterprise ecologically and socially sustainable. To imagine such a thing, we have to consult scattered pieces. Any conversation at home between grandparents and grandchildren is potentially the beginning of a local culture, even of a sustaining local culture, however it might be cut short and wasted.

Troy's practice of forestry, even when he was working mainly alone, was already one of the scattered pieces of a sustainable economy, and Troy knew that. But as he also knew, it was not nearly enough. He needed to be talking about his work with somebody, a younger somebody, at work with him. And so in 2005 he looked about in the Spartansburg area and hired Guy Dunkle, a bright young man just out of college, who, like Troy, was locally born and raised. I had gathered from my conversations with Troy that he had hired Guy as an apprentice or understudy, supplying the younger half of the necessary conversation. But when I asked the name of Guy's "position," hoping for better understanding, Troy said, not to my surprise, "We're not much on titles." He spoke of Guy simply as "a forester for Firth Maple Products." In the most practical terms, Guy is a younger forester coming behind an older forester and learning from him. "Everything depends on that," Troy said.

It is more than passingly significant, I think, that this complex enterprise of forestry, sawmilling, "maple products," and an emergent foundation is "not much on titles," and that its executive staff consists of two "foresters" involved in the fate of the same woodlands. It is significant also that Guy's college degree is in environmental science, not forestry, and that his job interview was conducted in the course of a week's work in the woods with Troy, tapping maple trees.

Guy Dunkle enjoys describing his job as whatever Troy doesn't much want to do. And so it was Guy who drove to Erie to meet my plane on May 16, 2012, when I came to walk and talk in the woods with the two of them and to take part in a couple of public events promoting the foundation. I had met Guy in October 2005 at one of Jason Rutledge's logging sites in Virginia, but our drive from the Erie airport to supper with Troy and Lynn Firth at their house was our first opportunity actually to become acquainted.

Guy is a tall, lean, pleasant man, impressive in several ways, not least for his apparent innocence of the too-common wish to "make an impression." He is not as quiet as Troy, but he is quiet enough. He seems, right away, to be a man of settled character, and you suppose he has answered several of the important questions about his life and his vocation. He and his wife, Wilma, have two young sons. They are living on their own farm, which is partly wooded, partly planted in Christmas trees, and partly in pasture. We spent a good many miles of our drive in a conversation about raising sheep. A young man with a farm, and fully interested in it, may be assumed to have some permanent intentions.

When we arrived at the house on Firth Road, Troy had come in from his day's work. Lynn gave us a good supper, most welcome to me, for my noon fare at the Cincinnati airport had been both meager and bad. I had asked Troy to show me as much of his practice of forestry as I would be able to see in the time available, and he had laid out a crowded itinerary. Soon after we had eaten, he and Guy and I used up the rest of the day on the first three stops of my tour of places important to see.

First we made a fairly quick drive-through at Troy's sawmill, most memorable to me for the ricks of beautiful large black cherry logs. To be profitable, a mill of this relatively small scale must have a specialty, or "niche," which here is the lucrative trade in veneer-quality cherry. That happens to be a proper niche for this area, which is uniquely productive of the finest cherry wood. To maintain the necessary stock of good logs, since he can't supply all he needs from his own relatively limited acreage, Troy regularly logs timber sales in the Allegheny National Forest or the Susquehannock State Forest.

We next went to the headquarters of Firth Maple Products, where the sap of thousands of trees is collected, reduced by boiling to the proper consistency of maple syrup, and bottled. My own country has a lot of sugar maples, but we have no tradition of syrup-making. We do, or did, have something of a tradition of sorghum molasses-making, and so I understand in principle the boiling down of a thin, mildly sweet liquid to a viscous liquid that is intensely sweet. But the equipment of Troy's maple syrup factory far exceeded in both scale and complexity any that I had seen before.

Troy's woodlands are full of pipelines. The black plastic pipe, most of it an inch and a quarter in diameter, is tightly stretched, reinforced by high-tensile, galvanized steel wire, and secured at about knee-height by other wires guyed to trees on either side. Small transparent tubes run the sap from the trees to the black pipes. I was impressed by the sturdiness of the pipelines, which are installed to last for several years, and also by the care with which they had been attached to the trees. Every wire that went around a tree was buffered by blocks of scrapwood to prevent injury to the bark. The pipelines carry the sap to tanks at well-sited collection points, from which it is trucked to the boilers. Troy is alert for any way to reduce costs. The maple syrup operation requires large tanks of various sizes and kinds, and many of these have been bought cheap—used or damaged but still good. The trucks are mostly stainless steel milk tankers, also used but good. If you have limited or no control of the markets you buy and sell on, thrift becomes an essential virtue.

Our third stop on the evening of my arrival was at a woodland that Troy had logged, he thought, ten times in the last forty-five years. And here we came to our principal subject and the purpose of my trip. Clearly, no patch of forest could have been so steadily productive for so many years if it had not been knowingly and carefully logged. And in fact this was an excellent demonstration of the results, over time, of worst-first single tree selection. The last cutting had been recent, the tops of the stumps were still bright and unweathered, and yet you could see immediately that the forest in that place was ecologically whole. There were no too-large openings in the canopy. The remaining trees were of a diversity of sizes, from large to small. None of the remaining trees had been damaged by the felling or skidding. Because the skidding had been done with horses, the ground had been only slightly scarred. The skid trails would mostly disappear by the next year.

My visit with Troy and Guy, and then with Jason Rutledge and his son Jagger, who arrived on Thursday afternoon, was from Wednesday evening until mid-morning on Sunday. Subtracting the several hours when we merely sat and talked, we probably spent two longish days walking, looking, and talking in the woods.

In the bright early sunlight of Thursday morning we walked into an open woodland of tall trees, and under them a uniform stand, virtually a garden, of graceful hay-scented ferns. This was a pleasant, "scenic" place where one might like to have a picnic, the sort of "natural" place that many a nature or wilderness enthusiast would like to "preserve." But there was a problem. There was no understory to speak of: no seedling trees, saplings, shrubs, or the smaller plants that in a thriving forest would occupy the spaces between the ground and the lower branches of the tallest trees. The succession of the trees had been broken. If one should be felled or blown down, there would be none to replace it. The culprit was the beautiful hay-scented fern, whose rhizomes and roots grow so thickly in the top layer of the soil that they cannot be penetrated by the roots of seedlings, and whose fronds steal the light from other plants.

This domineering population of ferns is caused by an over-population of white-tailed deer who eat and destroy the plants that compete with the ferns. Once the fern carpet is established, it remains dominant. Reducing the deer population then has no effect, and another remedy must be found. Among the ferns a few blackberry briars were scattered about, and these may offer some hope. It may be that where the blackberry roots have pierced the fern mat, the roots of other species may follow and take hold. Or it may be that the blackberries, because of their earlier leafing in the spring, will shade out the ferns. The interaction of the two species is as yet unclear. Alternatively, the ferns may have to be set back by the use of an herbicide. The immediate problem here for the long-term commercial forester—this is the point—is an ecological problem: how to restore to this impoverished woodland the highly diverse "vertical structure" indicative of ecological health.

The predominant and most valuable trees of this region are black cherry and sugar maple. In addition, there are northern red oaks, some white oaks, white pines, and red or soft maples. Because of their yield of sap, the sugar maples have a special status in the Firth woodlands. They are dual-purpose trees. Tapping them for syrup degrades them somewhat as timber, but at appropriate times they too are logged.

But when you are with him in the woods, you can feel Troy's partiality for the tall, veneer-grade cherry trees with their dark, thick, flawless trunks. Looking with him at these trees, you see one of the primary ways in which good forest management increases yield. We were discussing a tall, thick-boled cherry tree with a fork at about half its height: two long, large prongs with many smaller branches. This woodland, Troy says, was highgraded perhaps ninety years ago. "Highgrading," he says, "takes the top off the woods." This reduces the competition for space and light, with the result that the trees branch or fork lower than they normally or naturally would have done. This particular cherry is a valuable tree that will yield a log of high quality, but the main log will be half to two-thirds as long as it would have been in a properly managed forest. The tree, moreover, is vulnerable in a way that a single-stemmed tree is not. Cherry is a brittle wood. As the two great prongs of the fork become more top-heavy, and their leverage increases, it becomes more likely that they will cause the tree to break or split. A tree like this one must be closely watched. At some point, a skilled climber will top the tree so that it can be safely felled.

We visited, by contrast, a single-stemmed cherry tree in a woodland that had not been high-graded. This tree was perhaps ninety years old, about twenty inches in diameter at breast height, still flourishing, still making wood, and not nearly so old as, with right management, it may become. It was maybe a hundred and twenty feet tall, eighty feet to the first branch. Some day it will yield five sixteen-foot logs, the first two of veneer quality. It was by any measure a beautiful tree. We spent a good while looking up at it and complimenting it.

The great event of Friday morning was our encounter in the woods with two young Amish teamsters driving their two-horse teams from the high seats of the elegantly engineered carts known as "logging arches," which are used for skidding logs out of the forest. The carts are so constructed that, when the horses tighten against a log, the log's own weight causes its fore-end to lift off the ground, thus reducing friction, lightening the draft, and minimizing damage to the woods floor.

More remarkable were the two teams of horses that provide the traction power. These horses do the hardest sort of work daylong every weekday, drawing heavy loads over rough ground, and yet they had the appearance of show horses. They were sound, bright-eyed, in excellent flesh, their coats unflawed and shining. Their condition spoke well of everything involved, but principally of the intelligence, care, and skill of the two teamsters. These were friendly, articulate men in their late twenties or early thirties. In conversing with us, and particularly with Jason, who is himself an excellent horseman and teamster, they showed themselves to be constantly attentive to every point of contact between horse and harness. This comes from a fine kind of sympathy: A good horseman wants the horses to work as comfortably in their harness as he himself works in his clothes. Sometimes in horse logging one sawyer, felling the marked trees, will keep two or three teamsters busy skidding the logs out. Here, the teamsters themselves do the felling, and this gives an obvious advantage to the horses, who rest while the men work.

The use of horses in the woods is fundamental to Troy's way of logging. It may finally prove to be necessary to anybody's version of sustainable forestry. Horses are kinder to the forest than mechanical skidders. Their skid trails are narrower and require no digging or grading of the forest floor. Using less force, they are less violent. A horse's hooves may slip now and then, scarring the surface of the ground, but the wheels of a heavy skidder compact the soil and, in the roughest places, tend to dig. Horses, moreover, can work on wetter ground than skidders.

Horses and horse equipment also cost much less and work much cheaper than skidders, and this can yield a large advantage to the forest. A new skidder will cost in the neighborhood of $150,000, a used one something like $50,000. A teamster's outfit—horses, logging arch, chains, saw, etc.—might cost as little as $5,000. (The greatest "cost" of using horses is the knowledge of how to care for, breed, raise, train, and work them, which is now in short supply. But knowledge, once you know it, is free.) This is a significant and influential difference. The more expensive the logging equipment, the higher the operating costs, the greater must be the pressure to increase the daily volume of board feet, increasing in turn recklessness in the use of the forest and the likelihood of damage. The urge toward the highest possible volume works against selectivity, judgment, and forbearance. In forestry as in farming, low production costs can increase the quality of work and so of care for the land. The logger who is free of financial anxiety can stop and think.

Moreover, the use of horses in logging increases human employment. One person operating a skidder may replace three teamsters. We are taught to see that as an advantage, and it is, but only to the company that makes skidders, and to the suppliers of fuel. The advantages of horse logging, much to the contrary, accrue to the forest and the human community. The Amish horsemen we talked with earn a good living, working as "independent contractors." In a time officially obsessed with "job creation," perhaps we don't need to argue about the worth of three people in effect self-employed as opposed to one employee driving somebody else's machine. Horse logging, in addition, is principally maintained by the local economy, which in turn it helps to maintain, whereas the purchase of large machines and the fuel to run them siphons money out of the local economy to enrich the shareholders of remote corporations.

In keeping with his commitment to sustainability, Troy does what he can to favor continuity of employment. At present he has six teams at work in the woods. One of the teamsters, who happens to be one of Jason Rutledge's alumni, has worked for Troy for fifteen years, another for eight, two others for six. He tries to keep them busy all the year round. Except for times in the spring when the ground is too wet—and, when using horses, these times are not long—the logging is continuous. When it is too wet, Troy finds other work for the teamsters.

The connections and interactions among all the creatures in a thriving forest ecosystem are complex beyond the human ability to think. This is the starting point, the primary axiom. And so humility is the primary virtue of good forestry. One must get the scale right, so as not to put too much at risk. One must not use too much power, or be in too much of a hurry. Troy accordingly does not talk like an expert. Because the forest is complex, it requires a due complexity of knowledge and of thought. Ultimately it requires work that honors, not only the known complexity, but also the unknown, the mystery of the nature of any place. By now Troy has a lot of experience. He has thought and worked with care. But how does he know when he has worked well?

His summary statement to me was that he finds out from the songbirds. If one of his woodlands as a result of his care and management is ecologically healthy, if there is enough diversity of the species and ages of the plants from the lowliest flowers and shrubs to the tallest trees, if the "vertical structure" is right, then you will hear a lot of birds singing. This is not a Disney sentimentalism. The birds are not singing to compliment Troy. They are singing because they have found a diversity of places in which to nest and feed, as the nature of the forest requires. Their songs indicate that all, or at least enough, of the pieces are in place. They are singing specifically to Troy only in the sense that he knows at least this much of what they mean.

Another virtue essential to good forestry is generosity. You don't have to see much bad logging before you become unable to imagine that a selfish or greedy person could be a good forester. A good forester thinks first of the forest. I remember Troy saying at a forestry meeting several years ago, "A bad logger goes to the woods thinking of what he can take out. A good logger goes to the woods thinking of what he should leave." It is this generosity toward the forest that enables the necessary thought. Generosity involves the forester in the history of the forest, its past and its future.

Troy's knowledge of his woodlands is historical, extensive, highly particular, and intimate. If you follow him through the woods for several hours, you realize that you are passing through a succession of distinct places, each different from any other, in its community of plants, slope, exposure, soil quality, history, problems, and so on. About each place Troy is likely to know a more or less adequate history: from visible evidence, from local or personal memory, from dated aerial photographs. He will know that a given place, until a certain year, was cleared and farmed. Or he will know that another woodland was high-graded eighty years ago, or that another was carelessly logged fifty years ago. He is aware, of course, of general principles, ideas, and theories of forestry, some better than others, but he says, with emphasis, "Forestry is mainly observational, rather than theoretical." By "observation" Troy means walking and looking, paying attention, season after season, for many years. Eventually, a profound familiarity will grow between an observant forester and the places of the forest. Such knowledge is what we mean, maybe, by "sympathy" or "sixth sense" or "intuition." It is the knowledge that tells one, in a given situation, where to look or what to expect or how much is enough. It tells what to take and what to leave.

Though this certainly is knowledge, and though it certainly comes by a kind of education, it cannot be conveyed in courses or curricula or majors. This education is "observational" and it takes many years. To know competently a tract of forest, Troy says, "is going to take decades. That's all there is to it."

You can't learn one woodland by studying another, and you can load only a limited amount of competent or workable knowledge into one mind in one lifetime. This is why it is important for good foresters both to stay put and to have local successors. The United States Forest Service makes a practice of moving people around. We recognize this as an industrial ideal: The supposedly easy mobility of human populations is

an exploitable asset. Troy's idea, on the contrary, is to stay in
place himself, and to hire local people for life. There is a sound
economic reason for this, in addition to its obvious ecological
value: Local knowledge of the local forest, like the forest itself,
is an asset.

A local or personal economy, no matter how intelligently and
kindly managed, cannot be made entirely immune to changes
in the larger economy, no matter how false or fantastical the
larger economy may be. And so in the "recession" that began
in 2008, Troy says, "I lost a bundle." His inventory of standing timber "fell off a cliff." Now things are improving, and
he recently hired two more skidding teams. In general the
Firth economy, within a larger economy never favorable to
the economies of land use, has done remarkably well. Probably the most interesting sentence in the circular issued by the
Firth Family Foundation is this: "A 2006 study by LandVest
showed that the Firths averaged 11 percent annual return on
their timberlands for the previous 20 years."
    Troy Firth's years of work in the forest have carried him far
beyond the usual definitions of land ownership, of land use,
and even of forestry. Rather than requiring the forest to submit
to his economic demand, he has learned painstakingly to fit
his economy into the forest. He has taken his living from the
forest, not in defiance of the nature of the forest ecosystem,
but to the best of his ability cooperating with it. Under his
management, the production of timber has not reduced the
productivity of the forest. On the contrary, his management
has increased productivity—it can, he says, almost double the
production of poorly managed woodlands—at the same time
that it increases the number of human workers. It would be
hard to overestimate the importance of this, as an accomplishment and as an example: several human livelihoods taken from
the forest, to the forest's benefit, which becomes in turn and
complexly a further human benefit.
    Troy is an exemplary user of the land. At a time when such
exemplars are desperately needed, he is somebody to turn to
for confirmation of certain good possibilities. But the value of
this, as he well knows, is strictly limited. One or two good foresters here and there are not enough. Nor even would be some

widespread brand of "sustainable forestry," given an official label in the fashion of "organic agriculture." To develop and promote a forest ethic would do some good, but it would not do much good in the absence of working and paying examples of good practice. Governments may provide some checks against bad forestry by laws, regulations, restraints, "incentives," etc., but governments do not have the means to bring about good forestry. In a note of June 25, 1942, Aldo Leopold recorded his sense of this governmental limit: "I am skeptical about government timber production as a sole remedy for the apathy of private timber owners."

A sustainable local forest economy would supply as many as possible of its own needs, and it would perform most of the value-adding to most of its products. A country-wide project of sustainable forestry would require for many years an increasing number of such economies, which would in turn depend on the coincidence of available decent livelihoods from the forest with the passion for forestry that came to Troy in his student days and has continued with him ever since. If you are not squeamish about the word, you might say the "love" of forestry. To say that the good care of the forest, as of all the world's places, depends upon love is, sure enough, to define a difficulty. But not an impossibility. The impossibility is that humans would ever take good care of anything that they don't love. And we can take courage from the knowledge that millions of Americans once loved their vegetable gardens, cared well for them, and kept them dependably productive—and that a good many still do.

The Foundation for Sustainable Forests, at present, is the work of a few people. It has so far acquired in its own name only 740 acres. It is assured eventually of having, by bequest from Troy and Lynn, more than 6,000 additional acres. What seems most hopeful about it is its solid basis in practicality. Unlike most foundations, this one will be a viable commercial enterprise. It can use donated money to purchase woodlands, but these properties will become self-sustaining.

Like The Land Institute, Tilth, the Quivira Coalition, the Southern Agriculture Working Group, the Land Stewardship Project, and other organizations concerned with sustainable agriculture, the Foundation for Sustainable Forests will occupy the intersection of ecology and economy, which ought to be occupied by everybody but at present has a surplus of elbow room. These organizations promoting sustainable land economies are identifiable by their endless agendas of work, by their virtual invisibility to the worlds of policy and the news, and by the swarms of questions that hover about them—questions asked and waiting for answers, questions as yet unasked.

For example: If sustainable forestry depends upon sustainable local forest economies, how will we develop the necessary small-scale value-adding industries? I don't know the answer, and I doubt that anybody does. But then I remember hearing that small Amish furniture factories did more than a hundred million dollars worth of business last year, and I draw another breath. The necessary pieces may already exist, though widely scattered and perhaps lost from one another. No one person will be capable of putting them all together. Troy Firth knows this, but he also knows a further truth: One person can begin. "And can begin better with the help of others."

Local Economies to Save the Land and the People

As often before, my thoughts begin with the modern history of rural Kentucky, which in all of its regions has been deplorable. In my county, for example, as recently as the middle of the last century, every town was a thriving economic and social center. Now all of them are either dying or dead. If there is any concern about this in any of the state's institutions, I have yet to hear about it. The people in these towns and their tributary landscapes once were supported by their usefulness to one another. Now that mutual usefulness has been removed, and the people relate to one another increasingly as random particles.

To help in understanding this, I want to quote a few sentences of a letter written on June 22, 2013, by Anne Caudill. Anne is the widow of Harry Caudill. For many years she was involved in Harry's study of conditions in Eastern Kentucky and in his advocacy for that region. Since Harry's death, she has maintained on her own the long interest and devotion she once shared with Harry, and she is always worth listening to. She wrote:

> The Lexington Herald Leader last Sunday . . . published a major piece on the effects of the current downturn in the coal industry . . . Perhaps the most telling statement quoted came from Karin Slone of Knott County whose husband lost his job in the mines . . . finally found a job in Alabama and the family had to leave their home. Karin said, "There should have been greater efforts to diversify the economy earlier."
>
> [Fifty] years ago and more Harry tried . . . everything he could think of to encourage diversity. My heart goes out to those families who yet again are being battered by a major slump in available jobs. . . . Again they are not being exploited, but discarded.

This is a concise and useful description of what Anne rightly calls a tragedy, and "tragedy" rightly applies, not just to the present condition of Eastern Kentucky, but to the present condition of just about every part of rural Kentucky. The tragedy of Eastern Kentucky is the most dramatic and obvious because that region was so extensively and rapidly industrialized so early. The industrialization of other regions (mine, for example) began with the accelerated industrialization of agriculture after World War II, and it has accelerated increasingly ever since. The story of industrialization is the same story everywhere, and everywhere the result is ruin. Though it has developed at different rates of speed in different areas, that story is now pretty fully developed in all parts of our state.

To know clearly what industrialization is and means, we need to consider carefully some of the language of Anne Caudill's letter. We see first of all that she is speaking of a region whose economy is dependent upon "jobs." This word, as we now use it in political clichés such as "job creation," entirely dissociates the idea of work from any idea of calling or vocation or vocational choice. A "job" exists without reference to anybody in particular or any place in particular. If a person loses a "job" in Eastern Kentucky and finds a "job" in Alabama, then he has ceased to be "unemployed" and has become "employed," it does not matter who the person is or what or where the "job" is. "Employment" in a "job" completely satisfies the social aim of the industrial economy and its industrial government.

Perhaps there have always been "jobs" and "employees" to fill them. The point here is that the story of industrialization radically enlarges the number of both. It also enlarges the number of the unemployed and the unemployable. I can tell you confidently that the many owners of small farms, shops, and stores, and the self-employed craftspeople who were thriving in my county in 1945, did not think of their work as "a job." Most of those people, along with most skilled employees who worked in their home county or home town, have now been replaced by a few people working in large chain stores and by a few people using large machines and other human-replacing industrial technologies. Local economies, local communities, even local families, in which people lived and worked as members, have been broken. The people who once were members of mutually supportive memberships are now "human resources" in the "labor force," whose fate (to return to the language of Anne Caudill's letter) is either to be "exploited" by an employer or "discarded" by an employer when the economy falters or as soon as a machine or a chemical can perform their "job." The key word in Anne's letter is "discarded," which denotes exactly the meaning and the sorrow of our tragedy.

How can it be that the people of rural Kentucky can first become dependent upon officially favored industries, the "job-creating industries" that their politicians are always talking of "bringing in," and then by those industries be discarded? To answer that question, I need to refer again to Eastern Kentucky and something I learned there—or began consciously to learn there—nearly fifty years ago.

In the summer of 1965 I paid a visit of several days to my friend Gurney Norman, who was then a reporter for the *Hazard Herald*. At that time a formidable old man, Dan Gibson, armed with a .22 rifle, stopped a strip miner's bulldozer. The land Mr. Gibson was defending belonged to his stepson, who was serving with the Marines in Vietnam. Mr. Gibson's defiance and his arrest caused a considerable disturbance, and a crowd of troubled people gathered on a Friday night in the courthouse in Hindman. Gurney and I attended the meeting. That night Harry Caudill made a speech that recalled certain meetings in Philadelphia in the summer of 1776, for he spoke against the domestic successors of the British colonialists: "the mindless oafs who are destroying the world and the gleeful yahoos who abet them."

I am indebted to another speech of the same night. That speech was made by Leroy Martin, chairman of the Appalachian Group to Save the Land and the People. Mr. Martin bore witness to the significance of Dan Gibson's act, his loyalty, and his courage. He spoke impressively also of the forest that stood on the mountainside that Mr. Gibson had defended. He spoke the names of the trees. He reminded his hearers, many of whom were local people, that they knew the character and the value of such woodlands.

Three lines of thought have stayed with me pretty constantly from that time until now.

The first concerns the impossibility of measuring, understanding, or expressing either the ecological cost or the human heartbreak of the permanent destruction of any part of our only world.

The second consists of repeated returns to the impossibility, at least so far, of permanently stopping this permanent damage by confronting either actual machines or political machines. Dan Gibson's unlawful weapon was answered by the lawful weapons of thirteen state police, a sheriff, and two deputies. Our many attempts to confront the political machine that authorizes the industrial machinery have really not been answered at all. If money is speech, as our dominant politicians believe, then we may say that all our little speeches have been effectively answered by big money, which speaks powerfully though in whispers.

The third line of thought, the one I want to follow now, has to do with the hopefulness, and the correction, implied in the name of the Appalachian Group to Save the Land *and* the People. The name of that organization—and, if I have remembered it correctly, Leroy Martin's speech—assumed that we must not speak or think of the land alone or of the people alone, but always and only of both together. If we want to save the land, we must save the people who belong to the land. If we want to save the people, we must save the land the people belong to.

To understand the absolute rightness of that assumption, I believe, is to understand the work that we must do. The connection is necessary of course because it is inescapable. All of us who are living owe our lives directly to our connection to the land. I am not talking about the connection that is implied by such a term as "environmentalism." I am talking about the connection that we make economically, by work, by living, by making a living. This connection, as we see every day, is going to be either familiar, affectionate, and saving, or distant, uncaring, and destructive.

The loss of a saving connection between the land and the people begins and continues with the destruction of locally based household economies. This happens, whether in the United States after World War II or in present day China, by policies more or less forcibly moving people off the land. It happens also when the people remaining on the land are convinced by government or academic experts that they "can't afford" to produce anything for themselves, but must employ all their land and all their effort in making money with which to buy the things they need or can be persuaded to want. Leaders of industry, industrial politics, and industrial education decide, for example, that there are "too many farmers," and that the surplus would be "better off" working at urban "jobs." The movement of people off the land and into industry, away from local subsistence and into the economy of jobs and consumption, was one of our national projects after World War II, and it has succeeded.

This division between the land and the people has happened in all the regions of rural Kentucky, just as it has happened or is happening in rural places all over the world. The problem, invisible equally to liberals and conservatives, is that the forces that destroy the possibility of a saving connection between the land and the people destroy at the same time essential values and practices. The conversion of an enormous number of somewhat independent producers into entirely dependent consumers is a radical change that in many ways is immediately catastrophic. Without a saving connection to the land, people become useless to themselves and to one another except by the intervention of money. Everything they need must be bought. Things they cannot buy they do not have.

This great change is the subject of Harriette Arnow's novel *The Dollmaker*. In the early pages of this book we recognize its heroine, Gertie Nevels, as an entirely competent woman. Her competence does not come from any "success," political or social or economic. She is powerful because, within the circumstances of her agrarian life in the mountain community of Ballew, Kentucky, she is eminently practical. Among the varied resources of her native place, she is resourceful. She has, from her own strength and willingness and from her heritage of local knowledge, the means of doing whatever needs to be done. These are the means, for her, of being content in Ballew where she is at home. Her husband, Clovis, is not content or at home in Ballew. He is an off-and-on mechanic and coal hauler whose aspiration and frustration are embodied in a decrepit truck. This is during World War II. The world is changing, and people are being changed. Physically unfit for the draft, attracted to modern life and "big money," Clovis goes to Detroit and

finds a job as a "machine repair man." Gertie and their children
follow him to the city where, to Gertie, the cars seem to be
"driving themselves through a world not meant for people."
They find that Clovis has rented a disheartening, small, thin-
walled apartment, and is already in debt for a used car, a radio,
and other things that he has bought on credit.

In these circumstances, Gertie's practical good sense is depreciated nearly to nothing, except for the meaning it gives
to her grief. Back home, she had dreamed of buying, and had
almost bought, a small farm that would have given greater efficacy to her abilities and greater scope to her will. As her drasti-
cally narrowed life in Detroit closes upon her, she thinks: "Free
will, free will: only your own place on your own land brought
free will." (And now we should notice that those who have
lived in the saving way preferred by Gertie Nevels—and some
have done so—are solvent still, and Detroit is bankrupt.)

It is a small logical step from understanding that self-determination for an individual depends on "your own place on
your own land" to understanding that self-determination for a
community depends on the same thing: its home ground, and
a reasonable measure of local initiative in the use of it. This
gives us a standard for evaluating the influence of an "outside
interest" upon a region or a community. It gives us a standard
for evaluating the policy of "bringing in industry" and any industry that is brought in. Outside interests do not come in to
a place to help the local people or to make common cause with
the local community or to care responsibly for the local countryside. There is nothing at all to keep a brought-in industry in
place when the place has become less inviting, less exploitable,
or less profitable than another place.

We may not want to oppose any and all bringing in or coming in of industry, but localities and communities should insist
upon dealing for themselves with any outside interest that proposes to come in. They should not permit themselves merely
to be dealt *for* by state government or any other official body.
This of course would require effective, unofficial local organizing, and I believe we are developing the ability to do that.

But the most effective means of local self-determination
would be a well-developed local economy based upon the

use and protection of local resources, including local human intelligence and skills. Local resources have little local value when they are industrially produced or extracted and shipped out. They become far more valuable when they are developed, produced, processed, and marketed by, and first of all to, the local people—when, that is, they support, and are supported by, a local economy. And here we realize that a local economy, supplying local needs so far as possible from local fields and woodlands, is necessarily diverse.

As things now stand, the land and people of rural Kentucky are not going to be saved by the state and the federal governments or any of their agencies and institutions. All of those great official forces are dedicated primarily to the perpetuation of the corporate economy, not to new life and livelihood in small Kentucky communities. We must not make of that a reason to give up our efforts for better politics, better policy, better representation, better official understanding of our problems and needs. But to quit *expecting* the help we need from government bureaus, university administrations, and the like will give us an increase of clarity and freedom. It will give us back the use of our own minds.

For the fact is that if the land and the people are ever to be saved, they will be saved by local people enacting together a proper respect for themselves and their places. They can do this only in ways that are neighborly, convivial, and generous, but also, and in the smallest details, practical and economic. How might they do this? I will offer a few suggestions:

1. We must reject the idea—promoted by politicians, commentators, and various experts—that the ultimate reality is political, and therefore that the ultimate solutions are political. If our project is to save the land and the people, the real work will have to be done locally. Obviously we could use political help, if we had it. Mostly, we don't have it. There is, even so, a lot that can be done without waiting on the politicians. It seems likely that politics will improve after the people have improved, not before. The "leaders" will have to be led.

2. We should accept help from the centers of power, wealth, and advice only *if*, by our standards, it is actually helpful. The aim of the corporations and their political and academic

disciples is large, standardized industrial solutions to be applied everywhere. Our aim, to borrow language from John Todd, must be "elegant solutions predicated on the uniqueness of [every] place."

3. The ruling ideas of our present national or international economy are competition, consumption, globalism, corporate profitability, mechanical efficiency, technological change, upward mobility—and in all of them there is the implication of acceptable violence against the land and the people. We, on the contrary, must think again of reverence, humility, affection, familiarity, neighborliness, cooperation, thrift, appropriateness, local loyalty. These terms return us to the best of our heritage. They bring us home.

4. Though many of our worst problems are big, they do not necessarily have big solutions. Many of the needed changes will have to be made in individual lives, in families and households, and in local communities. And so we must understand the importance of scale, and learn to determine the scale that is right for our places and needs. Brought-in industries are likely to overwhelm small communities and local ecosystems because both the brought-in and the bringers-in ignore the issue of scale.

5. We must understand and reaffirm the importance of subsistence economies for families and communities.

6. For the sake of cultural continuity and community survival, we must reconsider the purpose, the worth, and the cost of education—especially of higher education, which too often leads away from home, and too often graduates its customers into unemployment or debt or both. When young people leave their college or university too much in debt to afford to come home, we need to think again. There can never be too much knowledge, but there certainly can be too much school.

7. Every community needs to learn how much of the local land is locally owned, and how much is available for local needs and uses.

8. Every community and region needs to know as exactly as possible the local need for local products.

9. There must be a local conversation about how best to meet that need, once it is known.

10. The high costs of industrial land-using technology encourage and often enforce land abuse. This technology is advertised as "labor-saving," but in fact it is people-replacing. The people, then, are gone or unemployed, the products of the land are taken by violence and exported, the land is wasted, and the streams are poisoned. For the sake of our home places and our own survival, we need many more skilled and careful people in the land-using economies. The problems of achieving this will be difficult, and probably they will have to be solved by unofficial people working at home. We can't expect a good land-based economy from people who wish above all to continue a land-destroying economy.

11. The people who do the actual work and take the most immediate risks in the land economies have almost always been the last to be considered and the poorest paid. And so we must do everything we can to develop associations of land owners and land users for the purpose of land use planning, but also of supply management and the maintenance of just prices. The nearest, most familiar model here in Kentucky is the federal tobacco program, which gave the same economic support to the small as to the large producers.

12. If we are interested in saving the land and the people of rural Kentucky, we will have to confront the issue of prejudice. Too many rural Kentuckians are prejudiced against themselves. They have been told and have believed that they are provincial, backward, ignorant, ugly, and thus not worthy to "stand in the way of progress," even when "progress" will destroy their land and their homes. It is hard to doubt that good places have been destroyed (as in the coal fields) or appropriated by hostile taking (as in Land Between the Lakes) because, in official judgment, nobody lived there but "hicks" or "hillbillies." But prejudice against other disfavored groups still is alive and well in rural Kentucky. This is isolating, weakening, and distracting. It reduces the supply of love to our needs and our work.

To end, I want to say how grateful I am to have this audience for this speech. I remember when there was no organization called (or *like*) Kentuckians for the Commonwealth, and so I know its worth. I am proud to be one of you. In speaking to

you, I've felt that I could reach, beyond several false assumptions, toward our actual neighborhoods and the actual ground
under our feet. If we keep faithful to our land and our people,
both together, never apart, then we will always find the right
work to do, and our long, necessary, difficult, happy effort will
continue.

Caught in the Middle

IN THE PRESENT political atmosphere it is assumed that everybody must be on one of only two sides, liberal or conservative. It doesn't matter that neither of these labels signifies much in the way of intellectual responsibility or that both are paralyzed in the face of the overpowering issue of our time: the destruction of land and people, of life itself, by means either economic or military. What does matter is that a person should choose one side or the other, accept the "thinking" and the "positions" of that side and its institutions and be so identified forevermore. How you vote is who you are.

We appear thus to have evolved into a sort of teenage culture of wishful thinking, of contending "positions," oversimplified and absolute, requiring no knowledge and no thought, no loss, no tragedy, no strenuous effort, no bewilderment, no hard choices.

Depending on the issues, I am often in disagreement with both of the current political sides. I am especially in disagreement with them when they invoke the power and authority of government to enforce the moral responsibilities of persons. The appeal to government is made, whether or not it is defensible, when families and communities fail to meet their prescribed moral responsibilities. Between the two moralities now contending for political dominance, the middle ground is so shaken as to be almost no ground at all. The middle ground is the ground once occupied by communities and families whose coherence and authority have now been destroyed, with the connivance of both sides, by the economic determinism of the corporate industrialists. The fault of both sides is that, after accepting and abetting the dissolution of the necessary structures of family and community as an acceptable "price of progress," they turn to government to fill the vacancy, or they allow government to be sucked into the vacuum. This, I think, explains both Prohibition and the War on Drugs, to name two failed government remedies. Another may be government-prescribed compulsory education.

To believe, as I do, that families and communities are necessary

615

despite their present decrepitude is to be in the middle and to be most uncomfortable there. My stand nevertheless is practical. I do not think a government should be asked or expected to do what a government cannot do. A government cannot effectively exercise familial authority, nor can it effectively enforce communal or personal standards of moral conduct.

The collapse of families and communities—so far, more or less disguisable as "mobility" or "growth" or "progress" or "liberation"—comes from or with the collapse of personal character and is a social catastrophe. It leaves individuals subject to no requirements or restraints except those imposed by government. The liberal individual desires freedom from restraints upon personal choices and acts, which often has extended to freedom from familial and communal responsibilities. The conservative individual desires freedom from restraints upon economic choices and acts, which often extends to freedom from social, ecological and even economic responsibilities. Preoccupied with these degraded freedoms, both sides have refused to look straight at the dangers and the failures of government-by-corporations.

The Christian or social conservatives who wish for government protection of their version of family values have been seduced by the conservatives of corporate finance who wish for government protection of their semireligion of personal wealth earned in contempt for families. The liberals, calling for too few restraints upon incorporated wealth, wish for government enlargement of their semireligion of personal rights and liberties. One side espouses family values pertaining to temporary homes that are empty all day, every day. The other promotes liberation that vouchsafes little actual freedom and no particular responsibility. And so we are talking about a populace in which nearly everybody is needy, greedy, envious, angry, and alone. We are talking therefore about a politics of mutual estrangement, in which the two sides go at each other with the fervor of extreme righteousness in defense of rickety absolutes that are indefensible and therefore cannot be compromised.

Nowhere has this callow politics asserted itself more thoughtlessly and noisily than on the so-called rights of abortion and homosexual marriage. The real issue here is the politicization of personal or private life, and inevitably, given the absence of authentic political discourse or dialogue, the reduction of the issues to two absolute positions. In addition to distracting from interests authentically public and political, the politicization of personal life, involving as it must the publicization of privacy, is inhumane and inherently tyrannical.

After Boris Pasternak's *Doctor Zhivago* was published in the West and Pasternak received the Nobel Prize in 1958, thus earning the Soviet government's reprisal, Thomas Merton wrote:

> Communism is not at home with nonpolitical categories, and it cannot deal with a phenomenon which is not in some way political. It is characteristic of the singular logic of Stalinist-Marxism that when it incorrectly diagnoses some phenomenon as "political," it corrects the error by forcing the thing to *become* political.

Now, after many decades of anticommunism, Merton's sentences have come remarkably to be descriptive of our own politics. Maybe people who focus their minds for a long time upon enmity finally begin to resemble their enemies. This has happened before. It is deeply embedded in the logic of warfare.

Whatever the cause, we seem to have become as adept as the old Soviet Union at politicizing the nonpolitical. Most notably, by the connivances of both political sides, we have invented a sexual politics, which, by the standards of our own political tradition, is a contradiction in terms. Or it is if there is to be a continuing political distinction between public life and private life. This distinction, after all, is the basis of the freedoms protected by the First Amendment, which holds essentially that people's thoughts and beliefs are of no legitimate interest to the government. The government is not in charge of our personal lives, our private affections, our prayers, or our political opinions. It is not in charge of our souls. Those who formed our government also limited it, forbidding it any freehold in our homes or in our minds.

I am as ready as any so-called conservative to worry about big government, though I would remember that government has gotten big in the much-needed effort to regulate big corporations and to help their victims. To my fear of big government I add my at least equal fear of unlimited government, which is to say total government. It is not entirely surprising that after our long, costly resistance to communist dictatorship, we should now see the rise of passions and excuses tending toward capitalist dictatorship. The most insidious of these passions tends toward state religion and government regulation of private behavior.

Sexual politics has to do with public disagreements about rights that, however valid, are newly proclaimed, obscure in origin, extremely controversial, and productive of conflicts that probably are not politically resolvable—the prime example being the apparently unendable conflict over abortion.

Not so long ago abortion was illegal in the United States. It was illegal, one must suppose, because of an innate aversion to a woman's destruction of her own child. And then the Supreme Court ruled in *Roe v. Wade* (1973) that abortion, within certain limits, was legal. The ruling is based on the right to privacy under the 14th Amendment and, more remarkably and controversially, on the proposition that a human fetus is not legally a person and therefore is not eligible for the protections guaranteed by that amendment. This distinction between a fetus and a person is, to common sense, arbitrary and therefore inevitably a source of trouble. *Fetus*, to begin with, is a technical term which once was rarely used by pregnant women, who had conventionally and naturally referred to the creature forming in their wombs as a *baby*, which is to say a human being, a person. The abortion debate involves endless, unendable disagreement about such issues as when a fetus becomes a human or a person, when life begins, when or whether abortion should be legal, whether we should call it "killing" or "termination." Some enlightened people hold in derision the idea that life begins at conception. But if life does not begin at conception, then we are at the beginning of a kind of sophistry: an argument about when life may be *said* to begin.

The right to have an abortion has been popularly justified as a woman's right to control her own body. Such a right seems to be implied by a number of other rights, but only recently has it been stated in this way. So stated, it is somewhat confusing, for many of our laws, legal and moral, *require* one to control one's body—to restrain it, for instance, from killing the body of another person, except of course when ordered to do so by the government. To say when and why a requirement may become a right, and when and why the requirement or the right should be suspended or opposed, needs a lot of spelling out—if such a spelling out is possible.

The facts remain, on one side, that abortions are still proscribed by some religious traditions and the old aversion is still felt by many people, and, on the other side, that the legalization of abortion answers a need desperately felt for real and pressing reasons by many women, and legal abortion would at least put an end to illegal abortions badly performed in bad circumstances by incompetent and disreputable people.

Also involved are questions of ultimate seriousness and importance: questions of life and death that exceed the competence of human intelligence and are forever veiled in mystery. The trains of causation run quickly out of sight. I know a man who said, plausibly, "Life begins with erection." Elders used to refer young people to a time "when you were just a look in your mother's eyes." But when I asked the geneticist Wes Jackson, "Does life begin at conception?" he replied, "Life *continues* at conception." This, I felt, was at last a statement sufficiently serious.

In making any choice, we choose for the future, and so all our choices involve us in mystery and in a kind of tragedy. To choose to have a baby, to abort a fetus, to save a life, to destroy a life is to make a whole change on the basis of partial knowledge. One chooses in light of what one knows now about the past and thus changes the future inevitably and forever. What would have been, had the choice been different, will never be known.

To reduce this complexity and mystery to a public contest between two absolutes seems to wrong everything involved. Some equivocation seems natural and appropriate because one is attending to two possibilities, both unknown. Saints, heroes, great artists and scientists began as fetuses. So did tyrants, torturers, and mass murderers. Choices do not invariably cut cleanly between good and evil. Sometimes we poor humans must choose between two competing goods, sometimes between two evils. Responsibility or circumstances will require us to choose. But we cannot choose to be unbewildered or not to grieve.

The theologian William E. Hull, worrying over the destructive animosities that divide religious organizations, asked, "How can we avoid the wrangling that breeds hostility?" And he answered: "By seeking clarity rather than victory." This sounds exactly right to me, and I find little clarity in the public argument about abortion. I know that both sides are made up of individual humans whose thoughts and feelings may differ in significant ways from the public positions of their sides. But the problem with those positions as they are generalized and vented into the political atmosphere is that they substitute simplicity for clarity. By separating the statistical facts of abortion from the lived experience—from the mystery, bewilderment, and suffering that attend it—the simplicity becomes obscure and heartless. To the proabortion side, abortion is simply a right, the creature to be aborted is a fetus, the act itself is termination of a pregnancy by a forthright medical procedure. To the antiabortion side, abortion is simply a wrong to be refused or opposed in obedience to a moral or religious law that ought to be the law of the land. The two sides seem about equally to disregard both the truth of human suffering and the possibility of human compassion. Sexual politics is overflowing with principles and abstractions, but otherwise seems deserted. No actual woman wanting or needing or refusing an abortion is present.

The issue, I think, can be clarified only by imagining a woman to whom an abortion is one of two heartbreaking alternatives, one of which she alone must choose, and between which, however she chooses, she will remain emotionally divided perhaps for the rest of her life. This woman, troubling as she is to the political atmosphere of opposed absolutes, cannot be admitted by either side into the public argument. But her example is starkly clarifying. Her absence from the argument stupefies both sides.

I am unsure of the whereabouts or even the possibility of truth in the abortion strife, but I, with perhaps a good many others, am somewhere in the middle, where I see no chance of a public reconciliation. In fairness, we have to acknowledge that within the experience and history of abortion there must be many shades and mixtures of right and wrong. As in the human condition generally, we are not dealing with a choice between a shadowless light and utter darkness.

I have said several times that I am opposed to abortion except when it is necessary to save the mother's life. I stick to that, for I still feel strongly the old aversion. Unlike the proabortion side, I think that abortion is killing. What else could it be? And I think that the creature killed is a human being, for it can be a being of no other kind, and it is not a nonbeing. But I feel just as strongly an aversion to our life-destroying economy and way of life, and every day increases our need to cherish life in all its forms. I oppose the official killings that bear the names of justice and defense and also the killing that is a cost or by-product of certain industrial enterprises. I do, however, recognize the cruelty that is inherent and inescapable in the life of this world, in which no creature lives but at the expense of other creatures, as I recognize right and wrong ways of exacting and recompensing such costs.

But when I have spoken of my opposition to abortion, I hope I have never neglected to say that I can imagine circumstances in which I would willingly aid and comfort a girl or a woman getting an abortion. And here I arrive at what is for me the moral difficulty, even the moral obscurity, of this problem: Though I can say that, in some circumstances, I would willingly help somebody get an abortion, I cannot say that I would willingly aid and abet a murder.

Whatever one may think of a woman's right to control her own body, the inexpressibly intimate involvement of her own body in a woman's decision to have an abortion is a real and urgent consideration, and for a man it is a special one. That it does not involve, and could never have involved, my body does not invalidate my belief that abortion is wrong, but it does require me to be carefully aware of the bodily difference. Whereas a person's demonstrated willingness to kill another person already born requires us to look upon that killer as a public menace, a woman's decision to kill the baby in her womb does not require us to look upon her as a menace to anybody else. In fact we *don't* look upon her in that way.

So far as I can see, there are four possible legislative choices in relation to the abortion controversy:

1. Abortion could be forbidden absolutely, with no exceptions.
2. It could be forbidden, with specified exceptions.
3. It could be permitted, with specified exceptions.
4. We could permit it without exception, which to me means that we would have no law related specifically to abortion.

The first of these would outrage the proabortion side, it would settle the controversy merely by ignoring it, it is immitigably harsh, and it makes little sense. Absolute forbidding would choose the life of the unborn child over any and all other considerations, including that of the mother's life. The government thus would abandon any obligation to protect the mother's life in order to protect the life of the child. If, for want of an abortion, mother and child *both* should die, then the state would accomplish no good at all except for the pacification of fanatics.

Any law forbidding abortion would be ineffective, and it could easily do harm. To forbid an established practice for which the demand is widespread and the supply dispersed and readily available would be virtually to license an illicit and lucrative economy that would reward the greed and enterprise of the worst people. It would repeat the futility of the War on Drugs and Prohibition. Such a situation undermines government authority and brings law enforcement into disrespect.

The two middle solutions, as opposed to an outright ban, would require niggling official regulation of the conduct of individual persons, conduct at least semiprivate. This would require an increase in police power that would be expensive and also a danger to everybody's freedom. We could, for example, make a law forbidding abortion except to save the mother's life, but what would we mean by "the mother's life"? Would it be denoted only by her vital signs or, more reasonably, by her ability to live and thrive in the world—in which case the definition of her life would include her economic life, the life of her family (if she has one), even the life of her community (if she has one). For another example, we could make a law permitting abortion except during the third trimester. But this would require a lot of official watching. And who is to say exactly when the third trimester begins? Such legislating can only strand everybody, including the government, in permanently painful and dangerous confusion.

The problem in forbidding or permitting with exceptions is that the exceptions apparently cannot be decided upon by precise determinants, but rather by "approximate" or "appropriate" judgments by experts. The language of *Roe v. Wade*, as the ruling implicitly acknowledges, is vague and uneasy. What exactly is meant, with respect to abortion, by *life*, *conception*, *viability*, *privacy*, and *person*? *Roe v. Wade* does not, to my mind, settle the meaning of any of those words. The legal definition of a person evaporated when the Supreme Court defined a corporation as a person. If a corporation is a person, contrary to all previous usage and to common sense, then personhood can be conferred upon virtually anything merely by decree. Issues are thus quickly carried not just into vagueness but beyond the bounds of language.

I am going to take the risk, therefore, of saying that there should be *no* law either for or against abortion. Like certain other wrongs—various addictions, let us say—this one is more personal than public and would be best dealt with by the persons immediately involved: the pregnant girl or woman, her family or her friends, her doctor.

This is my attempt to make a statement on abortion that is reasonably complete—and that, in result, may be necessarily incomplete. I should add that I may find further reasons that will require me to revise. To have a mind, I think, depends upon one's willingness to change it.

Regarding homosexual marriage, the fault that I again must acknowledge is that what I have said before has been incomplete. As far as I remember, I have made only two public statements about this issue. My argument, much abbreviated both times, was that sexual practices of consenting adults ought not to be subjected to the government's approval or disapproval and that domestic partnerships, in which people who live together and devote their lives to one another, ought to receive the spousal rights, protections, and privileges that the government allows to heterosexual couples.

In those two statements I was considering homosexual marriage as an issue of law—with reference to the contention from both sides that marriage is a right to be granted or withheld by the government. This puts me again in the middle but this time with more certainty of my whereabouts and with good reasons to object.

First, this "right to marriage" is still birth-wet. It exists only to be selectively withheld. Apart from its momentary political expediency there is no reason for it. Whatever one may think of all that is presently implied and entailed by the legalization of marriage, surely nobody can claim that marriage is either the government's invention or that the government has an inherent right to determine who may marry. A government that can forbid two women or two men to marry might with better reason forbid two bigots to marry.

Second, this right depends upon a curious agreement between liberals and conservatives that human rights originate in government, to be dispensed to the people according to their pleading and at the government's pleasure, implying inescapably that any right, being so expediently the government's gift, can just as expediently be withheld or withdrawn. This flatly contradicts the founding principle of American democracy that human rights are precedent to the government's existence, that the government is established to protect them, and that the government must be restrained from violating them.

Third, it cannot be allowable, under the above principles, for the government, on the pleading of *some* of the people, to establish a right solely for the purpose of withholding it from some other people. If this were to happen, it would amount to a punishment imposed on a disfavored group for no crime except their existence. I don't need to point out that this has happened before.

That the liberals, who so often have been rightly anxious about the protection of rights and liberties, should define those rights and liberties as the gifts of a generous and parental government is absurd.

The conservative program on this issue, promoting as it does a constitutional apportionment of rights according to sexual category, in implicit violation of the Fifteenth and Nineteenth Amendments, is more darkly absurd. The theory that accreditation of the sexual practices of individuals is a function proper to a "small" and noninterfering government is comical.

That homosexuals have been denied the right to marry, supposing for the moment that such a right can exist, surely violates the Fourteenth Amendment, which forbids the state to "deprive any person of life, liberty, or property, without due process of law; [or] to deny to any person within its jurisdiction the equal protection of the laws." There is no need for homosexuals to be granted a right to marry that is at all different from the right of heterosexuals to do so. There is no good reason for the government to treat homosexuals as a special category of persons.

To support their strategy of outlawing homosexual marriage, Christians of a certain disposition have found several ways to categorize homosexuals as a different kind from themselves, who are in the category of heterosexuals and therefore normal and therefore good. They are mindful that the Bible looks upon homosexual acts as sinful or perverse. But it is not clear to me why perversion should have been specifically assigned to homosexuality. The Bible has a lot more to say against fornication and adultery than against homosexuality. If one accepts the 24th and 104th Psalms as scriptural norms, then surface mining and other forms of earth destruction clearly are perversions. If we take the Gospels seriously, how can we not see industrial warfare and its unavoidable massacre of innocents as a most shocking perversion? By the standard of all scripture, neglect of the poor, of widows and orphans, of the sick, the homeless, the insane, is an abominable perversion. Jesus taught that hating your neighbor is tantamount to hating God, and yet some Christians hate their neighbors as a matter of policy and are busy hunting biblical justifications for doing so. Are they not perverts in the fullest and fairest sense of that term? And yet none of these offenses, not all of them together, has made as much political-religious noise as homosexual marriage.

Another way to categorize and isolate homosexuals from the general citizenry and the prerogatives of citizenship is to define homosexuality as a disease having a cause that can be discovered and removed or cured by some sort of therapy. This seems most promising as long-term job security for scientists and doctors. Ken Kesey once saw an inscription in a men's room: "My mother made me a homosexual." Under it somebody else had written: "If I gave her the yarn would she make me one?" My own speculation is that we will never do much better than that. We will discover that, like all the rest of us, homosexuals are made what they are by their mothers, their fathers, their genes, their germs, their upbringing and their education, by their friends and neighbors, their dwelling places, their time and its culture, by their economic and social status, their personal history, and by history.

Yet another such argument is that homosexuality is unnatural. If the nature in question is merely biological—the realm of the ape and the naked ape—that may prove too roomy and accommodating to be of much help. By the standard of that nature, monogamy is unnatural, an artifact of *some* cultures. If it is argued that homosexual marriage cannot be reproductive, is therefore unnatural and should be forbidden, must we not then argue that any childless marriage is unnatural and should be annulled?

Specifically *human* nature, by contrast, has always had a definition more complex and demanding than that of a naked ape. William Blake thought we are made human by being made in the image of God:

> For mercy, pity, peace, and love
> Is God our father dear;
> And mercy, pity, peace, and love
> Is man, his child and care.

Are homosexuals capable of mercy, pity, peace, and love? Some certainly are, as some heterosexuals certainly are. To deny that distinction to homosexuals is to deny categorically that they are human, which is hardly a proper employment for mercy, pity, peace, and love. Oversimplified moral certainties—always requiring hostility, always potentially violent—isolate us from mercy, pity, peace, and love and leave us lonely and dangerous. The only perfect laws are absolute, but perfect laws are only approximately fitted to imperfect humans. That is why we have needed to think of mercy, and of the spirit, as opposed to the letter, of the law.

One may find the sexual practices of homosexuals to be unattractive or displeasing and therefore unnatural. But anything that can be done in that line by homosexuals can be done, and is done, by heterosexuals. Do we need a political remedy for this? Would conservative Christians like a small government bureau to inspect, approve, and certify their sexual behavior? Would they like a colorful tattoo, verifying government approval, on the rumps of lawfully copulating persons? We have the technology, after all, to monitor everybody's sexual behavior, but so eager an interest in other people's most private intimacy is both prurient and totalitarian.

The oddest of the strategies to condemn and isolate homosexuals is to propose that homosexual marriage is opposed to and a threat to heterosexual marriage—as if the marriage market is about to be cornered and monopolized by homosexuals. If this is not industrial-capitalist paranoia, it at least follows the pattern of industrial-capitalist competitiveness, according to which you *must* destroy the competition. If somebody else wants what you've got—from money to marriage—you must not hesitate to use the government (small, of course) to keep them from getting it.

But if heterosexual marriage is so satisfying to heterosexual couples, why can they not just reside in their satisfaction? So-called traditional marriage, now mostly divested of a traditional household and traditional bonds to a community, is for sure suffering a statistical failure, but this is not the result of a homosexual plot. Heterosexual marriage does not need defending. It only needs to be practiced, which is pretty hard to do just now. But the difficulty is rooted mainly in the values and priorities of our industrial-capitalist system, in which every one of us is complicit.

It certainly is possible for a government to withhold the legal perquisites of marriage from any group that it may be persuaded to designate in our present civil cold war. That is mainly to say that a government can forbid its officers to license weddings for people in the designated group.

But a wedding is not a marriage. A wedding is traditionally an exchange of vows of fidelity and love in all circumstances until death. In some circumstances, for some people, a wedding may be a sacrament. But however complicated and costly the preparations, the costumes, the photography, and the reception, a wedding is over and done with in a few minutes.

A marriage, by contrast, is the *making* of marriage, by daily effort to live out the vows until death. The vows may be taken seriously or not, broken or not, but there is no way of withholding them from homosexuals. You cannot copyright the vows, which a homosexual couple is perfectly free to make. The government cannot forbid them to do so, nor can any church.

Conjugal love, Kierkegaard wrote,

> is faithful, constant, humble, patient, long-suffering, indulgent, sincere, contented, vigilant, willing, joyful. All these virtues have the characteristic that they are inward qualifications of the individual. The individual is not fighting with external foes but fights with himself. . . . Conjugal love does not come with any outward sign . . . with whizzing and bluster, but it is the imperishable nature of a quiet spirit.

No church can *make* a homosexual marriage, because it cannot make any marriage, nor can it withhold any degree of blessedness or sanctity from any pledged couple striving day by day to be at one. If I were one of a homosexual couple, the same as I am one of a heterosexual couple, I would place my faith and hope in the mercy of Christ, not in the judgment of Christians.

Condemnation by category is the lowest form of hatred, for it is cold-hearted and abstract, lacking the heat and even the courage of a personal hatred. Categorical hatred is the hatred of the mob, which makes cowards brave. And there is nothing more fearful than a religious mob overflowing with righteousness, as at the crucifixion, and before, and since. This sort of violence can happen only after we have made a categorical refusal of kindness to heretics, foreigners, enemies, or any other group different from ourselves.

Kindness is not a word much at home in current political and religious speech, but it is a rich word and a necessary one. There is good reason to think that we cannot live without it.

Kind is obviously related to *kin*, but also to *race* and to *nature*. In the Middle Ages *kind* and *nature* were synonyms. *Equal*, in the famous phrase of the Declaration of Independence, could be well translated by these terms: All men are created kin, or of a kind, or of the same race or nature.

Jesus saves the life of the woman taken in adultery by removing her from the category in which her accusers (another mob) have placed her and placing her within kindness, his own kindness first and then that of her accusers: "He that is without sin among you, let him first cast a stone at her."

The accusers take this kindness as a defeat, as we all are too likely to do, and they depart without another word. The brief dialogue that follows is wonderfully animated by Jesus's sense of humor:

> "Woman, where are those thine accusers? hath no man condemned thee?"
> "No man, Lord."
> "Neither do I condemn thee: go, and sin no more."

Good advice—but can we suppose he could have given it without smiling, knowing as he did the vast repertory of sins and the endless human susceptibility?

Within the larger story of the Gospels this story is not exceptional. It does show us Jesus's way of dealing with one of the biblically denominated sins, but he simply reaches out to the woman in her great need as he did many times to many others. In the Gospels the sinfulness of all humans is assumed. It is neediness that is exceptional, and in Jesus's ministry need clearly takes a certain precedence over sin. His kindness is best exemplified by his feedings and healings with no imputation at all of deserving.

But the wealth of this idea of kindness is not exhausted by kindnesses to humans. It is far more encompassing. From some Christians as far back as the twelfth century, certainly from farther back in so-called primitive cultures, and from some ecologists of our own time, we have the idea of a great kindness including and binding together all beings: the living and the nonliving, the plants and animals, the water, the air, the stones. All, ultimately, are of a kind, belonging together, interdependently, in this world. From the point of view of Genesis 1 or of the 104th Psalm, we would say that all are of one kind, one kinship, one nature, because all are *creatures*.

Much happiness, much joy, can come to us from our membership in a kindness so comprehensive and original. It is a shame, as I know from long acquaintance with myself, to be divided from it by the autoerotic pleasure of despising other members.

Our Deserted Country

IF WE ARE to understand the history of our landscapes, which mostly are economic—farms, ranches, working forests, mines —we will have to begin by understanding the impetus and motive of the Industrial Revolution. Many people, still, will regard this suggestion as odd or unthinkable, though even some economists have begun to acknowledge what has always been obvious: One result of replacing human workers by industrial technologies is "joblessness."

But joblessness has been exactly the aim of industrialization from its beginning, as the so-called Luddites understood immediately. The purpose of industrial technology has always been to cheapen work by displacing human workers, thus increasing the flow of wealth from the less wealthy to the more wealthy. We have dealt with the violence always implicit in these substitutions by disregarding it, or disguising it by an official, quasi-religious litany of synonyms: *labor-saving, efficiency, progress, convenience, speed, comfort,* even *creative destruction.* Maybe we have to grant the possibility of some degree of altruism. Painless dentistry is often invoked as a justification of technological progress, and surely we must concede that painlessness is preferable to pain, just as comfort is preferable to discomfort. But what might be the costs of the "God-wrought" painlessness and comfort that have come with industrialization? "There is no such thing as a free lunch," the hard-headed realists love to advise us, implying that they have completely "done the numbers." But actually in all the industrial world the least popular mathematical operation is subtraction. We habitually imply that all the gains of technological progress are entirely net. *Our* painlessness and comfort are *everybody's* painlessness and comfort. Nobody pays, nobody loses.

But obsolescent human workers, characteristically, have been both replaced and displaced. The costs of progress have routinely been borne by discarded workers, and often the costs have been exorbitant. Of the outmoded coal miners of eastern Kentucky, Harry Caudill wrote:

> Many voices have decried [mechanization], contending
> . . . that when machines relegate men to idleness their
> condition will be worse than before. There is a certain
> logic to this reasoning, but it is a logic hard to swallow
> when one watches a . . . bucket-excavator . . . dig-
> ging a ditch at a rate scores of men could not maintain.
> But what is the result when the machines sweep through
> an industry in a short time, replacing . . . fathers, sons
> and even grandsons . . . so that they simultaneously
> lose their livelihoods and all claim to status and stand-
> ing? What happens to entire communities leveled by such
> traumas? What befalls the psyche and spirit of people so
> afflicted . . . ?

Such questions have never burdened the consciences of coal
companies, and have seldom disturbed the sympathy of politicians. A society dominantly industrial has no effective means
either of democratizing the gains or of ameliorating the human
costs of industrialization. The fate of workers is abandoned to
"the market" and "the labor market."

Of the eastern Kentucky coal miners one needs to add that
when they moved into the "coal camps" they left not only their
homes but also the families, neighborhoods, and land-based
household economies that had supported them at home. They
thus became entirely dependent on the money economy and
their wages. Once progress had taken away their wages, they
became entirely dependent on "welfare," or they migrated
in search of jobs to the cities, where again they were wage-
dependent and now in circumstances for which their rural experience had in no way prepared them.

Similarly, when industrial machinery and chemicals came into
the sugarcane and cotton lands of the South, the mostly black
field hands and their families became immediately obsolete,
useless and wageless, with no recourse but to take their chances
in urban ghettos. They too had to move from rural homes,
communities, and a land-based subsistence—in which, however poor, they had lived with a long-established competence
—into situations alien to their experience and abilities.

Those who have been in this way uprooted, replaced and

displaced, and who by their sufferings pay far more than their share of "the price of progress," are then subjected to versions of political regard dependent on the same social disconnection, and about equally irresponsible. The "conservatives," eagerly serving both God and Mammon, have concluded much to their convenience that the poor are poor or "jobless" because they are lazy, requiring only the incentive of starvation, or the starvation of their children, to become "productive members of society." The "liberals," serving an abstracted and oblique political compassion, assume that poverty and joblessness can somehow be corrected by the tiding over or new start supposedly granted by public charity and government "programs."

These attitudes issue from the great blank in the political-industrial mind that has forgotten, if it ever knew, the public and political value of securing for all citizens a reasonable permanence of dwelling place and vocation, which depends upon the widespread ownership of small farms and other economically supportive small properties. Lacking that fundamental connection, individuals, families, and neighborhoods can originate nothing in their own support, but become helplessly dependent on the money economy and the government, as now in fact we all are. Lacking that connection, the public economy becomes little more than a financial system irrelevant to economic life and to need, as ours now has done.

The following sentences speak as well as any I have ever read of the essentially democratic connection between people and land. The writer praises

> the spirit of independence which is generated in countries where the free cultivators of the soil constitute the major part of the population. It can scarcely be imagined how proudly man feels, however small his property may be, when he has a spot of arable land and pasture . . . [H]is independence being founded on permanent property, he has an interest in the welfare of the state, by supporting which he renders his own property more secure, and, although the value of the property may not be great, it is every day in his view . . . Those who wish to see only the two castes of capitalists and day-labourers, may smile at this union of independence and poverty.

That statement is a part of Alexander Mackenzie's assessment of the "nineteenth-century clearances" of the small holders of the Scottish Highlands. It is a statement that Thomas Jefferson would have recognized and honored. So would Virgil and several of the authors of the Bible. So would have my father and many other rural Kentuckians whose minds were formed a generation before World War II. They and many others spoke for an authentic, ancient human need to have and to belong to a piece of land, however small or poor, to live on and from, and to care for, a place offering a significant measure of life-support to themselves and their families and, as needed, to their neighbors. This is in no way like the land-greed of "them that join house to house, that lay field to field . . . that they may be placed alone in the midst of the earth!"

The Scottish Highlanders who were "cleared" from their small holdings, to make way for landlords' sheep pastures and game preserves, were abandoned mostly to the choice between emigration and starvation. As I have already suggested, their fate was not unique. Such regardless harm was the deliberate result also of England's Enclosure Acts. The same dispossession of country people took place in Stalin's Russia, and is taking place now in China. The resemblance of these developments to our own clearances of the American Indians is plain enough. The modern American version, in addition to the casual substitution of machines for field hands, and also following World War II, was the semi-official agricultural policy of "too many farmers" or "Get big or get out." It was semi-official because no act of Congress gave it a legal basis. Its basis was the "expert" opinion of corporate, academic, and political leaders. The relatively self-sufficient producers on small farms, according to this opinion, needed to become members of the industrial "labor force" and consumers of industrial commodities. Reducing the number of farmers and farms became a devastatingly successful political goal. The "free market" was allowed to have its way, which meant, among much else, that the primary producers in the industrial food economy would buy dear and sell cheap. By now nearly all of the land-using population have left their farms and home places to be industrially or professionally employed, or unemployed, and to be entirely dependent on the ways and the products of industrialism. Or they have remained, as "farmers," to pilot enormous machines over thousands of acres continuously in annual row crops such as soybeans and corn.

From earliest times we have known, if we were willing to know, having learned by experience and example, that when people are disconnected from their land they suffer. But that is only half of the truth. The other half is that when its rightful people, the people who rightfully care for it, are absent from it, the land suffers. It is the mutual, indivisible suffering of land and people that sets in right perspective the suffering of either.

The first problem of a drastic reduction of the land-using population is to keep the land producing in the absence of the people. The expert solution followed unsurprisingly the expert doctrine of "too many farmers." At the end of World War II, the war industries conveniently could "gear up" to adapt the mechanical and chemical technologies of war to the "needs of agriculture." The departing farm families would be replaced by the re-rigged war technologies, depending upon a seemingly limitless bounty of "natural resources," mainly ores and fuels. Agriculture would become an industry. Farms would become factories, like other factories ever more automated and remotely controlled. Industrial land use thus has become a front in a war against the living world. For the remaining fewer and fewer farmers, this has required a shift of interest from husbanding the fertility of the land to the various means of consuming the fossil fuels—with consequences perhaps foreseeable even by the experts, had their eyes been open.

But there was a further problem that the experts did not recognize then, and have not recognized yet: Agricultural production without land maintenance can lead only to exhaustion. Land that is in use, if the use is to continue, must be used with care. And care is not, it can never be, an industrial product or an industrial result. It cannot be prescribed or enforced by a market, free or unfree. Care can come only from what we used to understand as the human heart, so called because it is central to human concerns and to human being. The human heart is informed by the history of care and by the need for it, also by the heritage of the skills of caring and of caretaking.

The replacement of our displaced farm families by technologies derived from warfare has involved, beyond appeal, a supposedly acceptable, and generally accepted, violence against land and people. By it we have established an analogy between land use and war that has remained remarkably consistent to our present wars with their transferable "precision" technologies of remote observation and control. The common theme is a terrible pragmatism that grants an absolute predominance of the end over the means, in oblivion or defiance of any natural or moral law that may stand in the way. In the industrial land economies, from agriculture to mining, anything coming from the land that cannot be sold is treated as an enemy, and this includes natural and human communities.

The quickest way is the best way: This is the industrial version of efficiency. The industrial economy thrives on the rapidest possible changes of technology and fashion. Industrial land use thrives, or its suppliers thrive, by ever-swifter passages of machines over the land. Anything obstructing or reducing speed must be cleared away. To realize this highest aim of industrial agriculture, everything must give way to the rule of the widest expanse and the straightest line. Every surviving woodland, every tree, every fencerow must be removed. So must the animals, their pastures and pens. So must the surplus people and their buildings. Streams must be straightened and ponds filled. Every acre that will support a tractor must be cropped. Such use of the land is determined entirely by the market, and is limited entirely by the capabilities of the available technology. Questions relating to ecological and human health, or to the health of the local economy, are easily ignored because there are no industrial answers to such questions.

The rhetoric and indignation of conservationists often leads to stereotyping and condemnation by category. I need now to be careful to avoid that. From the prevalence of villainous ways of land use, it does not necessarily follow that all industrial farmers and foresters are villains. For the sake of fairness and in the interest of valid remedies, we need to understand how economic systems and constraints, plus the availability of shortcutting technologies, incline or attract or force land users toward abuse. We need to understand how abuse is favored or rewarded in the absence of sound and conserving land-use policies, and of any informed public concern and discussion that might lead to such policies.

Moreover, though a large portion of our remnant of farmers are now fully industrialized and their "operations" more extensive than ever before, I believe that they have inherited the economic standing of farmers in general and nearly always. For reasons usually of scarcity or of crop failure elsewhere, some of them sometimes experience a good year or a few good years, but their future is never reasonably secure, and in the industrial era their children are less and less likely to become farmers. Though the corporations that supply "industrial inputs" to farmers and the corporations that purchase "farm products" may prosper exceedingly, the farmers themselves, the people who bear the primary financial burdens and who do the actual work, will be the last considered, the least respected, and the lowest paid.

And so however severe may be the current abuses of the land, and however urgent the need for conservation, we have got to bear in mind that the land will not be well used, because it cannot be, by people who cannot afford to use it well. I recently attended a meeting on agriculture and water quality, at which a thoughtful farmer made a point shockingly obvious. What he said ought, in reason, to have ended the meeting, though of course it did not. He said that keeping animal manure out of the streams makes sense, it makes both agricultural and economic sense, "but it is hard to think of your environmental responsibilities when you're wondering who will be the next family to live in your house."

To make as much sense as I can of our predicament, I must turn now to Wes Jackson's perception that for any parcel of land in human use there is an "eyes-to-acres ratio" that is right and necessary to save it from destruction. By "eyes" Wes means a competent watchfulness, aware of the nature and the history of the place, constantly present, always alert for signs of harm and signs of health. The necessary ratio of eyes to acres is not constant from one place to another, nor is it scientifically predictable or computable for any place, because from place to place there are too many natural and human variables. The need for the right eyes-to-acres ratio appears nonetheless to have the force of law.

Economic landscapes, in short, require the most careful watching. And "careful" is the right adjective here: People who don't care, or know enough to care, or care enough to know, don't watch. And here I need to add an indispensable comment from the biologist Robert B. Weeden, who read an earlier attempt at this essay:

> [T]he essential eye is both attentive and loving. I think you use the term caring, which can have the same meaning as loving. However, one may have a variety of motives for caring, including looking after an investment. The attentiveness needed is a whole and balanced thing, not allowed to skew over into mere analysis. Narrow focus is necessary but not sufficient. As to loving, it must never be left alone, at least in your context of relation to place. Romantics of the eighteenth and nineteenth centuries loved rural places, but as ideas and ideals—as projections of perfection. Someone has to check whether the lowing herd is chopping the life out of the saturated lea!

What is necessary and attractive here is the introduction of the idea of a practical and practicing love. Reading Bob Weeden's letter caused me to think again of something I have, from my own experience, begun to know: how intimately related, how nearly synonymous, are the terms "love" and "know," how likely impossible it is to know authentically or well what one does not love, and how certainly impossible it is to love what one does not know.

To anybody who takes seriously the eyes-to-acres ratio, and who is carefully and lovingly watching, it is apparent that the working landscapes of our country are now virtually deserted. In the vast, relatively flat acreages of the Midwest now given over exclusively to the production of corn and soybeans, the number of farmers is lower than it has ever been. I don't know what the average number of acres per farmer now is, but I do know that you often can drive for hours through those corn-and-bean deserts without seeing a human being beyond the road ditches, or any green plant other than corn and soybeans.

Any people you may see at work, if you see any at work anywhere, almost certainly will be inside the temperature-controlled cabs of large tractors, the connection between the human organism and the soil organism perfectly interrupted by the machine. Thus we have transposed our culture, our cultural goal, of sedentary, indoor work to the fields. Some of the "field work" is now done by airplanes.

This contact, such as it is, between land and people is now brief and infrequent, occurring mainly at the times of planting and harvest. The speed and scale of this work have increased until it is impossible to give close attention to anything beyond the performance of the equipment. The condition of the crop of course is of concern and is observed, but not the condition of the land. And so the technological focus of industrial agriculture, by which species diversity has been reduced to one or two crops, is reducing human participation ever nearer to zero. Under the preponderant rule of "labor-saving," the worker's attention to the work *place* has been effectively nullified even when the worker is present. The "farming" of corn-and-bean farmers—and of others as fully industrialized —has been brought down from the complex arts of tending or husbanding the land to the application of "purchased inputs" according to the instructions conveyed by labels and operators' manuals.

Almost suddenly in the last three years the Midwestern version of corn-and-bean farming has invaded the highly vulnerable sloping fields of my part of Kentucky—where the officially recommended and encouraged "no-till" farming, dependent upon herbicides, does not, as claimed, prevent erosion, and especially not under continuous cropping. (Until recently, weed control in crops was accomplished by plowing or other mechanical means of stirring and loosening the soil. The no-till method controls weeds, instead, by poisoning them with herbicides. Loosening the soil, of course, makes it vulnerable to erosion, the vulnerability increasing with the steepness of the land. The selling point that no-till stops erosion thus seems to justify planting on land that is too steep.) This expansion of the "cutting edge" has been caused by high grain prices, caused in turn by the officially recommended and encouraged production of "biofuels," supposedly sustainable but not so by ecological standards, and doubtfully so even economically.

We can suppose that the eyes-to-acres ratio is approximately correct when a place is thriving in human use and care. The sign of its thriving would be the evident good health and diversity, not just of its crops and livestock but also of its population of native and noncommercial creatures, including the community of creatures living in the soil. Equally indicative and necessary would be the signs of a thriving local and locally adapted human economy.

The great and characteristic problem of industrial agriculture is that it does not distinguish one place from another. In effect, it blinds its practitioners to where they are. It cannot, by definition, be adapted to local ecosystems, topographies, soils, economies, problems, and needs.

The sightlessness and thoughtlessness of the imposition of the corn-and-bean industry upon the sloping or rolling countryside hereabouts is made vividly objectionable to me by my memory of the remarkably careful farming that was commonly practiced in these central Kentucky counties in the 1940s and 1950s—though, even then, amid much regardlessness and damage. The best farming here was then highly diversified in both plants and animals. Its basis was understood to be grass and grazing animals; cattle, sheep, hogs, and, during the 1940s, the workstock, all were pastured. Grain crops typically were raised to be fed; the farmers would say, "The grain raised here must *walk* off." And so in any year only a small fraction of the land would be plowed. I knew an excellent farmer of that older kind who thought that only about 5 percent of most upland farms could be safely broken for row crops (best by rotating from sod and back to sod) in any year. This was farming fitted to the land, as J. Russell Smith said it should be. And the commercial economy of the farms was augmented and supported by the elaborate subsistence economies of the households. "I may be sold out or run out," the farmers would say, "but I'll not be *starved* out."

My brother recently reminded me how carefully our father thought about the nature of our home countryside. He had witnessed the ultimate futility—the high costs to both farmer and farm—of raising corn for cash during the hard times of the 1920s and 1930s. He concluded, rightly, that the only crop that could be raised here both abundantly and profitably in the long run was grass. That was because we did not have large acreages that could safely be used for growing grain, but our land was aboundingly productive of grass, which moreover it produced more cheaply than any other crop. And the grass sod, which was perennial, covered and preserved the soil the year round.

A further indication of the quality of the farming here in the 1940s and 1950s is that the Soil Conservation Service was more successful during those years than it would or could be again in the promotion of plowing and terracing on the contour to control soil erosion. Those measures at that time were permitted by the right scale of the farming and of the equipment then in use. Anybody familiar with topographic maps will know that contour lines, remaining strictly horizontal, over the irregularities of the land's surfaces, cannot be regularly spaced. This variability presents no significant problem to a farmer using one- or two-row equipment in relatively small lands or fields. And so for a while contour farming became an established practice on many farms, and to good effect. It was defeated primarily by the enlargement of fields and the introduction of larger equipment. Eventually many farmers simply ignored their terraces, plowing over them, the planted rows sometimes running straight downhill. Earlier a good many farmers had taken readily to the idea of soil conservation. A farmer in a neighboring county said, "I want the water to *walk* off my land, not run." But beyond a certain scale, the farming begins to conform to the demands of the machines, not to the nature of the land.

I should pause here to notice that within three paragraphs I have twice quoted farmers who used "walk" as an approving figure of speech: Grain leaving a farm hereabouts should *walk* off; and the rainwater fallen upon a farm should *walk*, not run. This is not accidental. The gait most congenial to agrarian thought and sensibility is walking. It is the gait best suited to paying attention, most conservative of land and equipment, and most permissive of stopping to look or think. Machines, companies, and politicians "run." Farmers studying their fields travel at a walk.

Farms that are highly diversified and rightly scaled tend, by their character and structure, toward conservation of the land, the human community, and the local economy. Such farms are both work places and homes to the families who inhabit them and who are intimately involved in the daily life of land and household. Without such involvement, farmers cease to be country people and become in effect city people, industrial workers and consumers, living in the country.

To understand the complex and demanding requirements of good agriculture, and to know the vast acreage now given over to bad agriculture (leaving aside for now the vast acreage consigned to bad forestry), is to recognize the utter futility of the notion, apparently still prevalent among conservation groups, that the health of the natural world, revealingly called "the environment," can be preserved in parks and "wilderness areas." This drastic abbreviation of land stewardship permits no competent concern for the effects of the lowing herd upon the lea or of corn and beans upon the slopes. It holds that the gated communities of "the wild" will somehow preserve the natural health of "the planet."

Such conservationists, one imagines, might take instruction from scientists about the need for ecological health in the food-producing landscapes. But such scientists are rare and scattered. The predominant agricultural science of the universities, the corporations, and the government is still almost unanimously promoting industrial agriculture despite the by now overwhelming evidence of its failure: soil erosion, salinization, aquifer depletion, nutrient depletion, dependence on fossil fuels and toxic chemicals, pollution of streams and rivers, loss of genetic and ecological diversity, destruction of rural communities and the cultures of husbandry. The agricultural scientists and experts go doggedly on in their "cutting edge" rut because they either are employed by agribusiness or because their universities are now helplessly dependent on grants from agribusiness.

Urban conservationists, university scientists and intellectuals, journalists, powerful officials and politicians moreover are unlikely to live in the economic landscapes. In general they don't like the "boondocks" and the "nowheres" of rural America, and they don't know anything about them. Most of them know nothing of the issues of land use, and they think them unimportant. The sentimentalized ignorance of the romantic wilderness lovers, and the institutionalized, fear-enforced ignorance of agricultural scientists, are thus in turn permitted and supported by the ignorance of the general public, most of whom see the economic landscapes only through the windows of their speeding automobiles. If, by some rare chance, they should get out of their cars and walk in the fields, pastures, and woodlands, they most likely would take the present look of things to be "normal." Knowing no history of places, having no memories of them, they could not distinguish the country as it is from the country as it ever was. They would not recognize the signs of deterioration, or the numerous alien species that have come in as a side effect of global trade.

Nowhere is this ignorance more poignantly manifest than in the common use of words such as "land" and "ecosystem" simply as ideas or metaphors. I recently read *Ill Fares the Land*, a mostly admirable book, by Tony Judt, an admirable man. The book's title comes from a line in "The Deserted Village," Oliver Goldsmith's poem of protest against the Enclosure Acts. As Goldsmith used the word, "land" meant land. But Tony Judt's book never mentions the land. It speaks of "environmental well-being," of climate change and "environmental effects," it quotes John Maynard Keynes on "the beauty of the countryside," but its author clearly had not thought of the land itself, the land-use economies, or the natural world. To him "the land" is merely a figure of speech denoting "the nation" or "the national economy."

In the absence of a widely practiced and capable attention to our use of the land, to the land-use economies, and to the natural sources of our life, we have a national, or global, economy consisting entirely of capital (rated at monetary value), minimal labor ("jobs," merely numbered, and the numbers always liable to reduction by technology), information (infinite perhaps, but never sufficient), marketing (seduction of the gullible), and consumption (conversion of goods into waste or poison). And so we have lost patriotism in the old sense of love for one's country, and have replaced it with an ignorant, hard-hearted military-industrial nationalism that devours the country.

Under such a dominance it is understandable that land use should be reduced to the application, at the greatest possible speed and with the least possible labor, of technology and information. Since no limit is implied in the economic assumptions of such use, its technology or its methods, its destructiveness is unlimited, far exceeding the reach of any envisionable public regulation or supervision, as in the mountaintop removal method of coal mining, which destroys absolutely and beyond remedy the original, invaluable forest ecosystem of the Appalachian coal fields. An economy operating on the basis merely of quantities runs oddly toward both infinity and nothing, limitless desire and final exhaustion. No billionaire, evidently, can be satisfied with one billion or with any conceivable number of billions. But the effect of so much wealth, uncontrollable because unaccountable and unknowable by any human mind, is to use up the world's real wealth, which is its ability to live and to renew its life.

I don't think we can set the terms of a restorative and conserving land-use economy simply by juggling and somehow fixing the technologies, methods, resources, assumptions, and regulations of the present economy. We seem to have exhausted the capacity of our present system to improve much of anything. Or it may be better to say that we improve things by means of costs that we never count or subtract from the supposed benefits, and so we relinquish any notion of net improvement, not to speak of net loss.

The possibility of actual improvement in our economic life, which is to say our way of living from our land, seems to lie, not just in stabilizing our occupation of our country on the basis of knowledgeable attachment to its localities, but also in studying the economic value of such intangible goods as knowledge, memory, familiarity, imagination, affection, sympathy, neighborliness, and so on. One might greatly lengthen this list simply by allowing the several goods to call forth their kindred, each of which, like the ones I have listed, would impose certain conditions and certain limits. An economy, of course, must deal in quantities, but an economy answerable to terms such as I have listed would be opposite, in its effects, to an economy merely of quantities. I resist the inclination to call such terms qualities or ideas, just as I resist the inclination to call them feelings. They certainly are informed by qualities, ideas, and feelings, but they also are mental powers capable of enforcing care in our treatment of persons, places, and things.

In thinking about the economy, not of agriculture or farming, but of a farm, we come quickly to see the value of knowledge. The knowledge of how to farm, how and when to do the work, has an obvious, and fairly reckonable, economic worth. But the knowledge that is limited to one's own particular farm —knowledge of its nature, character, history, limitations, and right use, gathered out of years of experience—also has an economic value that is far more specific and not so readily accounted. Its value can only be suggested by pointing out that it clearly is an asset to the farmer who has it, that it cannot be sold or adequately conveyed by instruction to a new owner, whose want of it will be an economic disadvantage requiring much time, and perhaps some costly mistakes, to remedy. The value of such knowledge may be further suggested by seeing that it is depreciated virtually to nothing by the continuous monocultures I have described.

The force and effect of such intangibles is nowhere better exemplified than in the better communities of the Amish, which, so far as my own observation goes, are the only communities in the United States that are successful by every appropriate standard. Some would argue that the fundamental power or principle of these communities is their religious faith, and some would argue, further, that no community can be successful without religious faith as a fundamental principle. I am inclined to accept their conclusion, though I doubt that it can be adequately demonstrated, let alone proved. And so I will deal here only with issues of economy—though, among the Amish, the economy is formed or informed at every point by religious faith.

I think it is reasonable to say that the fundamental principle of Amish economy is rightness of scale: rightness of the size of their farms, and the appropriateness of that size to the nature of the local landscape, to the community's way of farming, and to the scale of its work. Rightness of scale gives scope and efficacy to the powers I have listed, and to others. It is the principle that allows diversity, flexibility, and local adaptation of farm enterprises.

Amish economy is an economy dependent upon limits strictly understood and observed. And each of the limits produces an economic advantage. Traction for field work, for example, is limited to the use of horses or mules. And that limit implies a limit of farm size; it implies diversity, which implies a structure of interdependent and mutually supportive plant and animal enterprises; it makes the farm the source of most of its operating energy and fertility—all of which, together, lower the cost of production.

Another limit, of almost incalculable significance, is neighborliness. This means that you would rather have a neighbor than to have your neighbor's farm, and here is another limitation on the size of farms. If their farms are of the right size, neighbors can help one another in times of trouble or when the work calls for many hands. If one has a neighbor—in the sense derived from the Gospels, as it is here—then one has help, and help is an economic advantage. If neighbors exchange work, they don't have to hire help and pay in cash.

The Amish famously, or infamously, limit formal schooling to eight grades. (This is not at all to say that they limit learning. Some Amishmen, for example, have gone on to learn mechanical engineering.) They limit schooling in order to keep their children in the community. This makes sense if you *want* to keep your children in the community, and if you have understood that the purpose of mainstream education is to prepare children, and especially country children, to *leave* the community. If you contrive in general to keep the community's children in the community, there are two desirable results: 1) The children, from earliest childhood, learn the community's work, by observing it and, as they become able, by doing it; and 2) If you keep all or most of the community's children in the community, then as a matter of course you keep the brightest and most talented ones. To keep the community's own brightest and most talented members in the community, in the community's local circumstances, doing the community's work, is to have the best intelligence and talent applied to that work, producing good examples that can be followed by neighbors.

This is an economic advantage. Moreover, the community may thus gain the competence to deal with tools, to quote again my friend Robert B. Weeden, "that take more skill not to use, than to use."

If on a farm of a conscientiously limited size you balance row crops and forages with an appropriate number of animals, then as a matter of course you become familiar with those animals. You can know them individually, and so sympathy can enter into your association with them. The best-known example of this is Psalm 23, a shepherd's psalm in which the shepherd, identifying himself as one of God's sheep, identifies profoundly with his own sheep: "He maketh *me* to lie down in green pastures: he leadeth *me* beside the still waters" [my emphases].

Just so, if by the same order of limitation you use a team of horses for field work, you know by sympathy, by imagining yourself "in their place," when they are overheating or overtired or when you are asking too much. From there it is not far to sympathy for the field you are working in, which also can be too much demanded upon, can be overworked and in need of rest. Imagination leads to sympathy, sympathy leads to good care, and all three convey economic advantages.

Once this pattern of interdepending limits and advantages has been understood, it can be described perhaps endlessly. But I have said enough to show why, if you drive through one of the good Amish communities, you will see a lot of people outdoors and busy. You will see that they have honored their places with the visible signs of good work lovingly done. And you will think again, with sharpened sadness, of the nearly always deserted "farms" of thousands of acres of corn and beans.

II

I have spoken so far of the decline of country work, but the decline of country pleasures is at least equally significant. If the people who live and work in the country don't also enjoy the country, a valuable and necessary part of life is missing. And for families on farms of a size permitting them to be intimately lived on and from, the economic life of the place is itself the primary country pleasure. As one would expect, not every day or every task can be a pleasure, but for farmers who

love their livestock there is pleasure in watching the animals
graze and in winter feeding. There is pleasure in the work of
maintenance, the redemption of things worn or broken, that
must go on almost continuously. There is pleasure in the growing, preserving, cooking, and eating of the good food that the
family's own land provides. But around this core of the life and
work of the farm are clustered other pleasures, in their way also
life-sustaining, and most of which are cheap or free.

I live in a country that would be accurately described as
small-featured. There are no monumental land forms, no peaks
or cliffs or high waterfalls, no wide or distant vistas. Though
it is by nature a land of considerable beauty, there is little here
that would attract vacationing wilderness lovers. It is blessed
by a shortage of picturesque scenery and mineable minerals.
The topography, except in the valley bottoms, is rolling or
sloping. Along the sides of the valleys, the slopes are steep. It
is divided by many hollows and streams, and it has always been
at least partly wooded.

Because of the brokenness and diversity of the landscape,
there was never until lately a clean separation here between
the pursuits of farming and those of hunting and gathering.
On many farms the agricultural income, including the homegrown and homemade subsistence of the households, would
be supplemented by hunting or fishing or trapping or gathering provender from the woods and berry patches—perhaps by
all of these. And beyond their economic contribution, these
activities were forms of pleasure. Many farmers kept hounds or
bird dogs. The gear and skills of hunting and fishing belonged
to ordinary daily and seasonal life. More ordinary was the ram-
bling about and looking that kept people aware of the condition of the ground, the crops, the pastures, and the livestock,
of the state of things in the house yard and the garden, in the
woods, and along the sides of the streams.

My own community, centered upon the small village of Port
Royal, is along the Kentucky River and in the watersheds of local tributaries. Its old life, before the industrialization of much
of the farmland and the urbanization of the people, was under
the influence of the river, as other country communities of that
time were under the influence of the railroads. In the neighborhood of Port Royal practically every man and boy, some

girls and women too, fished from time to time in the Kentucky River. Some of the men fished "all the time" or "way too much." Until about a generation ago, there was some commercial fishing. And I can remember when hardly a summer day would pass when, from the house where eventually I would live, you could not hear the shouts of boys swimming in the river, often flying out into the water from the end of a swinging rope. I remember when I was one of them. My mother, whose native place this was, loved her girlhood memories of swimming parties and picnics at the river. In hot weather she and her friends would walk the mile from Port Royal down to the river for a cooling swim, and then would make the hot walk back up the hill to town.

Now all of that belongs to the past. The last of the habituated fishermen of the local waters are dead. They have been replaced by fishermen using expensive "bassboats," almost as fast as automobiles. This sport is less describable as "fishing" than as "using equipment." In the last year only one man, comparatively a newcomer, has come to the old landing where I live to fish with trotlines—and, because of the lack of competition, he has caught several outsize catfish. Some local people, and a good many outsiders, hunt turkeys and deer. There is still a fair amount of squirrel hunting. The bobwhite, the legendary gamebird of this region, is almost extinct here, and the bird hunters with them. A rare few still hunt with hounds.

Most remarkable is the disappearance of nearly all children and teenagers from the countryside, and in general from the out-of-doors. The technologies of large-scale industrial agriculture are too complicated and too dangerous to allow the participation of children. For most families around here, the time is long gone when children learned to do farmwork by playing at it, and then taking part in it, in the company of their parents. It seems that most children now don't play much in their house yards, let alone in the woods and along the creeks. Many now descend from their school buses at the ends of lanes and driveways to be carried the rest of the way to their houses in parental automobiles. Most teenagers apparently divide their out-of-school time between indoor entertainment and travel in motor vehicles. The big boys no longer fish or swim or hunt or camp out. Or work. The town boys, who used to hire themselves out for seasonal or part-time work on the farms, no longer find such work available, or they don't wish to do the work that is available.

Not so long ago, talking with one of my contemporaries about the time before school consolidation, when there were five high schools in Henry County, I said, "When we were boys, there were five basketball teams in this county."

He said, "I'll tell you something else. In my senior year, any of the five could have beaten this one we've got now."

"Well, you all were in shape," I said.

"Yes," he said. "We *worked*."

More recently I heard a woman complaining that her husband could not hire anybody to help him with farmwork, "not even young people."

Somebody asked, "What *are* the young people doing?"

She was indignant: "Nothing!"

They aren't doing nothing, of course, but nothing might be preferable to what they are likely to be doing. For young people accustomed (by habit and indoctrination) to indoor life, the dominant attractions are drugs (including alcohol), sex, and various digital devices. The drug use is merely predictable in a society in which drugs are everywhere recommended for every imaginable discomfort and "need." Sex (arranged by telecommunication, using various digital devices) is promoted everywhere as another cure-all, as an incentive to participate in the economy of spending and consuming, and no doubt as another of our new crop of "rights." The digital devices are recommended or required in order to prepare the young for "the world of the future." The cost of this expensive preparation is virtual exile from the present world that is available at no cost outside their front doors. And so they spend their liveliest years mostly sitting and looking at screens.

Local people who regularly hunted or fished or foraged or walked or played in the local countryside served the local economy and stewardship as inspectors, rememberers, and story tellers. They gave their own kind of service to the eyes-to-acres ratio. Now most of those people are gone or absent, along with most of the farming people who used to be at work here.

With them have gone the local stories and songs. When

people begin to replace stories from local memory with stories from television screens, another vital part of life is lost. I have my own memories of the survival in a small rural community of its own stories. By telling and retelling those stories, people told themselves who they were, where they were, and what they had done. They thus maintained in ordinary conversation their own living history. And I have from my neighbor, John Harrod, a thorough student of Kentucky's traditional fiddle music, his testimony that every rural community once heard, sang, and danced to at least a few tunes that were uniquely its own. What is the economic value of stories and songs? What is the economic value of the lived and living life of a community? My argument here is directed by my belief that the art and the life of settled rural communities are necessary to the sustainability of a life-supporting economy. But their value is incalculable. It can only be acknowledged and respected, and our present economy is incapable, and cannot on its own terms be made capable, of such acknowledgement and respect.

Meanwhile, the farmlands and woodlands of this neighborhood are being hurt worse and faster by bad farming and bad logging than at any other time in my memory. The signs of this abuse are often visible even from the roads, but nobody is looking. Or to people who are looking, but seeing from no perspective of memory or knowledge, the country simply looks "normal." Outsiders who come visiting almost always speak of it as "beautiful." But along this river, the Kentucky, which I have known all my life and have lived beside for half a century, there is a large and regrettable recent change, clearly apparent to me, and to me indicative of a drastic change in water quality, but perfectly invisible to nearly everybody else.

I don't remember what year it was when I first noticed the disappearance of the native black willows from the low-water line of this river. Their absence was sufficiently noticeable, for the willows were both visually prominent and vital to the good health of the river. Wherever the banks were broken by "slips" or the uprooting of large trees, and so exposed to sunlight, the willows would come in quickly to stabilize the banks. Their bushy growth and pretty foliage gave the shores of the river a distinctive grace, now gone and much missed by the few who remember. Like most people, I don't welcome bad news, and so I said to myself that perhaps the willows were absent only from the stretch of the river that I see from my house and work places. But in 2002 for the first time in many years I had the use of a motorboat, and I examined carefully the shores of the twenty-seven-mile pool between locks one and two. I saw a few old willows at the tops of the high banks, but none at or near the low-water line, and no young ones anywhere.

The willows still live as usual along other streams in the area, and they thrive along the shore of the Ohio River just above the mouth of the Kentucky at Carrollton. The necessary conclusion is that their absence from the Kentucky River must be attributable to something seriously wrong with the water. And so, since 2002, I have asked everybody I met who might be supposed to know: "Why have the black willows disappeared from the Kentucky River?" I have put this question to conservationists, to conservation organizations specifically concerned with the Kentucky River, to water-quality officials and to university biologists. And I have found nobody who could tell me why. Except for a few old fishermen, I have found nobody who knew they were gone.

This may seem astonishing. At least, for a while, it astonished me. I thought that in a state in which water pollution is a permanent issue, people interested in water quality surely would be alert to the disappearance of a prominent member of the riparian community of a major river. But finally I saw that such ignorance is more understandable than I had thought.

A generation or so ago, when fishing and the condition of the river were primary topics of conversation in Port Royal, the disappearance of the willows certainly would have been noticed. Fishermen used to tie their trotlines to the willows. That time, as I've said, is past, and I was seeking local knowledge from conservationists and experts and expert conservationists. But most conservationists, like most people now, are city people. They "escape" their urban circumstances and preoccupations by going on vacations. They thus go into the countryside only occasionally, and their vacations are unlikely to take them into the economic landscapes. They want to go to parks, wilderness areas, or other famous "destinations." Government and university scientists often have economic concerns or responsibilities, and some of them do venture into farmland or working forests or onto streams and rivers that are not "wild." But it seems they are not likely to have a particular or personal or long-term interest in such places, or to go back to them repeatedly and often over a long time, or to maintain an economic or recreational connection to them. Such scientists affect the eyes-to-acres ratio probably less than the industrial farmers.

It seems to me significant, and not a bit surprising, that among the many conservationists I have encountered in my home state, the most competent witness by far is Barth Johnson, a retired game warden who is a dedicated trapper, hunter, and fisherman, as he has been all his life. Barth has devoted much of his life to conservation. Like most conservationists he is informed about issues and problems. Unlike most, he is exceptionally alert to what is happening in the actual countryside that needs to be conserved. This is because he is connected to the neighboring fields and woods and waters by bonds of economy and pleasure, both at once. Because those bonds are long-sustained and continued from year to year, he knows those places well. He has, as a matter of course, a year-round interest in the health—which is, in the best sense, the productivity—of the countryside. Moreover, he has lived for thirty years in the same place at the lower end of the Licking River. This greatly increases the value of his knowledge, for he can speak of changes *over time*. People who stay put and remain attentive know that the countryside changes, as it must, and for better or worse.

In his grasp of ecological principles, Barth is completely up with the times, and in some ways maybe ahead of the times, but he also is genuinely a countryman of a kind that is old and now rare. As opposed to writers, the best story *tellers* are people who spend a lot of time outdoors. Barth is well supplied with stories, most of them funny, all of them interesting, but some of them could be better classed as "reports."

He tells a story about Harris Creek, a small stream along which he had trapped for many years. It was richly productive, and Barth was careful never to ask too much of it. But in 2007, confident that it would be as it always had been, he went there with his traps and discovered that the stream was dead. He could not find a live minnow or crawfish. There were no animal tracks. So far as he could tell, there could be only one reason for this: In the spring of that year, the bottom-land along the creek had been herbicided in preparation for a seeding of alfalfa. In 2008, the stream was still dead. In 2009, there was "a little coon activity." Finally, in 2013, the stream was "close to normal."

I have also learned from Barth that upstream as far as he has looked, to a point two and a half miles above the small town of Boston, the black willows are gone from the Licking River. And in October 2013, he wrote me that the river had turned a brownish "brine" color that he had never seen before.

What happened to the willows? Two young biologists at Northern Kentucky University are now at work on the question, and perhaps they will find the answer. But other scientists have led me to consider the possibility that such questions may never be answered. It may be extremely difficult or impossible to attach a specific effect to a specific cause in a large volume of flowing water.

What killed Harris Creek? Barth's evidence is "anecdotal," without scientifically respectable proof. I have read scientific papers establishing that the herbicide glyphosate and its "deg-radation products" are present in high concentrations in some Mississippi River tributaries, but the papers say nothing about the effects. I have called up scientists working on water quality, including one of the authors of one of the papers on glyphosate. What about the *effects*? Good question. Nobody knows the answer. It seems that the research projects and the researchers are widely scattered, making such work somewhat incoherent. And besides there is always the difficulty of pinning a specific cause to a specific effect. To two of these completely friendly and obliging people I told Barth's story of Harris Creek: Does that surprise you? One said it did not surprise him. The other said it was possible but unlikely that the stream was killed by an herbicide. Was an insecticide also involved?

What caused the strange discoloration of the Licking River? Since the discoloration was visible until obscured by mud in the water when the river rose, I suppose that, if it happens again, the odd color could be traced upstream to a source.*

*As of July 2014, it has not happened again.

Will somebody do that? I don't know. Is any scientist from any official body monitoring the chemical runoff from croplands and other likely sources? I have been asking that question too, and so far I have asked nobody who could answer.

In my search for answers, it may be that I have been making a characteristic modern mistake of relying on experts, which has revealed a characteristic modern failure: Experts often don't know and sometimes can never know. Beneficiaries of higher education, of whom I am one, often give too much credit to credentials.

By now, my interest has necessarily shifted from an attempt to find answers to an attempt to understand the implications, not just of my failure so far to find answers, but of my failure so far to find any reason to think that answers are likely to be found. And so I must back up to what I do actually know and start again.

I know that the willows have disappeared, apparently because of some toxic chemical in the water. And I know that this is the result of a scientific success in developing a chemical that serves some industrial purpose, but with water pollution as a side effect. It is also the result of a scientific failure to notice —or to care, if noticed—that this chemical has "accidentally" polluted the river.

The countryside and its waterways are now being significantly, recognizably, and sometimes measurably damaged by industrialized science. Because of scientific success and scientific failure, toxic chemicals are on the loose, the effects of which the available science does not know and therefore cannot hope to remedy. This obviously is beyond the competence of people who are not scientists. If anybody now is going to maintain an effective vigilance over the runoff of toxic chemicals from crop monocultures or surface mines or other industrial sites, that vigilance will have to be maintained by scientists (such as they are). But there is no indication that there is enough money anywhere to hire enough scientists to do this. And how could enough scientists be found to watch carefully enough over such an extent of country and so many miles of creeks and rivers?

To be fair, there are a lot of amateurs who voluntarily watch over some of our waterways, and some of these people become competent at testing for the presence of pollutants. But such vigilance, to be effective, must be constant and sustained over a long time, and I am unsure of the extent to which this has been, or can be, achieved. The problem is the "mobility" of our people, which obviously limits their ability to pay attention to places.

In the middle of the last century we had within easy reach the possibility of preserving a right ratio of eyes to acres. From people and ways then in place, we could have cultivated, educated, and encouraged a population of vigilant land stewards attached to our economic landscapes by bonds of economy, pleasure, affection, and long memory, possessing the cultural means and imperatives of good care. We not only abandoned that possibility but condemned it as of no worth, substituting a program of industrial innovation, scientific responsibility, and governmental regulation that appears definitively to have failed. Of the "spill" of 4-methylcyclohexane methanol that contaminated the water supply of 300,000 West Virginians, a *New York Times* editorial noticed that this was "the third major chemical accident in the region in five years," adding that "the EPA has tested just 200 of the roughly 85,000 chemicals in use today." The *Times* editors called for "meaningful reform." But by what conceivable reform could any agency locate, in all their inevitable wanderings, transformations, and combinations, and make harmless 85,000 chemicals already in use?

As another example, the Water Quality Branch of the Kentucky state government's Division of Water has nine biologists to monitor the state's 92,000 surface miles of streams. That is a biologist-to-miles ratio of 1 to more than 10,000. What, one has to ask, can be the adequacy of one biologist per 10,000 miles of streams? Or: What would be the value of one biologist per 10,000 miles compared to the value of one vigilant and stewardly farmer or trapper every 10 or even 100 miles—with, let us say, a staff of public biologists to help when there is a question or a problem?

Confronting industrial agriculture in particular, we are requiring ourselves to substitute science for citizenship, community membership, and land stewardship. But science fails at all of these. Science as it now predominantly is, by definition and on its own terms, does not make itself accountable for unintended effects. The intended effect of chemical nitrogen fertilizer, for example, is to grow corn, whereas its known effect on the Mississippi River and the Gulf of Mexico is a catastrophic accident. Moreover science of this kind is invariably limited and controlled by the corporations that pay for it.

In the economic landscapes, now so desperately in want of proper stewards and so little watched over, the industrial economy is free to do anything it can do, according to its wishes. This is a triumph of laissez-faire economics riding upon a triumph of its laissez-faire science—a science free to invent causes, and free of worry about effects. This triumph, highly complicated, highly profitable to a few corporations, intelligible (somewhat) only to experts, is nevertheless limited by its dependence on exhaustible materials, including, as now used, the soil. The bad ecological effects, even more complicated, and as now "regulated," may be limitable only by disease and death. This is the industrial version of the human predicament.

To this the only rational response is that it has been a mistake to allow industrialism, from the beginning, to measure its conduct exclusively by its own success, using such standards as mechanical efficiency or monetary profit, and ignoring all else. This great project of continuous technical innovation and obsolescence, substituting technologies for human workers, was intentionally wrong from the start in its evasion of the long-established moral requirement of neighborliness. Its ecological, and therefore its long-term economic, wrongs may at first have been to some degree innocent, for the earliest industrial abuses of nature were comparatively slight. No doubt the availability of long-distance transportation from "remote" areas encouraged the assumption that the world's supply of raw materials was all but infinite. And the availability in North America of "new land" to the west for three hundred years encouraged, and more or less subsidized, land abuses in the east. Now, though it certainly is possible to know better, and many people do, most industrialists still have not learned better. The frontier superstitions of the inexhaustibility of natural supplies and of the adequacy of human concern still prevail in the face of the overwhelming evidence against them.

Our original and continuing mistake has been to ignore the probability, even the inevitability, of a formal misfitting between the human economy and the economy of nature, or between economy and ecology. This misfitting has been dangerous and damaging at least since the beginnings of agriculture. The reason for this is the limited competence of the human mind, which will never fully comprehend the forms and functions of the natural world. With the development of industrialism, this misfitting has become increasingly a contradiction or opposition between industrial technologies and the creatures of nature, tending always toward the destruction of creatures, creaturely habitat, and creaturely life. We can respond rationally to this predicament only by honest worry, unrelenting caution, and propriety of scale. We must not put too much, let alone everything, at risk. We must never tolerate permanent damage to the ecosphere or to any of its parts. We must not, because we cannot for very long, tolerate compromises with soil erosion and agricultural poisons.

It is anyhow clear that if we want to do better, we will have to recognize the old mistake as a mistake: no more euphemisms such as "creative destruction," no more "sacrificing" of a present good for "greater good in the future." We will have to repudiate the too-simple industrial standards and replace them with the comprehensive standard of ecological health, realizing that this standard involves necessarily the humane obligation of neighborliness both to other humans and to other creatures. This means that all our uses of the natural world must be governed by our willingness to learn the nature of every place, and to submit to nature's limits and requirements for the use of every place. In short, agriculture and forestry must finally submit to ecology. Mining of course is not natural by any stretch of the term. Unlike agriculture and forestry, it cannot be made analogous to natural processes, because by method and purpose it exhausts its sources. But, at least, it must no longer be allowed to extract wealth from under the ground by destroying the invariably greater, self-renewing wealth that is on the surface. The human economy must operate, to the always extendable limits of its ability, as a good neighbor in both the natural and the human communities, because in the long run the health of one is the health of the other.

*

The history of industrialism has been an ever-ramifying series of substitutions. It began and has continued by substitutions of two kinds: the substitution of hotter to hottest fuels for the cooler energies of gravity, sunlight, wind, and food; and the substitution of ever more "sophisticated" technologies for human work and care.

The side effects of our use of fossil and nuclear fuels are too infamous to require much notice here. Their extraction has been increasingly, and often irreparably, damaging to the ecosphere, as well as to human communities and workers. Their use produces enormous volumes of earthborne and airborne poisons that human intelligence has never learned what to do with, probably because the safe disposal of permanently dangerous substances cannot be learned. The farce of such "disposability" is a major theme of the history of industrialism.

Another major theme is the disposability of people, and this is not a farce. It is one of the versions of "creative destruction," which is to say the theme of heartlessness, heartbreak, and permanent damage to people and their communities, endlessly repeated from the beginning, and with no proposed or theoretical end. We now use "Luddite" as a term of contempt, and this usage, often by people who consider themselves compassionate and humane, implies a sort of progressivist etiquette by which, in the interest of the future (of the more fortunate), we are to submit passively to our obsolescence, disemployment, displacement, and (likely enough) impoverishment. We smear this over with talk of social mobility, upward mobility, and retraining, but this is as false and cynical as the association of "safe" with the extraction, transportation, and use of fossil and nuclear fuels. Customarily we ignore the possibility that people's knowledge, intelligence, and skills may be needed in the places where their minds were formed, also the likelihood that such assets will become worthless as soon as the people leave home. We seem to have overlooked as well the possibility of using technologies to ameliorate the working conditions and the lives of workers while keeping them in place. Coal companies, for example, readily have used the newest technology to speed and cheapen the extraction of coal, but they often have used it reluctantly or too late to provide for the health and safety of their employees.

Replacement of workers by machines becomes more serious when it is enabled by the degradation of work. We have ignored the limits of compatibility between labor-saving and good work. And we have degraded almost all work by reducing the generously qualitative idea of "vocation" or "calling" to the merely quantitative integer, "a job." The purpose of education now is to make everybody eligible for "a job." A primary function of politics is "job creation." Persons deprived of work that they have loved and enjoyed and performed with pride are to consider their loss well-remedied by some form of "welfare" or "another job."

The idea of vocation attaches to work a cluster of other ideas, including devotion, skill, pride, pleasure, the good stewardship of means and materials. Here we have returned to intangibles of economic value. When they are subtracted, what remains is "a job," always implying that work is something good only to escape: "Thank God it's Friday." "A job" pretty much equals bad work, which can be performed as well or better by a machine. Once the scale and speed of farmwork have overridden any care for the health of the land community and any pride in the beauty of the farm, then we can talk, as we now are talking, of farming by remote control.

If such substitutions appear to work, we must consider the likelihood that they work only temporarily or according to criteria that are too simple or false. And we must acknowledge that some do not work at all: A "service economy" is immediately a falsehood when it is staffed by phone-answering robots. The computerization and robotification of the United States Postal "Service," far from improving the service, has impaired its ability to transport and deliver the mail on time or at any time. Somewhere along the line of industrial substitutions, it appears certain that we will find ourselves again confined within, and sharing the fate of, the natural, naturally-limited, industrially-depleted world that we have thought to transcend by more and more engineering. And then we will know, needing no experts to tell us, that our world, like our bodies, cannot survive unstanched bleeding and repeated doses of poison.

*

To replace industrial standards by the standard of ecological health would undo a failed substitution. It would not undo the history and the legacy of industrialism. We are not going to have the privilege of a clean slate or a new start. A change of standards would certainly change our ways of living and working, but those changes, even as they are chosen or forced upon us, cannot come with the speed of our various technological revolutions. They will require more patience than haste. (It may be that we can keep without harm some industrial comforts: warm baths in wintertime maybe, maybe painless dentistry.)

Though a clean slate is impossible, as it has always been, we are not destitute of instructions and examples. Though our present anxieties incline us toward theories and illustrations of the natural rapaciousness of humans, not all humans and not all human communities have been so. I don't think the present bunch of living humans can be allowed to make the (very restful) claim that there is nothing they can do, pleading the incorrigibility of their nature or their circumstances. And so I will end this essay with an inventory of the resources we have at hand that will support us in our effort to do better. This is *an* inventory, not *the* inventory. My obligation here is only to show that we do have resources, probably enough, if we would pay attention to them.

1. First and fundamental are the examples of nature's own ways of land care in the native ecosystems that precede the human economy, also in some humanly modified ecosystems that preceded the industrial economy. These ways have been carefully studied by agricultural scientists from Albert Howard, Aldo Leopold, and others before the middle of the last century to Wes Jackson and his colleagues at The Land Institute in Kansas right now. One of our most urgent needs is for ecologists, trained in the native forests and prairies, to apply their knowledge in the economic landscapes. We humans, because of our limited intelligence, are never going to understand perfectly the nature of any place, or fit our local economies perfectly into local ecosystems. But the right instructions, or the right "model," for our use of a place can come only from the nature of the place.

*

2. We have from all over the world, from F. H. King's *Farmers of Forty Centuries*, 1911, to the ongoing work of Vandana Shiva, examples of long-enduring traditional or peasant agricultures. These seem to have come from respectful and competent observation of the nature of places, and they have embodied natural processes or analogues thereof.

3. We have Thomas Jefferson's great principle: "The small landholders are the most precious part of a state." If we understand that the state and the state's economy depend upon the land, and if we understand the numerousness, diversity, uniqueness, and vulnerability of the land's small places, and so their dependence upon human care, we cannot doubt the truth of Jefferson's rule.

4. Scattered about the country, and concentrated in some Amish communities, we have, still surviving, farmers and foresters who have held to established good practices and tried for better. It is not altogether a secret that in many a rural community you will find farmers or memories of farmers who have survived, even prospered, in hard times by limiting their farms to a manageable size, by taking care of them, by avoiding or minimizing debt, by limiting or avoiding purchases of new equipment, by subsisting so far as possible from their own land, and in general by saving rather than spending.

5. Over the last two or three decades, there has been a growing national and international movement toward local economies, starting with economies of local food. This has been little noticed and poorly understood by the news media, mostly ignored in the state capitals and in Washington, D.C. Some city and county governments, however, have taken notice of this movement and understood its importance. For example, the local food effort is now well established as a part of the economic development plan of Louisville, Kentucky. To bring a local demand and a local supply into existence more or less simultaneously and on the principle of cooperation is a task obviously difficult, complicated, and long. But it has begun (there are now many examples in many places), and from nothing only a few years ago it has come a remarkable distance, though

the distance ahead is much longer. This project involves food production *in* cities as well as around them, and it is fostering a necessary urban agrarianism among gardeners and consumers.

6. We have a fairly long history of organic farming and gardening. We need to say, I think, that organic is as organic does. The term has often been too negatively defined (You *don't* use chemicals), and it can be too loosely defined, but it has always implied the standard of good health, and it seems by now to have escaped its old shadow of kookiness.

7. For half a dozen years we have had the 50-Year Farm Bill, which addresses the specifically agricultural problems of soil erosion, toxicity, loss of diversity, and the destruction of rural communities. It proposes to invert the ratio of 80 percent annual and 20 percent perennial crops at present to 20 percent annual and 80 percent perennial crops in fifty years. This bill certainly comes from beyond the present margin of farm policy, and of course the habitual objectors are happy to point out that there is at present no chance of its passage. But such a farsighted bill could not have been written by people foolish enough to believe in its immediate passage.

8. The 50-Year Farm Bill was published in 2009 by The Land Institute in Salina, Kansas, with the concurrence of other farm and conservation groups all over the country. That it came from The Land Institute should be no surprise, for that organization's great project for nearly forty years has been the perennialization of agriculture, arriving finally at perennial grainfields of mixed species, in analogy to the native prairie. This project of perennial grain crops has now spread from Kansas to other states and other countries. The first of the necessary medley of perennial grains, a domesticated intermediate wheatgrass, is already well developed and may be ready for distribution to farmers in eight to ten years. That in itself is a remarkable achievement, though much remains to be done. But I want to suggest that the kind of science practiced at The Land Institute is itself a great and necessary resource. It is by definition a local science, carried on conscientiously in the contexts of the local ecosystem and the local human community. Whatever

is developed in that place will require local adaptation, and the careful employment of many minds in other places. This science, moreover, is carried on with respect for local nature and local humanity. It is not going to produce a poison or an explosive.

9. This is somewhat speculative, since I am not an economist or an accountant, but I believe that our present systems of economy and accounting can be greatly and usefully improved simply by becoming more inclusive and more honest. My impression is that the prominent economists who advise and influence the government pay little attention to the so-called economy's basis in nature or to the use and produce of the land. The ruling assumption seems to be that a thriving economy can be sustained by abuse of the land and of the people who do the land's work. I remember, of course, my argument that some things of real economic worth cannot be quantified and accounted. But there should be a full and fair accounting of things that are accountable. Many of the costs of water pollution, for example, can be determined, and those costs should be charged to water-polluting industries and their customers. We need to be less interested in the "growth" of our gross income, and more interested in the subtraction of real costs and in net benefits.

10. Finally, and most necessarily, we have the ancient and long-enduring cultural imperative of neighborly love and work. This becomes ever more important as hardly imaginable suffering is imposed upon all creatures by industrial tools and industrial weapons. If we are to continue, in our only world, with any hope of thriving in it, we will have to expect neighborly behavior of sciences, of industries, and of governments, just as we expect it of our citizens in their neighborhoods.

On Being Asked for
"A Narrative for the Future"

S O FAR AS I can see, the future has no narrative. The future does not exist until it has become the past. To a very limited extent, prediction has worked. The sun, so far, has set and risen as we have expected it to do. And the world, I suppose, will predictably end, but all of its predicted deadlines, so far, have been wrong.

The End of Something—history, the novel, Christianity, the human race, the world—has long been an irresistible subject. Many of the things predicted to end have so far continued, evidently to the embarrassment of none of the predictors. The future has been equally, and relatedly, an irresistible subject. How can so many people of certified intelligence have written so many pages on a subject about which nobody knows anything? Perhaps we need a book—in case we don't already have one—on the end of the future.

None of us knows the future. Fairly predictably, we are going to be surprised by it. That is why "Take . . . no thought for the morrow . . ." is such excellent advice. Taking thought for the morrow is, fairly predictably, a waste of time.

I have noticed, for example, that most of the bad possibilities I have worried about have never happened. And so I have taken care to worry about all the bad possibilities I could think of, in order to keep them from happening. Some of my scientific friends will call this a superstition, but if I did not forestall all those calamities, who did? However, after so much good work, even I must concede that by taking thought for the morrow we have invested, and wasted, a lot of effort in preparing for morrows that never came. Also by taking thought for the morrow we repeatedly burden today with undoing the damage and waste of false expectations—and so delaying our confrontation with the actuality that today has brought.

The question, of course, will come: If we take no thought for the morrow, how will we be *prepared* for the morrow?

I am not an accredited interpreter of Scripture, but taking thought for the morrow is a waste of time, I believe, because all we can do to prepare rightly for tomorrow is to do the right things today.

The passage continues: "for the morrow shall take thought for the things of itself. Sufficient unto the day is the evil thereof." The evil of the day, as we know, enters into it from the past. And so the first right thing we must do today is to take thought of our history. We must act daily as critics of history so as to prevent, so far as we can, the evils of yesterday from infecting today.

Another right thing we must do today is to appreciate the day itself and all that is good in it. This also is sound biblical advice, but good sense and good manners tell us the same. To fail to enjoy the good things that are enjoyable is impoverishing and ungrateful.

The one other right thing we must do today is to provide against want. Here the difference between "prediction" and "provision" is crucial. To predict is to foretell, as if we know what is going to happen. Prediction often applies to unprecedented events: human-caused climate change, the end of the world, etc. Prediction is "futurology." To provide, literally, is to see ahead. But in common usage it is to look ahead. Our ordinary, daily understanding seems to have accepted long ago that our capacity to see ahead is feeble. The sense of "provision" and "providing" comes from the past, and is informed by precedent.

Provision informs us that on a critical day—St. Patrick's Day, or in a certain phase of the moon, or when the time has come and the ground is ready—the right thing to do is plant potatoes. We don't do this because we have predicted a bountiful harvest; history warns us against that. We plant potatoes because history informs us that hunger is possible, and we must do what we can to provide against it. We know from the past only that, if we plant potatoes today, the harvest *might* be bountiful, but we can't be sure, and so provision requires us to think today also of a diversity of food crops.

What we must *not* do in our efforts of provision is to waste or permanently destroy anything of value. History informs us that the things we waste or destroy today may be needed on the morrow. This obviously prohibits the "creative destruction" of the industrialists and industrial economists, who think that evil is permissible today for the sake of greater good tomorrow. There is no rational argument for compromise with soil erosion or toxic pollution.

For me—and most people are like me in this respect— "climate change" is an issue of faith; I must either trust or distrust the scientific experts who predict the future of the climate. I know from my experience, from the memories of my elders, from certain features of my home landscape, from reading history, that over the last 150 years or so the *weather* has changed and is changing. I know without doubt that to change is the nature of weather.

Just so, I know from as many reasons that the alleged causes of climate change—waste and pollution—are wrong. The right thing to do today, as always, is to stop, or start stopping, our habit of wasting and poisoning the good and beautiful things of the world, which once were called "divine gifts" and now are called "natural resources." I always suppose that experts may be wrong. But even if they are wrong about the alleged human causes of climate change, we have nothing to lose, and much to gain, by trusting these particular experts.

Even so, we are not dummies, and we can see that for all of us to stop, or start stopping, our waste and destruction today would be difficult. And so we chase our thoughts off into the morrow where we can resign ourselves to "the end of life as we know it" and come to rest, or start devising heroic methods and technologies for coping with a changed climate. The technologies will help, if not us, then the corporations that will sell them to us at a profit.

I have let the preceding paragraph rest for two days to see if I think it is fair. I think it is fair. As evidence, I will mention only that, while the theme of climate change grows ever more famous and fearful, land abuse is growing worse, noticed by almost nobody.

A steady stream of poisons is flowing from our croplands into the air and water. The land itself continues to flow or blow away, and in some places erosion is getting worse. High grain prices are now pushing soybeans and corn onto more and more sloping land, and "no-till" technology does not prevent

erosion on continuously cropped grainfields. Industrial agriculture, moreover, is entirely dependent on burning the fossil fuels, the most notorious of the alleged causes of climate
change.

Climate change, supposedly, is recent. It is apocalyptic, "big
news," and the certified smart people all are talking about it,
thinking about it, getting ready to deal with it in the future.

Land abuse, by contrast, is ancient as well as contemporary.
There is nothing futurological about it. It has been happening
a long time, it is still happening, and it is getting worse. Most
people have not heard of it. Most people would not know it if
they saw it.

The laws for conservation of land in use were set forth by
Sir Albert Howard in the middle of the last century. They were
nature's laws, he said, and he was right. Those laws are the basis
of the 50-Year Farm Bill, which outlines a program of work that
can be started now, which would help with climate change, but
which needs to be done anyhow. Millions of environmentalists
and wilderness preservers are dependably worried about climate change. But they are not conversant with nature's laws,
they know and care nothing about land use, and they have
never heard of Albert Howard or the 50-Year Farm Bill.

II

If we understand that Nature can be an economic asset, a help
and ally, to those who obey her laws, then we can see that she
can help us now. There is work to do now that will make us her
friends, and we will worry less about the future. We can begin
backing out of the future into the present, where we are alive,
where we belong. To the extent that we have moved out of the
future, we also have moved out of "the environment" into the
actual places where we actually are living.

If, on the contrary, we have our minds set in the future,
where we are sure that climate change is going to play hell
with the environment, we have entered into a convergence of
abstractions that makes it difficult to think or do anything in
particular. If we think the future damage of climate change to
the environment is a big problem only solvable by a big solution, then thinking or doing something in particular becomes
more difficult, perhaps impossible.

It is true that changes in governmental policy, if the changes were made according to the right principles, would have to be rated as big solutions. Such big solutions surely would help, and a number of times I have tramped the streets to promote them, but just as surely they would fail if not accompanied by small solutions. And here we come to the reassuring difference between changes in policy and changes in principle. The needed policy changes, though addressed to present evils, wait upon the future, and so are presently nonexistent. But changes in principle can be made now, by so few as just one of us. Changes in principle, carried into practice, are necessarily small changes made at home by one of us or a few of us. Innumerable small solutions emerge as the changed principles are adapted to unique lives in unique small places. Such small solutions do not wait upon the future. Insofar as they are possible now, exist now, are actual and exemplary now, they give hope. Hope, I concede, is for the future. Our nature seems to require us to hope that our life and the world's life will continue into the future. Even so, the future offers no validation of this hope. That validation is to be found only in the knowledge, the history, the good work, and the good examples that are now at hand.

There is in fact much at hand and in reach that is good, useful, encouraging, and full of promise, although we seem less and less inclined to attend to or value what is at hand. We are always ready to set aside our present life, even our present happiness, to peruse the menu of future exterminations. If the future is threatened by the present, which it undoubtedly is, then the present is more threatened, and often is annihilated, by the future. "Oh, oh, oh," cry the funerary experts, looking ahead through their black veils. "Life as we know it soon will end. If the governments don't stop us, we're going to destroy the world. The time is coming when we will have to do something to save the world. The time is coming when it will be too late to save the world. Oh, oh, oh." If that is the way our minds are afflicted, we and our world are dead already. The present is going by and we are not in it. Maybe when the present is past, we will enjoy sitting in dark rooms and looking at pictures of it, even as the present keeps arriving in our absence.

Or maybe we could give up saving the world and start to live savingly in it. If using less energy would be a good idea for the future, that is because it is a good idea. The government could enforce such a saving by rationing fuels, citing the many good reasons, as it did during World War II. If the government should do something so sensible, I would respect it much more than I do. But to wish for good sense from the government only displaces good sense into the future, where it is of no use to anybody and is soon overcome by prophesies of doom. On the contrary, so few as just one of us can save energy right now by self-control, careful thought, and remembering the lost virtue of frugality. Spending less, burning less, traveling less may be a relief. A cooler, slower life may make us happier, more present to ourselves, and to others who need us to be present. Because of such rewards, a large problem may be effectively addressed by the many small solutions that, after all, are necessary, no matter what the government might do. The government might even do the right thing at last by imitating the people.

In this essay and elsewhere, I have advocated for the 50–Year Farm Bill, another big solution I am doing my best to promote, but not because it will be good in or for the future. I am for it because it is good now, according to present understanding of present needs. I know that it is good now because its principles are now satisfactorily practiced by many (though not nearly enough) farmers. Only the present good is good. It is the presence of good—good work, good thoughts, good acts, good places—by which we know that the present does not have to be a nightmare of the future. "The kingdom of heaven is at hand" because, if not at hand, it is nowhere.

THE ART OF LOADING BRUSH
(2017)

The Thought of Limits in a Prodigal Age

> Is there, at bottom, any real distinction between esthetics
> and economics?
> —Aldo Leopold, *A Sand County Almanac & Other*
> *Writings on Ecology and Conservation*

I WANT TO SAY something about the decline, the virtual ruin,
of rural life, and about the influence and effect of agricultural surpluses, which I believe are accountable for more
destruction of land and people than any other economic "factor." This is a task that ought to be taken up by an economist,
which I am not. But economists, even agricultural economists,
farm-raised as many of them have been, do not live in rural
communities, as I do, and they appear not to care, as I do,
that rural communities like mine all over the country are either dying or dead. And so, only partly qualified as I am, I will
undertake this writing in the hope that I am contributing to a
conversation that will attract others better qualified.

I have at hand an article from the *Wall Street Journal* of
February 22, 2016, entitled "The U.S. Economy Is in Good
Shape." The article is by Martin Feldstein, "chairman of the
Council of Economic Advisors under President Ronald Reagan
. . . a professor at Harvard and a member of the Journal's
board of contributors." Among economists Prof. Feldstein appears to be somewhere near the top of the pile. And yet his
economic optimism is founded entirely upon current measures
of "incomes," "unemployment," and "industrial production,"
all abstractions narrowly focused. Nowhere in his analysis does
he mention the natural world, or the economies of land use by
which the wealth of nature is made available to the "American
economy." Mr. Feldstein believes that "the big uncertainties
that now hang over our economy are political."

But from what I see here at home in the watershed of the
Kentucky River, and from what I have seen and learned of
other places, I know that industrial agriculture is in serious
failure, which is to say that it is not sustainable. Projecting from
the damages of the comparatively brief American histories of

673

states such as Kentucky and Iowa, one must conclude that the present use of the farmland cannot be sustained for another hundred years: The rates of soil erosion are too high, the run-off is too toxic, the ecological impoverishment is too great, the surviving farmers are too few and too old. To anybody who knows these things, by witness of sight or by numerical measures, they would appear to qualify significantly the "good shape" of the economy. I conclude that Prof. Feldstein does not know these things, but is conventionally ignorant of them. Like other people of privilege for thousands of years, far more numerous now than ever before, he appears to take for granted the bounty of nature and the work that provides it to the human economy.

In remarkable contrast to the optimism of Prof. Feldstein, the *New York Times* of March 10, 2016, printed an article, "Who's Killing Global Growth?" by Steven Rattner, "a Wall Street executive and a contributing opinion writer." Mr. Rattner's downhearted assessment, like Prof. Feldstein's upbeat one, is based upon measures that are entirely economic or monetary, quantitative, and abstract: "financial markets," "projections for future growth," "wages," "consumer spending," "rising supply," "disappointing demand," etc. The "global growth" Mr. Rattner has in mind is purely financial and is without reference to the effect of such "growth" upon the health and the welfare of the globe's actual people and other creatures. Like Prof. Feldstein, Mr. Rattner appears to suppose, consciously or not, that the natural world and human workers will continue to supply their necessary goods without limit and to the allure simply of money.

I have at hand also a sentence from the *New York Review of Books*, September 24, 2015, by James Surowiecki, another highly credentialed economist. Mr. Surowiecki is reviewing among others a book by Joseph E. Stiglitz and Bruce C. Greenwald, *Creating a Learning Society: A New Approach to Growth, Development, and Social Progress.* This book, the reviewer says, "is dedicated to showing how developing countries can use government policy to become high-growth, knowledge-intensive economies, rather than remaining low-cost producers of commodities." I have kept this sentence in mind because of the problems it raises, all relating to my concern about the damages imposed by national and global "economies" upon land and people. Mr. Surowiecki's sentence seems to be highly condensed and allusive, a sort of formula for increasing economic growth—or, as it actually says, for turning countries into economies. The sentence no doubt is clear to economists, but it has put me to some trouble. My interest is not in the analyses and theories of these economists, since they seem mainly to ignore the natural world and the human communities that are my concern. I am interested here in their public language, by which they reveal what they accept, and expect most others to accept, as axioms—what one might call their lore or more accurately their faith.

I assume, then, that by "low-cost producers of commodities" Mr. Surowiecki means "poorly paid producers of cheap commodities," that these commodities are material goods or raw materials produced from the land, that "knowledge-intensive economies" are based upon the abilities to exploit, trade, add value to, and market the cheaply produced commodities. Apparently it is taken for granted that this improving formula applies to *all* developing countries, their people, their land, and their natural resources, without regard to differences or distinctions among them. Such disregard of local and personal differences is a major article of this faith. It takes for granted furthermore that a knowledge-intensive economy, by causing growth, development, and social progress, will change a developing country into a developed country, and that this will be an all-around improvement. From the standpoint of industrial economists and their clients, this apparently is self-evident and unquestionable. It becomes immediately and urgently questionable from the standpoint of a dweller in a rural countryside who is bound to the land and the community by ties of history, family, and affection.

Here we arrive at a fundamental division of interest and allegiance, as probably also at the difference between two kinds of mind. The attention of these economists and others like them is directed as a matter of course to the monetary economy and to what, according to their abstract measurements, is good or bad for it. The attention of settled dwellers, at home in their chosen or hereditary places, is directed partially to the monetary economy, of course, and often in fear or sufferance, but their attention is directed also, out of natural affection and solicitude, to their places, the particular, unique, and irreplaceable patches of ground under their feet. Another difference involved here, if the settled dwellers are farmers, is that between people whose livelihoods are primarily dependent upon salaries and people whose livelihoods are primarily dependent upon the weather.

The settled dwellers, then, in their natural desire to remain settled, and facing the "promises" of development, certainly are going to have questions for the developers, and the first would be this: What is the net good that industrial economists, their employers, and their clients appear customarily to credit to growth, development, and "social progress"? In the United States, since at least the Civil War, and ever more rapidly after World War II, we have achieved industrial versions of all of those goals. But almost nobody is asking what is the worth of that achievement after we deduct its ecological and human costs. We have, in fact, been turning our country into an economy as fast as possible, and we have been doing so by an unaccounted squandering of its actual, its natural and its cultural, wealth.

As a second question, we should ask why commodities, the material goods that support our life, and the work of producing them, should be "low cost" or significantly cheaper than the goods and services of a "knowledge-intensive economy." There is no reason to believe that the present market values of technological (developmental) knowledge and of commodities are absolute or in any way permanent. Nor is there reason to believe that such issues of value are, or can be, reliably settled by the free market of our present economy, or by any market. The good health of the land economies is a value that a market as such cannot consider and cannot protect. Moreover, agribusiness in all of its aspects is a knowledge-intensive system, which uses knowledge ruthlessly to control and exploit land and people.

Apparently it is assumed that a country's economy of commodity-production, which could be as diverse as the country's climate and soils permit, can safely be replaced or further depreciated by an economy of knowledge only. And so, as a third question, we must ask how secure and how beneficent is a one-product economy. Is the market for knowledge infinite in its demand, or can it be over-supplied and depressed, as the one-product economy of coal in the Appalachian coal fields has often been? And it hardly needs to be said that in the Appalachian coal fields the benefits of the coal economy to a rich and distant few has never adequately been measured against its impoverishment of the local people and their land.

Perhaps no outsider—no visiting expert, no dispassionate observer, certainly no outside investor—will notice the inherent weakness and cruelty of a one-product economy in a region or a country. But the adverse effects will certainly be visible and acutely feelable to the resident insiders. Those who live and must make their lives within the boundary of such an economy experience daily the readiness of their political leaders to endorse and excuse the destructiveness of "the economy," as well as the public unwillingness to remedy or compensate the damages to the land and the people. The Appalachian coal economy has not only inflicted immeasurable and immeasurably lasting ecological and social damage to its region, but it has also distracted attention and care from the region's other assets: its forests, soils, streams, and the (too often exported) talents of its people.

And so a fourth question: How, even in a knowledge-intensive economy, even unendingly "growing" and wealthy, are the people's needs for food, clothing, and shelter to be met? Does the development of a highly lucrative knowledge economy entirely eliminate the need for the fundamental economies of subsistence? Do people eat and wear knowledge? Do they sleep warm in it? I know very well what the far-seeing economists will answer: People earning large salaries from "high-growth, knowledge-intensive economies" will *buy* their material subsistence from "low-cost producers of commodities" at home or, if not at home, then elsewhere. It is assumed that where there is a demand, and enough money, there will be unfailingly a supply, and this is another article of the industrial economic faith: The land and its "resources" will be always

with us, and so will the poor who will dig, hack, and whittle an
everlasting supply of low-cost commodities until they can be
replaced by knowledge-intensive machines that will dig, hack,
and whittle, no doubt on solar energy, faster and at a lower
cost.

And so we come to question five: Do the economists of development ever attempt a fair assessment, or any assessment, of
the value to a knowledge-intensive economy of a dependable
local supply of life-supporting commodities? The answer, so
far as I have learned, is that the developmental economists do
no such thing. Their dream of human progress calls simply for
the *replacement* of the commodity economy by the knowledge
economy, and that is that. As evidence, let us consider a review
of our economic past and future, "Moving On from Farm and
Factory," by Eduardo Porter, in the *New York Times* of April
27, 2016.
   Mr. Porter's premise is that the economies of farming and
manufacturing, as a fixed and final consequence of historical
trends, are now obsolete, or nearly so, and his statistics are
sufficiently precipitous:

> Over the course of the 20th century, farm employment
> in the United States dropped to 2 percent of the work
> force from 41 percent, even as output soared. Since 1950,
> manufacturing's share has shrunk to 8.5 percent of non-
> farm jobs, from 24 percent.

To this state of things Mr. Porter grants something like half
an approval. Whereas nearly all of the "work force" once
employed in farming have been "liberated . . . from their
chains," he thinks that "The current transition, from manufacturing to services, is more problematic." Though for workers
in the United States there are "options: health care, education
and clean energy, just to name a few," these options "present
big economic and political challenges." The principal challenges will be to get the politicians to abandon their promises
of an increase of employment in manufacturing, and then to
provide the government help necessary to make "the current
transition from manufacturing to services" without too much

rebellion by workers "against the changing tide." Mr. Porter's conclusion, despite these challenges, is optimistic:

> Yet just as the federal government once provided a critical push to move the economy from its agricultural past into its industrial future, so, too, could it help build a postindustrial tomorrow.

Mr. Porter's article, which clearly assumes the agreement or consent of a large number of his fellow economists and fellow citizens, rests upon the kind of assumptions that I have been calling articles of faith. Though it is certain that a lot of people, economists and others, are putting their faith in these assumptions, they are nonetheless entirely groundless. The assumptions, so far as I can trace them out, are as follows:

1. The economy of a country or a nation needs only to provide employment, it does not matter at what. And so of course no particular value can be assigned to the production of commodities.
2. So long as there is enough employment at work of some kind, a country or a nation can safely dispense with employment in sustainable farming and manufacturing, which is to say a sustainable dependence upon natural resources and the natural world.
3. Farming has little economic worth. It is of the past, and better so. Farm work involves no significant responsibilities, and requires no appreciable intelligence, knowledge, skill, or character. It is, as is often said, "mind-numbing," a servile condition from the "chains" of which all workers, even owners who work on their own farms, need to be "liberated."
4. The "output" of industrial agriculture will continue to "soar" without limit as ever more farmers are "liberated from their chains" by technology, and as technologies are continuously succeeded by more advanced technologies.
5. There is no economic or intrinsic difference between agriculture and industry: A farm is no more than a factory, a plant or an animal is no more than a machine or a substance.

6. The technological advances that have disemployed so many people from farming and manufacturing will never take away the jobs of service or postindustrial workers.

7. History, including economic history, is a forward motion, a progress, made up of irreversible changes. These changes can be established absolutely and forever in so little time as a century or two. Thus the great technological progress since perhaps the steam engine—a progress enabled by the fossil fuels, war, internal combustion, external combustion, and a sequence of poisons—will carry us right on into the (climactic? everlasting?) "postindustrial tomorrow."

8. The need for "health care, education and clean energy, just to name a few," will go securely on and on, supplying without limit the need for jobs, whereas the need for food, clothing, shelter, and manufactured goods will be supplied by *what*?

Now I must tell why, as a comparatively prosperous and settled resident of my home country in the United States, I should be as troubled as I am by the faith or superstition or future-fantasy of the economists of so-called development. My family and I live, as we know and fear, in what the orthodox economists consider a backward, under-developed, and to-be-developed country. This is "rural America," the great domestic colony that we have made of our actual country, as opposed to the nation, the government, and the economy. This particular fragment of it is called "The Golden Triangle," a wedge of country bounded by the three interstate highways connecting Louisville, Lexington, and Cincinnati. The three are connected also by rail and by air. The Triangle is bounded on its northwest side also by the Ohio River. Because it is so fortunately located with respect to transportation and markets, this area is thought (by some) to be "Golden," which is to say eminently suited to (future) development.

The landscapes within the Triangle are topographically diverse —rolling uplands, steep valley sides, fairly level bottomlands —all, though varyingly, fragile and vulnerable to various established abuses. The soils are fertile, productive, responsive to good treatment, but much diminished by erosion and misuse in the years of "settlement," severely eroded in some places, still eroding in others. The native forest is predominantly hardwood, much diminished and fragmented, suffering from diseases and invasive species, largely undervalued, neglected or ill-used, but potentially of great economic worth if well used and cared for. The watercourses are numerous, often degraded, mostly polluted by silt or chemicals or both. There is, in most years, abundant rainfall.

The three cities seem generally to be prospering and expanding, but are expectably troubled by social disintegration, drugs, poverty, traffic congestion, and violence. The towns, including county seats, are in decay or dead, preyed upon by the cities and chain stores, diseased by urban and media culture, cheap energy, family disintegration, drugs, and the various electronic screens.

Especially during the early decades of the tobacco program, the farming here was highly diversified and, at its best, exemplary in its husbanding of the land. Because of the program, tobacco was the basis of a local agrarian culture that was both economically and socially stabilizing. The farms were mostly small, farmed by their resident families and neighborly exchanges of work. In addition to tobacco and provender for the households, they produced (collectively and often individually) corn, small grains, hogs, chickens and other poultry, eggs, cream, milk, and an abundance of pasture for herds of beef cattle and flocks of sheep.

The tobacco program with its benefits ended in 2004. Though it served growers of a crop that after the Surgeon General's report of 1965 could not be defended, the program itself was exemplary. Both the people and the land benefitted from it. By the combination of price supports with production control, limiting supply to anticipated demand, the program maintained the livelihoods of the small farms, and so maintained the livelihoods of shops and stores in the towns. It gave the same protection on the market to the small producer as to the large. By limiting the acreage of a high-paying crop, it provided a significant measure of soil conservation. Most important, it supported the traditional family and social structure of the region and its culture of husbandry.

For once and for a while, then, the farmers of this region

stood together, stood up for themselves, and secured for themselves prices reasonably fair for one of their products. The tobacco program, once and for a while, gave them an asking price, with results in every way good. Before and after the program, which was *their* program, they have had simply to accept whatever the buyers have been pleased to offer. When producers of commodities have no asking price, the result is plunder of both land and people, as in any colony. By "asking price" I mean a fair price, as determined for example by "parity," which would enable farmers to prosper "on a par with" their urban counterparts; a fair price, then, supported by bargaining power.

After the demise of the tobacco program, and with it the economy and way of life it had preserved and stood for, this so-favored Triangle and its region have declined economically, agriculturally, and socially. The tobacco that is still grown here is grown mainly in large acreages under contracts written by the tobacco companies, and primarily with migrant labor. Most of the farms that are still working are mainly or exclusively producing beef cattle—which is good, insofar as it gives much of our vulnerable countryside the year-round protection of perennial pastures and hay crops, but it is a far cry from the old diversity of crops and livestock. Of much greater concern is the continuous planting of large acreages of soy beans and corn, a way of farming unsuited to our sloping land (or, in fact, to any land), erosive, toxic, requiring large expenditures of money for uncertain returns. For such cropping the fences are removed, making the land useless for grazing. Farms are being subdivided and "developed," or cash-rented for corn and beans.

The land is no longer divided and owned in the long succession, by inheritance or purchase, of farmer after farmer. It has now become "real estate," ruled by the land market, owned increasingly by urban investors, or by urban escapees seeking the (typically short-lived) consolation and relaxation of "a place in the country." The government now subsidizes land purchases by some young farmers, "helping" them by involving them in large long-term debts and in ways of farming that degrade the land they may, late in their lives, finally own. For many young people whose vocation once would have been farming, farming is no longer possible. You have to be too rich to farm before you can afford a farm in my county.

Only a few years ago, I received a letter from a man extraordinarily thoughtful, who described himself as an ex-addict whose early years were spent under the teaching and influence of a family elder, in the tobacco patches of a neighborhood of small farms. Caught up by the centrifugal force of a disintegrating community and way of life, he drifted into addiction, from which, with help, he got free. He wrote to me, I think, believing that I should know his story. People, he said, were wondering what comes after the tobacco program. He answered: drug addiction. He was right. Or he was partly right. His answer would have been complete if he had added screen addiction to drug addiction.

As long as the diverse economy of our small farms lasted, our communities were filled with people who needed one another and knew that they did. They needed one another's help in their work, and from that they needed one another's companionship. Most essentially, the grownups and the elders needed the help of the children, who thus learned the family's and the community's work and the entailed duties, pleasures, and loyalties. When that work disappears, when parents leave farm and household for town jobs, when the upbringing of the young is left largely to the schools, then the children, like their parents, live as individuals, particles, loved perhaps but not needed for any usefulness they may have or any help they might give. As the local influences weaken, the outside influences grow stronger.

And so the drugs and the screens are with us. The day is long past when most school-age children benefitted from work and instruction that gave them in turn a practical assurance of their worth. They have now mostly disappeared from the countryside and from the streets and houseyards of the towns. In this new absence and silence of the children, parents, teachers, church people, and public officials hold meetings to wonder what to do about the drug problem. The screen problem receives less attention, but it may be the worse of the two because it wears the aura of technological progress and social approval.

The old complex life, at once economic and social, was fairly coherent and self-sustaining because each community was focused upon its own local countryside and upon its own people, their needs, and their work. That life is now almost entirely gone. It has been replaced by the dispersed lives of dispersed individuals, commuting and consuming, scattering in every direction every morning, returning at night only to their screens and carryout meals. Meanwhile, in a country everywhere distressed and taxed by homelessness, once-used good farm buildings, built by local thrift and skill, rot to the ground. Good houses, that once sheltered respectable lives, stare out through sashless windows or have disappeared.

I have described briefly and I am sure inadequately my home country, a place dauntingly complex both in its natural history and in its human history, offering much that is good, much good also that is unappreciated or unrecognized. Outsiders passing through, unaware of its problems, are apt to think it very beautiful, which partly it still is. To me, and to others known to me, it is also a very needy place. When I am wishing, as I often do, I wish its children might be taught thoroughly and honestly its own history, and its history as a part of American history. I wish every one of its schools had enough biologists and ecologists to lead the students outdoors, to show them where they are in relation to drainages, soils, plants, and animals. I wish we had an economy wisely kind to the land and the people.

A good many years ago somebody, or several somebodies, named this parcel of land "The Golden Triangle." Like I assume most people here, I don't know who the somebodies were. I don't know how or what they were thinking or what their vision was. I know that the name "The Golden Triangle" is allied to other phrases or ideas, equally vague and doubtful, that have been hovering over us: the need for "job creation," the need to "bring in industry," the obligation (of apparently everybody) to "compete in the global economy," the need (of apparently everybody) for "a college education," the need for or the promises of "the service economy," and "the knowledge economy." None of these by now weary foretellings has anything in particular to do with anything that is presently here. They and the thinking they represent all gesture somewhat heroically toward "the Future," another phrase, obsessively repeated by the people out front in politics and education, signifying not much. Perhaps the most influential "future" right now is that of "the knowledge economy," as yet not here but surely expected. This means that in order to get jobs and to compete in the global economy, our eligible young people need to major in courses of science, technology, engineering, and mathematics while they are still in high school. This is the so-called STEM curriculum, dear to the hearts of our several too expensive, overadministered, underfunded, and ravenous state universities. And STEM is promoted by slurs, coming from the highest offices of state government, against such studies as literature and history.

The advantage of the STEM-emphasis to the education industry is fairly obvious. And if the great corporations of the global knowledge economy settle in The Golden Triangle or somewhere nearby, they surely will be glad to have a highly trained workforce readily available. But no supreme incarnation of the knowledge economy has yet arrived. If such an arrival is imminent or expected, that has not been announced to the natives. No doubt for that reason, the authorities have not predicted how many STEM graduates the future is going to need (and, as predicted, pay well). The possibility that the schools may turn out too many expensively educated, overspecialized STEM graduates evidently is not being considered. Nor evidently is the possibility that a surplus of such graduates, like their farming ancestors, will have no asking price, and so will come cheap to whomever may hire them. Maybe someday the people living here will have a fine, affluent Scientific, Technological, Engineered, and Mathematized Future to live in. Or, of course, maybe not.

That, anyhow, is development as we know it in The Golden Triangle. Meanwhile, our land is going to the devil, and too many of our people are addicted to drugs or screens or to mere distraction.

For a person living here, it is possible to imagine an economic project that would be locally appropriate and might actually help us. This likely is a project that could not be accomplished by economists only, but economists surely will be needed. The project would be to define a local or regional economy that, within the given limits, would be diverse, coherent, and lasting. If they were not so fad-ridden, economists might see that a knowledge economy, or any other single economy, cannot and should not occupy a whole region or a whole future. They might consider the possibility of a balance or parity of necessary occupations.

I am assuming a need for any locality, region, or nation to provide itself so far as possible with food, clothing, and shelter. Such fundamental economic provision, one would think, should be considered normal or fitting to human inhabitation of the earth. In addition to the economic benefits to local people of local supplies, a future-oriented society such as ours ought to consider the possibility that any locality might become stranded by lasting interruptions of long-distance transportation. Since for many years I have been trying to think as a pacifist, I feel a little strange in addressing issues of military strategy. But it seems preposterous to me that we should maintain an enormous, enormously expensive armory of weapons, including nuclear bombs, ready at every moment to defend a country in which most people live far from the sources of their food, clothing, and other necessities. Arguing from our leaders' own premises, then, the need for balanced local economies is obvious. From the recognition of local needs, both visible and supposable, the people of this or any region might reasonably proceed to a set of questions needing to be answered. Eventually, I think, there would be many such questions. I am sure that I don't know them all, but it is easy to foresee some of them:

1. After so long a history of diminishment and loss, what remains here, in the land and people of this place, that is valuable and worth keeping? Or: What that is here do the local people need for their own use and sustenance, and then, the local needs met, to market elsewhere?
2. What is the present use or value of the local land and its products to the local people?
3. How might we earn a sustainable income from the local land and its products? This would require adding value locally to the commodities—the goods!—coming from our farms and woodlands, but how might that be done?
4. What kinds of work are necessary to preserve and to live from the productivity of our land and people?

5. What do our people need to know, or learn and keep in mind, in order to accomplish the necessary work? The STEM courses might help, might be indispensable, but what else is needed? We are talking of course about education for livelihood, but also for responsible membership, citizenship, and stewardship.

6. What economic balances are necessary to reward adequately, and so to maintain indefinitely, the necessary work?

To answer those questions, close and patient study will be required of economists and others. The difficulty here is that, within the terms and conditions of the dominant economy over the last century and a half, the communities and economies of land use have been increasingly vulnerable. The effort to make them something like sustainable would have to begin with attention to the difference between the industrial economy of inert materials and monetary abstractions and an authentic land economy that must include the kindly husbandry of living creatures. This is the critical issue. As for many years, we are still hearing that almost any new technology will "transform farming." This implies an almost-general approval of the so far unrestrainable industrial prerogative to treat living creatures as comprising a sort of ore, and the food industry as a sort of foundry. If farming is no more than an industry to be unendingly transformed by technologies, as is still happening, then farmers can be replaced by engineers, and engineers finally by robots, in the progress toward our evident goal of human uselessness. If, on the contrary, because of the uniqueness and fragility of each one of the world's myriad of small places, the land economies must involve a creaturely affection and care, then we must look back fifty or sixty years and think again. If, as even some scientists have recognized, there are natural and human limits beyond which farming (and forestry) cannot be industrialized, then we need a more complex and particularizing language than the economists so far use.

The six questions I have proposed for my or any region do not derive from a wished-for or a predicted future. They have to do with what I would call "provision," which depends upon being attentively and responsibly present in the present. We do not, for example, love our children because of their potential to become well-trained workers in a future knowledge economy. We love them because we are alive to them in this present moment, which is the only time when we and they are alive. This love implicates us in a present need to *provide*: to be living a responsible life, which is to say a responsible economic life.

Provision, I think, is never more than caring properly for the good that you have, including your own life. As it relates to the future, provision does only what our oldest, longest experience tells us to do. We must continuously attend to our need for food, clothing, and shelter. We must care for the land, care for the forest, plant trees, plant gardens and crops, see that the brood animals are bred, keep the house and the household intact. We must teach the children. But provision does not foresee, predict, project, or theorize the future. Provision instructs us to renew the roof of our house, not to shelter us when we are old—we may die or the world may end before we are old—but so we may live under a sound roof now. Provision merely accepts the chances we must take with the weather, mortality, fallibility. Perhaps the wisest of the old sayings is "Don't count your chickens before they hatch." Provision accepts, next, the importance of diversity. Perhaps the next-wisest old saying is "Don't put all your eggs in one basket." When the bad, worse, or worst possibility presents itself, provision only continues to take the best possible care of what we have, or of what we have left.

The answers to my questions of course will affect the future. They might even bring about the "better future for our children" so famous with some politicians. But the answers will not come from the future. We must study what exists: what we know of the past, what we know now, what we can see now, if we look. It is likely that, if we look, we will see a need for the STEM disciplines, for we know already their capacity to serve some good purposes. But we will see that the need for them is limited by, for one thing, the need for other disciplines. And we will see a need also not to allow the value of highly technical knowledge to depress the value of the equally necessary and respectable knowledge of land use and land husbandry.

From its beginning, industrialism has depended on a general

willingness to ignore everything that does not serve the cheapest possible production of merchandise and, therefore, the highest possible profit. And so to look back and think again, we must acknowledge real needs that have continued through the years to be real, though unacknowledged: the need to see and respect the inescapable dependence even of our present economy, as of our lives, upon nature and the natural world; and upon the need, just as important, to see and respect our inescapable dependence upon the economies—of farming, ranching, forestry, fishing, and mining—by which the goods of nature are made serviceable to human good.

And now, because it seems to be somewhat conventionally assumed that we are "moving on from farm and factory," we need to recognize again our inescapable dependence upon manufacturing. This does not imply that we must be dependent always and for every product upon large corporations and a global economy. If manufacturing as we have known it is in decline, then that gives room to the thought of a genuinely domestic and conserving economy of provision. This would be a national economy made up of local economies, which, to an extent naturally and reasonably possible, would be complete, self-sustaining, and local in scale. For example, in a town not far from where I am writing, we have recently gained a small, clean, well-equipped, federally inspected slaughtering plant, which completes locally the connection between local pastures and local kitchens, while providing work to local people. There is no reason for this connection and provision to be more extensive. To make the connection between pasture and kitchen by way of the industrial food system is to siphon livelihoods and life itself out of the rural communities.

We also have woodlands here that could even now produce a sustainable yield of valuable hardwoods. But trees cut here at present leave here as raw lumber or saw logs, at the most minimal benefit to the community. Other places and other people may prosper on the bounty of our forest, but not our place and, except minimally for the sellers and a few workers, not our people. It is not hard, considering this, to imagine a local forest economy, made up of small enterprises that would be, within the given limits, complete and coherent, yielding local livelihoods from the good use and care of the living forest to the production of lumber for buildings to finished cabinetry. The thought of such economies is of the nature of provision, not of projection, prediction, or contingency planning. The land and the people are here now. The *present* economic questions are about the work by which land and people might thrive mutually in the best health for the longest time, starting now.

To think well of such enterprises, and of the possibility of combining them in a diverse and coherent local economy, is to think of the need for sustaining all of the necessary occupations. Because a local, a *placed*, economy would be built in sequence from the ground up, from primary production to manufacturing to marketing, a variety of occupations would be necessary. Because all occupations would be necessary, all would be equally necessary. Because of the need to keep them all adequately staffed, it would be ruinous to prefer one above another by price, custom, or social prejudice. There must be a sustained economic parity among them.

In such an economic structure the land-using occupations are primary. We must be mindful of what is, or should be, the fundamental difference between agriculture or forestry and mining, but until the farmers, ranchers, foresters, and miners have done their work, nothing else that we count as economic can happen. And unless the land users do their work *well*— which is to say without depleting the fertility of the earth's surface—nothing we count as economic can happen for very long.

The land-using occupations, then, are of primary importance, but they are also the most vulnerable. We must notice, to begin with, that almost nobody in the supposedly "higher" occupational and social strata has ever recognized the estimable care, intelligence, knowledge, and artistry required to use the land without degrading or destroying it. It is as customary now as it was in the Middle Ages to regard farmers as churls— "mind-numbed," backward, laughable, and dispensable. Farmers may be the last minority that even liberals freely stereotype and insult. If farmers live and work in an economic squeeze between inflated purchases and depressed sales, if their earnings are severely depressed by surplus production, if they are priced out of the land market, it is assumed that they deserve no better: They need only to be "liberated from their chains."

*

The problem to be dealt with here is that the primary producers in agriculture and forestry do not work well inevitably. On the contrary, in our present economy there are constraints and even incentives that favor bad work, the result of which is waste of fertility and of the land itself. Good work in the use of the land is work that goes beyond production to maintenance. Production must not reduce productivity. Every mine eventually will be exhausted. But where the laws of nature are obeyed in use—as we know they can be, given sufficient care and skill—a farm, a ranch, or a forest will remain fertile and productive as long as nature lasts. Good work also is informed by traditional, locally adapted ways that must be passed down, taught and learned, generation after generation. The standard of such work, as the lineages of good farmers and of agrarian scientists have demonstrated, cannot be established only by "the market." The standard must be partly economic, for people have to live, but it must be equally ecological in order to sustain the possibility of life, and if it is to be ecological it must be cultural. The economies of agriculture and forestry are vulnerable also because they are exceptional, in this way, to the rule of industry.

To obtain the best work in the economies of land use, those who use the land must be enabled to afford the time and patience necessary to do the best work. They must know how, and must desire, to do it well. Owners and workers in the land economy who grow their own food will not likely be starved into mistreating their land. But they can be taxed and priced into mistreating it. And so the parity of necessary occupations must be supported by parity of income.

Parity in this sense is not a new thought, although new thinking may be required in applying it to the variety of crops and commodities produced in a variety of regions. But we do fortunately have some precedence for such thinking. The Agricultural Adjustment Act defines parity as "that gross income from agriculture which will provide the farm operator and his family with a standard of living equivalent to those afforded persons dependent upon other gainful occupation." Perhaps the idea of parity does not need much explanation or defense. If, as now and always, a sufficient staff of land-users is necessary to the health of the land and therefore to the lives of all of us, then they should be assured a decent livelihood. And this the so-called free market cannot provide except by accident.

The concept of parity, as fair-minded as it is necessary, addresses one of the problems of farming and farmers in the industrial economy. Another such problem, more fundamental and most in need of understanding, is that of overproduction. "Other gainful employment" in the cities escapes this problem because the large industrial corporations have not characteristically overproduced. Overproduction moreover is not a problem of subsistence farming, or of those enterprises of any farm that are devoted to the subsistence of the farm family. The aim of the traditional economy of the farm household—a garden, poultry, family milk cow, meat animals, vines, fruit trees—was *plenty*, enough for the family to eat in season and to preserve, plus some to share or to sell. Surpluses and scraps were fed to the dogs or the livestock. There were no leftovers.

Surplus production is a risk native to commercial agriculture. This is because farmers individually and collectively do not know, and cannot learn ahead of time, the extent either of public need or of market demand. Given the right weather and the "progressive" application of technologies, their failure to control production, even in their own interest, is thus inevitable. This is not so much because they won't, but because, on their own, they can't. Either because the market is good and they are encouraged, or because the market is bad and they are desperate, farmers tend to produce as much as they can. They tend logically, and almost by nature, toward overproduction. In the absence of imposed limits, overproduction will fairly predictably occur in agriculture as long as farmers and the land remain productive. It has only to be allowed by a political indifference prescribed by evangels of the "free market." For the corporate purchasers the low price attendant upon overproduction is the greatest benefit, as for farmers it is the singular cruelty, of the current agricultural economy. Farm subsidies without production controls further encourage overproduction. In times of high costs and low prices, such subsidies are paid ultimately, and quickly, to the corporations.

This version of a farm economy pushes farmers off their farms. By increasing the wealth of urban investors and shoppers for "country places," it increases the price of farmland, making it impossible especially for small farmers, or would-be small farmers, to compete on the land market. The free market lays down the rule: Good land for investors and escapists, poor land or none for farmers. Young people wishing to farm are crowded to the economic margins and to the poorest land, or to no land at all. Meanwhile overproduction of farm commodities always implies overuse and abuse of the land.

The traditional home economies of subsistence, while they lasted, gave farmers some hope of surviving their hard times. This was true especially when the chief energy source was the sun, and the dependence on purchased supplies was minimal. As farming became less and less subsistent and more and more commercial, it was exposed ever more nakedly to the vagaries and the predation of an economy fundamentally alien to it. When farming is large in scale, is highly specialized, and all needs and supplies are purchased, the farmer's exposure to "the economy" is total.

It ought to be obvious that an economy that works against its sources will finally undercut the law of supply and demand in the most fatal way, that is, by destroying the supply. A food economy staffed by producers who are always fewer and older, whose increasing dependence on industrial technologies puts them and their land at ever greater risk, obviously cannot feed without limit an increasing population. But the reality of such an increasing scarcity is unaccounted for by the doctrine of the free market as applied to agriculture. Even less can this version of freedom comprehend the need for strict limits upon land use in order to preserve for an unlimited time the land's ability to produce. In a natural ecosystem, even on a conservatively managed farm, the fertility cycle may turn from life to death to life again to no foreseeable limit. By opposing to this cycle the delusion of a limitlessness exclusively economic and industrial, the supposedly free market overthrows the limits of nature and the land, thus imposing a mortal danger upon the land's capacity to produce.

When agricultural production is not controlled by a marketing cooperative such as the tobacco program, the market becomes, from the standpoint of the farmers, a sort of limitless commons, the inevitable tragedy of which is inherent in its limitlessness. In the absence of any imposed limit that they collectively agree to and abide by, all producers may have as large a share of the market as they want or can take. Only in this sense is the market, to them, "free." To limit production as a way of assuring an equitable return to producers is assuredly an abridgement of freedom. But freedom for what? For producers, it is the freedom to produce themselves into bankruptcy—to fail, that is, by succeeding. For the purchaser, it is the freedom to destroy the producers as a normal and acceptable expense. The only solution to the tragedy of the limitless market is for the producers to divide their side, the selling side, of the market into limited fair shares by limiting production, which is exactly what the tobacco program accomplished here in my region. By preventing the farmers' overuse of the market, it prevented as well the overuse of the land.

Agribusiness corporations of course don't openly advocate overproduction. They don't have to assume visibly the moral burden of their bad motive. All they have to do is stand by, praising American agriculture's record-breaking harvests, while either hope or despair drives the remaining farmers to produce as much as they can. The agribusinesses then are glad to sell the very expensive surplus of seeds, chemicals, and machines needed for surplus production.

The agricultural tragedy of the market is in part political. And how was the by-now entirely dominant political position on the agricultural free market defined? In the middle of the twentieth century, think tanks containing corporate and academic experts laid down the decree that there were too many farmers. They decreed further that the excess should be removed as rapidly as possible, and that the instrument of this removal should be the free market, with all price supports and production controls eliminated. The assumption evidently was that the removed farmers would be replaced by industrial technologies, recommended by the land-grant universities, and supplied by the corporations. The surplus farmers would increase the industrial labor force, and they and their families would enlarge the population of consumers of industrial products. It was proposed of course that all of society, including the displaced farm families and farm workers, would benefit from this. There would be no costs, social or agricultural, no problems, no debits, nothing at all to subtract from the accrued economic and social assets. This would institute an evolution-ary process that would unerringly eliminate "the least efficient producers." Only the fittest would survive.

In short, by granting a limitless permission and scope to the free market and technological progress—which is assumed to work invariably for the best—politicians, by doing merely nothing, could rid themselves of any concern for farmers or farmland. The representatives of the people and the guardians of the common good were thus able to "free" the market to promote the (allegedly) inefficient farmers to (supposedly) the suburbs while subjecting the countryside to limitless progress and modernization. Against this heartless determinism, it is useful to remember that it was the aim of the program for burley tobacco in my region to include and help every farmer, even the smallest, who wanted to grow the crop. The differ-ence was in the minds of the people whose work during four decades at last shaped an effective program. Those people, un-like the experts of the midcentury think tanks, were thoughtful of the needs of farming and farmers as opposed to the needs of the corporate free market known as the economy. The doctrine of "too many farmers" has never been revoked. No limit to the attrition has been proposed.

As evidence of the persistence of this doctrine, here is a pas-sage from a letter of October 3, 2016, from John Logan Brent, Judge Executive of Henry County, Kentucky:

I have taken a couple of afternoons to work on the ac-counting for farming cattle under the current terms. Enclosed you will find that product based upon a real example, which is our 100 acre farm . . . and its ap-proximately 25 cow herd. . . . The good news is that for a young man wishing to earn a middle, to slightly be-low middle class annual salary of $45,000 farming cattle full-time, he only has to have $3,281,000 in capital to get started. If he can find 780 acres to rent, he only has to have $551,000 for used cows and equipment. I say this is the good news, because the reality is that this was based on a weaned calf price of $850 from June of this year. According to today's sales reports, that same calf is now $650 at best.

That alone, forgetting other adverse agricultural markets, would be an excellent recipe for the elimination of farmers. And conservationists should take note, as mostly they have not done, that in the absence of the eliminated farmers and with the consequent increase of agricultural dependence on the fossil fuels and toxic chemicals, there will be more pollution of water and air.

The related problems of low prices and overproduction of a single but significant crop were solved for about sixty years, in my part of the country and in others, in the only way they could be solved: by a combination of price supports and production controls. This was the purpose and the work of the tobacco program. I want now to look more closely at the Burley Tobacco Growers Co-operative Association, not this time as the brightest public occurrence in the history of my home countryside, but in terms of the suitability of its economic strategy to farming everywhere.

Here I must acknowledge that this organization and, more important, its economic principles have had the allegiance and the service of members of my family for three generations. Beginning in the winter of 1941, when the "Burley Association" renewed its work under the New Deal, my father, John M. Berry, Sr., served as vice president for sixteen and as president for eighteen years, retiring in 1975 but serving as an advisor until a few years before his death in 1991. My brother, John M. Berry, Jr., served as president from 1987 to 1993. My daughter, Mary Berry, started the Berry Center in New Castle, Kentucky, in 2011 for the purpose mainly of remembering, advocating, and applying the Association's proven economic strategy and its purpose of assuring a decent livelihood for small farmers. My son serves on the Berry Center board.

Under this program, support prices for the various grades of tobacco were set according to a formula for assuring a fair return on the cost of production. Production was controlled by allotting to each farm, according to its history of production, at first an acreage, and later a poundage, that would be eligible for price supports under the program. The total of the allotments for each year was determined by the supply, worldwide, that was available for manufacture. The rule was that the supply on hand should be sufficient for 2.4 years. If the supply was less than the predicted demand for 2.4 years, allotments would be increased; if more, allotments would be reduced. I don't know why the factor was set at 2.4. Its significance, however, is that production was limited according to an established measure of expected demand.

To buy a crop or a portion of a crop protected by the program, a purchaser had to bid a penny a pound above the support price. The government's assistance to the program consisted of a loan, made annually "against the crop," which permitted the program to purchase, store, and resell the portion of any year's crop that did not earn the extra penny a pound—which, thanks to the loan, would be purchased by the Association and the grower paid at the warehouse. The cost to the government was only administrative until, in response to protests, this cost was charged to the farmers, and the program then operated on the basis of "no net cost" to the government. This program succeeded remarkably well in doing what it was designed to do, and a part of its success is that it still provides a pattern for the thought and hope of those who are working for the survival of land and people.

The tobacco program is an example of a necessary service that government can provide to people who cannot provide it to themselves. The point most needing to be made now is that parity of pricing under the tobacco program was in no sense a subsidy. It did not involve a grant of money, a government giveaway, or a public charity. The concept of parity was used, by intention, to *prevent* government subsidation. Its purpose was to achieve fair prices, fairly determined, and with minimal help from the government. My father defended parity as an appropriate incentive: "It accords with our way of life, and it gives real and tangible meaning to the philosophy of 'equal opportunity.'" He thought of "direct subsidy payments" as virtually opposite to parity and an "abominable form of regimentation."

The tobacco program in all of its versions was finally defeated and destroyed in 2004 by the political free marketers who had always opposed it, and who had resented it in proportion to its success. During the six decades of its life, the Burley Tobacco Growers Co-operative Association helped keep farm families on their farms and gainfully employed in Kentucky, Missouri, Indiana, Ohio, and West Virginia. One measure of its success was the decrease of farm tenancy among the growers from 33 percent in 1940 to 9 percent in 1970. During those years some of the population of tenant farmers undoubtedly died, and some left farming, but most of them ceased to be tenant farmers by becoming owners of farms. This was a defining event in the lives of a considerable number of worthy people whom I knew. The farmer-members of the Association overwhelmingly renewed their support in referendum after referendum.

The Burley Association was thus truly a commons and a common good, based not only upon correct political and economic principles, but also upon the common history and culture, and thus upon the understanding consent of its sharers. So complete was the understanding of the members that in 1955, because of an oversupply of tobacco in storage, they voted for a 25 percent reduction of their allotments. On April 8, 2016, my neighbor Thomas Grissom, by far the best historian of the Association, wrote in a personal letter to me:

> After years of research, I have concluded that the most distinctive characteristic of the Kentucky [Burley] Tobacco Program is its design and application of an industrial agriculture commodity program to the cultivation and production of an agrarian crop indigenous to an agrarian society.

I think that Tom's perception is exactly right and that he found the right and necessary terms to describe it.

Burley tobacco, despite the dire health problems that it was found to cause and the consequent disfavor, was very much an agrarian crop. It was characteristically and mainly the product of small family farms, produced mainly by family labor and exchanges of work among neighbors. It was for a long time the staple crop in a highly diversified way of farming on landscapes that typically required considerate and affectionate care. As long as the market paid highly for high quality (which it finally ceased to do), the production of burley tobacco demanded, and from its many highly competent producers it received, both conscientious land husbandry and a fine artistry.

Industrialism and agrarianism are almost exactly opposite and opposed. Industrialism regards mechanical or technical functions as ideal. It rates its accomplishments by quantitative measures. Though it values the prestige of public charity, it is motivated necessarily by the antisocial traits that assure success in competition. Agrarianism, by contrast, arises from the primal wish for a home land or home place—the wish, in the terms of our tradition, for the freedom and independence that come with dependence on a parcel of land, however small, that one owns and is owned by or has at least the use of. Agrarianism grants its highest practical value to the good husbandry of the land. It is motivated, to an extent effective and significant, by neighborliness, family loyalty, and devotion to the coherence and longevity of communities.

As long as it has a sufficiency of "natural resources" and remains free of imposed political or economic restraint, an industrial economy will dominate and destroy an agrarian economy —no matter that the agrarian economy is indispensable for a continuing supply of resources. This defines precisely the need for the "design and application of an industrial agricultural commodity program to the cultivation and production of an agrarian crop indigenous to an agrarian society." For a while the Burley Tobacco Growers Co-operative Association—never mind the deserved infamy of tobacco—did preserve a sort of balance between the interests of industrialism and agrarianism, which prevented their inherent difference and opposition from becoming absolute, and thus absolutely destructive of the agrarian society. This balance was fair enough to the industry and it permitted the growers to prosper. The program worked in fact to the best interest of both economies.

From the perspective of this balance during the decades when it worked as it should have, it is possible to see that a step too far toward industrialism was probably taken by the Burley Association itself when, in 1971, it permitted the "lease and transfer" of production quotas away from the farms to which they had been assigned. This change, made under pressure from industrializing members, permitted the accumulation of allotments finally into very large acreages dependent upon more extensive technology and migrant labor. The program then was obliged to "balance" a reduced agrarianism against an increased industrialism.

With the demise of the program in 2004, the region's indigenous agrarianism could survive only as a history, a memory, and a set of vital principles that someday may be revived and reincarnated in reaction against the damages of industrialism.

*

For the past six decades, except for such a remnant of the New Deal, the government has done nothing for farmers except to quiet them down by subsidizing uncontrolled production, which really is worse than nothing. But this "policy," in the minds of the dominant politicians, signified that they were "doing something for agriculture" and so relieved them of thinking or knowing about agriculture's actual requirements. For example, the Democratic platform preceding President Clinton's first term initially contained no agricultural plank. My brother, John M. Berry, Jr., who was on the platform committee, was dismayed by this innovation, and he said so. He was then told that a plank was being drafted. When he saw the result, he laughed. He asked if he might draft a more meaningful plank. After much resistance, he was allowed to do so. He then "spent the next six hours redrafting the amendment so as to satisfy the Clinton staff." I am quoting his letter of June 29, 1992, to Dr. Grady Stumbo, chair of the Kentucky Democratic Party. The letter goes on to say that Clinton's staff refused to permit any reference to

> "supply management," "price support" or any government guarantee of a fair price for farmers. They also refused to permit any reference to agribusiness control of farm policy or the level of agribusiness profits. They also refused to permit any language that could be construed as a commitment . . . to anything specific for agriculture or the rural community. . . .
>
> I had already been advised that Chairman Ron Brown had formed an agriculture task force and sold seats to its members for $15,000* contributions to the Democratic National Committee.
>
> Those seats went to representatives of agribusiness and other interests that have traditionally written farm policy for the Republicans.

The doctrine of "too many farmers" thus had become the established orthodoxy of the leaders of both parties. My brother was then president of the Burley Tobacco Growers

---

*My brother told me not long before his death in 2016 that the "contributions" actually were $30,000.

Co-operative Association, which was still a major life support of our state's small farmers. By 1992 tobacco had become indefensible as a product, and it bore too great a public stigma to be touchable by a national candidate. My brother understood that, and he did not expect approval specifically of the tobacco program. But he knew that the working principles of that program would protect farmers who produced commodities other than tobacco everywhere in the country—and would also protect our own farmers when they no longer produced tobacco. He knew that Mr. Clinton, if he wanted to, could endorse the program's principles without endorsing its product. The agricultural plank of the 1992 Democratic platform, as published, gave a general approval to "family farmers receiving a fair price," to "a sufficient and sustainable agricultural economy . . . achieved through fiscally responsible programs," and to "the private–public partnership to ensure that family farmers get a fair return for their labor and investment." And of course it condemned "Republican farm policy." It committed Mr. Clinton and his party to do nothing. And nothing was what they did.

In 1995 President Clinton spoke to an audience of farmers and farm leaders in Billings, Montana. He acknowledged that the farm population by then was "dramatically lower . . . than it was a generation ago." But, he said, "that was inevitable because of the increasing productivity of agriculture." Nevertheless, he wanted to save the family farm, which he held to be "alive and well" in Montana. He believed we had "bottomed out in the shrinking of the farm sector." He said he wanted to help young farmers. He spoke of the need to make American agriculture "competitive with people around the world." And so on.

He could not have meant what he said, because he was speaking without benefit of thought. And why should he have thought when he was not expected to do so? He was speaking forty or fifty years after politicians and their consulting experts had abandoned any effort to think about agriculture. "Inevitable" is a word much favored by people in positions of authority who do not wish to think about problems. When and why did Mr. Clinton in 1995 think that the inevitable "shrinking of the farm sector" had ceased? In fact, "the farm sector" had not bottomed out in 1995; there is no good reason to think that it has bottomed out, at less than 1 percent of the population, in 2016. And how could he have helped young farmers except by giving them the protections against the free market that my brother had recommended three years before? Mr. Clinton was talking nonsense in Billings in 1995 because he did not have, and could not have had from his advisers, the means to think about what he thought he was talking about. The means of actual thought about the use and care of the land had been intentionally discounted and forgotten by people such as themselves.

It appears to be widely assumed by politicians, executives, academics, public intellectuals, industrial economists, and the like that they have a competent understanding of agriculture because their grandparents were farmers, or they have met some farmers, or they worked on a farm when they were young. But they invoke their understanding, which they do not have, only to excuse themselves from actual thought about actual issues of agriculture. These people have found "inevitability" a sufficient explanation for the deplorable history of industrial agriculture. They see the reason for the present discontent of "blue-collar" voters as low or "stagnant" wages. They don't see, in back of that, the dispossession that made many of them wage-workers in the first place. The loss everywhere of small farms and small towns and the respectable livelihoods that they provided was ruled "inevitable" and thus easily explained and forgotten. In their perceived worthlessness and dispensability, at least, the people of the farms and small towns were in effect racially equal. If, for instance, black small farmers were helped to prosper, as some liberals would have liked, then white small farmers would have had to be helped to prosper, which would have pleased neither liberals nor conservatives.

It was, then, "inevitable" that the independent livelihoods in the old economies of the countryside and the small towns should be replaced by the mainly subservient livelihoods in industry, or by unemployment. But if the "working class" counted for nothing and were dispensable as small farmers or farmhands or as small independent keepers of shops and stores or as independent tradespeople and craftspeople, why then should they count for something and be more than dispensable as "blue-collar workers"? In the corporate and urban economy the blue-collar workers were just as "inevitably" replaceable by technologies as they had been before. They were then notified that they were losing out because they were "uneducated." They needed "a college education," in default of which they were offered "retraining" and "job creation." But these were only political baits, which left the blue-collared ones to their "inevitable" fate of low or stagnant wages or unemployment.

This doctrine of inevitability, also known as technological progress, is in fact a poor excuse for an economic and technological determinism, as heartless as it is ignorant, which has belonged about equally to the political establishment of both parties. Realizing that they were the broken eggs of an omelet that others would eat, the blue-collar workers became angry. Their anger turned them to Donald Trump, who at least recognized their existence and the political usefulness of their anger.

In the pre-Trump version of the history of progress, determinism and inevitability overruled any need for actual knowledge and actual thought. But with the ascendancy of Mr. Trump, at least some of the determinists seem to be reverting to free will. While the conservatives, who have strained at a gnat and swallowed a camel, endeavor to digest their dinner, the liberals talk of "connecting" with the blue-collar workers of rural America, to whom they have given not a substantive thought since Ezra Taft Benson, Eisenhower's secretary of agriculture, pronounced to their grandparents the political death sentence, "Get big or get out."

Let us remember also the workers, white and black, who in their thousands became simply obsolete at the instant when "efficient" machines were brought into the coal mines, the factories, and the fields of sugarcane and cotton. I thought of them when I read in a column by Roger Cohen in the *New York Times* of November 19, 2016, that "the very essence of the modern world" is "the movement of people and ever greater interconnectedness, driven by technology." Mr. Cohen approves of this "essence" and is afraid that Mr. Trump will stop it. What he has in mind surely must be the *voluntary* movement of people. The movements of people *actually* "driven by technology" are outside Mr. Cohen's field of vision, surely only because of his political panic. Millions of people, as we know, have been driven away from their homes in the modern world by the similarly imperative technologies of industrial production and industrial war.

I am uncertain what value Mr. Cohen assigns to "interconnectedness," but he cannot be referring to the interconnectedness of families in their home places, or of neighbors in their neighborhoods. How the loss of those things can be compensated by movement, driven or not, is far from clear. The same obscurity clouds over any massive "movement of peoples," as over the arguments by which these movements are excused or justified. It does not require a great refinement of intellect to see the harm that is in all of them.

The experts who decided in the middle of the twentieth century that there were too many farmers had in fact no agricultural knowledge or competence upon which to base such a judgment. They and their successors certainly had not the competence to assume any responsibility for, or in any way to mitigate, the totalitarian displacement of about twenty million farmers.

Farming is one of the major enactments of the connection between the human economy and the natural world. In the industrial age farming also enacts the connection, far more complicated and perilous than industrialists admit, between industrial technologies and living creatures. Some science certainly needs to be involved, also more and better accounting. But good farming is first and last an art, a way of doing and making that involves human histories, cultures, minds, hearts, and souls. It is not the application by dullards of methods and technologies under the direction of a corporate-academic intelligentsia.

If we should want to revive, or begin, in a public way the actual thinking about agriculture that has actually taken place in some cultures, that is still taking place in some small organizations and on some farms, what would we have to do? We would have to begin, I think, by giving the most careful attention to issues of carrying capacity, scale, and form, to issues of production, of course, but also and just as necessarily to issues of maintenance or conservation. The indispensable issue of conservation would apply, not just to the farm's agricultural "resources," but also to the ecosystem that includes the farm and to the waterways that drain it. I think, moreover, that this attention to issues must be paid always outdoors in the presence of examples. The thing of greatest importance is to think about the land with the land's people in the presence of the land. Every theory, calculation, graph, diagram, idea, study, model, method, scheme, plan, and hope must be caught firmly by the ear and led out into the weather, onto the ground.

It is obvious that this effort of thinking has to confront everywhere the limits both of nature and of human nature, limits imposed by the ecosphere and ecosystems, limits of human intelligence, human cultures, and the capacities of human persons. Such thought is authenticated by its compatibility with limits, its willingness to accept limits and to limit itself. This will not be easy in a time overridden by fantasies of limitlessness. A market limitlessly usable by sellers and limitlessly exploitable by buyers is merely normal in such a time. And limitlessness is the common denominator of the dominant political sides, both of which tend to refer to limitlessness as "freedom."

We have the liberal freedom of unrestrained personal behavior, and the conservative freedom of unrestrained economic behavior. These two freedoms are more alike, more allied, and more collaborative than either side would like to admit. Opposition to the industrial economy's ravaging of the landscapes of farming and forestry now comes from a small and scattered alliance of agrarians, not from liberals or conservatives.

Conservatives and liberals disagree passionately about climate change, for example, yet liberal protests against climate change far exceed protests against the waste and pollution that occur locally in industrial agriculture and are its reputed causes. And neither the conservatives who esteem the fossil fuels nor the liberals who deplore them have advocated rationing their use, either to make them last or to reduce their harm. For these people the old ideals of *enough* and *plenty* have been overruled by the ideals of *all you want* and *all you can get*. They cannot imagine that for farmers a limitless market share, like a limitless appetite, can lead only to the related diseases of *too much* and *too little*.

Science, apart from moral limits in scientists, seems to be limitless, for it has produced nuclear and chemical abominations that humans, with their very limited intelligence, can neither limit nor safely live with. "Anything goes" and "Stop at nothing" are the moral principles that some scientists have borrowed apparently from the greediest of conservatives and the most libertine of liberals. The faith that limitless technological progress will finally solve the problems of limitless contamination seems to depend upon some sort of neo-religion.

The good care of land and people, on the contrary, depends primarily upon arts, ways of making and doing. One cannot be, above all, a good neighbor without such ways. And the arts, all of them, are limited. Apart from limits they cannot exist. The making of any good work of art depends, first, upon limits of purpose and attention, and then upon limits specific to the kind of art and its means.

It is a formidable paradox that in order to achieve the sort of limitlessness we have begun to call "sustainability," whether in human life or the life of the ecosphere, strict limits must be observed. Enduring structures of household and family life, or the life of a community or the life of a country, cannot be formed except within limits. We must not outdistance local knowledge and affection, or the capacities of local persons to pay attention to details, to the "minute particulars" *only* by which, William Blake thought, we can do good to one another. Within limits, we can think of rightness of scale. When the scale is right, we can imagine completeness of form.

The first limit to be encountered in making a farm—or a regional or national economy—is carrying capacity: How much can we ask of *this* land, this field or this pasture or this woodland, without diminishing the land's response? And then we come to other limits, perhaps many of them, each one addressing directly our imagination, sympathy, affection, forbearance, knowledge, and skill. And now I must call to mind Aldo Leopold, who, unlike most conservationists since John Muir, could think beyond wilderness conservation to conservation of the country's economic landscapes of farming and forestry. His conception of humanity's relation to the natural world was eminently practical, and this must have come from his experience as a hunter and fisherman, his study of game management, and his and his family's restoration of their once-exhausted Sand County farm. He knew that land-destruction is easy, for it requires only ignorance and violence. But the obligation to restore the land and conserve it requires humanity in its highest, completest sense. The Leopold family renewed the fertility and health of their land by their work, their pleasure, and their love for their place and for one another.

Aldo Leopold thought carefully about farming and forestry because he knew that far more land would be put to those uses than ever could be safeguarded in wilderness preserves. In an essay of 1945, "The Outlook for Farm Wildlife," he laid side by side "two opposing philosophies of farm life" (the italics being his):

1. *The farm is a food-factory*, and the criterion of its success is salable products.
2. *The farm is a place to live.* The criterion of success is a harmonious balance between plants, animals, and people; between the domestic and the wild; between utility and beauty.

This is a statement about form, contrasting a form that is too simple and too exclusive with a form that may be complex enough to accommodate the interest of what is actually involved. Under the rule of the first form, "the trend of the landscape is toward a monotype." This form can be adequately described as the straightest, shortest line between input and income. All else is left out or denied. Such a form concedes nothing to its whereabouts. It is placed upon whatever landscape merely by imposition, as a cookie cutter is imposed upon dough. In its simplicity and rigidity, such a form is bad art, but also, as Leopold knew and as we now know better than he could have, it is bad science.

The second form is described as "a harmonious balance" among a diversity of interests. On such a farm, made whole by the high artistry of farming, every part is both limited and enabled by the others. This harmonious balance, I should not need to say, cannot be prefabricated. It can be realized only uniquely within the boundary of any given farm, according to the natures and demands of its indwelling plants and animals, and according to the abilities, needs, and wishes of its resident human family. Wherever this is fully accomplished, it is a grand masterpiece to behold.

The Presence of Nature in the Natural World: A Long Conversation

THE GREAT TROUBLE of our age, involving the whole human economy from agriculture to warfare, is in our relationship to the natural world—to what we call "nature" or even, still, "Nature" or "Mother Nature." The old usage persists even seriously, among at least some humans, no matter how "objectivity" weighs upon us. "Of all the pantheon," C. S. Lewis wrote, "Great Mother Nature has . . . been the hardest to kill." With Nature we have, properly speaking, a relationship, for the responses go both ways: Nature is fully as capable of responding to us as we are of responding to her. In the age of industrialism, this relationship has been radically brought down to a pair of hopeless assumptions: that the natural world is passively subject either to unlimited pillage as a "natural resource," or to partial and selective protection as "the environment."

We seem to have forgotten that there might be, or that there ever were, mutually sustaining relationships between resident humans and their home places in the world of Nature. We seem to have no idea that the absence of such relationships, almost everywhere in our country and the world, might be the cause of our trouble. Our trouble nonetheless exists, is severe, and is getting worse. Instead of settled husbanders of cherished home places, we have become the willing parasites of any and every place, destroying the source and substance of our lives, as parasites invariably do.

This critical state of things has not always been explicitly the subject of my writing, but it has been constantly the circumstance in which I have written, for I have had constantly and consciously in sight the progressive decline of my home countryside and community. I have been perforce aware that this is a local manifestation of a decline that is now worldwide, affecting not only every place but also the oceans and the air, and I have of course felt a need to understand, and to oppose so far as I have been able, this downslope of all creation. In this effort of thought, I have been always in need of teachers, friends, and allies among the living and the dead. The mercy or the generosity accompanying this effort has been that I have found perhaps not all but many of the teachers, friends, and allies I have needed, often when I have needed them most.

By that I do not mean to suggest that my looking for help has been easy or carefree. To begin with, I never trusted, and after a while I rejected, the hope that many people have invested in what we might call the industrial formula: Science + Technology + Political Will = The Solution. This assumes that the science is adequate or soon will be, that the technology is adequate or soon will be, and therefore that the only essential task is to increase political pressure favoring the right science and the right technology.

An outstanding example of the industrial formula in action is the present campaign against global warming. If the scientific calculations and predictions about global warming are correct (which I am willing to assume that they are, though I have no science of my own by which to know), then its causes are waste and pollution. But "global warming" is a ravenously oversimplifying phrase that gobbles up and obscures all the myriad local instances of waste and pollution, for which local solutions will have to be worked out if waste and pollution are ever to be stopped. The phrase "global warming" suggests no such thing. Global warming, the language insists, is a *global* emergency: a global problem requiring a global solution. To solve it we have the science, we have the technology, and now we have only to prevail upon the world's politicians to enforce the recommended solution: burning less fossil fuel.

But from where I live and watch, I see the countryside and the country communities being wrecked by industrial violences: a heartless gigantism of scale and power, massive and irreparable soil erosion, pollution by toxic chemicals, ecological and social disintegration—and of course an immense burning of fossil fuels, making in turn an immense contribution, as alleged by scientists, to global warming.

I have no doubt at all that even if the global climate were getting better, our abuses of the land would still be the disaster most seriously threatening to the survival of humans and other

creatures. Land abuse, I know, is pretty much a global phenomenon. But it is not happening in the whole world as climate change happens in the whole sky. It is happening, because
it can happen, only locally, in small places, where the people
who commit the abuses also live. And so my question has been,
and continues to be, What can cause people to destroy the
places where they live, the humans and other creatures who are
their neighbors, and ultimately themselves? How can humans
willingly turn against the earth, of which they are made, from
which they live? To treat that as a scientific and technological
or political question is not enough, is even misleading. The
question immediately and at least is economic: What is wrong
with the way we are keeping house, the way we make our living, the way we live? (What is wrong with our minds?) And to
take the economic question seriously enough is right away to
ask another that is also but not only economic: What is happening to our souls?

There is no industrial answer to such questions. Industrialism
has never provided a standard by which such questions can be
answered. I long ago hatched out of the egg in which I could
believe that industrialism is capable of competent judgments
of its effects, let alone competent solutions to the problems it
has caused. Its "solutions," on the contrary, tend to increase
the problems, as in the desperate example of industrial war,
in which more never produces less; or the example, equally
desperate, of agricultural pesticides, which must become more
toxic and diverse as immunities develop among the pests. Since
industry has no language with which to speak to us as living
souls and children of Nature, but only as interchangeable employees, customers, or victims, by what language can we, in the
fullness of our being, speak back to industrialism?

Seeing that my need for help was defined by that question,
I have faced another difficulty: I am inescapably a product of
"Western" culture, first as I was born and grew up in it, and
then as I, by my work, have made myself able to know it and
more responsibly to inherit it. The difficulty was that Western
culture, especially when it is understood as Christian culture,
however decayed, has been for many years in disfavor among
writers and intellectuals, some of whom have seen it as the
very origin of our unkindness toward "the environment." At

its best, this disfavor has produced useful criticism, for professedly Christian institutions, nations, and armies have much cruelty and violence to answer for. At its worst, it is fashionably attitudinal and dismissive, so that we now have "Shakespeare scholars" who cannot recognize Shakespeare's frequent references to the Gospels.

For some, it has been possible and useful to turn away from the Western or Christian inheritance to find instruction and sustenance in, mainly, oriental or tribal cultures. I am glad to acknowledge my own considerable indebtedness to the little I have managed to learn of oriental, American Indian, and other cultures, which have been often confirming and sometimes clarifying of the cultural lineage and faith that I consider my own. From Gary Snyder's Buddhism, for instance, I understood more clearly than from any other source that the practice of a religion is necessarily economic: how we live on and from the earth.

But I am too completely involved in Western culture by the history of my mind, my people, and my place to be capable of a new start in another tradition. My need to make as much sense as I could of my history and experience, as I began fifty years ago to think of my task, clearly depended on my willingness to do so, not only as a native of a small patch of country in Henry County, Kentucky, but also as an heir and inevitably a legator of Western culture. If I hoped to make sense, my culture would have to be always, at least implicitly, my subject, and I would have to be its critic. If I was troubled by our epidemic mistreatment of the natural or given world, especially in the economic landscapes that we live from, then I would have to search among the artifacts and records of my culture, in the time I had and so far as I was able, to find probable causes, appropriate standards, and possible corrections. This search began, as I would later realize, in my school years, before I could have explained, even to myself, my need and purpose.

Eventually, provoked by attacks on Western culture as founded upon a supposedly biblical permission to humans to use the earth and its creatures in any way they might please, I read through the Bible to see how far it might support any such permission, or if, on the contrary, it might impose on humans the obligation to take good care of a world both given to them as a dwelling place, as to the other creatures, and to which they had been given as caretakers.

I found of course Genesis 1:28 ("Be fruitful and multiply, and replenish the earth, and subdue it: and have dominion over the fish of the sea, and over the fowl of the air, and over every living thing that moveth upon the earth"), which is the verse the detractors have found, and is pretty much the sum of their finding. But I also found that the Bible as a whole is a context that very sternly qualifies even the "dominion" given in Genesis 1:28. And I found many verses and passages that, even out of context, require humans to take the best possible care of the earth and its creatures. To speak to Genesis 1:28, for instance, there is the first verse of Psalm 24: "The earth is the Lord's, and the fullness thereof: the world and they that dwell therein." To that verse, singly and without context, if one takes it seriously, there are only two sane responses: (1) fear and trembling, and (2) a human economy that would conserve and revere the natural world on which it obviously depends. But it is possible for materialist environmentalists (as for many professing Christians) to read and not see, even to memorize a passage of language and not know what it says. It is easy, even intellectually permissible, to quote from a book that one has not read. And now most people are too distracted automotively and electronically to know what world they are in, let alone what the Bible might say about it.

In a world allegedly endangered by biblical religion, it is dangerous to be ignorant of the Bible. I must say, now, that I am not an uncritical reader of the Bible. There are parts of it that I dislike, but there are parts of it also in which I place my faith. From those parts, or some of them, I gather an old belief that God is not merely above this world, or behind it, but is *in* it, and it lives by sharing His life and breathing His breath. "If he gather unto himself his spirit and his breath," Elihu says to Job, "All flesh shall perish together, and man shall turn again unto dust" (Job 34:14–15). Like Psalm 24, these fearful verses clearly imply a mandate for the good care of all creation, and this becomes explicit in the Gospels' paramount moral commandment (Matthew 22:39) that we must love our neighbors as ourselves, even when our neighbors happen to be our enemies. This neighborly love cannot be a merely human transaction, for you cannot love your neighbor while you destroy the earth and its community of creatures on which you and your neighbor mutually depend.

The Bible, then, is not defined by misreading, or Western culture by economic violence, any more than a body is defined by a disease.

Over the years, for my soul's sake and for the sake of my work, I have returned many times to the Bible, testing it and myself by such understanding, never enough, as I have been able to bring to it.

I have also returned repeatedly to the pages of several English poets, who lived and worked of course under the influence of the Bible, and whose interest in the natural world and the magisterial figure of Nature has been instructive and sustaining to me.

Both Chaucer and Spenser testify that "this noble goddesse Nature" or "great dame Nature" comes to them from a Latin allegory written in about the seventh decade of the twelfth century by Alanus De Insulis, or Alan of Lille. This book, *De Planctu Naturae* (*Plaint of Nature*), is composed of alternating chapters of verse and prose. It tells the story of Alan's dream-vision in which he converses with Nature, whom he recognizes both as his "kinswoman" and as the Vicar of God. The Latin original apparently is extremely hard to translate. According to James J. Sheridan, whose translation I have used,

> The author revels in every device of rhetoric. . . . He so interweaves the ordinary, etymological and technical signification of words that, when one extracts the meaning of many a section, one despairs of approximating a satisfactory translation.

Despite its intricacy even in English, and the strangeness in our time of Alan's leisurely pleasure in rhetorical devices—he elaborates, for example, a sexual symbolism of grammar—this is a book extraordinarily useful to a reader interested in the history of our thought about the natural world as well as the history of conservation.

At the beginning of his book, Alan's happiness has been overcome by grief because of humanity's abandonment of Nature's laws. His own immediate complaint is against homosexuality, which seems to have been widespread and fairly openly practiced and acknowledged in his time. This he understands as a consequence that comes "when Venus wars with Venus." For my own sake in my own time, I am sorry that he begins this way, for I am much troubled by our current politics of private life. Some of us now may reasonably object to Alan's view of homosexuality as unnatural, or anti-natural, but it would be a mistake to dismiss him on that account at the end of his first sentence.

As Alan continues the story of his vision, his objection to homosexuality becomes incidental to his, and Nature's, objection to lust, which is one of the whole set of sins that are opposed both to the integrity of Nature and to the integrity of human nature. The two Venuses that are at war are the Venus of lust or mere instinct and the Venus of responsible human love. This duality is analogous to, and dependent upon, the double nature of humankind. In the order of things, as in the Chain of Being, humans are placed between the angels and the animals. Strongly attracted in two conflicting directions, they have the hardship of a moral obligation to keep themselves in their rightful place.

Nature, then, as "faithful Vicar of heaven's prince," makes two requirements of us. She requires us to be natural in the general sense that, like other animals, we are born into physical embodiment, we reproduce, and we die. But she most sternly conditions her first requirement by requiring us also to be natural in the sense of being true to our specifically human nature. This means that we must practice the virtues—chastity, temperance, generosity, and humility, as they are named, their procession led by Hymenaeus,* in the *Plaint*—that keep us in our rightful place in the order of things. We must not be prompted by lust, avarice, arrogance, pride, envy, and the other sins, prodigality being the sum of them all, either to claim the prerogatives of gods or to lapse into bestiality and monstrosity.

---

*Hymenaeus was once, until forsaken by her, the husband of Venus. The son of that marriage is Desire. In Alan's mythology, Hymenaeus is "closely related" to Nature and has "the honor of her right hand." He represents the sanctity, and the bad history, of marriage.

This is Nature's requirement because her own integrity and survival depend upon it, but she herself is unable to enforce it. She very properly acknowledges the limit of her power when she explains to Alan that although "man is born by my work, he is reborn by the power of God; through me he is called from non-being into being, through Him he is led from being into higher being."

To put it another way: Humanity as an animal species—"the only extant species of the primate family Hominidae," says *The American Heritage Dictionary*—is a limitless category in which anything we may do can be explained, even justified, as "natural." But the necessity of our communal life as fellow humans, as well as our shared and interdepending life as a species among species, imposes limits, defining, for the sake of our survival and that of the living world, a higher or moral human nature, specified by Nature herself in her conversation with Alan. The nature of every species is uniquely specific to itself. Human nature in its fullest sense is the most complex of all, for it involves standards and choices.

To speak of this from the perspective of our own time and the fashionable environmentalist prejudice against "anthropocentrism," it is our mandated human nature that allows us, by understanding the legitimacy of our own self-interest, or self-love as in Matthew 22:39, to understand the self-interest of other humans and other creatures. A human-centered and even a self-centered point of view is inevitable—What other point of view can a human have?—but by imagination, sympathy, and charity *only* are we able to recognize the actuality and necessity of other points of view.

It is impossible to speak of such things in a positivist or "objective" language, pruned of its upper branches and presided over by professional or specialist consciences embarrassed by faith, hope, and charity. And so it is a relief to resume the freedom and completeness of the language in which we can say that Nature, the Vicar of God, is the maker or materializer of a good world, as affirmed by Genesis 1:31, entrusted to the keeping of humans who must prove worthy of it by keeping true to their own, specifically human, nature, which is defined by that worthiness, which they must choose.

According to Alan's great instructor, the integrity of the

natural world depends upon the maintenance by humans of *their* integrity by the practice of the virtues. The two integrities are interdependent. They cannot be separated and they must not be separately thought about. This is the moral framework of *Plaint of Nature*, and I would argue that it persists in a lineage of English poets from Chaucer to Pope, whether or not they can be shown to have read Alan. The same set of traditional assumptions can be shown also to support the work of a lineage of agricultural writers and scientists of about the last hundred years, as I will later demonstrate. So far as my reading, observation, and experience have informed me, I believe that Nature's imperative as set forth in the *Plaint* is correct, not just for the sake of morality or for the sake of Heaven, but in its conformity to the practical terms and demands of human life in the natural world. The high standards of Nature and of specifically human nature obviously will not be comforting to humans of the industrial age, among whom I certainly must include myself, and they are not in fashion, but that hardly proves them false.

As Alan grieves over the degeneracy of humankind, Nature appears to him. She is "a woman" whose hair shines with "a native luster surpassing the natural." He proceeds to describe in patient detail her hair, her face, and her neck. He devotes seven fairly substantial paragraphs, in Sheridan's translation, to her crown, the jewels of which are elaborately allusive and symbolic. He describes with the same attention to detail and reference her dress, on which is depicted "a packed convention" of all the birds; her mantle, which displays "in pictures . . . an account of the nature of aquatic animals," mainly fish; and her embroidered tunic, a part of which has been damaged by man's abuse of his reason, but on the remainder of which "a kind of magic picture makes the land animals come alive." He discreetly refrains from looking at her shoetops and her undergarments, but is "inclined to think that a smiling picture made merry there in the realms of herbs and trees"—which he then describes.

The presence of Nature is so numinous and exalted, of such "starlike beauty," that Alan faints and falls facedown. Though my interest is fully invested in this book, my own reaction here is much less excited, and I doubt that any reader now could find much excitement in this exhaustive portraiture, though I agree with C. S. Lewis that "the decorations do not completely obscure the note of delight." The problem is that Alan's description of Nature is too distracted by his rhetorical extravagances to give us anything like a picture of her.

We get a better sense of her, I think, in his later invocation, in which he describes Nature's character rather than her appearance:

> O child of God, mother of creation, bond of the universe and its stable link . . . you, who by your reins guide the universe, unite all things in a stable and harmonious bond and wed heaven to earth in a union of peace; who, working on the pure ideas of [Divine Wisdom], mould the species of all created things, clothing matter with form . . .

It is no doubt wrong, or at least pointless, to ask Alan for a "realistic" portrait, as if he were dealing merely with a personification. His effort to describe Nature's appearance, exaggerated as it is, failing as it does, speaks nonetheless of his need to realize her, not merely as an allegorical person, but as a presence actually felt or known.

The really useful question, it seems to me, is not how Alan's Nature functions in a medieval allegory, but what she means historically as a figure very old and long-lasting, of whom people have needed, of whom some people need even now, to speak. How might one explain this need? My reading is sufficient only to assure me that Alan's vision of Nature draws upon a variety of sources, classical and Christian. She came to him as a figure of attested standing and power, he did imagine her, she did appear to him in something like a vision.

In Alan's *Plaint of Nature* and in Christian poems much later, the classical deities continue to appear, still needed by imagination, no longer perhaps as vital gods and goddesses, but at least as allegorical figures standing for their qualities or powers. That Nature takes her place prominently among them in the work of later poets appears to be owing largely to Alan. Sheridan says, "It is in Alan of Lille that she reaches full stature." And according to Sheridan it was Alan who "theorized that God first created the world and then appointed Natura as His substitute and vice-regent . . . in particular to ensure that by like producing like all living creatures should increase and multiply . . ."

And so Nature comes into the plot of Alan's Christian allegory reasonably enough, to keep company with Venus, Cupid, Genius, and a cast of personified vices and virtues. But if allegory tends toward simplification and a kind of shallowness —Truth is going to be purely true, and Greed purely greedy —Alan's representation of Nature tends in virtually the opposite direction: She is complex, vulnerable, mysterious, somewhat ghostly, less a personification than the presence, felt or intuited, of the natural world's artificer. She comes from an intuition of order and harmony in creation that is old and independent of empirical proofs. That we have personified her for so long testifies that we know her as an active, purposeful, and demanding force. Her domain, in the *Plaint* and elsewhere, is both natural and supernatural. As Vicar of God, she joins Heaven and Earth, resolving the duality of spirit and matter.

To say, as I just did, that we know her is obviously to raise the question of what we mean, and what we ever have meant, by saying that we "know." In our time an ideological tide has been carrying us toward a sort of apex at which none of us will claim to know anything that has not been proved and certified by scientists. But to read Alan of Lille is to realize that there was a time, a *long* time, when such knowledge was neither available nor wanted, but when many necessary things were nonetheless known. Thomas Carlyle, writing from his own perch in the modern world about the twelfth-century monk Jocelin of Brakelond, had to confront this very question: "Does it never give thee pause . . . that men then had a *soul*, not by hearsay alone, and as a figure of speech; but as a truth that they *knew* . . ."

A century and a half later in the same darkening age, I asked my friend Maurice Telleen, who had much experience of livestock shows: "Does a good judge measure every individual by the breed standards, or does he go by intuition and use the breed standards to check himself?" In reply Maury said something remarkable: "He goes by intuition! A slow judge is always wrong." I thought his reply remarkable, of course, because it confirmed my belief that, among people still interested in the qualities of things, intuition still maintains its place and its standing as a way to know.

Intuition tells us, and has told us maybe as long as we have been human, that the nature of the world is a great being, the one being in which all other beings, living and not-living, are joined. And for a long time, in our tradition, we have called this being "Natura" or "Kind" or "Nature." And if we forget, our language remembers for us the relation of "natural" (by way of "kind") to "kindness" and "kin," and to "natal," "native," "nativity," and "nation." Moreover, as understood by Alan of Lille and the poets who descend from him, the being and the name of "Nature" also implicates the history of human responsibility toward the being of all things, and Nature's continuing requirement of that responsibility.

Contrast "Nature," then, with the merely clever poeticism "Gaia" used by some scientists to name the idea of the unity of earthly life, and perhaps to warm up and make congenial the term "biosphere." "Gaia" is the name of the Greek goddess of the earth, in whom no modern humans, let alone modern scientists, have even pretended to believe. Her name was used, no doubt, because, unlike "Nature," it had no familiar or traditional use, could attract no intuitive belief, or appeal at all to imagination.

By name Nature familiarly belongs to us, and she has so belonged to us for many hundreds of years, but considered from a viewpoint strictly biblical and doctrinal, she still may seem an intruder. Like the other classical deities, she joins the biblical tradition somewhat brazenly. The writers needed her perhaps because she has persisted in oral tradition, and because there remained not a Christian only but a general human need for her. That she usurps or threatens the roles of all three persons of the Trinity suggests that there was also a felt need for her to some degree in spite or because of them. Speaking only for myself, I will say that for me it has always been easy to be of two minds about the Trinity. The Father, the Son, and the Holy Spirit, as they appear in their places, as "characters" so to speak, in the Gospels, I have found simply recognizable or imaginable as such, but I think of them as members of "the Trinity" only deliberately and without so much interest. As an idea, the Trinity, the three-in-one, the three-part godhead, seems to me austerely abstract, complicated, and cold. The more it is explained, the less believable it becomes.

Perhaps it is this aloofness of the Trinity that calls into being and causes us to need this other, more lowly presence, holy yet familiar, matronly, practical, concerned, and eager to teach: Nature, the mother of creatures as the Virgin is the mother of God. And here I remember that Thomas Merton, in his prose poem "Hagia Sophia," sees both Nature and the Virgin as sharing in the identity or person of Holy Wisdom: "There is in all visible things an invisible fecundity, a dimmed light, a meek namelessness, a hidden wholeness. This mysterious Unity and Integrity is Wisdom, the Mother of all, *Natura naturans*." And later: "*Natura* in Mary becomes pure Mother." In Merton's treatment these three—Natura, Mary, Sophia—seem to fade into one another, shadowing forth an always evasive reality. And this, I think, is what we must expect when human thought reaches toward the mystery of the world's existence: an almost visible, almost palpable presence of a reality more accessible to poetry than to experiment, never fully to be revealed by any medium of human knowledge. And yet it is to be, it can be, learned and known, not by peering through a lens or by assemblies of data, but perhaps only by being quietly observant for a long time. Merton's vision does not explain or clarify Alan of Lille's, but between the two there is a sort of recognition and a mutual verification.

I am not entirely confident of my grasp of Alan's *Plaint of Nature*, partly because of my very reasonable doubt of my scholarship and understanding, but also because I am looking back at Alan from my knowledge of later writers who seem to be his successors, and from my own long preoccupation with the issues he addresses. For I do not wish, and I have not tried, to read Alan as a writer of historical or literary interest only. Despite the always considerable differences of times and languages, I have thought of him as a fellow writer with immediately useful intelligence of the world that is both his and mine.

His concept of the integrity of the natural world, and of the dependence of the world's integrity upon the integrity of human nature, leads by the most direct and simple logic to our own more scientific recognition of the integrity of ecosystems, the integrity ultimately of the ecosphere, and to the recognition (by not enough of us) of the necessity of an ecological ethic.

But a large part of the value of the *Plaint* is that, though I suppose it is as full as it could be of the biology and taxonomy of its time, it cannot be reduced to the sort of knowledge that we call scientific. It cannot, for that matter, be reduced to the sort of knowledge that we call poetic. It belongs instead to the great Western family of writings that warn us against what we now call reductionism, but which traditionally we have called the deadly sin of pride or hubris: the wish to be "as gods," or the assumption that our small competence in dealing with small things implies or is equal to a great competence in dealing with great things. In our time we have ceased to feel the traditional fear of that equation, and we have a world of waste, pollution, and violence to show for it.

Plaint of Nature cannot be comprehended within the bounds of any of our specialties, which exclude themselves by definition from dealing with one another. But once we acknowledge, once we permit our language to acknowledge, the immense miracle of the existence of this living world, in place of nothing, then we confront again that world and our existence in it, forever more mysterious than known. And then the air swarms with questions that are scientific, artistic, religious, and all of them insistently economic. Some of the questions are answerable, some are not. The summary questions are: What are our responsibilities? and What must we do? The connection of all questions to the human economy is finally not escapable. For our economy (how we live) cannot leave the world or any of its parts alone, as the ideal of the wilderness preserve seems to hope. We have only one choice: We must either care properly for all of it or continue our lethal damage to all of it.

That this is true we may be unable to know until we have understood how, and how severely, we have been penalized by the academic and professional divorces among the sciences and the arts. Division into specialties as a necessity or convenience of thought and work may be as old as civilization, but industrialism certainly has exaggerated it. We could hardly find a better illustration of this tendency than Shelley's "Defence of Poetry," in which he makes, with characteristic passion, a division between imagination and reason, assigning things eternal and spiritual to imagination, and things temporal and material to reason. *Plaint of Nature*, as if in answer, resolves exactly that duality in the person of Nature herself, who joins Heaven and Earth, and whose discourse is of the vices that break her laws and her world, and of the virtues by which her laws would be obeyed, her rule restored, and her work made whole.

Nearly all of the *Plaint* occurs in Alan's dream-vision, which is to say in his mind. It is an intensely mental work, if only because he worked so hard to render the style of it, and it becomes rather stuffy. Going from Alan to Chaucer is like stepping outdoors. Nature, the "noble emperesse" herself, is presented twice in Chaucer's work, once with acknowledgement of "Aleyn" and "the Pleynt of Kynde." But in Chaucer, as never in Alan, the natural world itself is present also.

It has been said that in Chaucer nature is mainly idealized in dreams or paradisal visions. And I suppose it could be said that in his treatment especially of birds, he is a mere fabulist or a protocartoonist. But to me, his attitude toward the outdoor world seems in general to be familiar, affectionate, sympathetic, and, as only he could be, humorous. This we would know if he had written only the first twelve lines of *The Canterbury Tales*. A teacher required me to memorize those lines, and I did, more or less, sixty-some years ago. Since then I have found no other writing that conveys so immediately the *presence* (the freshness of the sight, feel, smell, and sound) of an early spring morning, as well as the writer's excitement and happiness at the thought of it.

Chaucer was a man of the city and the court, but I think he was not a "city person," as we now mean that phrase. London in Chaucer's time was comparatively a small city. Inundated as we are by the commotions of internal combustion, I doubt that we can easily imagine the quietness of fourteenth-century London, from any part of which, above or beyond the street noises, Chaucer probably could have heard the birds singing.

Travel on horseback through the countryside was then an ordinary thing. This involved familiar relationships with horses,
and on horseback at a comfortable gait anybody at all observant
participates in the life of the roadside and the countryside with
an intimacy impossible for a traveler in a motor vehicle. City
people then would have had country knowledge as a matter of
course. If they dreamed or imagined idealized landscapes, that
may have been a "natural" result of their close knowledge of
real ones.

And so in *The Parliament of Fowls* we have an ideal or visionary garden with real trees. And Chaucer's list of its trees
is not just an inventory. He clearly enjoys sounding their
names, but he also introduces them, so to speak, by their per-
sonalities: "The byldere ok, and ek the hardy asshe, / The
piler elm," and so on. A little later he identifies the birds in
the same way. *The Knight's Tale* also has its list of trees, but
this one names the membership of a grove that has been cut
down, "disinheriting" its further membership of gods, beasts,
and birds, and leaving the ground "agast" at the novelty of
the sunlight falling upon it. Ronsard too felt this shock, this
sympathy and dismay, at the exposure and vulnerability of the
ground after the "butchery" of the Forest of Gastine (Elegy
XXIV), and some of us are feeling it still at the sight of our
own clear-cut forests. There is nothing fabulous or visionary
about the Knight's small elegy for the fallen grove. It is informed by observation of such events, and by a real regret.

That the ground under the fallen grove was "agast" (aghast,
shocked, terrified) at its new nakedness to the sunlight is not
something that the Knight or Chaucer knew by any of the
"objective facts" that industrial foresters would use to deny
that such a thing is possible. The much older, long-enduring
knowledge came from sympathy and compassion.

If Chaucer heard the birds singing in his own English, that
was owing to his sympathy for them. Birds and animals use
human speech by convention in literary fables, but that usage came into literature, I am sure, from the conversation of
country people who lived, as some of them live still, in close
daily association with the birds and animals of farmsteads and
with those of the natural world. People who are in the habit of
speaking to their nonhuman neighbors and collaborators are
likely also to have the habit of translating into their own speech

the languages and thoughts of those other creatures. There are practical reasons for this, and obviously it can be amusing, but it comes from sympathy, and in turn it increases sympathy.

Some years ago I wrote the introduction for a new printing of Theodora Stanwell-Fletcher's *Driftwood Valley*, a book I had loved since I was in the seventh or eighth grade. When I sent my typescript to the author, her strongest response was to my anticipation of the objection of some readers "to Mrs. Stanwell-Fletcher's anthropomorphizing of animal thoughts and feelings." One of her degrees was in animal ecology; she was "scientific" enough to know that I was right. But her marginal note, as I remember, said: "How else could we understand them?" She spoke with authority, for she and her husband had spent three years in a remote part of British Columbia, where their nearest neighbors were the native plants and animals they had come to study.

From *The Nun's Priest's Tale* we know that Chaucer was a perfect master of the literary fable, but that tale signifies also that he was a close observer of the manners of household poultry, and he no doubt had listened with care to the conversation of country people—several of whom, after all, were in his company of Canterbury pilgrims. That his attention to them and their kind had been deliberate and lively we know from two of his portraits in *The Canterbury Tales*: those of the Plowman in the *General Prologue* and the "poor widow" of *The Nun's Priest's Tale*. The descriptions of those two characters give us a sort of compendium of agrarian values—that are, by no accident, agreeable to Nature's laws.

The Plowman is said to be one of Chaucer's "ideal characters," and I suppose he is, but his idealization does not make him in any way simple. His description, though brief, is ethically complex. The Plowman is an exemplary man and countryman because he is an exemplary Christian. He loves God best, and then he loves his neighbor as himself. He is a hard worker, good and true, and a man of peace. And those virtues enable his charity, for he will work without pay for those who need his help. If I am not mistaken, Chaucer's sense of humor is at work here, pointed at those who think of goodness as a sacrifice paid here for admission to Heaven. The Plowman's virtues are understood as solidly practical and economic. He is a good neighbor, and good neighbors are likely to *have* good neighbors. The payoff, to complete the joke, may be Heavenly, but it is also earthly: A good neighborhood is an economic asset to all of its members.

The portrait of the elderly widow in *The Nun's Priest's Tale* attends to other practical and necessary virtues: patience, frugality, and good husbandry. These are congenial with the virtues of the Plowman and complete them, as his virtues complete hers. That the whole set of virtues is divided between the two characters is a matter only of appropriateness. The widow lives with her two daughters in a small cottage. She owns (God has "sente" her) three large sows, three cows, a sheep named Malle (Mollie?), seven hens, and a rooster. She works "out" as a "dairymaid," but it is clear that her economy is most securely founded upon her own small holding and her few head of livestock, also upon needing little. She needs no "poignant sauce" for her food because the food is good and hunger seasons it well. Her charity is to need no charity, another recognized way of being a good neighbor in a country community. This widow is a first cousin to the old Corycian farmer in Virgil's fourth Georgic. She practices exactly the "cottage economy" later praised and advocated by William Cobbett (and others).

My approach to Chaucer's first representation of Nature has been backward, from late work to early, as a way of knowing how he understood her. If we read *The Parliament of Fowls* simply as an exemplary "dream-vision" by a master poet and courtier, then perhaps we are free to regard the figure of Nature as a pleasing, even a beautiful, "picture" borrowed from Alan's *Plaint* and requiring little more of us than to place the poem historically and to appreciate it critically. But if we would like to take seriously this appearance or apparition of Nature, then we have to ask how seriously Chaucer took it, and that is not so easy. It helps, I think, if we conceive of Chaucer not only as the great poet and sophisticated "man of the world" that he certainly was, but also, and on the evidence of his writing, a man on easy speaking terms with the countryside and all of its inhabitants.

The poem is a dream-vision, a lighthearted fantasy, above all a comedy. It also takes place in a landscape in which the poet is well acquainted with the trees, the flowers, and the birds. The birds speak English, sometimes at length, but they speak also in their own tongues. We hear from the goose, the cuckoo, and the duck all together and all in the same line: "Kek kek! kok-kow! quek quek!" When they speak English, they use images that might have been used at court or by a farmer's hearth, but they certainly came from people who had spent time outdoors at night: "There are more stars, God knows, than a pair" and "You fare by love as owls do by light."

Faced with the possibility of copying Alan's vision of Nature in her extremely elaborate finery, Chaucer politely declines by referring us to the *Plaint*, where we can find her in "such array." This is a part of the comedy, but it leaves him free to describe her in his own way, which he does by a single graceful compliment which serves to acknowledge her sanctity and her standing: She is so far fairer "than any creature" as the summer sun's light surpasses that of any star. And then, it seems, he draws aside a curtain:

She is sitting in a forest glade on a flowery hill. She is "set off" by halls and bowers of branches—evergreen branches, as Chaucer's readers must always have assumed—that have been "wrought" according to her design. This is Nature in her English vicarage. She sits before us as a distinctly hieratic figure, yet made familiar by the homely setting that combines art and nature. This is clearly akin to the scene in which, two and a half centuries later, Robert Herrick's Corinna will go a-Maying:

> Come, my Corinna, come; and comming, marke
> How each field turns a street; each street a Parke
> Made green, and trimm'd with trees: see how
> Devotion gives each House a Bough,
> Or Branch: Each Porch, each doore, ere this, An arke a
>   Tabernacle is
> Made up of white-thorn neatly enterwove . . .

Chaucer's "noble emperesse, ful of grace" seems perfectly to belong here, in the center of such a mating ritual as he must have witnessed many times. She is there, not in May but on Saint Valentine's day, as in every year, to see to the match-making of all the birds, according to her judgment. To see that they in their great numbers find mates and reproduce after their kinds is her high office. Following Alan's doctrine, Chaucer has her presiding as

> the vicaire of the almyghty Lord,
> That hot, cold, hevy, light, moyst, and dreye
> Hath knyt by evene noumbres of acord . . .

Her work is to reconcile the world's opposites and contentions into a lasting, self-renewing composure.

From Nature enthroned our attention is drawn to the assembly of all the birds, to the clamorous low "common sense" of the goose, the cuckoo, and the duck, and finally to the elegant "roundel" that the mated birds all sing together at the end. These incongruities take nothing away from Nature's dignity and they don't need to be justified or explained. They notify us of the range of Chaucer's art and his knowledge, authenticated by the world itself, which often puts high seriousness and low comedy into the same event or the same instant.

But let us go ahead and ask the modern question: Did Chaucer "believe in" this Nature? Did he "know" her? Did he "actually see" her seated in her glade "upon an hil of floures"? Well, he seems to have seen her there, he seems to have invited me to see her, and I too seem to have seen her. Where are the proofs? There are of course no proofs, no photograph, no second witness. But we are talking here about the imaginative life of a country—not a nation, a *country*—which, in its apprehension of the natural world and its "invisible fecundity," its "hidden wholeness," always must outreach its proofs, its sciences, its mechanic arts, its political economy, its market. If it fails in that, as it has with us, then we get probably what we have got: a country, mainly unknown to its occupying humans, rapidly melting into a toxic slurry and flowing away through its rivers.

That Alan's Nature appears again, years later, in *The Canterbury Tales* suggests that Chaucer thought her important enough to keep her in mind to the end of his life. At the beginning of his tale, the Physician "quotes" what he imagines she would say of the singular beauty of Virginia, his chaste heroine. Though the Physician is a man whose honesty needs watching, we need not doubt his characterization of Nature. Following his professional habit of learned speech, he has Nature describe herself in keeping with Alan's description, and with Chaucer's in *The Parliament of Fowls*. She is God's "vicar general," in charge of the sublunar creation, whose "forming and painting" of the creatures is work done "to the worship of my lord." What may be unique here is her insistence that her work cannot be "counterfeited." In this, she takes the side of William Carlos Williams* and others in our own time, against Hamlet et al. who have argued that the "end" of art "was and is to hold . . . the mirror up to Nature . . ."

This disagreement is of interest now because it clearly defines the problem confronted by scientists of the last hundred or so years who have, against the industrial declivity, taken an ecological approach to our use of the land, primarily in agriculture. To those scientists, whose work I will discuss later in this essay, two truths have been obvious: first, that we humans cannot live in unaltered natural settings, for we passed up our chance to be "rational apes" too long ago; and, second, though we cannot make mirror images of natural places, even if we could do so, even if we could live in them if we could make them, we nonetheless are obliged to obey Nature's laws, which are imposed absolutely and will never change.

From Alan's vision of Nature, and then from Chaucer's, came Spenser's (as he testifies) in the fragmentary seventh book of *The Faerie Queene*, and of these three Spenser's is the finest. It is the most fully developed and detailed and the richest in meaning. The story he places Nature in, as he tells it, is the most dramatic. And the problem he sets for her is one that was urgent for him then, and that seems still more urgent for us.

Book VII of *The Faerie Queene* contains only the sixth and seventh cantos and two stanzas of an eighth. These are known as "the Mutability cantos," for they tell how "the Titanesse," Mutability, rebels against the classical deities and attempts to establish her sovereignty over them, and over all creation. She disdains the rule of Jove, and refuses to accept his judgment, appealing instead to "the highest him . . . Father of Gods and men . . . the God of Nature." And so great Nature herself, as God's deputy, comes to preside over a trial to determine the justice of Mutability's claim. For this all the creatures are assembled on Arlo hill, where they are "well disposed" by Order, who is "Natures Sergeant."

*In especially some of the most lucid prose passages of *Spring and All.*

And then Nature enters with the great processional dignity
that Spenser seems to have learned from the twenty-fourth
Psalm,* as maybe of all poets only he could have learned from it:

Then forth issewed (great goddesse) great dame *Nature*,
        With goodly port and gracious Maiesty;
        Being far greater and more tall of stature
        Then any of the gods or Powers on hie:
        Yet certes by her face and physnomy,
        Whether she man or woman inly were,
        That could not any creature well descry:
        For, with a veile that wimpled euery where,
Her head and face was hid, that mote to none appeare.

That some doe say was so by skill deuized,
        To hide the terror of her vncouth hew,
        From mortall eyes that should be sore agrized;
        That eye of wight could not indure to view:
        But others tell that it so beautious was,
        And round about such beames of splendor threw,
        That it the Sunne a thousand times did pass,
Ne could be seene, but like an image in a glass.

That well may seemen true: for, well I weene
        That this same day, when she on *Arlo* sat,
        Her garment was so bright and wondrous sheene,
        That my fraile wit cannot deuize to what
        It to compare, nor finde like stuffe to that,
        As those three sacred *Saints*, though else most wise,
        Yet on mount *Thabor* quite their wits forgat,
        When they their glorious Lord in strange disguise
Transfigur'd sawe; his garments so did daze their eyes.

. . . .

This great Grandmother of all creatures bred
        Great *Nature*, euer young yet full of eld,
        Still moouing, yet vnmoued from her sted;
        Vnseene of any, yet of all beheld . . .

*This assumption is fairly obvious if one is thinking also of the twelfth stanza
of his "Epithalamion." Only a poet of the greatest skill, and confidence, would
have attempted this.

This representation of Nature clearly derives from those of Alan of Lille and Chaucer, but the resemblance, though unmistakable, is distant. Here there is nothing at all of Alan's relentless accumulation of details. And in our own time, when poets are supposed or expected to disown their forebears, it is a relief to come upon Spenser's filial devotion to Chaucer, but nothing here reminds us of the at-home conviviality of Nature's precedence over the Parliament of Fowls.

Spenser's Nature is altogether hieratic and luminous. In a way that recalls Dante's frustration in the *Paradiso*, Spenser describes this "great goddesse" mostly by describing his inability to describe her. Though she is heavily veiled like perhaps a nun and he refers to her always by feminine pronouns, she may be "inly" either a man or a woman, and her face is either unendurably terrible or so radiantly beautiful that it could not be seen except "through a glass darkly." And in describing her, Spenser, unlike Alan and Chaucer, recalls the Gospels, for her garment was "so bright and wondrous sheene" as to recall the garments of Christ at the Transfiguration. And here Spenser must be remembering also the Gospel of John 1:3: "All things were made by him; and without him was not any thing made that was made."

That she is veiled we may take as Spenser's admonishment to the scientists, from Francis Bacon to the highest-funded prophets and oracles of our time, who have proposed, by their merely human wits and devices, to wring from Nature her ultimate Truth, and thus to mate every problem with its ultimate solution. There is nothing so "progressive" in Spenser's view of Nature—or, hence, of nature. Knowing as he knew, he would have recognized Bacon's arrogance as old under the sun. He could not have expected Science, our own allegorical giant, to quiz Nature face to face.

Furthermore, Mutability, in pleading her case before Nature, addresses her as "greatest goddesse, only great"

> Who Right to all dost deale indifferently,
> Damning all Wrong and tortuous Iniurie,
> Which any of thy creatures doe to other
> (Oppressing them with power, vnequally)
> Sith of them all thou art the equall mother,
> And knittest each to each, as brother vnto brother.

*

Any twenty-first-century reader familiar with the formal principle of interdependence, as it operates in ecosystems, will recognize this "knitting" for what it is and means, as they will recognize also Nature's "equal" motherhood of "all" the creatures. Her "indifference" is not apathetic; she is merely impartial, preferring no single species over any other, just as the realists of present-day biology know her to be. And this supposedly modern perception is much older than Spenser, for he took it from the *Plaint* in which, four hundred years earlier, Nature had warned Alan that "my bounteous power does not shine forth in you alone individually but also universally in all things." We have our lives by no right of our own, but instead by the privilege of sharing in the life that sustains all creatures. This great convocation is the work of Nature, its "equall mother," which makes her not only, as Alan saw, our teacher, but also, as Spenser was first to see, our judge.

And so, submitting to Nature as the supreme worldly authority, Mutability presents her argument, and summons a procession of witnesses: the earth, the four elements, the seasons, the months, day and night, and finally life and death, all of whom support her argument. Jove then argues that worldly change is ruled by the "heavenly" gods. But Mutability charges and proves the changingness of the planetary deities, of the sun, which is sometimes eclipsed, and of Jove himself, who once was not and then was born. She is an unbluffable, brilliant lawyer, and her case is nearly perfect, as is the poetry that Spenser gives to it.

Finally, after a long considering pause, Nature gives her verdict:

> I well consider all that ye haue sayd,
>> And find that all things stedfastnes doe hate
>> And changed be: yet being rightly wayd
>> They are not changed from their first estate;
>> But by their change their being doe dilate:
>> And turning to themselues at length againe,
>> Doe worke their owne perfection so by fate:
>> Then ouer them Change doth not rule and raigne;
> But they raigne ouer change, and doe their states maintaine.

C. S. Lewis speaks, with justice, of the "deep obscurity" of those lines. But we may clarify them somewhat by placing beside them these earlier lines in which Mutability argues that

 For, all that from [Earth] springs, and is ybredde,
           How-euer fayre it flourish for a time,
           Yet see we soone decay; and, being dead,
           To turne again vnto their earthly slime:
           Yet, out of their decay and mortall crime,
           We daily see new creatures to arize;
           And of their Winter spring another Prime,
           Vnlike in forme, and chang'd by strange disguise . . .

Mutability here seems to argue, rather craftily, that change is absolute, leading invariably to something "unlike" and entirely new. Nature, if at the end she is remembering those lines, sees them as an imperfect description of a natural cycle capable of endlessly repeating itself—with, we would now say, occasional variations or "mutations," and depending on appropriate human cooperation. In her verdict, Nature readily acknowledges the ceaselessness of change, but she confirms, if not quite clearly, its cyclicality as greater, and as a form, or *the* form, of stability. She would have been confirmed in this by the *Plaint* where, in Alan's vision of her, Nature says (more clearly) that

> it was God's will that by a mutually related circle of birth and death, transitory things should be given stability by instability, endlessness by endings, eternity by temporariness, and that the series of things should ever be knit by successive renewals of birth.

This will be better understood about three and a half centuries later by scientists aware of the biology, and the supreme economic importance, of the fertility cycle.

   At the end of Canto VII, Nature, having completed her work for the time being, "did vanish, whither no man wist." *The Faerie Queene*, as we have it, ends two stanzas later with Spenser's prayer for the "Sabaoths sight"

           Of that same time when no more *Change* shall be,
           But stedfast rest of all things firmely stayd
           Vpon the pillours of Eternity . . .

Nature's standing in the order of things, as Spenser understood it, is exalted, well above that of humanity, and she has about her the nimbus of sanctity. Her equitable motherhood of all the creatures and her judgeship over them impose upon humans a responsibility that is both worshipful and relentlessly practical. But the order in which she is placed is firmly Christian, and her jurisdiction is limited to the incarnate world. Reassuring as may be her verdict against Mutability, it offers little comfort to individual humans in their suffering of their own mortality and that of their loved ones. And so *The Faerie Queene* as we have it, though incomplete, ends appropriately by invoking our so far undying hope for a "time," beyond Nature's world and all of its stories, "when no more change shall be."

Within the compass of my reading, Spenser's vision of Nature is the highest and fullest, the most responsibly imagined, the most complete, and the most instructive. And this, I think, is because it is the most thoughtful. In the Mutability cantos, Spenser confronts a question serious enough to have no definitive or final answer: On what terms are we to live with the perpetual changingness of this world? And he answers with an argument meticulously constructed. That his stanzas on Nature's appearance are so complex and beautiful must be partly the result of his thoughtfulness. I do not mean that he used his poetry as a vehicle to express or communicate his finished thought, but rather that his poetry was the vital means by which his thinking was done. Strange as it may seem to say this after the division of the mental functions into departments, it is clear that some poets have recognized that poetry in its way, like prose in its way, can be serviceable to thought, and when they have needed to do so they have used it as a way to think.

After Spenser, so far as I have read or remember, no other English poet acknowledged the influence of Alan of Lille or thought so carefully about Nature—probably because, after Spenser, no poet needed to think so carefully about her. But there certainly have been other English poets who appear to have been influenced by the earlier visions of Nature, and who have contributed to a line of thought about the proper human use of the natural world.

Passages in the eleventh and twelfth parts of *Piers Plowman* suggest that William Langland, for one, had read *Plaint of Nature*. C. S. Lewis assumed that he had. A remarkable difference is that, to Langland, Nature or "Kynde," rather than the Vicar of God, is God Himself. Langland, anyhow, was a better naturalist than either Chaucer or Spenser. Kynde instructs him in a dream to study the creatures, "the wonders of this world," to gain understanding and to learn to love his creator. And he observes carefully the variety of the creatures and their ways of mating, the skills of the nesting birds, the woodland flowers and their colors. His intimate knowledge of these things authenticates his wonder at them and his sense of the miraculousness of their existence. His wonder involves a tenderness unlike any other that I know: Of the flowers, the stones, and the stars only Kynde "himself" knows the causes; He is the magpie's patron and tells her, "putteth it in hir ere," to build her nest where the thorns are thickest.

Langland and Chaucer both died in 1400. After them, and after Spenser, I have in mind poems by Milton and Pope, in which they seem to have remembered Alan's Nature or Chaucer's or Spenser's and called her, so to speak, by name.

John Milton's *Comus* is a masque, an elegant play in verse, presented at Ludlow Castle in 1634 when the poet was twenty-five years old. It tells the story of a temptation, remembering the story of Satan's temptation of Christ (Matthew 4:1–11), and anticipating *Paradise Regained*, first published in 1671. As Spenser in the Mutability cantos had asked on what terms we are to live with the perpetual changing of this world, a question that had become urgent for him in the latter part of his life, so Milton in *Comus* was asking a question no doubt urgent in his youth but approximately parallel to Spenser's: On what terms are we to live with the material abundance of this world? Human nature, by any honest measure, is limited strictly and narrowly—we don't live very long, and we don't know very much—whereas the nature of the world at large by comparison seems a limitless plenitude. The two poets, then, were asking how to make human sense, a *little* sense, of an immensity.

In *Comus* a young maiden, identified simply as "The Lady," becomes lost in the woods. Alone and vulnerable, she meets Comus, perhaps the Tempter himself, disguised as a shepherd, who offers to guide her to safety. Instead, he takes her to "a stately palace, set out with all manner of deliciousness: soft music, tables spread with all dainties." He then proves himself the masterful and eloquent seducer he really is. He makes the conventional argument of *carpe diem*, "seize the day," which Jonson, Waller, Herrick, and Marvell also made in famous poems of elegant wit, giving to their dire and perfect logic a characteristic lightness of heart. The poet, as would-be lover, reminds his lady, as Marvell would put it, that "The grave's a fine and private place, / But none, I think, do there embrace."

This is exactly Comus's argument, but he makes it with a philosophic impudence and gravity that greatly enlarges its bearing. Milton's poem is sometimes described as a defense of "the sage / And serious doctrine of Virginity." It is that, but also far more than that. The poem's great question, as Comus himself raises it, is about the proper use and care of natural gifts:

> Wherefore did Nature pour her bounties forth
> With such a full and unwithdrawing hand,
> Covering the earth with odors, fruits, and flocks,
> Thronging the seas with spawn innumerable,
> But all to please and sate the curious taste?
>
> . . . .
>
>   If all the world
> Should in a pet of temperance feed on pulse,
> Drink the clear stream, and nothing wear but frieze,
> The All-giver would be unthanked, would be unpraised,
> Not half his riches known, and yet despised . . .

He eventually takes up the conventional *carpe diem* theme of the transience of mortal beauty and roses that wither, but he is arguing, as the Lady quickly understands, not for using the world but for using it up. His ideology goes beyond mere personal gluttony and lust to a modern avarice and utilitarianism: the assumption, laid bare in our own time, that all of the natural world that we humans do not consume either is worthless or is wasted.

The Lady, threatened by Comus's passionate intensity and his power of enchantment, is protected only by her inner light —even Comus can see that "something holy lodges in that breast"—and her hope of rescue, which eventually comes, but she easily overmatches Comus as a debater, arguing from a better premise and with sufficient courage. Nature, she says,

> Means her provision only to the good,
> That live according to her sober laws
> And holy dictate of spare Temperance.
> If every just man that now pines with want
> Had but a moderate and beseeming share
> Of that which lewdly-pampered luxury
> Now heaps upon some few with vast excess,
> Nature's full blessings would be well dispensed
> In unsuperfluous even proportion . . .

Comus and the Lady are too allegorical in character to allow for much in the way of drama, but I don't think Milton can be accused of rigging their debate. He seems to have taken care to make Comus's argument as attractive as a vital man of twenty-five would have known it to be. The Lady's argument, attractive in a way perfectly opposite, more soundly appreciative of Nature's abundance, and approving temperance as the only safeguard of abundance, is Christian and democratic. (The "Christian conservatives" of our day would call it socialism.) Her argument serves my own by what I take to be its completion of the poets' long-evolving characterization of Nature. The Lady enlarges the import of Nature's demand upon humanity by making it, at last, explicitly economic. It remained for Milton to perceive clearly that Nature requires of us a *practical* reverence. Temperance in the use of natural gifts is certainly a religious obligation, but it is also an economic virtue. *Comus* requires us to think of the right use of gifts. To be in one's right mind is to know the right use of gifts. The Lady reacts so fiercely to Comus's proposition, not just because it assaults her personal virtue, but because it disdains and destroys the idea of economy. The word "economy," taken literally, as I am taking it, does not designate a financial system, but rather the management and care of the given means of life.

Alexander Pope was a poet in many ways unlike John Milton, and yet from the Lady's rebuke to Comus it is only a step to certain poems and passages of Pope. In spite of his physical debilities, Pope had more fun than Milton, but he certainly would have recognized his kinship to the Lady in *Comus*. As the Lady puts it, the human obligation to Nature is defined by obedience "to her sober laws" and the "holy dictate" of temperance. To Pope, that obligation is defined by "Sense" or "Good Sense," which in his use of those terms is pretty much the same. As he understood her, Nature requires of us certain proprieties, not only of manners but also of work. To him, everything depended on a proper sense of scale. We must act and work with the awareness always of the magnitude of Nature's work and of our own comparative smallness as individuals and as a species. And so Pope was another enemy of prodigality, of ostentation and the utter silliness of every kind of extravagance or waste.

In the "Epistle to Burlington," his satire is against the owners of country estates who surround themselves with houses and gardens magnificent to the point of ugliness and discomfort, far exceeding in dimensions and cost any use or pleasure: "huge heaps of littleness" built to display the wealth of the owner, who by comparison is "A puny insect, shiv'ring at a breeze." This is contrary to "Good Sense, which only is the gift of Heaven" and is worth as much as all the sciences. Against the Prodigals who waste their wealth on expensive, fashionable things they don't even like, and especially against their grotesque extravagances in what we now call "landscaping," Pope lays down two rules—"In all, let Nature never be forgot" and "Consult the Genius of the Place in all"—that my friend Wes Jackson and I have quoted back and forth for years in confirmation of our efforts for good husbandry of the land.

I need to confess, however, that I have often wondered how seriously I ought to take, not Pope's rules, which I think are sound, but Pope himself as a critic of land use. In the "Epistle to Burlington," after all, he is talking about the country houses and pleasure gardens of wealthy gentlemen, not working farms and forests. He goes so far as to say that even the worst examples, which he curses for "lavish cost, and little skill," are pardonable at least for giving employment to the poor—too much as polluting industries now are justified by "job creation." I forgive him for that because of his prophecy, clearly hopeful and immediately following, that these show places of extravagant littleness in "another age" will be wheat fields. Moreover, his understanding of the relation of art and nature is authentically complex and practical, appropriate to land use of any kind. Good sense leads to a proper mindfulness of Nature, which leads to collaboration between Nature and the gardener. The gardener's intention or design is completed by Nature's gathering of the parts into wholeness. By her gift, moreover, the land is made useful:

> 'Tis Use alone that sanctifies expense,
> And Splendor borrows all her rays from Sense.

The lawns of the estate should not be ashamed to be grazed by livestock, or the beautiful forests to yield timber. Though with less passion and not so explicitly, Pope thus consents to Milton's argument in *Comus*, that the human economy should be appropriate to the human dependence on Nature.

Pope is the last of the English poets to be mindful of Nature as mother, maker, teacher, giver of patterns and standards, and judge—so far as I know. I have repeatedly acknowledged the limits of my knowledge, first as a duty, but also with the hope that my deficiencies will be supplied by better scholars. What I am sure of is that we have lost the old apprehension of Nature as a being accessible to imagination, linking Heaven and Earth, making and informing the incarnate creation, and requiring of humanity an obedience at once worshipful, ethical, and economic. Her stern instruction, never disproved, that we humans have a rightful but responsible place in the order of things, has disappeared, and has been absent a long time from our working consciousness and our formal schooling.

Nature, as she appeared to Chaucer, Spenser, Milton, and Pope, does not appear in the "nature poetry" of the Romantic poets, and she is absent from the history of their influence upon both poetry and the conservation movement. By the time we come to Wordsworth, who often wrote about the natural world and often was on foot in it, there is already a powerful sense of being alienated from it, with a concomitant longing to escape into it from "the din / Of towns and cities," "the heavy

and the weary weight / Of all this unintelligible world," "the
fretful stir / Unprofitable, and the fever of the world."

I am quoting now from the poem familiarly called "Tintern
Abbey," which displays pretty fully our modern love for nature, our often-lamented distance from it, and the vacationer's
sensibility and economy that bring us occasionally "close" to
it again, allowing us to feel more or less a religious sense of
beauty and peace.

In "Tintern Abbey," addressed to his sister, Dorothy, who
accompanied him, the poet has returned after a five-year absence to the Wye valley, "a wild secluded scene" of "beautiful
forms," where the weight of the busy world is lightened and

> with an eye made quiet by the power
> Of harmony, and the deep power of joy,
> We see into the life of things.

In the quiet and beauty of that "wild" place he feels or perhaps
recalls

> A presence that disturbs me with the joy
> Of elevated thoughts; a sense sublime
> Of something far more deeply interfused,
> Whose dwelling is the light of setting suns . . .

There can be no doubt of the strength of his emotion here, or
of the loftiness of his language. Perhaps the presence he feels is
that of Nature as the older poets imagined her, but these lines,
however intense, are vague by comparison, and his thought
is entirely dissipated by his resort to "something." His claim,
soon following, that he recognizes

> In nature and the language of the sense
> The anchor of my purest thoughts, the nurse,
> The guide, the guardian of my heart, and soul
> Of all my moral being

is devoid of any particular thought or any implication of a practical responsibility. His version of "nature" thus lacks altogether
the intelligence and moral energy of Nature as she appeared to
the older poets. Of Wordsworth she seems to have required
nothing at all in particular, except perhaps his admiration.

In line 122 the poet personifies Nature by capitalizing her name, but he then also sentimentalizes her:

> Nature never did betray
> The heart that loved her; 'tis her privilege,
> Through all the years of this our life, to lead
> From joy to joy . . .

and thus she becomes the poet's guardian against human evils and "The dreary intercourse of daily life . . ." To perceive in Nature this favoritism is clearly more self-indulgent and less true than Spenser's characterization of her as the "equall mother" of all creatures, which conforms exactly to Jesus's reminder that God "maketh his sun to rise on the evil and on the good, and sendeth rain on the just and on the unjust." Since Nature is so exceptionally kind to Wordsworth and his sister, it is in a manner logical that he declares himself a "worshipper of Nature." Here he departs maybe as far as possible from the Nature of his predecessors, to whom she was God's Vicar and thus forever subordinate to Him. The older poets were, as C. S. Lewis said of Spenser, Christians, not pantheists.

In comparison to the imaginative force and complexity of the earlier poets, this poem looks simpleminded and slack. Nature is understood merely as the purveyor of a sort of consolation or what we now call "mental health." Nobody could take from this version of Nature any sense of our economic dependence upon her, much less of her dependence upon our virtue. The wonder is that this poem contains in lines 11–17 a fine and moving description of an economic landscape—"pastoral farms / Green to the very door," "plots of cottage ground," orchards, and hedge rows—but he makes no approach to the economic life of that place or to its farmers, who certainly could have enlightened him about Nature's special preferences and favors. In this poem, the farms rate only as scenery, as they do for nature lovers of our own time.

A further wonder is that two years later Wordsworth wrote "Michael," a poem that penetrates the scenery. In that poem, the poet imagines in plenitude of detail the lives of a "pastoral" family of the Lake District: Michael, an elderly shepherd, Isabel, his wife, younger by twenty years, and their late-born only son, Luke. They live in difficult country, in weather that can be harsh. Like the many generations of their forebears, they live by endless work, from daylight to dusk and on into the night, the aging parents

> neither gay perhaps,
> Nor cheerful, yet with objects and with hopes,
> Living a life of eager industry.

There is something of the quality of legend in the telling, for their life was old in the poet's memory, and old beyond memory. But I think there is little if any idealization, no "romanticizing." Above all, Wordsworth passed unregarding the temptation to present these people as "clowns" or, as we would say, "hicks." Michael's mind "was keen, / intense, and frugal." As a shepherd, he is "prompt / And watchful." He knows "the meaning of all winds."

The poem tells of Luke's upbringing, during which he learned his people's world and their work by accompanying his father almost from infancy, by being "Something between a hindrance and a help" at the age of five in helping his father to manage their sheep flock, and by becoming his father's "companion" in the work by the age of ten.

Theirs is a "world" in itself almost complete and everlasting. But when Luke is eighteen trouble comes from the outside, as trouble is apt to come from the outside in, for instance, the novels of Thomas Hardy. Michael has mortgaged his land "in surety for his brother's son," to whom misfortune has come, and Michael is called upon to pay the debt, which amounts in value to about half his property. To preserve the land undivided as Luke's inheritance, his old parents decide to send him away, to work for another kinsman, "A prosperous man, / Thriving in trade." Working for this man, Luke would earn enough to lift the debt from the land. As Michael says to Isabel,

> He quickly will repair this loss, and then
> He may return to us. If here he stay,
> What can be done? Where everyone is poor,
> What can be gained?

For a time, Luke does well, the kinsman is pleased with him, he writes "loving letters" home. And then doom falls with terrible swiftness upon the family and their long history:

> Luke began
> To slacken in his duty; and, at length,
> He in the dissolute city gave himself
> To evil courses: ignominy and shame
> Fell on him, so that he was driven at last
> To seek a hiding-place beyond the seas.

Thus an ancient story receives almost abruptly its modern version: The Prodigal, the far-wandering son, does not now finally make his way home to continue the family lineage in its home place. Now he is gone forever. The forsaken parents live on alone, and die, and the land, at such great cost held to, is sold "into a stranger's hand."

In "Tintern Abbey," without of course intending to do so, Wordsworth laid out pretty fully the model of industrial-age conservation, which reduces too readily to the effort to preserve "wilderness" and "the wild," in certain favored places, as if to compensate or forget the ongoing industrial devastation of the other landscapes. This version of conservation, industrial and romantic, orthodox and dominant for at least a century, simplifies and sentimentalizes nature as friendly, wild, virgin, spectacular or scenic, picturesque or photogenic, distant or remote from work or workplaces, ever-pleasing, consoling, restorative of a kind of norm of human sanity. Conservationists of this order have thus established and ratified a division, even a hostility between nature and our economic life that is both utterly false and limitlessly destructive of the world that they are intent upon "saving." Such conservationists are no threat at all to the economy of industry, science, and technology, of recreational equipment and vacations, which threatens everything those conservationists think they are defending, including "wilderness" and "the wild." Meanwhile the absolute dependence, even of our present so-called economy, even of our lives, upon the natural world is ignored. In my many years of advocacy for better care of farms and working forests, the silence of conservationists and their organizations has been conspicuous. They oppose sensational abuses such as global warming or fracking or (sometimes) surface mining, but they don't oppose bad farming. Most of them would not recognize bad farming if they saw it, and they see plenty of it even from the highways as they drive toward the virgin forests and the snowcapped mountains. It seems never to have occurred to them that soil erosion and stream pollution in agricultural lands threaten all of this natural world, even "the wild," and that such abuses are caused by an economy that ruins farmers and farms by policy.

"Tintern Abbey" is an archetypal poem, for it gives us the taste, tone, and "spiritual" justification of the escapist nature-love of the many romantic nature poems that descended from it, and of the still-prevailing mentality of conservation. "Michael" also is an archetypal poem, but in a sense nearly opposite. Wordsworth understood it as no more than a family tragedy. Two centuries later, we must see it as uncannily predictive of millions of versions of its story all over the world: the great and consequential tragedy, little acknowledged or understood, of the broken succession of farm families, farm communities, and the cultures of husbandry. Generation after generation the children of farm couples have moved away, leaving the land's human memory, often of great ecological and economic worth, to die with their parents. Whether they have gone away to fail or succeed by the measures of their time, they still have gone away, and their absence is a permanent and enormous loss. And so "Michael" is great both because of its achievement and stature as a work of art and because of its rarity and significance as a cultural landmark. It stands high among the poems that have meant most to me. I first read it more than sixty years ago.

The second tragedy of "Michael" has to do with the history of the poem itself. Unlike "Tintern Abbey," it has had, so far as I am aware, no influence on our thinking about the natural world and our use of it, and little or no influence on the subsequent history of poetry. It stands almost alone. The only poem I know that I think worthy of its company is "Marshall Washer," by my friend Hayden Carruth. Marshall Washer was for many years Hayden's neighbor, a dairy farmer near Johnson, Vermont. Like Michael, Marshall lived a life of hard work on a small farm in a demanding place and climate. As, like Michael, the master and artist of his circumstances, Marshall earned Hayden's love and respect, and from his companionship and example Hayden learned many things that he valued. Hayden, moreover, saw Marshall in perspective of the hard history of such farmers before and after the time of Wordsworth. In the poem, Marshall has lived and worked and aged into a loneliness known to millions of his kind and time. His wife has died, and his sons have

> departed, caring little for the farm because
> he had educated them—he who left school
> in 1931 to work by his father's side
> on an impoverished farm in an impoverished time.

Beyond Marshall's life, the life of the farm had become unknowable and unimaginable, and all such farms were coming under the influence of "development." Land prices and taxes were increasing. It was becoming less possible for a small farmer to own a small farm. And Hayden imagined Marshall's sorrow:

> farming is an obsolete vocation—
> while half the world goes hungry. Marshall walks
> his fields and woods, knowing every useful thing
> about them, and knowing his knowledge useless.

This describes the breaking of what Hayden called "the link of the manure"—"manure" in the senses both of fertilizing and caring for, hand-working, the land—which cannot be ignorantly maintained. The "link of the manure" is the fundamental economic link between humans and the natural world. No matter the plain necessity of this link, it is breaking, or is broken, everywhere.

Ignorance certainly will break it. But so also will forces imposed upon it by what we falsely and too readily think of as the "larger" economy. There obviously can be no economy larger than its own natural sources and supports. Less obvious, farming being "an obsolete vocation" as far as most of us are concerned, is the impossibility that any economy can be larger or more important or more valuable than the economies of land use that connect us practically to the natural sources.

Nevertheless, we have this small contrivance we call "the economy," utterly detached from our households and our need for
food, clothing, and shelter, in which people "put their money
to work for them" and sit down to await the increase, in which
money interbreeding with money enlarges itself to monstrosity, glutting on the world's goods. This small economy, centralized and concentrated in the larger cities, imposes in its great
equation of ignorance and power a determining and limitlessly
destructive influence upon the economies of land use, of farming and forestry, which are large, dispersed, and weak.

Those of us who would like to understand this could do worse
than consult with poets, or with the too few of them who have
taken an interest and paid attention. Those two widely separated and lonely poems, "Michael" and "Marshall Washer," are
tragedies of the modern world. The stories they tell become
tragic because the interest of the land, the human investment
of interest and affection in the land, becomes subordinated to
the interest of a "larger" economy that removes the human
interest native to a place and replaces it with its own interest
in itself.

After the story of the westward migration, the dominant
American story so far is that of the young people who have
departed from their rural birthplaces, "humble" or "small" or
"backward" or "poor," to find success or failure in the big city.
The story of the loneliness of the elders left behind, though
surely as common, is rarely told. The implication of its rarity is
that it does not matter, but is only a small sorrow incidental to
the quest for a greater happiness. But sometimes a backward
look will occur, bringing a recognition of loss and suffering
that matter more than expected. The example I have in mind is
a Carter Family song, "The Homestead on the Farm":

> I wonder how the old folks are at home,
> I wonder if they miss me when I'm gone,
> I wonder if they pray for the boy who went away
> And left his dear old parents all alone.

The song, which I like, carries the story only as far as nostalgia:
The departed boy looks back with fondness for his old home,
which his memory no doubt has improved, and with some

regret, but he clearly has no plans to return. Maybe we can do no better than this, having as yet no common standards or a common language to deal with social disintegration, much less with diminishments of culture and the loss of local memory, all of which will enter into an accounting indivisibly cultural, ecological, and economic.

I don't know if Ezra Pound ever actually knew such people as Michael or Marshall Washer. But his poems on usury, "sin against nature," cast on the stories of these men what has been for me an indispensable light. Maybe nobody ever gave more passionate attention than Pound to the ability of a monetary system by means of usury to drive the cost of land and its products beyond any human measure of their worth, and thus to prey upon and degrade the work and the health that sustain us. And nobody has ever instructed us about this with more economy or grace or beauty:

> With usury has no man a good house
> made of stone, no paradise on his church wall
> With usury the stonecutter is kept from his stone
> the weaver is kept from his loom by usura
> Wool does not come into market
> the peasant does not eat his own grain
> The girl's needle goes blunt in her hand
> The looms are hushed one after another
>
> . . . .
>
> Usury kills the child in the womb
> And breaks short the young man's courting
> Usury brings age into youth; it lies between the bride
> and the bridegroom
> Usury is against Nature's increase.

Pound had the misfortune, self-induced, of becoming more notorious for his prejudices and political mistakes than famous for sanity. But he did have in him a broad streak of good sense. He could see all the way to the ground, not invariably a talent among poets, and he had moments of incisive agrarianism. He wrote in these lines a clarifying history of modern land husbandry:

> Dress 'em in folderols
> > and feed 'em with dainties,
> In the end they will sell out the homestead.

He praised with perfect soundness "Chao-Kong the surveyor," who "Gave each man land for his labour"—wages being always vulnerable, whereas the value of land, like the value of a life, is unreckonable and absolute. And he wrote these lines,

> > Pull down thy vanity, I say pull down.
> Learn of the green world what can be thy place
> In scaled invention or true artistry . . .

in which he stood before Nature much in the posture of Alan of Lille eight hundred years before.

Whoever would think at the same time of the home ecosystem on the one hand, and on the other of the home community (ecosystem plus humans), is all but forced to think of the local economy—and its tributary local economies elsewhere. Very few poets—very few people—have thought of both at once, because, I suppose, of the intensity of the stretch. It is something like standing with one foot on shore and the other in a loosely floating small boat. It requires a big heart and a strong crotch.

 Of the poets I know of my own time, Gary Snyder is the one whose thought and work (in poetry and prose) have most insistently inclined toward the daily and practical issues of our economic life, which is to say our life: the possibility of bringing our getting and spending into concordance with terrestrial reality. This has been, for Gary, a lifelong effort, involving the events and materials of his life, much travel, much reading, much study and thought of his home geography. At a loss as to how to represent this effort in its complexity, I will say only that I think it has begun and renewed itself again and again in realizations of the profound, indissoluble link, the virtual identity, between the world's life and the lives of creatures. How is it with us who live our ever-changing lives as parts or members of the ever-changing world? Or to use language Gary has borrowed from Dogen: How is it with us who are walking on mountains that are walking? This question is not comfortable,

and we sometimes would like to ignore or wish against it. But
it means at least that the world is always with us, new and fresh:

> Clear running stream
> > clear running stream
>
> Your water is light
> > to my mouth
>
> And a light to my dry body
> > your flowing
>
> Music,
> > in my ears. free,
>
> Flowing free!
> With you
> > in me.

It is not hard to imagine thoughts of economy and of "the
economy" starting from the tenderness, intimacy, and inherent
delight of this kindness, this kinship. But the poem also is a
thanksgiving, the proper conclusion to such thoughts.

    And I ask myself, Can it be that, by way of Buddhism and
American Indian anthropology, Gary Snyder is another who
has come face to face with the great goddess, great dame Nature, and he makes her this cordial greeting? Maybe.

It must be clear enough by now that I am a reader who reads
for instruction. I have always read, even literature, even poetry, for instruction. I am a poet, and so I have read other
poets as a poet, to learn about poetry, and, just as important,
to learn from poetry. I am also in a small way a farmer devoted
to farming, as my father and his father made me, and I have
read poetry, and everything else, also as a farmer. As a farmer I
have lived daily with the inherent hardships and pleasures, but
also with an ever-fascinating and utterly intimidating question:
What should I be doing to care properly for this (as it happens)
very difficult and demanding place? And a further one: How
can I make myself a man capable of seeing what is the right
thing to do, and of doing it?

    Such questions now are typically addressed to experts, and
are answered by a letter or a pamphlet replete with statistics,

graphs, diagrams, and instructions, all presumed to be applicable to all persons and to any and every place within a designated region or zone. This may in fact prove helpful, but it is not enough. It is not enough because it includes no knowledge of the particular and unique place where the expertise is to be applied. Moreover, expertise which is credentialed and much dependent on the demeanor and language of authority is typically unable to acknowledge either its inner ignorance or the immense mystery that surrounds equally the asker and the answerer. And so askers are in effect left alone with the expert answer on their singular small places within the mostly uninformative universe.

Supplements and alternatives are available. One can talk with one's neighbors and elders, who never are oracles but are usually worth listening to. One can consult with other experts. I have always liked to consult older or earlier experts, whose knowledge may be seasoned with affection and humility. Or one can read, following one's nose as good hounds and readers do, always on watch for what people in other times and places have known or learned, what penalties they paid for ignorance, what satisfactions they gained by knowledge. These benefits can come from books of every kind.

Such broad and even random conversation is necessary because the present industrial world is not an isolated empire fortified against history or incomparable by its improvements with other times and ways. As I have conversed and read for the last fifty years, I have been reminding myself that many new ways of doing work have been adopted not because they were better for land and people, but simply because the old ways were technologically outmoded. Industrial agriculture, as probably the paramount example, is better than pre-industrial agriculture by the standard mainly of corporate profitability, excluding by a sort of conventional or fashionable blindness the paramount standard of mental, bodily, and ecological health. One can learn this by the study of rural landscapes, but conversation and reading also are necessary. To think responsibly about land use, the whole known spectrum of means and ways must be available to one's thoughts. If, for example, one cannot compare a tractor (and its attendant economy and ecological effects) with a team of horses (and its attendant

economy and ecological effects), then one has the use of less
than half of one's mind.

Do I, then, think that farmers, or persons of other vocations,
can find actual help in reading Chaucer or Milton or Spenser or
Pope? Yes, that is what I think. To know Chaucer's plowman
is not to know something merely of historical interest. It is to
know, to recognize immediately, something that one needs to
know. It would be too much to say that *all* farmers, foresters,
economists, ecologists, and conservationists should know what
Chaucer, Spenser, Milton, and Pope wrote about Nature, but
it is a pity and a danger that none or only a few of them do.

Having argued that it is possible to learn valuable, even useful
and necessary things from poetry, I now have to answer two
further questions. First: Is it necessary for a poem to be instructive in order to be good? I hope not, and I don't think so.
There is nothing very instructive, for example, in hearing that
"the cow jumped over the moon," but who is not delighted by
that poem's exuberant indifference to the possibility of making
sense? It is a masterpiece. Even so, I am happy to know that
some poems delight *and* instruct, which is a richer possibility.

Second: How does an instructive poem instruct? The answer seems obvious—by containing something worth knowing
—but there is one condition: It must teach without intending
to do so. In support of this I offer a sentence by Jacques Maritain, who said of the cathedral builders: "Their achievement
revealed God's truth, but without *doing it on purpose*, and because it was not done on purpose." The point, I believe, is
that what the cathedral builders were doing on purpose was
building a cathedral. Any other purpose would have distracted
them from the thing they were making and spoiled their work.
Teaching as a purpose, as such, is difficult to prescribe or talk
about because the thing it is proposing to make is usually
something so vague as "understanding." My own best teachers, as I remember, sometimes undertook deliberately to teach
me something: "You handed me that board wrong, and made
my work harder. Now I'm going to hand it back to you *right*.
Now you hand it back to me *right*." But this was rare. When
I was most learning from them, I think, they were attempting
something besides teaching, if only to make a sentence that

made sense. They taught me best by example, unaware that they were teaching or that I was learning. Just so, an honest poet who is making a poem is doing neither more nor less than making a poem, undistracted by the thought even that it will be read. Poets, or some poets, bear witness as faithfully as possible to what they have experienced or observed, suffered or enjoyed, and this inevitably is instructive to anybody able to be instructed. But the instruction is secondary. It must be embodied in the work.

One remarkable thing I have noticed in my reading is that agrarianism, readily identifiable by certain themes (the importance of the small holding, the relationship to Nature, etc.), appears throughout the written record from as far back as Homer and the Bible. A second remarkable thing is that these appearances are intermittent, sometimes widely separated. When this vital strand of human thought and concern has disappeared from writing—as it does, in poetry, from Wordsworth's "Michael" to Hayden Carruth's "Marshall Washer"—it has continued in the conversation of farmers. I was (I am still) Hayden Carruth's friend. I knew him from much conversation and many letters. In Hayden's company I met and talked at some length with Marshall Washer. Though Hayden was far better read than I am, I am confident that the agrarianism of Hayden's poem did not come from his reading. It came straight from Marshall himself, who had it from his father, who had it from a lineage of living voices going back and back, more or less parallel to, but rarely, maybe never, intersecting with the lineage of writings. I know this also from my own long participation in such conversation. The appearances of agrarian thought in the written record are like stepping stones, separate and irregularly spaced, but resting on a firm and (so far) a continuous bottom.

Who were, and to a much-reduced extent still are, the people who have passed this tradition by living word from generation to generation for so long? The answer, obviously, is the country people who have done the work of field and forest, generation after generation. They are the people once known as peasants, serfs, or churls, and now, still doing the same work, as farmers or country people, when "farmer" and "country" are still as readily used as terms of abuse as "peasant" and "churl" ever were. My friend Gene Logsdon says that when he was a schoolboy in Wyandot County, Ohio, "*Dumb farmer* was one word." Here I need to remember two sentences from G. G. Coulton's *Medieval Panorama*. The first is this: "Four-fifths, at least, of the medieval population had grown their own food in their own fields; had spun and woven wool from their own sheep or linen from their own toft; and very often it was they themselves who made it into clothes." I should add that they also kept themselves housed, and they fed and sheltered their animals. Anybody who has done any part of such work knows that it involves knowledge of the highest order and value, for the human species has survived by it, and it is neither simple nor easy to learn.

But how were these people with their elaborate and indispensable knowledge valued in the Middle Ages? Here is the second sentence from *Medieval Panorama*: "Froissart shows us plainly enough how, after the bloody capture of a town or castle, the gentles [knights and nobles] were spared, but the common soldiers [churls] were massacred without protest from their more exalted comrades in arms." They were thought to be, in a word, dispensable.

I don't believe that this thought has changed very substantially from then until now. We, whose humane instincts have so famously evolved and improved, are firmly opposed to outmoded forms of massacre, but we still regard farmers as dispensable, and by various economic constraints and social fashions we have dispensed with many millions of them over the last sixty or so years. If the twenty-first-century American farmer is farming several thousand acres and employing a million dollars' worth of equipment, he still is a member of a disparaged, and therefore a vanishing, population. He does not control any part of his economy. He has no more influence over the markets on which he buys and sells than he has over the weather. No prominent politician, economist, or intellectual is thoughtful either of him or of the condition of the land he works, or of the health of the ecosystem that includes his land. In the entire food industry he will be dependably the most at risk, the least valued, and the lowest paid, except, of course, for the migrant laborers he may at times be constrained to hire. If he miscalculates or has a bad year and is ruined, that will register among the professional onlookers as a minor instance of "creative destruction."

In the course of this progress of industrialization and depopulation, farming itself has become so radically simplified as to be unworth the name. Most damaging has been the division between the field crop industry and the meat-animal industry. To remove the farm animals from farming is to remove more than half of the need for knowledge, skill, and intelligence, and nearly all of the need for sympathy. To crowd the animals into the tightest possible confinement to live and function exclusively as meat-makers is to do away with sympathy as a precondition, to reduce mindfulness to routine, and to replace all the free helps of Nature and natural health with purchased machinery and medicines.

But when plants and animals, croplands and pastures, are gathered into the care of a single farmer, this calls for a mind versatile and accomplished, competent to deal with the fairly stable natures of the kinds of plants and animals, as well as the variable and sometimes surprising circumstances of mortality, the weather, and the economy. Among the good farmers I have known, I have found also a formal intelligence by which they ordered the spatial arrangement of fences, fields, and buildings to be most conserving of the land and most usable, and by which they formed also the temporal structures of their crop rotations and the days, seasons, and years of their work.

Good farmers, whose minds are comparable to the minds of artists of the "fine arts," have been instructed so often for so long that they are "dumbfarmers," that they only half believe how smart and capable they actually are. I have heard too many of them describe themselves as "just a farmer." It is, at any rate, impossible for highly credentialed professionals and academics to appraise justly the intelligence of a good farmer. They are too ignorant for that. You might as well send a bird dog to judge the competence of a neurologist.

About such things there are no "objective" measures. I can offer only my testimony. In my by-now long life, I have known well and observed closely the work of a good many farmers whom I have respected. I have thought carefully about them for a long time, and some of them I have written about. In general: I have found them alert, observant, interested, interesting, thoughtful of their experience, conversant with the experience of others. By midlife or sooner they have come to know many things so completely as to be unconscious of knowing them. They have had, in general, the humorous intelligence that recognizes natural limits and their own limits. They have had also a commonplace sense of tragedy that permitted them to accept their helplessness in the face of loss and suffering—as Marshall Washer faced his burning barn with thirteen heifers trapped inside, and kept on, still working, by the testimony of his friend Hayden Carruth, within weeks of his death at eighty-eight. That is not indicative either of "dullness" or "stoicism," nor does it mean that farmers in general are better or more virtuous than other persons. It means simply that farmers live and work in circumstances unremittingly practical, and for this certain strengths of mind are necessary.

In general, the good farmers I have known have had no taste for self-display, did not parade their knowledge, and would say little where their knowledge might not be respected. But if you had a mind and ear for their conversation, and some ground of friendship or common knowledge, you were likely to hear sound sense that came from their experience of the natural world, and of farming and its local history. As late as my own generation, because the young then were still working with and listening to their elders, the talk of farmers was still carrying the traditional themes and attitudes. Many of the things I heard from them I have kept always in my mind, their words and their voices.

Probably in the summer of 1965, after my family and I had moved here to the "twelve acres more or less" where we still live, I was not equipped for some mowing that needed to be done. I hired a neighbor of my parents' generation, a man I had known from my childhood, liked, and respected. When he had finished the work, he turned off the tractor engine and we talked a while, he much in the spirit of welcoming me to my new place. He had been renewing his own acquaintance with the place as he worked. We spoke of the ways it had been used (not always kindly), and of the uses I might make of it.

When he was ready to go, he started the tractor and, to end the conversation, said cheerfully, "Well, try a little of everything, you'll hit on something."

He knew, and he knew I knew, how little of "everything" the old place was actually capable of producing on its steep slopes and its one pretty good "garden spot" by the river. But he had said a good thing, and we both knew that. He had stated one of Nature's laws: the Law of Diversity, vital both to ecological and to economic health. One of my father's father's rules was "Sell something every week," a different version of the Law of Diversity.

Another of my older friends, a fine farmer whose friendship I inherited from my father, once told me a crucial part of the story of his beginning on his own farm. This, I think, would have been in 1940 or a little earlier. He and his wife had gone into debt to buy their place, and they had no money. My friend went early in the year to a grocery store in town and asked the owner if he would "carry" him until he sold his crop. The owner knew him, trusted him, and agreed. When my friend sold his crop at the year's end and went to settle up at the store, his bill came to eleven dollars and some cents. He and his wife had been thrifty and careful, living so far as possible from their place, thus obeying another of Nature's laws, the Law of Frugality: Don't be prodigal.

A man of about the same age, a dear friend who was conscientiously my teacher for thirty years, once told me with an emphasis amounting to passion: "If you've got grass and room to keep a milk cow [for family use] and you don't do it, you've *lost* one milk cow." The absent cow he saw, not as a neutral matter of mere preference, but as a negative economic force, a subtraction from thrift and thriving. This was in respect for the Law of Diversity and the Law of Frugality.

A steadying influence on my mind for most of my life has been the metaphor by which my father referred to a patch of abused land that had been healed and grassed as "haired over." He thus perceived a wound to the earth as a wound to a living creature, perhaps a collar gall on the shoulder of a workhorse. We could say that this phrase observes another of Nature's laws—Keep the ground covered—but much else of value is concentrated in it. By it my father wakened my mind to the thought of the system of kinships by which the world survives, and to the thought of the whole significance of the smallest healing. I don't know whether or not this language originated with my father, but I knew *his* father and so I know that it was not the thought of one mind. Such a thought is as far as possible unlike the thought of the earth as an inert material mass to be shoved about, poisoned, and blasted at human discretion.

None of those sayings comes from a note I made at the time or from any record I kept, but only from memory, and to me this signifies their importance and value. I have remembered them and many others, I believe, because I recognized them not quite consciously as parts of a whole, a kind of mind which, like my mentors, I inherited, which probably is as old as farming and necessary to it. This mind, which I think has never been fully conscious or coherent in any one person, never perfect, always in some manner failing, almost never having the incentive of public appreciation or adequate economic support, has nonetheless cohered, coming to consciousness as needed, and expressing itself in an articulate local speech that takes its heft and inflection from the reconciliations between people and their circumstances, their work and their feelings about it. This mind, much to the credit of its inherent good sense, has survived all its adversities until now, living on, talking to itself, so to say, in the conversation of its local memberships. It still speaks, where it survives, of the importance of the well-doing of small tasks that the dominant culture has always considered degrading but which are nonetheless essential and worthy of care: "Don't think of the dollar," my friend and teacher said to me. "Think of the *job*." And it still speaks of our dependence on the particular natures of creatures and on the natural world, and, more practically, of a necessary respect and deference toward Nature. In this speech, Nature customarily is personified: the great dame herself, who knows best and will have her way. Always there is the implication that Nature helps when you work *with* her, with knowledge of her ways, their value, and their ultimate dominance, and that she does not help but works against you when you work against her.

Recently my son brought me an issue of a conventional industrial farming magazine, containing an article, by a "semi-retired" rancher, Walt Davis, about the need to manage a cow

herd so as "to be in sync with nature." Though some of the
vocabulary ("in sync") is new, this advice is old. I know nothing
that suggests it is wrong. The point, as often, is economic: To
be in sync with Nature is to make full use of her helps that are
free or cheap, as opposed to the use of industrial substitutes that
are expensive. By observance of Nature's laws, the land survives
and even thrives in human use. By the *same* laws, farmers can
hope at least to survive in the almost conventional adversity of
the farm economy.

Almost typical of young farmers is the impulse to intervene
in natural processes with some "latest" expertise, requiring
typically some "latest" merchandise in order to increase production or fend off some perceived threat. Older farmers are
more likely to be suspicious of anything that costs money, and
to rely instead on the gifts of their own intelligence and the
nature of creatures. Young shepherds, in their eager and self-
regarding sympathy, may try to help a newborn lamb to stand
and suck—may even succeed, or seem to. Older shepherds will
know that such "help" is most likely a waste of time. They
know when to walk away and "leave it to Nature." They are
likely to know also that Nature bids them to get rid of a ewe
whose lamb can survive its birth only with human help.

Going on forty years ago, because I knew his grand reputation as a breeder of Southdown sheep, I had the honor of becoming a little acquainted with Henry Besuden of Vinewood
Farm in Clark County, Kentucky. Mr. Besuden was seventy-
six years old then and not in good health; it had been several
years since he had owned a sheep. But when I visited him,
which I did at least twice, and he showed me his place and
we talked at length, I found that his accomplishments as a
farmer excelled in fact and in interest his accomplishments as
a sheepman.

In 1927, at the age of twenty-three, he inherited a farm of
632 acres, a large holding for that country, but it came to him
as something less than a privilege. The land had been ruined
by the constant row-cropping of renters: "Corned to death."
Gullies were everywhere, some of them deep enough to hide a
standing man. Between Mr. Besuden then and an inheritance
of nothing was the stark need virtually to remake his farm.

To do this he had to return the land from its history of human carelessness to the care of Nature: Every gully, through which the land was flowing away, had to be transformed into a grassed swale that would check and retain the runoff. This required a lot of work and a long time, but by 1950 all of the scars at last were grassed over.

To this effort the sheep were not optional but necessary. They would thrive on the then-inferior, weedy and briary pastures, and, rightly managed, they were "land builders." He became a sheepman in order to become the farmer Nature required for his land. The constant theme of his work was "a way of farming compatible with nature" or, as Pope would have put it, with "the Genius of the place." I have not known a farmer in whose mind the traditional agrarianism was more complete or more articulate.

He wrote in one of his several articles entitled "Sheep Sense": "It is good to have Nature working for you. She works for a minimum wage." Soil conservation, he wrote, "also involves the heart of the man managing the land. If he loves his soil he will save it." He wrote again, obedient to Nature's guidance, of the need "to study the possibilities of grass fattening." He wrote of the importance of "little things done on time." He told me about a farmer who would wait until he came to a spot bare of grass to scrape the manure off his shoes: "That's what I mean. You have to keep it in your mind."

In any economy that becomes exploitive of land or people or (as usual) both, it appears that some of the people working within it will recognize that its special standards of judgment are inadequate. They will see the need for a more comprehending standard by which the constrictive system can be judged, so to speak, from the outside. At present, for example, some doctors and others who work in the medical profession have the uneasy awareness that their industrial standards are failing, and their thoughts are going, as always in such instances, toward Nature's laws, which is to say, toward the health that is at once bodily, familial, communal, economic, and ecological. The outside standard invariably will turn out to be health, and perceptive people, looking beyond industrial medicine, see that health is not the painlessness of a body part, or the comfort of parts of a society. They see that the idea of a healthy individual in an unhealthy community in an unhealthy place is an absurdity that, by the standards of industry, can only become more absurd.

In agricultural science the same realization was occurring at the beginning of the twentieth century. The first prominent sign of this, so far as I know, was a book, *Farmers of Forty Centuries*, in which F. H. King, who had been professor of agricultural physics at the University of Wisconsin, recounted his travels among the small farms of China, Korea, and Japan. As a student of soil physics and soil fertility, King must be counted an expert witness, but he was also a sympathetic one. He was farm-raised, and his interest in agriculture seems to have come from a lifelong affection for farming, farmers, and the details of their work.

The impetus for his Asian journey seems to have been his recognition of the critical deficiencies of American agriculture, the chief one being the relentless exploitation of soil and soil fertility by practices that were supportable only by the importation to the farms of "cargoes of feeding stuffs and mineral fertilizers." This, he knew, could not last. It was, we would say now, unsustainable, as it still is. It was possible in the United States because the United States was still a thinly populated country.

Such extravagance had never been possible in China, Korea, and Japan, where "the people . . . are toiling in fields tilled more than three thousand years and who have scarcely more than two acres per capita, more than one-half of which is uncultivable mountain land." This long-enduring agriculture was made possible by keeping the fertility cycle intact and in place. The "plant food materials" that we were wasting, "through our modern systems of sewage disposal and other faulty practices," the Asian farmers "held . . . sacred to agriculture, applying them to their fields."

As a sample of the economic achievement of the peasant farms he visited, King says in his introduction that "in the Shantung province [of China] we talked with a farmer having 12 in his family and who kept one donkey, one cow, both exclusively laboring animals, and two pigs on 2.5 acres of cultivated

land where he grew wheat, millet, sweet potatoes and beans."
The introduction ends with this summary:

> Almost every foot of land is made to contribute material
> for food, fuel or fabric. Everything which can be made
> edible serves as food for man or domestic animals. What-
> ever cannot be eaten or worn is used for fuel. The wastes
> of the body, of fuel and of fabric worn beyond other use
> are taken back to the field . . .

The chapters that follow contain finely detailed descriptions
of one tiny farm after another, every one of them exemplify-
ing what I have called Nature's Law of Frugality, or what Sir
Albert Howard would later call her "law of return." Fertility,
we might say, was understood as borrowed from Nature on
condition of repayment in full. "Nothing," King wrote, "jars
on the nerves of these people more than incurring of needless
expense, extravagance in any form, or poor judgment in making purchases." No debt was to be charged to the land and left
unpaid, as we were doing in King's time and are doing still.

Farmers of Forty Centuries was published in 1911. In 1929,
J. Russell Smith, then professor of economic geography at
Columbia University, published *Tree Crops*, another necessary
book, another of my stepping stones. I don't know whether or
not Smith had read King's book, but that does not matter, for
Smith's book was motivated by the same trouble: our senseless
waste of fertility and the prospect of land exhaustion.

What Smith saw was not only our fracture of the fertility cycle, but also another, an opposite, cycle: "Forest—field—plow
—desert—that is the cycle of the hills under most plow agricultures . . ." Smith was worried about soil erosion. He had seen
a part of China that King had not visited:

> The slope below the Great Wall was cut with gullies,
> some of which were fifty feet deep. As far as the eye could
> see were gullies, gullies, gullies—a gashed and gutted
> countryside.

And Smith knew that there were places similarly ruined in the
"new" country of the United States. The mistake, everywhere,
was obvious: "Man has carried to the hills the agriculture of

the flat plain." The problem, more specifically, was the agri-culture of annual plants. "As plants," he wrote, "the cereals are weaklings." They cannot protect the ground that their cultiva-tion exposes to the weather.

As Henry Besuden had realized on his own farm two years before the publication of Smith's book, the reaction against the damage of annual cropping could only go in the direction of perennials: Slopes that were not wooded needed to be per-manently grassed. So Mr. Besuden rightly thought. J. Russell Smith took the same idea one step further: to what he called "two-story agriculture," which would work anywhere, but which he saw as a necessity, both natural and human, for hill-sides. The lower story would be grass; the upper story would consist of "tree crops" producing nuts and fruits either as for-age for livestock or as food for humans.

The greater part of Smith's book is devoted to species and varieties of trees suitable for such use, and to examples (with photographic plates and explanatory captions) of two-story agriculture in various parts of the world. He was, he acknowl-edged, a visionary: "I see a million hills green with crop-yielding trees and a million neat farm homes snuggled in the hills." But he was careful to show, by his many examples, that two-story agriculture *could* work because in some places it was working, and in some places it had been working for hundreds of years. He was promoting a proven possibility, not a theory.

Like Alan of Lille, Smith understood Nature's condemnation of prodigality, and he accepted her requirement that we should save what we have been given. The passion that informs his book comes from his realization that Nature's most precious gift was given only once. The Prodigal Son could squander his inheritance of money, repent, and be forgiven and restored to his father's favor. But Nature, "the equall mother" of all the creatures, has no bias in favor of humans. She gave us, along with all her other children, the great gift of life in a rich world, the wealth of which reduces finally to its thin layer of fertile soil. When we have squandered that, no matter how we may repent, it is simply and finally gone.

Smith was above all a practical man, and he stated one of Nature's laws in terms exquisitely practical: "*farming should fit the land*." The italics are his. He said that this amounted to "*a new point of view,*" and again his italics indicate his sense of the urgency of what he was saying. He was undoubtedly right in assuming that this point of view was new to his countrymen, or to most of them, but of course it was also very old. It was implicit in Nature's complaint to Alan of Lille, it was explicit in Virgil's first Georgic, more than a thousand years before Alan, and it certainly informed Pope's instructions to consult "the Genius of the Place." The point is that, in using land, you cannot know what you are doing unless you know well the place where you are doing it.

In the long lineage that I am discussing, the fundamental assumption appears to be that Nature is the perfect—and, for our purposes, the exemplary—proprietor and user of any of her places. In our agricultural uses of her land, we are not required to imitate her work, because, as Chaucer's physician says, she is inimitable, and in order to live we are obliged to interpose our own interest between her and her property. We are required instead to do, not *as* she does, but *what* she does to protect the land and preserve its health. For our farming in our own interest, she sets the pattern and provides the measure. We learn to farm properly only under the instruction of Nature: "What are the main principles underlying Nature's agriculture? These can most easily be seen in operation in our woods and forests."

Those are the key sentences of *An Agricultural Testament*, by Sir Albert Howard, published in 1940. Howard had read *Farmers of Forty Centuries*, and his testimony begins with the worry that had sent King to Asia:

> Since the Industrial Revolution the processes of growth have been speeded up to produce the food and raw materials needed by the population and the factory. Nothing effective has been done to replace the loss of fertility involved in this vast increase in crop and animal production. The consequences have been disastrous. Agriculture has become unbalanced: the land is in revolt: diseases of all kinds are on the increase: in many parts of the world Nature is removing the worn-out soil by means of erosion.

From this perception of human error and failure he made the turn to Nature, that we can now recognize, across a span of many centuries, as characteristic and continuing in a lineage of some poets, some intelligent farmers and farm cultures, and some scientists. Howard's study of agriculture rests solidly upon his study of the nature of the places where farming was done. If the land in use was originally forested, as much of it was and is, then to learn to farm well, the farmer should study the forest.

Industrial agriculture, far from consulting the Genius of the Place or fitting the farming to the land or remembering at all the ecological mandate for local adaptation, has instead and from the beginning forced the land to submit to the capabilities and the limitations of the available technology. From the ruinous and ugly consequences, now visible and obvious everywhere the land can be farmed, one turns with relief to the great good sense, the mere sanity, the cheerful confidence of Howard's advice to farmers: Go to the woods and see what Nature would be doing on your land if you were not farming it, for you are asking her, not just for her "resources," but to accept you as her student and collaborator. And Howard summarizes the inevitable findings in the following remarkable paragraph:

> The main characteristic of Nature's farming can therefore be summed up in a few words. Mother earth never attempts to farm without livestock; she always raises mixed crops; great pains are taken to preserve the soil and to prevent erosion; the mixed vegetable and animal wastes are converted into humus; there is no waste; the processes of growth and the processes of decay balance one another; ample provision is made to maintain large reserves of fertility; the greatest care is taken to store the rainfall; both plants and animals are left to protect themselves against disease.

Howard was familiar with the work of F. H. King and other scientific predecessors. *An Agricultural Testament*, he says, was founded upon his own "work and experience of forty years, mainly devoted to agricultural research in the West Indies, India, and Great Britain." He had the humility and the good sense to learn from the peasant farmers his work was meant to serve. He came from a Shropshire farm family "of high reputation locally," and so I assume he had grown up hearing the talk of farmers. I don't know what, if anything, he had learned from the poets I have learned from, but of course I am delighted that he called Nature personally by name as Chaucer, Spenser, Milton, and Pope had done.

As Milton in *Comus* had enlarged the earlier characterizations of Nature by recognizing the economic significance of her religious and moral demands, making it explicit and practical, so the work of Sir Albert Howard completed their ecological significance. The agricultural economy, and the economies of farms, as determined by the economies and technologies of industrialism, would by logical necessity run to exhaustion. Agriculture in general, and any farm in particular, could survive only by recognizing, respecting, and incorporating the integrity of ecosystems. And this, as Howard showed by his work, could be understood practically and put into practice.

Spenser's problem of stability within change, moreover, we may think of as waiting upon Howard for its proper solution, which he gives in *The Soil and Health*, published in 1947:

It needs a more refined perception to recognize throughout this stupendous wealth of varying shapes and forms the principle of stability. Yet this principle dominates. It dominates by means of an ever-recurring cycle, a cycle which, repeating itself silently and ceaselessly, ensures the continuation of living matter. This cycle is constituted of the successive and repeated processes of birth, growth, maturity, death, and decay. An eastern religion calls this cycle the Wheel of Life and no better name could be given to it. The revolutions of this Wheel never falter and are perfect. Death supersedes life and life rises again from what is dead and decayed.

As Howard of course knew, religion would construe that last sentence analogically or symbolically, but he does not diminish it or make it less miraculous by treating it as fact. The Wheel of Life, or the fertility cycle, is not an instance of stability as opposed to change, as Spenser may have hoped, but rather an instance, much more interesting and wonderful, of stability dependent upon change. The Wheel of Life is a religious principle of which Howard saw the scientific validity, or it is a principle of Nature, eventually of science, which has long been "natural" to religion. This suggests that beyond the experimental or empirical proofs of science, there may be other ways of determining the truth of a solution, a very prominent one being versatility. Is the solution valid economically as well as ecologically? Does it serve the interest both of the land and of the land's people? The Wheel of Life certainly has that double validity. But we can go on: Does it fit both the farm and the local ecosystem? Does it satisfy the needs of both biology and religion? Is it imaginatively as well as factually true? Can it reconcile utility and beauty? Is it compatible with the practice of the virtues? As a solution to the problem of change and stability, the Wheel of Life answers affirmatively every one of those questions. It is complexly, and joyously, true.

Such a solution could not have originated, and cannot be accommodated, under the rule of industrial or technological progress, which does not cycle, which recycles only under compulsion, and makes no returns. Waste is one of its definitive products, and waste is profitable so long as you are not liable for the cost or replacement of what you have wasted.

Waste of the gifts of Nature from the beginning of industrialism has subsidized our system of corporate profits and "economic growth." Howard, judging by the unforgiving standard of natural health, described this "economy" unconditionally: "The using up of fertility is a transfer of past capital and of future possibilities to enrich a dishonest present: it is banditry pure and simple."

The falsehood of the industrial economy has been disguised for generations by the departmentalization and over-specialization of the essential academic and intellectual disciplines. In the absence of any common or unifying standard of judgment, the professionals of the schools have been free to measure their work by professional standards exclusively, submitting only to the "peer review" of fellow professionals. As in part a professional himself, Howard understood the ethical blindness of this system. Against it he raised the standard of the integrity of the Wheel of Life, which is comprehensive and universal, granting no exemptions.

Whether because of his "instinctive awareness of the importance of natural principle" or because of respect for his own intelligence or because of an evident affection for Nature's living world and the life of farming, Howard kept the way always open between his necessarily specialized work as a scientist and his responsibility to the Wheel of Life. His personal integrity was to honor the integrity of what he called the "one great subject": "the whole problem of health in soil, plant, animal, and man." The practical effect of this upon his work was his insistence upon thinking of everything in relation to its context. In his "revolt against 'fragmentation' of knowledge," his investigation of a crop would include its "whole existence": "the plant itself in relation to the soil in which it grows, to the conditions of village agriculture under which it is cultivated, and with reference to the economic uses of the product." As a researcher and teacher, he would not offer an innovation to his farmer clients that they could not afford, or that they lacked the traction power to use.

There clearly is no break or barrier between Howard's principles and his practice and the "land ethic" of his contemporary, Aldo Leopold, who wrote:

Health is the capacity of the land for self-renewal. Conservation is our effort to understand and preserve this capacity.

And:

A thing is right when it tends to preserve the integrity, stability, and beauty of the biotic community. It is wrong when it tends otherwise.

Moreover, there is no discontinuity between the ethics of Howard and Leopold and Nature's principle, given to Alan of Lille, that her good, the good of the natural world, depends upon human goodness, which is to say the human practice of the human virtues. And here I will say again that the needs of the land and the needs of the people tend always to be the same. There is always the convergence of what Nature requires for the survival of the land with what economic demand or economic adversity requires for the survival of the farmer.

From Alan's *Plaint of Nature*, and surely from the conversation of farmers and other country people before and after, it is possible to trace a living tradition of deference, respect and responsibility to Nature and her laws, carried forward by many voices speaking in agreement and in mutual help and amplification. They speak in fact with more agreement and continuity than I expected when I began this essay. And they speak, as also I now see, without so much as a glance toward moral relativism.

For many years I have known that the books by King, Smith, Howard, and Leopold, supported by other intellectually respectable books that I have not mentioned, constitute a coherent, sound, and proven argument against industrial agriculture and for an agricultural ecology and economy that, had it prevailed, would have preserved both the economic landscapes and their human communities. It is significant that all of those books, and others allied with them, were in print by the middle of the last century. Together, they might have provided a basis for the reformation of agriculture according to principles congenial to it. If that had happened, an immeasurable, and so far an illimitable, damage to the land and the people could have been prevented.

The opposite happened. All specifically agricultural and ecological standards were replaced by the specifically industrial standards of productivity, mechanical efficiency, and profitability (to agri-industrial and other corporations). Meanwhile, the criticisms and the recommendations of King, Smith, Howard, Leopold, et al., have never been addressed or answered, let alone disproved.

That they happen to have been right, by any appropriate standard, simply has not mattered to the academic and official forces of agriculture. The writings of the agrarian scientists, the validity of their science notwithstanding, have been easily ignored and overridden by wealth and power. The perfection of the industrial orthodoxy becomes clear when we remember that King, Smith, and Howard traveled the world to search out and study examples of sustainable agriculture, whereas agri-industrial scientists have traveled the world as evangelists for an agriculture unsustainable by every measure.

The truth of agrarian scientists, and their long cultural tradition, casts nevertheless a bright and exposing light upon the silliness and superficiality of the industrial economy, the industrial politicians and economists, and the industrial conservationists. Industrial politicians and economists ignore everything that can be ignored, mainly the whole outdoors. Industrial conservationists ignore everything but wilderness preservation ("Give us the 'wild' land; do as you please with the rest") and the most sensational and fashionable "environmental disasters."

In opposition to industrial and all other sorts and ways of subhuman consumption of the living world, the tradition of cooperation with Nature has persisted for many centuries, through many changes, sustained by right principles, the proofs of experience and eventually of science, and good sense. Its diminishment in our age has been tragic. If ever it should be lost entirely, that would be a greater "environmental disaster" than global warming. It is reassuring therefore to know that it has lasted until now, in the practice of some farmers, and in the work of some scientists. I now have in mind particularly the scientists of The Land Institute, which to me has been a source of instruction and hope for nearly half my life, and which certainly belongs in, and is in good part explained by, the cultural lineage I have been tracking.

The Land Institute was started forty years ago. It is amusing, amazing, and most surely confirmative of my argument, that when Wes Jackson laid out in his mind the order of thought that formed The Land Institute, he had never heard of Sir Albert Howard. The confirming gist of this is that Wes's ignorance of the writings of Sir Albert Howard at that time did not matter. What matters, and matters incalculably, is that Wes's formative thought developed exactly according to the pattern followed by the thinking of King, Smith, and Howard: perception of the waste of fertility and of soil, recognition of the failure of the current standards, and the turn to Nature for a better standard.

Wes had read *Tree Crops*, which had given him the fundamental principle and pattern, which he must more and more consciously have needed. He had in himself the alertness and the sense to look about and see where he was. And like King and Howard he had grown up farming—in, as I know, an agrarian family of extraordinary intelligence and attentiveness. He grew up loving to farm. He knew, and probably long before he knew he knew, the essential truth spoken by Henry Besuden: that soil conservation "involves the heart of the man managing the land. If he loves his soil he will save it." The entire culture of husbandry is implicit in that last sentence. The practical result was, again exactly, Albert Howard's comparison of human farming to Nature's farming—or, as Wes has phrased it in many a talk, a comparison of "human cleverness" to "Nature's wisdom." Whereas Howard had looked to the woods for this, Wes, a Kansan, looked to the native prairie, a possibility that Howard himself had foreseen and approved. If you want to know what is wrong with a Kansas wheat field, study the prairie. And then, taking his own advice, Wes conceived the long-term and ongoing project of The Land Institute: to replace the monocultures of annual grains with polycultures of grain-bearing perennials, which would have to be developed by a long endeavor of plant-breeding.

Among those who are interested, this project is well known. I want to say only two things about it: First, it is radical, for it goes to the roots of the problem and to the roots of the plants. Second, there is nothing sensational about it. To Wes, having thought so far, it was obviously the next thing to be done. To any agricultural scientist whose mind had been formed as Wes's had been formed—and Wes, unique as he is, could not have been *that* unique—it ought to have been obvious. If one of only two possibilities has failed, the alternative is not far to seek. That the need for an agriculture of perennials was obvious to nobody but Wes, and that it is still by principle unobvious to many agricultural scientists, suggests that the purpose of higher education has been to ignore or obscure the obvious.

The work of The Land Institute is as dedicated to the best interests of the land and the people as if it, rather than the present universities, had sprung from the land-grant legislation. And so the science of The Land Institute is significantly different from the industrial sciences of product development and product addiction—also from the "pure" science that seeks, with the aid of extremely costly and violent technology, the ultimate truth of the universe.

I am talking now of a science subordinate and limited, dedicated to the service of things greater than itself, as every science and art ought to be. There are some things it won't do, some dangers it won't risk. It will not, I think, commit "creative destructions," for the sake of some future good or higher truth. It is a science founded upon the traditional respect for Nature, the natural world, the farmland, and those farmers whose use of the land enacts this respect.

This old respect, really one of the highest forms of human self-respect, provides an indispensable basis, unifying and congenial, for work of all kinds. It might unify a university. In evidence, I offer the friendship that for nearly forty years has kept Wes Jackson and me talking with each other. According to the departmentalism of thought and the other fragmentations that beset us, a useful friendship or a conversation inexhaustibly interesting ought to be impossible between a scientist and a poet. That such a conversation is possible is in fact one of its most interesting subjects. Different as we necessarily are, we have much in common: an interest in the natural world, an interest in farming, an inherited agrarianism, an unresting concern about the problems of farming, of land use in general, and of rural life. Making a common interest of these subjects depends of course on speaking common English, which we can do. But that is not all.

I have heard Wes say many times that "the boundaries of causation always exceed the boundaries of consideration." The more I have thought about that statement, the more interesting it has become. The key word is "always." Mystery, the unknown, our ignorance, always will be with us, to be dealt with. The farther we extend the radius of knowledge, the larger becomes the circumference of mystery. There is, in other words, a boundary that may move somewhat, but can never be removed, between what we know and what we don't, between our human minds and the mind of Nature or the mind of God. To ignore or defy that division, wishing to be as gods, believing that the human mind is so capacious as to contain the whole universe and its whole truth, is characteristic of a kind of science that is at once romantic and industrial, ever in search of new worlds to conquer. From its work, I fear, we can expect only a continuing spillover of violence, to the world and to ourselves.

But a scientist who knows that the boundary exists and accepts, even welcomes, its existence, who knows that the boundary has a human side and elects to stay on it, is a scientist of a kind opposed to the would-be masters of the universe. A poet, too, can choose the human side of that boundary. Any artist, any scientist can so choose. Having so chosen, they can speak to one another congenially, in good faith and friendship. When we know securely the smallness of our minds relative to the immensity of our ignorance, then a certain poise and grace may become possible for us, and we can think responsibly of the circumstances in which we work, of the issues of limits and of the proprieties of scale. If we can talk of limits and of scale, we can slack off our obsession with quantities and immensities and take up the study of form: of the forms of Nature's work, of the forms by which our work might be adapted to hers. We may then become capable of the hope, that Wes and I caught years ago from our friend John Todd, for "Elegant solutions predicated upon the uniqueness of place."

And now the long journey of my essay is ended. I have written it in order to encounter once again, and more coherently than before, the writings and the thoughts among which my own writings have formed themselves. The newest to me of the predecessors I have discussed is the oldest: Alan of Lille's *Plaint of Nature*. My reading and re-reading of that book over the last four or five years helped me to see the thread that connects the others: the books and the voices—to borrow Alan's happiest metaphor—that through many years have been in conversation in "the little town of my mind."

Chronology

1934   Born Wendell Erdman Berry on August 5 in Henry County, Kentucky, the first child of John Marshall and Virginia Erdman Perry Berry. Father, born November 8, 1900, near Lacie, Kentucky, earned a bachelor's degree at Georgetown College in central Kentucky in 1922, and then worked as secretary to Congressman Virgil Chapman while earning a law degree from The George Washington University in Washington, D.C. He graduated in 1927 and later that year opened a private legal practice in Henry County. Mother born November 6, 1907, in Port Royal, in Henry County. She attended Randolph-Macon Woman's College in Lynchburg, Virginia, a Methodist women's college tied to the all-male Randolph-Macon College in Ashland, Virginia, graduating with a degree in English in 1930. Both families can trace roots back to early years of Henry County's history; both families also owned slaves before the Civil War (at time of Berry's birth, there are many children of both slaves and slaveholders in the community). Parents married on July 14, 1933.

1935   Brother John Marshall Jr. born October 13.

1936   Family moves to New Castle, Kentucky, where father practices law and helps with paternal grandfather's farm, the "Home Place."

1938   Sister Mary Jo born March 18.

1939   Sister Martha Frances born July 22.

1940   Enters New Castle Elementary School in Henry County. Over the next few years, mother introduces Berry to books about King Arthur's knights and Robin Hood, as well as *The Swiss Family Robinson*, *Treasure Island*, and *The Yearling*. Later will read Mary O'Hara's *My Friend Flicka*, *Thunderhead*, and *Green Grass of Wyoming*.

1941   The Burley Tobacco Growers Cooperative Association, representing burley tobacco farmers in Kentucky, Missouri, Indiana, Ohio, and West Virginia with an average farm size of 150 acres, which had been founded in 1921, is revived under the New Deal, establishing parity prices and production control as protections for farmers. Father is one of the principal authors of the program and serves as counsel and vice-president and later president.

1942   Uncle Morgan Perry serves in the navy medical corps during World War II, and because of his knowledge of agriculture, is placed in charge of the garden of the naval hospital at Pearl Harbor.

1944   On July 3, uncle Wendell Holmes Berry, with whom Berry was extremely close, is shot at a defunct lead mine property, where he had been working with a crew salvaging lumber and roofing, after a quarrel with Floyd Martin, one of the owners of the property, and dies the same day.

1948   Finishes the eighth grade at the New Castle School. Enters Millersburg Military Institute in Millersburg, Kentucky, at the same time as his brother, John, Jr. Berry will remember, "It was very confining and military, of course. I never liked the military part of it. I did have some good teachers there, and I began there to read more seriously than I had before. I began to have a favorite subject: literature." Father speaks before Congress on behalf of the Tobacco Program. In high school, begins tentatively to write poems.

1952   Graduates from Millersburg Military Institute. In the fall, enters University of Kentucky, in Lexington, Kentucky. Co-edits freshman magazine, *Green Pen*.

1953   Takes "Introduction to Literature" from Thomas Stroup. Begins bringing Stroup poems for criticism. Later takes classes from Hollis Summers and Robert Hazel. Reads T. S. Eliot and Ezra Pound. In May, publishes first essay, "The Wings of the Future," in *Green Pen*, about the survival of society in the light of two world wars and the entrance of the U.S. into the Korean conflict; he writes that it is the "American debt to tomorrow" to ensure that a democratic form of government succeeds over Soviet-style communism. During a composition class, the instructor Robert D. Jacobs introduces Berry to the term "agrarian" and to the 1930s manifesto of the twelve Southern Agrarian writers (who include Allen Tate, John Crowe Ransom, and Robert Penn Warren), *I'll Take My Stand*, a work that will be significant (both in influence and in reaction) to the development of Berry's thought.

1954   Publishes first poem, "Spring," and first short story, "Summer Crop," in the spring issue of *Stylus*, the university literary magazine. Becomes editor of *Stylus*. At a writers' conference at Morehead State College in Kentucky meets writer James Still, whose work will become an influence (Berry will later write, "His short stories, I think, are as near to perfect as any writing I know"). In fall, meets James Baker Hall, who will become a close friend, in a creative writing class taught by Hollis Summers, for which Berry writes story "The Brothers" (later part of *Nathan Coulter*).

1955   Wins *Stylus*'s Dantzler Award for short story "The Brothers." In fall, meets Tanya Amyx (born April 30, 1936, in Berkeley, California), a French and music major and daughter of University of Kentucky art professor Clifford Amyx and textile artist Dee Amyx. "The first time I ever saw Tanya," Berry would later remark, "she was standing by [a] wooden newel post in Miller Hall at the University of Kentucky. Years later, they started to remodel the place. I went over and said, 'Look. When you tear that post out, I want it.'" (The post is now in the Berrys' home.) With fellow student Edward M. Coffman, travels to Kenyon College to meet John Crowe Ransom.

1956   Wins *Stylus*'s Dantzler Award for short story "The Chestnut Stud." Meets fellow student Gurney Norman. On May 16, submits term paper on *I'll Take My Stand*: "The Regional Context: A Consideration of the Southern Agrarians as Southerners." Graduates from the University of Kentucky with a bachelor's degree in English. In summer, Berry and James Baker Hall join literature seminars at Indiana University School of Letters; Berry takes course in Joyce and Yeats from Richard Ellman, and course in modern poetry from Karl Shapiro. "The Brothers" wins first prize in the *Carolina Quarterly*'s 6th Annual Fiction Contest and is published in its summer issue. In the fall, enters the master's program in English at the University of Kentucky.

1957   In February, poem "Rain Crow" is published in *Poetry* magazine; in the next five years, his work will be published in *Poetry* five times. Meets writer Ed McClanahan during a graduate course. Completes MA in English. "Elegy" wins the Farquhar Award for Poetry from the University of Kentucky. Two short stories, "Whippoorwills" and "Apples," are published in *Coraddi: Arts*

Forum 1957. Marries Tanya Amyx, May 29. In summer, they live in a cabin built by Berry's great-uncle Curran Mathews, called "the Camp," on the Kentucky River near Port Royal, which was used as a family retreat from the 1920s. In the fall, they move to an apartment in Georgetown, Kentucky; Berry begins teaching freshman composition and sophomore literature at Georgetown College, father's alma mater, while Tanya continues to study at the University of Kentucky.

1958   Daughter Mary Dee born May 10 in Lexington. Awarded a Wallace Stegner Fellowship at Stanford University; in August, family moves to Mill Valley, California, to live while he studies creative writing at Stanford University under Stegner (about whom Berry will later write, "My debt to him is probably greater than I know") and Richard Scowcroft; fellow members of the fiction seminars include Ernest J. Gaines (also a Stegner fellow), Ken Kesey, and Nancy Packer. Lives with family in small house owned by Tanya's aunt and uncle, Ann and Dick O'Hanlon, which would later become part of the O'Hanlon Center. Works on novel *Nathan Coulter*: "I had about half of it written when I went out there, and I just continued to work on it." Introduced by Holman Hamilton, one of Berry's history professors at the University of Kentucky, to Craig Wylie of Houghton Mifflin, to whom Berry submits the manuscript of *Nathan Coulter*. Houghton Mifflin purchases an option on the book for $250.

1959   Accepts appointment as Edward H. Jones Lecturer in Creative Writing at Stanford University for one year.

1960   In January, begins writing second novel, *A Place on Earth.* First novel, *Nathan Coulter*, published in April by Houghton Mifflin. In spring, family moves back to Kentucky, living on the "Home Place" in Henry County, and also farming with Berry's friend and neighbor Owen Flood.

1961   Receives a Guggenheim Foundation Fellowship. Will later write of Henry County that summer: "I began to understand that so long as I did not know the place fully, or even adequately, I belonged to it only partially. That summer I began to see, however dimly, that one of my ambitions, perhaps my governing ambition, was to belong fully to this place, to belong as the thrushes and the herons and the muskrats belonged, to be altogether at home here." In August, leaves with family for a year in Tuscany and southern France to study and write: "the land in Tuscany has had about 2,000 years of good care. And it looked like it when I was there. . . . The sight of that changed my mind about what was possible in land use."

1962   Meets critic and translator Wallace Fowlie when both stay at La Napoule Arts Foundation near Cannes. Awarded Vachel Lindsay Prize by *Poetry* magazine. Returns to Kentucky from Europe in July. Son Pryor Clifford (Den) born August 19 in Lexington. In the fall, moves with family to New York to take up post as assistant professor of English and director of freshman writing at New York University's University Heights campus in the Bronx. Rents apartment in New Rochelle, New York. Meets Bobbie Ann Mason.

1963   Family moves to loft at 277 Greenwich Street, New York City, downstairs from poet Denise Levertov (whom Berry had met the previous fall, although they had corresponded since 1958, when Levertov wrote Berry about a poem he had published in *Poetry*) and writer Mitchell Goodman. Tanya arranges playdates for daughter Mary and a fellow student at St. Luke's School, the daughter of James Parks Morton, later dean of St. John the Divine. Spends summer in Kentucky. Dismantles the Camp, which has flooded several times over the years, and rebuilds it higher on the bank of the river. Works on novel *A Place on Earth*. Reads Harry Caudill's *Night Comes to the Cumberlands*, an important influence ("It showed me what it might mean to be a responsible Kentucky writer living in Kentucky, and it affected me deeply"). Returns to New York for fall semester. Meets poet Donald Hall, who becomes friend and correspondent, at a cocktail party on Riverside Drive in Manhattan. Offered and accepts position as professor of English at the University of Kentucky to begin in the fall of 1964, where he will teach creative writing and other courses.

1964   In May, visits Levertov and Goodman's Maine farm, where he reads in typescript Hayden Carruth's poetry collection *North Winter* (Berry will later begin a correspondence with Carruth that will last until Carruth's death in 2008). Leaves New York for Kentucky soon after. A poem about Kennedy's assassination, *November Twenty Six Nineteen Hundred Sixty Three* (with drawings by Ben Shahn) published by George Braziller in May. Begins commuting weekly from Lexington (where family lives) to the Camp to write. Meets Kentucky artist Harlan Hubbard and his wife Anna by chance while on a canoe trip on the Ohio River; later writes that their homestead "differed from Thoreau's economy radically in some respects, and also advanced and improved upon it. The main differences were that, whereas Thoreau's was a bachelor's economy, the Hubbards' was that of a married couple." First poetry collection *The Broken Ground* published by Harcourt, Brace & World in September (a *New York Times* review will begin: "The quiet but sure and melodious voice of Wendell Berry makes 'The Broken Ground' an immediate pleasure"). In November, purchases the Lanes Landing property as a weekend place, a twelve-acre hillside farm adjacent to the Camp and bordering his maternal grandfather's farm on the west side of the Kentucky River. Berry's editor at Harcourt, Dan Wickenden, introduces him to the work of organic farming pioneer Sir Albert Howard.

1965   Awarded a Rockefeller Foundation Fellowship. On July 4, moves with family to Lanes Landing Farm, where they will continue to live and farm year round. "I remember a moment—in 1965, or a little after—when I realized that I didn't have to be a writer; there were other kinds of work also that required artistry and offered satisfaction. From here, looking back, I can see what a defining moment that was. I had, in effect, decided not to be a 'professional' writer, but instead, in the literal sense, an amateur: I would work for love." Over time, expands the farm until it consists of about 117 acres in two tracts; at first, raises only food for the family. In July, sees first strip mine above Hardburly, Kentucky, while visiting writer Gurney Norman; Norman introduces him to Kentucky lawyer, writer, and environmentalist Harry Caudill and Caudill's wife Anne, who will become lifelong friends, at a meeting of the Appalachian Group to Save the Land and People. Will one day write, "Gurney has been my Virgil on the upper end (the mountain end) of the Kentucky River watershed, where he is at home. I'm at home on the lower end."

1967   In summer, makes first notes for a work that will become *The Unsettling of America* after reading about the report of President Johnson's "special commission on federal food and fiber policies" that defined the nation's greatest agriculture problem as a surplus of farmers. Novel *A Place on Earth* published by Harcourt, Brace & World in September. Receives Bess Hoken Prize from *Poetry* magazine.

Becomes president of the Cumberland Chapter of the Sierra Club in Kentucky. Begins work with farmers and landowners to oppose a dam across the Red River Gorge that would destroy the gorge's unique topography and ecosystem. In December, visits Trappist monk Thomas Merton at Abbey of Our Lady of Gethsemani near Bardstown, Kentucky, with Ralph Eugene Meatyard and his wife Madelyn, Tanya, and Denise Levertov.

1968   On February 10, delivers speech, "A Statement Against the War in Vietnam," during the Kentucky Conference on War and the Draft in Lexington, Kentucky: "I have come to the realization that I can no longer imagine a war that I would believe to be either useful or necessary. I would be against any war." In fall, takes leave from University of Kentucky to become visiting professor of creative writing at Stanford University during fall 1968 and winter 1969 quarters; lives with family in Menlo Park. Colleagues at Stanford include Wallace Stegner, Ed McClanahan, and Ken Fields. Meets writer John Haines when he gives a reading at Stanford. During the Christmas holiday, writes *The Hidden Wound*, an extended meditation on race and memories of two older black people who were his friends during his childhood. Poetry collection *Openings* published by Harcourt, Brace & World in October; *The New York Times* reviewer calls it a book "to win the respect of anyone who cares about contemporary verse."

1969   Returns with family to Kentucky. Essay collection *The Long-Legged House*, of which the title essay is about the history of the Camp, published by Harcourt, Brace & World in April; other essays concern place and belonging, agriculture, community, small farming, the Vietnam War, and consumerism. The same month, poetry collection *Findings* published by The Prairie Press. Wins first prize in the Borestone Mountain Poetry Awards. Receives grant from the National Endowment for the Arts.

1970   On March 7, speaks at the march at the state capitol in Frankfort, Kentucky, to protest the war in Vietnam; his speech notes government disdain for the will of the people and the huge expense of the war: "With the very earth dying under our feet, with the air full of pollution; we are spending billions of dollars and thousands of lives to assure the success of tyranny in Vietnam." At the first Earth Day celebration at the University of Kentucky, delivers speech "Think Little." Publishes essay "The Regional Motive" in the autumn issue of *The Southern Review*, a response to the Southern Agrarians; in it, he is critical of those Agrarians who had moved to northern universities, saying that it "invalidated their thinking, and reduced their effort to the level of an academic exercise." (Receives letter from Allen Tate in response, and when the essay is included in *A Continuous Harmony* in 1972, Berry appends an apologetic footnote; he later writes that "whatever the amount of truth in that statement, and there is some, it is also a piece of smartassery.") Poetry collection *Farming: A Hand Book* published by Harcourt Brace Jovanovich in September. Meets writer and farmer Gene Logsdon, who becomes a friend and ally, after Logsdon reads *Farming: A Hand Book*. *The Hidden Wound* published by Houghton Mifflin in September.

1971   *The Unforeseen Wilderness: An Essay on Kentucky's Red River Gorge* published by University Press of Kentucky in April, with photographs by Ralph Eugene Meatyard; it is a response to the plan to dam the Red River, which Berry had opposed since 1967. (Congress will withdraw funds for the project in 1976. In 1993 the river is designated by Congress in the National Wild and Scenic Rivers System Act.) Elected Distinguished Professor of the Year by the University of Kentucky. Receives the Arts and Letters Literary Award from the American Academy of Arts and Letters.

1972   With his brother John Berry Jr. and the Save Our Land Committee, helps in the successful opposition to the Jefferson County Air Board's plan for an international jetport in Henry and Shelby Counties. In the fall, takes a one-year sabbatical from the University of Kentucky; his courses are covered by Ed McClanahan. Essay collection *A Continuous Harmony: Essays Cultural and Agricultural* published by Harcourt Brace Jovanovich in September. Meets bookseller and future publisher Jack Shoemaker when Shoemaker visits Port Royal after reading *The Long-Legged House*.

1973   *The Country of Marriage* published by Harcourt Brace Jovanovich in February, his fifth poetry collection, with thirty-five poems, including the title poem, "Prayer After Eating," and "Manifesto: The Mad Farmer Liberation

Front." Buys neighboring forty-acre farm that had been sold to a developer, bulldozed, and graveled; begins process of repair and restoration. Invites writer and photographer James Baker Hall to take photographs during a tobacco harvest (they will be published with an essay by Berry in 2004). Begins correspondence with poet and essayist Gary Snyder, and after a few months Snyder travels to Port Royal to visit Lanes Landing. In November, brother elected to the Kentucky senate, in which he will serve until 1981.

1974   Serves as Elliston Poetry Lecturer at the University of Cincinnati for winter 1974. Novel *The Memory of Old Jack* published by Harcourt Brace Jovanovich in February (*The New York Times Book Review* calls it "a slab of rich Americana" and *Library Journal* says the novel is "worthy of a place among the best pieces of prose written by American writers of this century"). Meets Maurice Telleen at a draft horse sale in Waverly, Iowa, whose *Draft Horse Journal* will publish many of Berry's stories. Gives speech on July 1 at "Agriculture for a Small Planet Symposium" in Spokane, Washington, which will form the basis for the first chapters of *The Unsettling of America*: "I was asked to talk about 'Labor Intensive Micro-Systems Agriculture.' That's not my language, and it's not the sort of language I wish to use because it's the way people speak when they don't want to be understood by most people. I'm not sure what to make of these particular phrases, but they seem to suggest a very methodological or technological approach to agriculture. Part of my purpose here is to suggest that any such approach will necessarily be too simple." Mentions critically Secretary of Agriculture Earl Butz's "adapt or die" policy. Speech helps encourage the Tilth movement, a regional network of organic farmers in the Northwest (still active today). Poetry collection *An Eastward Look* published by Sand Dollar Press in the fall, Berry's first publication with editor Jack Shoemaker. Wins the Emily Clark Balch Prize from *The Virginia Quarterly Review*.

1975   *The Memory of Old Jack* wins the Friends of American Writers Award. A poetry pamphlet, *Horses*, published by Larkspur Press in Kentucky in April; another, *To What Listens*, published by Best Cellar Press. Poetry collection *Sayings and Doings* published by Gnomon Press in December.

1977   Serves as writer-in-residence for the winter quarter at
            Centre College in Danville, Kentucky. Poetry collec-
            tion *Clearing* published by Harcourt Brace Jovanovich
            in March. *The Unsettling of America: Culture and Ag-
            riculture* published by Sierra Club Books in August and
            dedicated to Maurice Telleen, the editor of *Draft Horse
            Journal*. Donald Hall reviews both volumes for *The New
            York Times*, writing, "Berry is a prophet of our healing, a
            utopian poet-legislator like William Blake." Resigns from
            the University of Kentucky to become contributing editor
            at Rodale Press, including its magazines *Organic Garden-
            ing* and *The New Farm*.

1978   With Tanya, buys first flock (six ewes and one buck) of Bor-
            der Cheviot sheep, a Scottish breed, which they will raise for
            the next four decades. In November, debates Secretary of
            Agriculture Earl Butz on the agricultural crisis at Manches-
            ter University, North Manchester, Indiana: "As I see it, the
            farmer standing in his field is not simply a component of a
            production machine. He stands where lots of cultural lines
            cross. The traditional farmer, that is the farmer who first fed
            himself off his farm and then fed other people, who farmed
            with his family, who passed the land on down to people who
            knew it and had the best reasons to take care of it—that
            farmer stood at the convergence of traditional values, our
            values: independence, thrift, stewardship, private property,
            political liberties, family, marriage, parenthood, neighbor-
            hood—values that decline as that farmer is replaced by a
            technologist whose only standard is efficiency."

1979   On June 3, opposes and with eighty-nine others is arrested
            during nonviolent protests against the construction of the
            Marble Hill nuclear power plant on the Ohio River near
            Madison, Indiana. The company eventually abandons the
            effort to finish it. Poetry collection *The Gift of Gravity*
            published by Deerfield Press.

1980   Fired from Rodale Press: "I think this was because I was
            more for small farmers than I was for organic farmers. Also,
            I don't think I've ever been a very good employee." Ernest
            J. Gaines visits the Berrys in Kentucky. Begins working
            with editor Jack Shoemaker, who had left bookselling to
            cofound North Point Press in Berkeley, California, in 1979,
            with William Turnbull. (Most of Berry's books will hence-
            forward be published with Shoemaker.) Poetry collection

A Part published in October. Joins with brother and other residents of Henry County along with the group Kentuckians for the Commonwealth in a successful effort to oppose the building of a hazardous chemical waste incinerator in the county. Publishes poetry pamphlet *The Salad*. On November 11, writes to Wes Jackson, founder of The Land Institute, after reading Jackson's book, *New Roots for Agriculture*. Visits Jackson in December to write an article about The Land Institute for *The New Farm* magazine. The two begin a long correspondence and friendship.

1981   *Recollected Essays: 1965–1980* published in August. Essay collection *The Gift of Good Land: Further Essays Cultural and Agricultural*, dedicated to Gene Logsdon, published in November. Both volumes are reviewed in *The Washington Post* by Larry Woiwode, who calls them "reference works of the body and soul." Granddaughter Katie Jean Smith born, December 16.

1982   Poetry collection *The Wheel* published in October.

1983   Speaking at Oberlin College in February, meets David and Elsie Kline, an Amish couple from Holmes County who had read *The Unsettling of America* and came to hear him speak. Kline (with Logsdon, Telleen, and Jackson) is one of the only four agrarian friends Berry has from outside of Henry County for the next decade. Essay collection *Standing by Words* published in October, which *The Christian Science Monitor* review calls "nothing short of splendid." Publishes substantially revised and shortened version of 1967 novel *A Place on Earth* in March; in a new introduction, writes that the original book was "clumsy, overwritten, wasteful."

1985   Co-edits with Wes Jackson and Bruce Coleman *Meeting the Expectations of the Land: Essays in Sustainable Agriculture and Stewardship*, published by North Point in February; Berry's contribution is the essay "Whose Head is the Farmer Using? Whose Head is Using the Farmer?" Granddaughter Virginia Dee Smith born, March 6. *Collected Poems 1957–1982* published in May; *The New York Times* reviewer writes that Berry "can be said to have returned American poetry to a Wordsworthian clarity of purpose." Publishes substantially revised version of *Nathan Coulter* in May. Helps found Community Farm Alliance, a group of Kentucky farmers, bankers, businessmen, and clergy

members organized to help small farmers shift from tobacco to other products. Begins correspondence with
writer John Haines.

1986   *The Wild Birds: Six Stories of the Port William Membership*
published in March. Receives honorary doctorate from
the University of Kentucky. In November, delivers lecture
"Preserving Wildness" at the first Temenos Academy Conference in Devon, England.

1987   Returns to the English Department of the University of Ken-
tucky, now teaching courses for "future teachers and farmers
or anyone preparing to work in practical and comprehensive
ways with young minds or with nature"; courses include
"Composition for Teachers," aimed at public school English
teachers; "Readings in Agriculture," which assigns Spenser,
Milton, Shakespeare, Pope, and Wordsworth, as well as Sir
Albert Howard, J. Russell Smith, Wes Jackson, and Gene
Logsdon; and "The Pastoral." Publishes *Home Economics:
Fourteen Essays* in June; the *Christian Science Monitor*, in reviewing it, calls Berry "*the* prophetic American voice of our
day." Serves as writer-in-residence for the interim term at
Bucknell University. Receives the American Academy of Arts
and Letters Jean Stein Award and the Kentucky Governor's
Milner Award in the Arts. In the fall, "Why I Am Not Going
to Buy a Computer" published in the *New England Review
and Bread Loaf Quarterly* and reprinted in *Harper's*; the
essay inspires many critical responses. *Sabbaths: Poems* published in September, the first of what will become a sustained
project of poems on the themes of work and rest, fields and
woods. *The Landscape of Harmony: Two Essays on Wildness
and Community*, including the lecture "Preserving Wildness," published by Five Seasons Press, United Kingdom, in
September. Poetry collection *Some Differences* published by
Confluence Press in December.

1988   Novel *Remembering* published in October and reviewed
favorably in the *Los Angeles Times*.

1989   Receives Lannan Foundation Award for nonfiction. Poetry
collection *Traveling at Home* published in November. Delivers the Blazer Lecture on the life and work of Kentucky
artist and author Harlan Hubbard at the University of
Kentucky.

1990   Granddaughter Tanya Christine Smith born February 19.

Essay collection *What Are People For?*, dedicated to Gurney Norman, published in March; it includes the oft-repeated line, "Eating is an agricultural act." (Berry later calls this quote, as repeated without context, "an oversimplification he is now damned sorry to have written.") Bill McKibben publishes a major profile of Berry in *The New York Review of Books*, writing that "wherever we live, however we do so, we desperately need a prophet of responsibility; and although the days of the prophets seem past to many of us, Berry may be the closest to one we have. But, fortunately, he is also a poet of responsibility. He makes one believe that the good life may not only be harder than we're used to but sweeter as well." Inducted into the Fellowship of Southern Writers. Berry's Blazer lecture, *Harlan Hubbard: Life and Work*, published by University Press of Kentucky in November.

1991   *Standing on Earth: Selected Essays* published in the United Kingdom by Brian Keeble's Golgonooza Press in April. Publishes poetry collection *Sabbaths 1987* with Larkspur Press in October. Father dies October 31. North Point Press closes, and Berry follows Shoemaker to Pantheon.

1992   Receives the Victory of Spirit Ethics Award from the University of Louisville and the Louisville Community Foundation. From late August to late September, teaches a course at Schumacher College in South Devon, England, on "Nature as Teacher: The Lineage of Writings Which Link Culture and Agriculture"; writers assigned include Shakespeare, Sir Albert Howard, and Wes Jackson. On September 8, meets Charles, Prince of Wales. "It was a much easier meeting than I expected, for he is intelligent, considerate, and talks and listens well. He is an ally, is deeply concerned about what is happening to rural life and to civilization," Berry wrote in a letter to Wes Jackson. *Fidelity: Five Stories* published in October to favorable reviews, including in *The New York Times* and *The New Criterion*.

1993   Granddaughter Emily Rose Berry born, May 2. *Sex, Economy, Freedom and Community: Eight Essays* published in October; *The New York Times* writes that the book "in its eight essays distills the author's radically conservative views (in the good senses of both words)." Receives The Orion Society's John Hay Award for nature writing. Quits his job at the University of Kentucky.

1994   Receives T. S. Eliot Award for creative writing from the

Ingersoll Foundation. Poetry collection *Entries* published in May. *Watch With Me and Six Other Stories of the Yet-Remembered Ptolemy Proudfoot and His Wife, Miss Minnie, Née Quinch* published in August. Berry follows his editor to Counterpoint Press, which Shoemaker cofounds in Washington, D.C., with Frank H. Pearl.

1995   Grandson Marshall Amyx Berry born June 22. In October, long poem *The Farm* (one of the *Sabbath* series) published by Larkspur Press and the essay collection *Another Turn of the Crank* published by Counterpoint.

1996   Co-authors *Three on Community* with Gary Snyder and Carole Koda, published by Limberlost Press in April. Receives the Harry M. Caudill Conservationist Award from the Cumberland Chapter of the Sierra Club. Novel *A World Lost* published in October.

1997   Mother dies, January 3. Father-in-law Clifford Amyx dies, July 30. *Two More Stories of the Port William Membership* published by Gnomon Press. Receives the Lyndhurst Prize for individuals making a significant contribution to the arts. Speaks at a benefit for The Garden Project in San Francisco, November 10.

1998   Gives keynote address, "In Distrust of Movements," at the Northeast Organic Farming Association conference. *A Timbered Choir: The Sabbath Poems 1979–1997* published in April. *The Selected Poems of Wendell Berry* published in October.

1999   Receives the Thomas Merton Award from the Thomas Merton Center for Peace and Social Justice in Pittsburgh. On November 10, joins Gary Snyder and Jack Shoemaker in reading and conversation at a Lannan Foundation event in Santa Fe, New Mexico.

2000   Wins Poets' Prize for *The Selected Poems of Wendell Berry*. In May publishes *Life Is a Miracle: An Essay Against Modern Superstition*; Bill McKibben, reviewing it for *The Washington Monthly*, writes, "it's hard to imagine that the millennium has seen a more important book than this slim volume from our finest essayist." Novel *Jayber Crow* published in September; *The New York Times* reviewer writes that "by the end this melancholy barber has won both our attention and our hearts." In October, teaches one-week course at Schumacher College in England on "Community, Sustainability and Globalisation."

2001   In response to the events following 9/11, writes "Thoughts in the Presence of Fear," which is published on October 30 by *The Land Report*; in December it is joined by two other essays, "The Idea of a Local Economy" and "In Distrust of Movements," and published by The Orion Society in paperback as *In the Presence of Fear: Three Essays for a Changed World*. Publishes *Sonata at Payne Hollow: A Play* with Larkspur Press.

2002   *The Art of the Commonplace: The Agrarian Essays of Wendell Berry*, edited and introduced by Norman Wirzba, published in April. Berry follows his editor to Shoemaker & Hoard, which Shoemaker cofounds with Trish Hoard.

2003   On February 9, essay "A Citizen's Response to the National Security Strategy of the United States," critical of the Bush administration's post-9/11 strategy, published as a full-page advertisement in *The New York Times*. Essay collection *Citizenship Papers* published in August. *Citizen's Dissent: Security, Morality, and Leadership in an Age of Terror: Essays*, co-authored with David James Duncan, published by The Orion Society. Receives Lifetime Achievement Award from the Cathedral Heritage Foundation in Louisville, Kentucky.

2004   *That Distant Land: The Collected Stories* published in February. Receives the Eli M. Oboler Memorial Award in Orlando in June, shared by David James Duncan, awarded by the American Library Association Intellectual Freedom Round Table, for *Citizen's Dissent*. Mother-in-law Dee Rice Amyx dies, July 3. Awarded the Charity Randall Citation from the International Poetry Forum, as well as the Soil and Water Conservation Society Honor Award from the Kentucky Bluegrass Chapter for dedicated efforts toward the preservation of unique rural and cultural resources. Writes essay for *Tobacco Harvest: An Elegy*, to accompany photographs by James Baker Hall; published by University Press of Kentucky in September. Novel *Hannah Coulter* published in September. Poetry collection *Sabbaths 2002* published by Larkspur Press.

2005   Receives O. Henry Prize for short story "The Hurt Man." *Given: Poems* published in May. Inducted into the University of Kentucky Arts and Sciences Hall of Fame. *Blessed Are the Peacemakers: Christ's Teachings about Love, Compassion & Forgiveness*, selections of the gospels compiled and introduced by Berry, published in October. *The Way of Ignorance and Other Essays* published in October. Essay "Not a Vision of our Future, But of Ourselves," about the earth's destruction due to coal mining, included in *Missing Mountains: We Went to the Mountaintop but It Wasn't There*, published by Wind Publications in October. Receives Lifetime Achievement Award from the Conference on Christianity and Literature, December 29.

2006   In August, speaks at "One Thing to Do About Food: A Forum" with Alice Waters, Eric Schlosser, Marion Nestle, Peter Singer, and others. In October, gives the keynote address for the thirtieth anniversary of The Land Institute in Salina, Kansas. Novel *Andy Catlett: Early Travels* published in November.

2007   Visits Ernest J. Gaines in Louisiana. Shoemaker and Charlie Winton acquire both Counterpoint Press and Soft Skull Press; merge both with Shoemaker & Hoard to re-form Counterpoint.

2008   On February 14, at the Kentuckians for the Commonwealth's annual I Love Mountains Day, delivers speech at the Kentucky state capitol in Frankfort calling for non-violent resistance and civil disobedience to protest mountaintop removal in Kentucky. In May, awarded honorary doctorate by Duke University. *The Mad Farmer Poems* published by Counterpoint in November. Children's book *Whitefoot: A Story from the Center of the World* published in December. Receives the Cynthia Pratt Laughlin Medal from the Garden Club of America.

2009   On January 4, publishes an op-ed in *The New York Times* with Wes Jackson, "A 50-Year Farm Bill," calling for legislation that would address soil loss, pollution, fossil fuels, and the preservation of rural communities. On January 23, releases public statement against the death penalty through the Kentucky Coalition to Abolish the Death Penalty: "As I am made deeply uncomfortable by the taking of a human life before birth, I am also made deeply uncomfortable by the taking of a human life after birth." On March 2, joins protests in Washington, D.C., against mountaintop removal and burning of fossil fuels. On April 3, awarded the Fellowship of Southern Writers' Cleanth Brooks Medal for Excellence in Southern Letters. Essay collection *Bringing It to the Table: On Farming and Food* with an introduction by Michael Pollan, published in August. Poetry collection *Leavings* published in October. In November, joins public letter signed by forty Kentucky writers to Kentucky governor and attorney general asking for a moratorium on the death penalty. On December 20, removes personal papers on loan to the University of Kentucky Archives as a result of the University's too close relationship with the coal industry. (In August 2012, those papers will be donated to the Kentucky Historical Society in Frankfort.)

2010   Receives O. Henry Prize for short story "Stand By Me." Essay collections *Imagination in Place* (including reflections on Wallace Stegner, Gurney Norman, Hayden Carruth, Donald Hall, Jane Kenyon, John Haines, James Still, Gary Snyder, and Kathleen Raine) published in January, and *What Matters? Economics for a Renewed Commonwealth*, with a foreword by Herman E. Daly, published in May.

2011   In mid-February, joins nineteen other activists protesting mountaintop removal in eastern Kentucky by occupying the office of Governor Steve Beshear in Frankfort for a weekend. The governor agrees only to tour coal communities to see the damage. *The Poetry of William Carlos Williams of Rutherford*, a personal tribute to the poet, published in February and is reviewed in *The New York Review of Books*. On March 2, awarded National Humanities Medal by President Barack Obama at the White House. On May 4, speaks at the Future of Food conference at Georgetown University: "the future of food is not distinguishable from the future of the land, which is indistinguishable, in turn, from the future of human care." Daughter Mary Berry establishes The Berry Center in New Castle, Kentucky, to continue the work of Berry and his father and brother toward healthy and sustainable agriculture.

2012   Great-granddaughter Charlcye Anne Johnson born February 7. Receives O. Henry Prize for short story "Nothing Living Lives Alone." *New Collected Poems* published in March; *Christian Science Monitor* hails it as "a welcome complement to the rest of his work." Receives the Steward of God's Creation Award at the National Cathedral in Washington, D.C., April 22. Delivers the 41st Jefferson Lecture, the highest honor the federal government confers for distinguished intellectual achievement in the humanities, at the John F. Kennedy Center for the Performing Arts on April 23: "We cannot know the whole truth, which belongs to God alone, but our task nevertheless is to seek to know what is true. And if we offend gravely enough against what we know to be true, as by failing badly enough to deal affectionately and responsibly with our land and our neighbors, truth will retaliate with ugliness, poverty, and disease." *It All Turns on Affection: The Jefferson Lecture and Other Essays* published in September. *A Place in Time: Twenty Stories of the Port William Membership* published in October. On October 17, receives the James Beard Foundation Leadership Award for efforts toward a healthier, safer, and more sustainable food system, in New York City. On October 30, receives the inaugural Green Cross Award, given by the Bishop of California, for increasing cultural engagement in environmental issues.

2013   In April, elected into the American Academy of Arts and Sciences. The Berry Center and St. Catharine College in Springfield, Kentucky, establish The Berry Farming Program, an undergraduate, multidisciplinary degree inspired by Wes Jackson and Berry's philosophy and work, and designed for students from generational farm families. The program accepts its first students in the fall (St. Catharine College will close in 2016, and the following year the Berry Farming Program will move to partnership with Sterling College in Vermont). *This Day: Sabbath Poems Collected & New 1979–2013* published in October. On October 17, awarded the Freedom Medal from the Roosevelt Institute's Four Freedoms Awards in New York City. Daughter Mary marries Steve Smith, October 26.

2014   *Distant Neighbors: The Selected Letters of Wendell Berry and Gary Snyder*, edited by Chad Wriglesworth, published in June with reviews in the *Los Angeles Review of Books*, *Commonweal*, and other publications. *Terrapin and Other Poems*, with illustrations by Tom Pohrt, published in November. Poetry collection *Roots to the Earth*, with woodcuts by Wesley Bates, published by Larkspur Press.

2015   On January 28, inducted into the Kentucky Writers Hall of Fame, the first living writer to be so honored. *Our Only World: Ten Essays* published in February. Poetry collection *Sabbaths 2013* published by Larkspur Press.

2016   Receives O. Henry Prize for short story "Dismember-
        ment." On March 17, receives the Ivan Sandrof Lifetime
        Achievement Award from the National Book Critics Circle
        in New York City, presented by Nick Offerman. *A Small
        Porch: Sabbath Poems 2014 and 2015* together with *"The Pres-
        ence of Nature in the Natural World: A Long Conversation"*
        published in June. On September 25, delivers the Strachan
        Donnelley Lecture on Conservation and Restoration at
        the 2016 Prairie Festival in Salina, Kansas, celebrating the
        fortieth anniversary of The Land Institute and the ten-
        ure and retirement of Wes Jackson as president. Brother
        John M. Berry Jr. dies, October 27. Speaks in conversa-
        tion with Wes Jackson, and moderated by Mary Berry,
        at the 36th Annual Schumacher Lectures, October 22.
        Great-granddaughter Wendy Jean Johnson born, October
        29. In December, speaks with Eric Schlosser at the twenti-
        eth anniversary of the Center for a Livable Future at Johns
        Hopkins University.

2017   *The World Ending Fire: The Essential Wendell Berry*, se-
        lected by Paul Kingsnorth, published by Allen Lane in
        the United Kingdom in January. Contributes to *Letters to
        a Young Farmer: On Food, Farming, and Our Future* by
        Stone Barns Center for Food and Agriculture, published
        by Princeton Architectural Press. *The Art of Loading Brush*
        published by Counterpoint in October.

Note on the Texts

This volume—the second of a two-volume set, *What I Stand On: The Collected Essays of Wendell Berry 1969–2017*—gathers essays from the latter half of Berry's career. It contains the whole of *Life Is a Miracle* (2000) and forty-two essays culled from nine other books by Berry published from 1993 to 2017: *Sex, Economy, Freedom & Community* (1993), *Another Turn of the Crank* (1995), *Citizenship Papers* (2003), *The Way of Ignorance* (2005), *What Matters?* (2010), *Imagination in Place* (2010), *It All Turns on Affection* (2012), *Our Only World* (2015), and *The Art of Loading Brush* (2017). The contents were chosen by the author and his longtime editor and publisher, Jack Shoemaker, and are arranged in chronological order by first book publication. In accord with Shoemaker and Berry's wishes, and as described below, this volume prints the texts of the essays from the most recent Counterpoint Press printings, which reflect corrections made by the author. The companion volume presents selections from books published from 1969 to 1990.

After the publication of *The Unsettling of America* (1977) by Sierra Club Books, Berry began to look for a new publisher. In 1980 he embarked on his long collaboration with Jack Shoemaker, then a thirty-four-year-old editor who, with William Turnbull, had recently cofounded North Point Press in Berkeley, California. From that point forward most of Berry's books would be edited by Shoemaker. Berry followed his editor first to Pantheon when North Point closed in 1991; then to Counterpoint Press, which Shoemaker cofounded with Frank H. Pearl in Washington, D.C., in 1994; then to Shoemaker & Hoard in 2002, which Shoemaker cofounded with Trish Hoard; and finally back to Counterpoint, which merged with Shoemaker & Hoard when Shoemaker and Charlie Winton purchased both Counterpoint Press and Soft Skull Press in 2007. None of the books mentioned above appeared in British editions.

Sex, Economy, Freedom & Community was published by Pantheon Books on October 19, 1993; the most recent Counterpoint Press paperback printing, the source for the texts printed here, was published in 2018. The first publication of the essays selected from *Sex, Economy, Freedom & Community* is given below.

> Conservation and Local Economy. Lecture delivered at the Southern Baptist Theological Seminary in Louisville, Kentucky, during a conference sponsored by the Clarence Jordan

Center for Christian Ethical Concerns, on March 23, 1992, published in *Sex, Economy, Freedom & Community* (New York and San Francisco: Pantheon, 1993).

Conservation Is Good Work. Lecture delivered at the Southern Baptist Theological Seminary in Louisville, Kentucky, on March 25, 1992, published in *Wild Earth*, Spring 1992, and in *The Amicus Journal*, Winter 1992.

Christianity and the Survival of Creation. Lecture delivered at the Southern Baptist Theological Seminary in Louisville, Kentucky, on March 24, 1992, published in *CrossCurrents*, Summer 1993.

Sex, Economy, Freedom, and Community. Lecture given at the University of Louisville School of Business, on April 8, 1992, Parts I and II printed as a pamphlet under the title "Sex, Economy, and Community" by the Louisville Community Foundation, 1992. The essay also appeared prior to its book publication in *Pantheon Previews*, Fall 1993.

Another Turn of the Crank was published by Counterpoint Press on October 18, 1995; the most recent Counterpoint paperback printing, the source for the texts printed here, was published in 2011. The first publication of the essays selected from *Another Turn of the Crank* is given below.

Farming and the Global Economy. *Another Turn of the Crank* (Washington, D.C.: Counterpoint Press, 1995).

Conserving Forest Communities. Speech from the Kentucky Forest Summit and Governor's Conference on the Environment, September 29–31, 1994, published by the State of Kentucky as a proceedings paper in 1994 and excerpted in *The Land Report*, Summer 1995.

Health Is Membership. Speech at a conference, "Spirituality and Healing," at Louisville, Kentucky, on October 17, 1994, published in *Utne Reader*, September/October 1995.

Life Is a Miracle was published by Counterpoint Press on May 16, 2000. The text printed in this volume is that of the Counterpoint paperback printing, published in 2001.

Citizenship Papers was published by Shoemaker & Hoard on August 8, 2003; the most recent Counterpoint Press paperback printing, the source for the texts printed here, was published in 2014. The first publication of the essays selected from *Citizenship Papers* is given below.

A Citizen's Response to the National Security Strategy of the United States. Originally published in the United States on Sunday, February 9, 2003, in *The New York Times* in a full-page ad taken out by *Orion Magazine* and on www.Orion Online.org; and in the United Kingdom in *Irish Pages*, Autumn–Winter, 2002/2003. It appeared as the cover article in the March/April 2003 issue of *Orion Magazine*.

Thoughts in the Presence of Fear. Originally published October 30, 2001, on www.OrionOnline.org, in a feature entitled "Thoughts on America: Writers Respond to Crisis."

The Failure of War. Originally published online at www. yesmagazine.org, November 5, 2001.

In Distrust of Movements. *Orion Magazine*, Summer 1999, adapted from a keynote address delivered at the 1998 Northeast Organic Farmers' Association Conference.

The Total Economy. *Orion Magazine*, Winter 2001.

Two Minds. *Citizenship Papers* (Washington, D.C.: Shoemaker & Hoard, 2003).

The Whole Horse. Originally published in a shorter version in *The Amicus Journal*, Winter 1998.

The Agrarian Standard. *Orion Magazine*, Summer 2002.

Conservationist and Agrarian. Originally published as "For Love of the Land" in *Sierra*, May–June 2002.

The Way of Ignorance was published by Shoemaker & Hoard on October 21, 2005; the Counterpoint Press paperback printing, the source for the texts printed here, was published in 2006. The first publication of the essays selected from *The Way of Ignorance* is given below.

Secrecy vs. Rights. *The Way of Ignorance* (Washington, D.C.: Shoemaker & Hoard, 2005).

Contempt for Small Places. *Farming*, Spring 2004.

Compromise, Hell! Speech for the annual Earth Day celebration of the Kentucky Environmental Quality Commission at Frankfort, Kentucky, April 22, 2004, published on May 3, 2004, in the *Lexington Herald-Leader* as "People Can't Survive If the Land Is Dead" and, under its present title, in *Orion Magazine*, November/December 2004.

Charlie Fisher. Originally published in the spring 1996 issue of *The Draft Horse Journal* as "Trees for My Son and Grandson to Harvest."

The Way of Ignorance. *New Letters*, Summer 2005.

Quantity vs. Form. Originally published in the Fall 2005 issue of *Southern Arts Journal* as "Quantity and Form."

Renewing Husbandry. *Crop Science: A Journal Serving the International Community of Crop Scientists*, May–June 2005.

The Burden of the Gospels. Speech for the first joint convocation of the Lexington Theological Seminary and the Baptist Seminary of Kentucky, Lexington, Kentucky, August 30, 2005, printed in *The Christian Century*, September 20, 2005.

What Matters? was published by Counterpoint Press on May 10, 2010, as a paperback original. The texts of the essays selected from *What Matters?* are taken from that edition. The first publication of the essays chosen for inclusion is given below.

> Money Versus Goods. Originally published in two parts in *The Progressive*: "Inverting the Economic Order" (September 2009) and "The Cost of Displacement" (December 2009/ January 2010).
>
> Faustian Economics. *Harper's Magazine*, May 2008.

Imagination in Place was published by Counterpoint Press on January 19, 2010; the Counterpoint paperback printing, the source for the texts printed here, was published in 2011. The first publication of the essays selected from *Imagination in Place* is given below.

> Imagination in Place. H. L. Weatherby and George Core, eds., *Place in American Fiction: Excursions and Explorations* (Columbia: University of Missouri Press, 2004).
>
> American Imagination and the Civil War. *The Sewanee Review*, Fall 2007.
>
> Sweetness Preserved. Bert G. Hornback and Peter Lang, eds., *"Bright Unequivocal Eye": Poems, Papers, and Remembrances from the First Jane Kenyon Conference* (New York: Peter Lang Publishing Group, 2000).
>
> The Uses of Adversity. *The Sewanee Review*, Spring 2007.
>
> God, Science, and Imagination. *The Sewanee Review*, Winter 2010.

It All Turns on Affection was published by Counterpoint Press on September 11, 2012, as a paperback original. The texts of the essay selected from *It All Turns on Affection* are taken from that edition. Their first publication is given below.

> It All Turns on Affection. Jefferson Lecture delivered April 23, 2012, published in *It All Turns on Affection* (Berkeley, CA: Counterpoint Press, 2012).
>
> About Civil Disobedience. *The Progressive*, January 2012.

Our Only World was published by Counterpoint Press on February 10, 2015; the Counterpoint paperback printing, the source for the texts printed here, was published in 2016. The first publication of the essays selected from *Our Only World* is given below.

> Paragraphs from a Notebook. *Our Only World* (Berkeley, CA: Counterpoint Press, 2015).
>
> The Commerce of Violence. *The Progressive*, June 2013.
>
> A Forest Conversation. *Our Only World* (Berkeley, CA: Counterpoint Press, 2015).
>
> Local Economies to Save the Land and the People. *Our Only World* (Berkeley, CA: Counterpoint Press, 2015), based on a

speech to Kentuckians for the Commonwealth in Carrollton, Kentucky, on August 16, 2013.

Caught in the Middle. *Christian Century*, April 2013.

Our Deserted Country. *Our Only World* (Berkeley, CA: Counterpoint Press, 2015).

On Being Asked for "A Narrative for the Future." *Our Only World* (Berkeley, CA: Counterpoint Press, 2015).

The Art of Loading Brush was published by Counterpoint Press on October 31, 2017; the Counterpoint paperback printing, the source for the texts printed here, was published in 2019. The first publication of the essays selected from *The Art of Loading Brush* is given below.

The Thought of Limits in a Prodigal Age. *The Art of Loading Brush* (Berkeley, CA: Counterpoint Press, 2017).

The Presence of Nature in the Natural World. *A Small Porch: Sabbath poems 2014 and 2015 together with The Presence of Nature in the Natural World* (Berkeley, CA: Counterpoint Press, 2016).

This volume presents the texts of the printings chosen for inclusion here, but it does not attempt to reproduce features of their typographic design. Original citations have been altered to conform to Library of America practices; however, original footnotes that take the form of personal observation or argument have been preserved verbatim. In a few instances quotations have been corrected with the author's approval. Original in-text citations have been removed where both the author and title of the work under discussion are evident. The texts are presented without change, except for the correction of typographical errors and minor changes necessitated by the formatting of notes. Spelling, punctuation, and capitalization are often expressive features, and they are not altered, even when inconsistent or irregular. The following is a list of typographical errors corrected, cited by page and line number: 6.8, urbanindustrial; 8.7, L'il; 83.18, dreary ,; 121.19, book,; 133.16–17, "Whenever; 133.17, physiochemical; 172.34, agriculturalist,; 180.33, The *ABC*; 181.26, generation; 217.20, the Land; 227.34, entomologist,; 265.25, 1930's; 311.37, September,; 344.32, becomes; 383.11, discreet; 384.2, them to do; 447.11, cost.; 499.17, wife Tanya; 505.22, brother Oliver,; 506.10, says are; 545.7 (and *passim*), Howard's; 553.21, anthology,; 561.14, essays,; 561.15, Ethic,"; 561.31, novel,; 563.5, geometer,; 576.25, something.; 577.35, life upon; 596.4, Tory; 626.27, care."; 639.15, agriculture by; 640.28, famer; 641.10, 1950s, is; 670.6, prophesies; 675.3, Surowieki's; 683.25, worst; 712.18, western; 723.3, Defense; 731.8, presidence; 737.1, breast–; 742.13, keen, intense,; 752.15, sometime.

Notes

In the notes below, the reference numbers denote page and line of this volume (the line count includes headings). No note is made for material included in standard desk-reference books. Quotations from Shakespeare are keyed to *The Riverside Shakespeare*, ed. G. Blakemore Evans (Boston: Houghton Mifflin, 1974). Biblical quotations are keyed to the King James Version. Original citations have been altered to conform to Library of America practices; however, original footnotes that take the form of personal observation or argument have been preserved verbatim. For references to other studies and further information than is included in the Chronology, see *Wendell Berry: Life and Work*, ed. Jason Peters (Lexington: The University Press of Kentucky, 2007); and *Conversations with Wendell Berry*, ed. Morris Allen Grubbs (Jackson: University Press of Mississippi, 2007).

From SEX, ECONOMY, FREEDOM & COMMUNITY

7.13–20   "the county . . . earned it."] From an article by editor William Allen White (1868–1944) in the *Emporia Gazette* (1912), a small-town paper with a national reputation.

7.39–40   the General Agreement on Tariffs and Trade] Also known as GATT, an international trade agreement signed by multiple nations in Geneva on October 30, 1947, that was designed to reduce trade barriers. Negotiations between GATT members in the years 1986 to 1994 would lead to trade liberalization in agricultural products and the establishment of the World Trade Organization (WTO). In the early 1990s, Berry argued against proposed changes to GATT and has written subsequently on how trade liberalization has been damaging to American farmers.

8.7   Snuffy Smith or Li'l Abner] Snuffy Smith, a gun-toting, moonshine-swilling hillbilly in the still-running American comic strip *Barney Google and Snuffy Smith*, which debuted in 1919; Li'l Abner Yokum, a gullible hillbilly who lives in Dogpatch, USA, in the American comic strip *Li'l Abner*, which ran from 1934 to 1977.

8.9   "The business of America is business,"] Frequently attributed to President Calvin Coolidge (1872–1933), who in a speech of January 17, 1925, before the American Society of Newspaper Editors, Washington, D.C., remarked, "After all, the chief business of the American people is business."

10.26–27   what Edward Abbey called . . . cancer cell"] Edward Abbey (1927–1989), American essayist, novelist, ecologist, and a dedicatee of Berry's *Sex, Economy, Freedom & Community* (1993). See Abbey's essay "Arizona: How Big Is Enough?," reprinted in his collection *One Life at a Time* (1987): "Growth for the sake of growth is the ideology of the cancer cell."

15.18–20   Wilderness preserves . . . marketable timber.] Dave Foreman, *Confessions of an Eco-Warrior* (1991).

15.21–23   Wes Jackson] American plant geneticist, advocate for sustainable practices, and cofounder of The Land Institute in Salina, Kansas (b. 1936). Jackson was one of the four agrarian allies of Berry for several decades (outside his family), along with Gene Logsdon, Maurice Telleen, and David Kline.

17.4–5   Love Canal, Bhopal, Chernobyl, the burning oil fields of Kuwait.] Love Canal, near Niagara Falls, New York, the site of a chemical waste dump in the 1940s and 1950s that forced the evacuation and resettlement of more than a thousand families in the Love Canal neighborhood, after President Jimmy Carter (twice) declared a state of emergency in the area, in 1978 and 1981; Bhopal, the capital city in Madhya Pradesh, India, which gained international attention in 1984 when a pesticide-manufacturing plant leaked toxic gas, killing thousands; Chernobyl, an uninhabitable city in northern Ukraine, synonymous with the nuclear disaster of 1986, when the Chernobyl nuclear power station emitted large amounts of radiation into the atmosphere; the burning fields of Kuwait, more than 600 oil wells set aflame by the Iraqi military army as it withdrew from Kuwait in 1991, in the Persian Gulf War.

17.17   the Sacramento River is now dead] On July 14, 1991, a Southern Pacific tanker car spilled 19,000 gallons of metam-sodium (a soil fumigant and weed killer) into the Upper Sacramento River.

28.1   *Christianity and the Survival of Creation*] Originally delivered as a lecture by Berry at the Southern Baptist Theological Seminary in Louisville, Kentucky, in 1992.

30.13–14   "The earth . . . dwell therein."] Psalms 24:1.

30.25–26   "The land . . . with me."] Leviticus 25:23.

30.29–30   "All things . . . was made."] John 1:3.

31.2   John 3:16] "For God so loved the world, that he gave his only begotten Son, that whosoever believeth in him should not perish, but have everlasting life."

31.13–14   "gather unto . . . perish together."] Job 34:14–15.

31.17–18   "Creation is . . . hidden Being."] Philip Sherrard, *Human Image: World Image* (1992).

31.22–23   Thou art . . . and all.] George Herbert, "Providence," stanza 11, first published in *The Temple* (1633).

31.29–30   "despising Nature . . . against God.] Dante, *Inferno*, XI.46–48, Charles S. Singleton translation (1970). All subsequent quotations of *The Divine Comedy* are from the Singleton translation.

31.36–37   "condemns Nature . . . hope elsewhere."] Dante, *Inferno*, XI.109–111.

31.39   "everything that lives is holy."] The concluding line of Willian Blake's *The Marriage of Heaven and Hell* (1790–93).

32.1–2   "the sense . . . human norm."] See *Golgonooza: City of Imagination* (1991), by Kathleen Raine (1908–2003), English poet, essayist, and scholar whose expertise was William Blake and W. B. Yeats.

33.24–26   "Behold, the heaven . . . have builded?"] 1 Kings 8:27.

33.29–33   God that made . . . have said.] Acts 17:24 and 28.

34.10–11   "where two . . . my name."] Matthew 18:20.

34.30–39   Your new moons . . . the widow.] Isaiah 1:13–17.

35.4   "hypaethral book,"] "Hypaethral" means roofless, open to the sky. Writing in his *Journal* on June 29, 1851, Henry David Thoreau (1817–1862) characterized *A Week on the Concord and Merrimack Rivers* (1849) as "a hypaethral or unroofed book, lying open under the ether and permeated by it, open to all weathers, not easy to be kept on a shelf."

35.36–38   "The invisible things . . . are made."] Romans 1:20.

36.30–39   And of Joseph . . . the bush.] Deuteronomy 33:13–16.

37.32–33   "For what . . . own soul?"] Matthew 18:20.

38.17   "bruised to pleasure soul."] W. B. Yeats, "Among School Children," stanza VIII, published in his collection *The Tower* (1928).

39.33–34   "It is a tale . . . Signifying nothing,"] *Macbeth*, V.v.26–28.

39.35   "supp'd . . . the sun."] *Macbeth*, V.v.13, 48.

40.22–25   "the normal view . . . of artist."] Ananda K. Coomaraswamy, "What Is the Use of Art, Anyway?," collected in his book of essays *Christian and Oriental Philosophy of Art* (1956). Ananda Kentish Coomaraswamy (1877–1947) was a Ceylonese historian of Indian art and culture, art theorist, and scholar of comparative religious thought who was instrumental in bringing ancient Indian art and culture to the attention of the English-speaking world.

40.33–34   "the plowman . . . cosmic function."] Walter Shewring, *Artist and Tradesman* (1984).

41.9–11   Mother Ann Lee's famous instruction . . . die tomorrow."] In his original note Berry cites as his source for Lee's instruction June Sprigg's *By Shaker Hands* (1990). Ann Lee (1736–1784), also known as Mother Ann Lee, was the spiritual leader and founder of the Shaker movement in America.

41.12–26 "the perfection . . . reformation must begin."] Coomaraswamy, *Selected Papers*, vol. 1 (1977).

42.9–11 "Genius inhabits . . . has neighbors."] Coomaraswamy, "What Is the Use of Art, Anyway?," collected in *Christian and Oriental Philosophy of Art*.

42.23–25 "far worse . . . the art."] Dante, *Paradiso*, XIII.121–123.

43.34 "I made . . . try for it."] See the beginning of *The Adventures of Huckleberry Finn* (1884) by Mark Twain (1835–1910).

44.31–33 "Love your enemies . . . persecute you."] Matthew 5:44.

46.5–6 the Clarence Thomas hearing] In 1991, during Clarence Thomas's U.S. Supreme Court confirmation hearings, Thomas's former subordinate Anita Hill testified that Thomas had sexually harassed her at the Department of Education and the EEOC. Thomas denied any wrongdoing, and the Senate confirmed Thomas on October 15, 1991.

52.27–53.18 The . . . reason for . . . afterwards absorbed.] From *Hawkshead (The Northernmost Parish of Lancashire): Its History, Archaeology, Industries, Dialect, Etc.* (1899) by English antiquary Henry Swainson Cowper (1865–1941).

53.21–23 The industrialization . . . pattern exactly.] Berry, in his original note, indicates his indebtedness to Harry M. Caudill's *Theirs Be the Power* (1983) for this observation. Caudill (1922–1990), from Letcher County, Kentucky, was an author, legislator, opponent of strip-mining in the Appalachian coalfields, and professor of Appalachian Studies at the University of Kentucky. He is a dedicatee of Berry's *Sex, Economy, Freedom & Community*.

53.37–54.13 In the traditional culture . . . its value.] Helena Norberg-Hodge, *Ancient Futures: Learning from Ladakh* (1991).

55.12–18 Increasingly, people . . . longer afford.] Norberg-Hodge, *Ancient Futures*.

55.25 Luddites] Disaffected nineteenth-century English artisans—primarily textile workers and weavers—who destroyed new labor-saving machinery that was replacing them, in a widespread rebellion that lasted from 1811 to 1817.

55.33–56.8 By the adoption . . . his hire.] In his original note Berry cites as his source for Byron's speech Humphrey Jennings's *Pandaemonium, 1660–1886: The Coming of the Machine as Seen by Contemporary Observers* (1985).

56.9–16 The Luddites did . . . hangings and transportations."] Berry's source is *Encyclopedia Britannica*, 14th ed., s.v., "Luddite."

57.25–26 the Community Farm Alliance] Founded in 1985, an alliance of Kentucky farmers dedicated to helping independent family farmers across Kentucky and to promoting sustainable farming methods and traditional practices.

58.23–24 "fearfully and wonderfully made."] Psalms 139:14.

59.32–33   If sex . . . wished-for words.] Wallace Stevens, "Le Monocle de Mon Oncle" (1918), stanza XI, reprinted in Stevens's first collection of poems, *Harmonium* (1923).

61.10–12   "Of a surety . . . would return."] Dante, *La Vita Nuova*, Chapter XIV, translation by D. G. Rossetti (1828–1882), first published in *The Early Italian Poets* (1861).

62.22–23   "Losing kindness . . . to justness."] *The Way of Life According to Lao-tzu*, translation by Witter Bynner (1962).

67.26–27   "Ye shall know . . . you free,"] John 8:32.

68.15   "no ethical . . . an indictable one."] Mary McGrory, "Ed Meese on the Stand," the Louisville *Courier-Journal* (March 31, 1989).

70.11–13   I am indebted to Judith Weissman . . . the individual.] Berry, in his original note, directs readers "especially" to the Introduction and Chapter I of Judith Weissman's *Half Savage and Hardy and Free: Women and Rural Radicalism in the Nineteenth-Century Novel* (1987).

70.33–38   The Liberation . . . feminist writers. . . .] Weissman, *Half Savage and Hardy and Free.*

73.10–16   I am immodestly . . . American life."] William Mootz, "Arthur Kopit Plans to Offend Almost Everyone," the Louisville *Courier-Journal* (February 26, 1989).

77.26–27   "the purpose . . . *to move*"] Coomaraswamy, *Selected Papers*, vol. 2 (1977).

81.37–38   "certainly it . . . will die"] Dante, *La Vita Nuova*, Chapter XXIII, translation by D. G. Rossetti.

83.17–22   People were bathing . . . like water-lilies.] D. H. Lawrence, "The Gods! The Gods!," published in the posthumous collection *Last Poems* (1932).

From ANOTHER TURN OF THE CRANK

94.32   the Stamp Act] To recoup expenses incurred as a result of the French and Indian War, in 1765 the British parliament imposed a tax on printed matter in the American colonies. After protests and colonial boycotts, the Stamp Act was repealed the following year.

96.1   *Conserving Forest Communities*] Delivered as a speech at the Kentucky Forest Summit (in conjunction with the Nineteenth Governor's Conference on the Environment) at Louisville, Kentucky (September 29, 1994).

98.13–14   the Burley Tobacco Growers Co-operative Association] Founded in 1921, the Association resumed its work under the New Deal's Federal Tobacco Program starting in 1941. By a combination of production controls and price supports, it gave vital help to small farmers in Kentucky and other burley-growing states for six decades.

98.18–20 "is owned . . . 26 acres."] From "Kentucky's Forest Resources," an unpublished paper by William H. Martin, Mark Matuszewski, Robert N. Muller, and Bradley E. Powell (1994). In his original note Berry says, "I have taken statistics and other information on Kentucky forests from this paper and also from William H. Martin, 'Sustainable Forestry in Kentucky,' *In Context* (Center for Economic Development at Eastern Kentucky University, winter 1993) . . . and William H. Martin, 'Characteristics of Old-Growth Mixed Mesophytic Forests,' *Natural Areas Journal* 12, no. 3 (July 1992)."

103.10   an article in *California Forestry Notes*] Dave McNamara, "Horse Logging at Latour," *California Forestry Notes* (September 1983).

106.39   the Menominee Indians] Algonquian-speaking tribe of Native American people who originally lived along the Menominee River on the present-day border between Wisconsin and Michigan; the Menominee Indian Reservation in Wisconsin is approximately 60 miles west of the mouth of the Menominee. Berry visited the Menominee tribal forest in June 1994.

108.21–23 "they could . . . secondary processing."] Scott Landis, "Seventh-Generation Forestry," *Harrowsmith Country Life* (November/December 1992).

110.1   *Health Is Membership*] Delivered as a speech at a conference entitled "Spirituality and Healing," in Louisville, Kentucky, on October 17, 1994.

112.17   Sir Albert Howard . . . *The Soil and the Health*] Albert Howard (1873–1947), British botanist and agriculturist. Howard's publications drew on extensive practical research in India, where, working as an agricultural advisor, he came to see the wisdom of traditional Indian farming practices. *The Soil and Health: A Study of Organic Agriculture* was originally published under the title *Farming and Gardening for Health or Disease* (1945) before its 1947 reissue.

121.19–22 "Machines . . . calculate with."] Neil Postman, *Technopoly: The Surrender of Culture to Technology* (1992).

122.25–27 "Are not . . . before God?"] Luke 12:6.

122.28–31 "If a clod . . . diminishes me."] John Donne, *Devotions upon Emergent Occasions* (1623), Meditation XVII.

LIFE IS A MIRACLE

128.2   *Lionel Basney (1946–1999)*] American poet, essayist, environmental activist, and professor of English whose interests included the history of the Luddite movement and the writings of Wendell Berry; a personal friend of Berry.

131.34   "I stumbled when I saw."] *King Lear*, IV.i.19.

132.5–6   "'Twixt two . . . joy and grief. . . ."] *King Lear*, V.iii.199.

132.12   "Falls forward and swoons."] *King Lear*, IV.vi.41 s.d. Berry quotes the stage directions in the Pelican Shakespeare edition of the play.

132.18   "Away, and let me die."] *King Lear*, IV.vi.48.

132.32   "Thy life's . . . yet again."] *King Lear*, IV.vi.55.

133.16–18   "Wherever one . . . as such."] See *Physics and Philosophy: The Revolution in Modern Science* (1958) by German theoretical physicist Werner Heisenberg (1901–1976), a pioneer of quantum mechanics.

133.26–28   What seems . . . seems to Be . . .] William Blake, *Jerusalem* (1804–20), plate 32 [36], lines 51–53.

134.36–37   Kathleen Raine was correct . . . being experienced.] See Kathleen Raine, *The Inner Journey of the Poet* (1982).

135.17–19   O you mighty . . . affliction off.] *King Lear*, IV.vi.34–36.

135.27–29   You ever-gentle gods . . . you please] *King Lear*, IV.vi.217–219.

135.33   "Well pray you, father."] *King Lear*, IV.vi.219.

140.26–27   Dante saw it from a higher level of accomplishment] See Dante, *Paradiso*, XXII.133–154.

143.4–5   C. P. Snow spoke . . . "automatic corrective"] In *The Two Cultures and the Scientific Revolution* (1959), based on his Rede Lecture of the same year. Charles Percy Snow (1905–1980), English novelist and chemist.

145.13–14   *Consilience* by Edward O. Wilson] E. O. Wilson (b. 1929), American biologist, sometimes called the "father of sociobiology." *Consilience* was published in 1998 by Alfred A. Knopf.

152.17   so-called Y2K problem] Also known as the Y2K bug, the Millennium bug, the Year 2000 problem, or simply Y2K, the Y2K problem arose because many computer programs used two digits to signify the calendar year, raising the possibility of catastrophic consequences on January 1, 2000, if computers were unable to distinguish between 2000 and 1900. Estimates vary, but hundreds of billions of dollars were spent around the globe to make computers and application programs Y2K-compliant.

153.21–22   "We murder to dissect."] William Wordsworth, "The Tables Turned," published in Wordsworth and Coleridge's *Lyrical Ballads* (1798).

161.20–21   "Life," wrote Erwin Chargaff . . . the inexplicable."] Erwin Chargaff, *Heraclitean Fire: Sketches from a Life Before Nature* (1978).

162.36–37   Tarzan theory of the mind . . . Edgar Rice Burroughs.] Burroughs (1875–1950), American author of more than sixty works, including *Tarzan of the Apes* (1912) and other books in the Tarzan series. In *Tarzan of the Apes*, Tarzan as a young boy teaches himself to read and write English by puzzling over alphabet books and primers he finds in the ruins of his parents' home.

169.8–9   Cortés . . . Montezuma] Spanish conquistador Hernán Cortés (1485–1547) held Aztec emperor Montezuma II (1466–1520) prisoner after

Montezuma welcomed him to the Aztec capital Tenochtitlán; Montezuma was killed in fighting for control of the city, largely destroyed by the Spanish and their native allies.

169.20–27 The intellect . . . the night's remorse.] W. B. Yeats, "The Choice," published in *The Winding Stair and Other Poems* (1933).

170.1 Dante . . . found this Ulysses in Hell] Dante encounters Ulysses in the eighth trench of the eighth circle of hell, a place reserved for evil counsellors. See the *Inferno*, XXVI.

179.30–32 "all great scientific discoveries . . . to lose."] Chargaff, *Heraclitean Fire*.

179.35–39 "The Nazi experiment . . . modern science."] Chargaff, *Heraclitean Fire*.

181.7–12 "There can be . . . ideals or opinions."] Coomaraswamy, "The Christian and Oriental, or True Philosophy of Art," collected in his book of essays *Christian and Oriental Philosophy of Art* (1956).

181.13–15 "when there . . . new or old."] Coomaraswamy, *The Transformation of Nature in Art* (1936), Chapter 1.

182.12 "dark Satanic mills"] William Blake, *Milton* (1810), Preface.

182.13 "we murder to dissect"] See note 153.21–22.

182.20 Hitler's troops marched into Prague] March 15, 1939.

182.20–21 the Scottish poet Edwin Muir] Edwin Muir (1887–1959) and his wife, Willa Muir (1890–1970), translated the works of the German-speaking Jewish Prague writer Franz Kafka into English.

182.23–30 "So many goodly cities . . . oh base conquest."] Montaigne, "Of Coaches," *The Essayes of Michael Lord of Montaigne*, John Florio translation (1603).

182.30–41 "Think of all . . . merely new."] Edwin Muir, *The Story and the Fable* (1940), revised and reissued in 1954 as *An Autobiography*.

183.2–7 "Dreams of . . . for progress."] C. S. Lewis, *That Hideous Strength* (1945), Chapter 9.

183.9–21 "It is a severe question . . . photographing the corpse . . ."] Albert Howard, *The Soil and Health* (1947), Chapter 6. See also note 112.17.

183.25–28 "The wonderful . . . be recalled."] Chargaff, *Heraclitean Fire*.

183.38–39 "deep-set repugnances"] C. S. Lewis, *That Hideous Strength*, Chapter 9.

184.8–9 "We ought to stay out of the nuclei."] Wes Jackson in conversation with Berry.

187.6–8   a panel discussion . . . published in the *Authors Guild Bulletin*.]
"Whose Life Is It, Anyway?," the *Authors Guide Bulletin* (Winter 1999).

198.8   "Thou'lt come . . . never never."] *King Lear*, V.iii.308–309.

198.13–14   General Lee was overheard saying . . . too bad!"] Berry's source
is *Lee* (1961), Richard Harwell's abridgment of Douglas Southall Freeman's
four-volume biography *R. E. Lee* (1934).

198.24–27   "I know . . . shall I see God. . . ."] Job 19:25–26.

201.25–27   "I could be bounded . . . bad dreams."] *Hamlet*, II.ii.254–255.

202.16–17   All flesh . . . breath of God.] Cf. Job 34:14–15.

202.17   "live, and move, and have our being"] Acts 17:28.

202.20   the fall of every sparrow] Cf. Matthew 10:29.

202.21   "the very hairs of your head"] Matthew 10:30.

202.22   "Thy life's a miracle,"] *King Lear*, IV.vi.55.

202.23–24   "everything that lives is holy."] See note 31.39.

202.25–27   "want us . . . simple things as well."] See *Revelations of Divine
Love* (1395), Chapter 32, by English anchoress and Christian mystic Julian of
Norwich (1342–c. 1416).

202.28–29   "*A Sand County Almanac* . . . distinct beings. . . ."] Stephanie Mills, *In Service of the Wild: Restoring and Reinhabiting Damaged Land*
(1995).

203.12–14   "The gods' presence . . . the gods."] See Guy Davenport's
translation of the fragments of pre-Socratic Greek philosopher Herakleitos (c.
535–c. 475 B.C.) in *Herakleitos and Diogenes* (1979).

206.30–31   "assert Eternal Providence . . . to men."] Milton, *Paradise Lost*,
I.25–26.

211.6–7   To see a World . . . Wild Flower.] The opening lines of William
Blake's "Auguries of Innocence" (1866).

213.10   Gerard David's *Annunciation*] Oil on wood painting, by Netherland-
ish painter Gerard David (c. 1460–1523), of the announcement of the Incarnation by the angel Gabriel to Mary, commissioned in 1506.

213.20   Luke 1:28–35] The Annunciation is narrated in Luke 1:28–35.

217.19–20   Wes Jackson . . . co-founder of The Land Institute] Wes Jackson
cofounded The Land Institute in 1976 with Dana Jackson. The Land Institute is dedicated to the development of robust locally adapted, perennial grain
crops, to reducing or eliminating soil erosion and the use of herbicides, and
to the overall reduction of dependence on industrial agriculture. See also note
15.21–23.

221.24   Wallace Stegner] American novelist, short story writer, historian, environmentalist, teacher, and founder of Stanford University's creative writing program (1909–1993). Stegner won the Pulitzer Prize for his novel *Angle of Repose* (1971) and the National Book Award for the novel that followed, *The Spectator Bird* (1977).

221.27–30   "boomers . . . made it in."] Wallace Stegner, *Where the Bluebird Sings to the Lemonade Springs: Living and Writing in the West* (1992), Introduction.

226.19–20   "the tones given off by the heart."] From *The Great Digest*, translation by Ezra Pound (1928), reprinted in his anthology *Confucius* (1951).

227.35   French entomologist Jean Henri Fabre] Largely self-taught, Fabre (1823–1915) conducted important research on bees and wasps, beetles, and grasshoppers.

227.38   "a patch of pebbles . . . four walls."] Jean Henri Fabre, "The Laboratory of Open Fields," in *The Insect World of J. Henri Fabre* (1949), translated by Alexander Teixeira de Mattos, with an introduction and comments by Edwin Way Teale.

228.20   *The Wheelwright's Shop*] Autobiography published in 1923 by English writer George Sturt (1863–1927), who after his father's death managed the family wheelwright shop in the town of Farnham, in Surrey, where he spent the remainder of his life.

230.5–11   Richard Strohman has made clear . . . predictability] In *The Daily Californian* (April 1, 1999), and *The Wild Duck Review* (Summer 1999).

235.1–2   I learned from the theologian Philip Sherrard to ask this question] In an embedded citation in the original text, Berry points readers to Sherrard, *Human Image: World Image* (1992).

From CITIZENSHIP PAPERS

239.20   The new "National Security Strategy"] Issued by the executive branch of the U.S. government, the "National Security Strategy" is a report to Congress that articulates national security concerns and how an administration plans to address them. The "new" report to which Berry refers is the one published on September 20, 2002, by the George W. Bush administration.

241.8   the events of September 11, 2001] Coordinated attacks by al-Qaeda on U.S. targets using four hijacked commercial planes; 2,996 people including the 19 hijackers died at the World Trade Center, the Pentagon, and in a field near Shanksville, Pennsylvania, where one of the planes crashed before reaching its intended target (believed to be either the U.S. Capitol or the White House).

243.22   "There is none . . . God,"] Matthew 19:17.

243.23–24   "He that is . . . at her."] John 8:7.

243.25–27   "for nothing . . . superintendence."] Thomas Jefferson, letter to Mann Page, August 30, 1795.

245.21–24   "Laws are commanded . . . to uphold."] Edmund Burke, *Reflections of the Revolution in France* (1790).

246.32   Saddam Hussein] President of Iraq from 1979 to 2003, when Hussein's Baathist regime was overthrown by U.S. and coalition forces in the invasion of Iraq. Found guilty of crimes against humanity by an Iraqi court, Hussein was hanged on December 30, 2006.

247.4   America Whose Business Is Business] See note 8.9.

249.3–4   the Alien and Sedition Laws of 1798] Four laws passed by U.S. Congress in 1798 amid growing fears of war with France. Aimed at French and Irish immigrants, the laws extended naturalization from five to fourteen years, authorized the imprisonment and expulsion of any noncitizens determined to be dangerous, and restricted speech critical of the U.S. government. Negative public reaction to the laws contributed to the victory of Thomas Jefferson over the incumbent, John Adams.

251.24   Monsanto] On June 7, 2018, Bayer AG purchased the Monsanto Company for $66 billion; shortly before the purchase was completed Bayer announced its intention to drop the Monsanto name.

254.21   the Gulf War of 1991] Also known as the First Gulf War, the First Iraq War, the Kuwait War, and the Persian Gulf War, a war conducted by coalition forces led by the United States against Iraq in response to Saddam Hussein's invasion of Kuwait.

259.7–8   The modern doctrine of such warfare . . . General William Tecumseh Sherman] During the Civil War, Sherman (1820–1891) conducted a scorched earth campaign in his march through Georgia and the Carolinas.

264.38–265.2   "Both . . . answered fully."] Lincoln's Second Inaugural Address, March 4, 1865.

269.24–29   "Love your enemies . . . the unjust."] Matthew 5:44–45.

275.40–276.1   in Mary Austin's phrase "summer and winter with the land."] See the Preface to *The Land of Little Rain* (1903) by Mary Hunter Austin (1868–1934), a writer of the American Southwest.

283.14   "Love thy neighbor as thyself"] Matthew 22:39.

283.15   "Sentient beings . . . to save them."] One of the Four Bodhisattva Vows in Mahayana Buddhism, an expression of commitment to the enlightenment of all sentient beings.

284.39–40   the Committee for Economic Development] Public policy organization created in 1942 by leaders of American corporations, in anticipation of World War II's conclusion and transition to a peace-time economy. The CED championed American economic internationalism.

286.38–39   the World Trade Organization] Under the Marrakesh Agreement of April 15, 1994, signed by 124 member nations, the WTO replaced the General Agreement on Tariffs and Trade (GATT). See also note 7.39–40.

290.7–10   "the principle . . . fitted by nature. . . ."] Ananda K. Coomaraswamy, "What Is Civilization?" (1946), in *What Is Civilization and Other Essays* (1989).

290.12   "the mammon of injustice,"] Cf. Luke 16:9.

290.15   "to work is to pray"] Benedictine motto (*laborare est orare*).

293.22–25   "Whenever the timber . . . as possible."] See Albert Schweitzer's *Out of My Life and Thought: An Autobiography* [*Aus meinem Leben und Denken*], first published in German in 1931. Berry quotes from the English translation by Antje Bultmann Lemke (1990). An Alsatian-German theologian, philosopher, organist, and philanthropist, Schweitzer (1875–1965) was awarded the 1952 Nobel Prize in part for his medical missionary work in west-central Africa.

293.27–29   "These people . . . own needs."] Schweitzer, *Out of My Life and Thought*.

295.12–13   Thomas Kuhn described this process] In place of the gradual accumulation of scientific knowledge, American physicist Kuhn (1922–1996) saw alternating phases of "normal science" and revolutionary invention that brought "paradigm shifts."

301.33–34   "the instinctive integrations . . . for living."] Wallace Stevens, "Imagination as Value," *The Necessary Angel: Essays on Reality and Imagination* (1951).

303.7–12   "If a man . . . not astray."] Matthew 18:12.

308.17   Bertie Wooster] Character and narrator in the Jeeves fictions by English writer P. G. Wodehouse (1881–1975); Wooster, an English gentleman, relies on his valet, Jeeves, to assist him in the simplest tasks and to rescue him from more complicated situations.

309.1   James Laughlin's poem "Above the City,"] James Laughlin (1914–1997), American poet and the founder of the publishing house New Directions. "Above the City" first appeared in *New Directions in Prose & Poetry 9* (1946) and was reprinted in the collection *A Small Book of Poems* (1948).

311.33–35   "let us build . . . the whole earth."] Genesis 11:4.

320.3–4   "I think . . . with animals"] Walt Whitman, "Song of Myself," section 32, initially published without sections in the 1855 edition of *Leaves of Grass*.

323.12–13   J. Russell Smith, Liberty Hyde Bailey, Albert Howard, Wes Jackson, John Todd] Joseph Russell Smith (1874–1966), American economic geographer, conservationist, and writer whose books include *Tree Crops: A Permanent Agriculture* (1929); Liberty Hyde Bailey (1858–1954), American botanist, horticulturalist, and cofounder of the American Society for Horticultural Science; Albert Howard, see note 112.17; Wes Jackson, see note 15.21–23; John Todd (b. 1939), Canadian ecological designer whose design innovations address challenges of food production, waste treatment, and environmental repair through ecosystem technologies that emulate nature.

323.16–17   "let the forest judge"] *As You Like It*, III.ii.121.

323.17   "consult the Genius of the Place"] Alexander Pope, "Epistle to Burlington" (1731), line 57.

323.29–33   "the eternal task . . . lived in."] John Haines, "On a Certain Attention to the World," published in *Fables and Distances: New and Selected Essays* (1996). An American poet and essayist, Haines (1924–2011) was poet laureate of Alaska from 1969 to 1973.

324.29–30   the Twelve Southerners of *I'll Take My Stand*] The Twelve Southerners, also known as the Southern Agrarians, were a circle of Southern writers and academics formed in the 1920s and united in the cause of preserving an agrarian way of life against the forces of industrialism. Among the group's most prominent members were John Crowe Ransom, Robert Penn Warren, Donald Davidson, and Allen Tate. Each of the Twelve Southerners contributed an essay to the landmark anthology *I'll Take My Stand: The South and Agrarian Tradition* (1930).

332.21–23   The purpose . . . "food dictatorship."] Vandana Shiva, "Food Democracy v. Food Dictatorship: The Politics of Genetically Modified Food," *Z Magazine* (April 2003).

336.35–337.6   I saw . . . unbought delicacies.] Virgil, *Georgics*, IV.132. Berry quotes from L. P. Wilkinson's translation (1982).

338.13–16   "it is not . . . a state. . . ."] Thomas Jefferson, letter to James Madison Fontainebleau, October 28, 1785.

339.19–20   "the ideology of the cancer cell"] See note 10.26–27.

341.32   "a false nostalgia . . . never existed."] Richard Lewontin, "Genes in the Food!," *The New York Review of Books* (June 21, 2001).

348.10–11   "fit the farming to the land"] See J. Russell Smith, *Tree Crops: A Permanent Agriculture* (1929), which describes the traditional use of tree crops and how trees such as the carob, honey locust, mulberry, chestnut, and black walnut can be an important source for food, soil conservation, and sustainable agriculture.

From THE WAY OF IGNORANCE

362.4–10   The contradictions . . . to live."] James R. Carroll, "Mountain-top Mining at Center of Controversy," the Louisville *Courier-Journal* (June 14, 2003).

363.1   *Compromise, Hell!*] Written originally as a speech for the annual Earth Day celebration of the Kentucky Environmental Quality Commission at Frankfort, Kentucky, April 22, 2004.

368.33   Joe Begley] Joseph Taylor Begley (1919–2000), storekeeper, longtime resident of Blackey, Kentucky, and activist who was instrumental in mounting opposition to strip-mining in eastern Kentucky.

377.1   *The Way of Ignorance*] Written as a preliminary paper for a Land Institute conference of the same title at Matfield Green, Kansas, June 3–5, 2004.

377.9–10   Richard Dawkins's assertion, in an open letter to the Prince of Wales] Dated May 21, 2000.

378.13–17   "we cannot imagine . . . not possess."] Kathleen Raine, "On the Symbol," published in her essay collection *Defending Ancient Springs* (1967).

385.39   "man as artist . . . after form."] David Jones, "Art in Relation to War," published in his *The Dying Gaul and Other Writings* (1978).

387.32–38   But Eliot . . . one part.] Kathleen Raine, "The Poet of Our Time," in *T. S. Eliot: A Symposium* (1948), edited by Tambimuttu and Richard March.

388.19–20   "As to diseases . . . no harm."] Hippocrates, *Of the Epidemics*, Book I.

389.21–22   "the way of ignorance,"] T. S. Eliot, "East Coker," first published in *New English Weekly* (March 1940) for its "Easter Number" and reprinted as the second poem in *Four Quartets* (1943).

391.1   *Quantity vs. Form*] Written originally for the conference "Evidence-Based, Opinion-Based, and Real World Agriculture and Medicine," convened by Charlie Sing at Emigrant, Montana, October 10–15, 2004.

392.11–12   Tiresias' foretelling of the death of Odysseus] Recounted in *The Odyssey*, Book XI.

392.23–24   Robert Southey's . . . biography of Lord Nelson.] *The Life of Horatio, Lord Nelson* (1813).

394.14–15   I wrote in a poem . . . infinite form,"] Wendell Berry, "From the Crest," first published in *The Virginia Quarterly Review* (Summer 1974). See Berry's *New Collected Poems* (2012).

395.4–6   "our common responsibility . . . can recognize."] Jeffery L. Amestoy, "Uncommon Humanity: Reflections on Judging in a Post-Human

Era," *New York Law Review* (November 2003), first delivered as a speech on February 6, 2003, at New York University School of Law for the annual Justice William J. Brennan, Jr. Lecture on State Courts and Social Justice.

399.3–4   McCormick High Gear No. 9] An animal-powered mower manufactured by International Harvester from 1939 to 1951.

399.5   Farmall A] A compact row-crop tractor manufactured by International Harvester from 1939 to 1947. In 1948, Harvester introduced the Super A, the model purchased by Berry's father.

405.33–34   "health in soil . . . one great subject."] Albert Howard, *The Soil and Health* (1947), Introduction.

406.16   "a dispenser of the 'Mysteries of God.'"] Liberty Hyde Bailey, *The Holy Earth* (1915).

412.1   *The Burden of the Gospels*] Written originally as a speech for the first joint convocation of the Lexington Theological Seminary and the Baptist Seminary of Kentucky, Lexington, Kentucky, August 30, 2005.

413.15–20   "Don't resist . . . children of God."] Cf. Matthew 5:39, 5:44–45.

413.31   "If ye love . . . my commandments."] John 14:15.

414.4   "one of the least of these my brethren"] Matthew 25:40.

415.7–8   "They have their reward"] Matthew 6:2.

415.31   parable of the Samaritan] Told in Luke 10:25–37.

415.33   Buddhist vow to "save all sentient beings,"] See note 283.15.

415.40–416.1   "Love thy neighbor as thyself."] Matthew 22:39.

416.28–30   "And there are . . . be written."] John 21:25.

417.35–37   "I am come . . . more abundantly."] John 10:10.

419.27   remembering Genesis 1:27] "So God created man in his own image, in the image of God created he him; male and female created he them."

From WHAT MATTERS?

432.16–18   He classes usurers . . . wrong sources."] Aristotle, *Nicomachean Ethics*, Book IV. Berry quotes from the English translation of 1908 by Sir William David Ross.

438.4–5   "Greed is . . . known to man."] William Safire, "Ode to Greed," *The New York Times* (January 5, 1986).

438.29–31   "I hold no brief . . . or Sloth."] Safire, "Ode to Greed."

439.11   Luddites] See note 55.25.

442.2–3 a stranger . . . a strange place] Cf. Exodus 2:22.

443.2–4 the difference, defined by Paul Goodman . . . between competent poverty and abject poverty] American psychologist, author, and film-
maker Paul Goodman (1911–1972) advocated "decent poverty," as opposed to
misery, as a way of life for those interested in pursuits not concerned primarily
with making money. Speaking of himself, Goodman claimed not to see a significant difference between being decently poor and modestly rich.

448.33–34 Menominee Forest . . . the Pioneer Forest] Berry's visit in June
1994 to the Menominee Indians of northern Wisconsin to learn about sustainable forest management is described in "Conserving Forest Communities," on
pages 96 to 109 in this volume; another, similar information-gathering trip to
Missouri's Pioneer Forest, made by Berry in December 2007, is described in
his op-ed piece, "Better Ways to Manage, Study Robinson Forest," the *Lexington Herald-Leader* (December 28, 2007), which also appeared as "Sustainable
Forestry: Lessons for Kentucky," Louisville *Courier-Journal* (December 28,
2007).

450.31 "I am that I am."] Exodus 3:14.

451.10–13 "The United States . . . rate of consumption."] Bill Wolfe, "Ky.
Urged to Be Power Player," the Louisville *Courier-Journal* (October 21,
2006).

453.35–36 "all nature's treasury . . . new-found world . . ."] Christopher
Marlowe, *The Tragical History of Doctor Faustus* (1604), I.i.77, 85–86. Cita-
tions of Marlowe's play correspond to those in *English Renaissance Drama: A
Norton Anthology* (2002), edited by David Bevington et al.

454.3–5 "How comes . . . out of it." *Faustus*, I.iii.77.

454.6–8 Hell hath . . . ever be.] *Faustus*, II.i.122–124.

454.12–13 "Tut, Faustus . . . more of it."] *Faustus*, II.i.152–153.

454.22–23 "to answer . . . *within bounds*] Milton, *Paradise Lost*, VII.119–
120.

454.25–29 Knowledge is . . . to wind.] Milton, *Paradise Lost*, VII.126–30.

457.19–20 "myself am Hell."] Milton, *Paradise Lost*, IV.75.

457.29 "Faustus, and a man."] *Faustus*, I.i.23.

From IMAGINATION IN PLACE

466.6–7 James Still, Harlan Hubbard, and Harry Caudill] James Still (1906–
2001), of Knott County, Kentucky, a poet, novelist, and story writer who spent
most of his writing life in a log house near the Dead Mare Branch of Little Carr
Creek; Harlan Hubbard (1900–1988), of Trimble County, Kentucky, a painter
and writer who with his wife, Anna, lived a life of subsistence in remote Payne

Hollow on the Ohio River (Berry's portrait of Hubbard, *Harlan Hubbard: Life and Work*, was published in 1990); Harry Caudill, see note 53.21–23.

466.15   old Cilician of *Georgics*] Virgil's aged gardener, who works a few acres of land on otherwise unproductive soil near Taranto in Apulia.

466.17–18   Ananda Coomaraswamy, Titus Burckhardt, Kathleen Raine, and Philip Sherrard] Ananda Coomaraswamy, see note 40.22–25; Titus Burckhardt (1908–1984), Swiss historian of sacred art, cultural anthropologist, and publisher; Kathleen Raine, see note 32.1–2; Philip Owen Arnould Sherrard (1922–1995), English poet, translator, theologian, and interpreter of Orthodox Christianity.

467.15   the Southern Agrarians] See note 324.29–30.

467.23–25   John Crowe Ransom] American poet, critic, one of the founders of the magazine *The Fugitive*, and a prominent member of the Southern Agrarians (1888–1974). His name remains closely associated with the New Critics, who argued for the autonomy of the literary work.

468.11   Allen Tate] American poet, critic, novelist, and biographer of Stonewall Jackson and Jefferson Davis who was a member of the Southern Agrarians (1899–1979). Like Ransom, Tate was a New Critic.

469.3–5   only Andrew Lytle and John Donald Wade . . . knowledge of actual farming] Novelist, essayist, and teacher Andrew Lytle (1902–1995) grew up on the family farm near Guntersville, Alabama; noted biographer, essayist, scholar, and editor John Donald Wade (1892–1963) lived in the family home in Macon County, Georgia, where farming remains an important part of the economy.

470.40   The Land Institute] See note 217.19–20.

471.11–12   Gene Logsdon, Maurice Telleen, Wes Jackson, and David Kline.] Gene Logsdon (1931–2016), American farmer and author of many books including the posthumously published *Letter to a Young Farmer: How to Live Richly Without Wealth on the New Garden Farm* (2017), with a foreword by Wendell Berry; Maurice Telleen (1928–2011), American farmer, writer, and founder of *The Draft Horse Journal*, to whom Berry dedicated *The Unsettling of America*; Wes Jackson, see note 15.21–23; David Kline (b. 1945), American author, farmer, editor of *Farming Magazine*, and a member of the Old Order Amish community of Fredericksburg, Ohio.

473.9–10   woman with six arms in the picture by . . . Emmeline Grangerford] Emmeline Grangerford, Buck Grangerford's deceased sister in *The Adventures of Huckleberry Finn* (1884), a juvenile artist whose crayon drawings include an unfinished "picture of a young woman in a long white gown, standing on the rail of a bridge all ready to jump off, with her hair all down her back, and looking up to the moon, with the tears running down her face, and she had two arms folded across her breast, and two arms stretched out in front, and two more reaching up towards the moon—and the idea was to see which pair would look best, and then scratch out all the other arms; but, as I was saying, she died before she got her mind made up, and now they kept this picture over the head of the bed in her room, and every time her birthday come they hung flowers on it."

474.1–4   "Realism . . . aspire and strive."] Kathleen Raine, "The Use of the Beautiful," published in her collection of essays *Defending Ancient Springs* (1967).

475.4–9   Parochialism and provincialism . . . his parish.] Patrick Kavanagh, "The Parish and the Universe," *Collected Pruse* (1967).

478.17–19   As Guy Davenport saw . . . replace it.] Throughout his writing career, American essayist and fiction writer Guy Davenport (1927–2005) was a fierce critic of money-valued society. Davenport began teaching at the University of Kentucky in 1964, the same year Berry joined the faculty as a creative writing instructor.

479.8–13   . . . we give . . . soonest winner.] *Henry V*, III.vi.108–113.

479.14–15   Edmund Burke's *Speech on American Taxation*] Delivered in the British House of Commons on April 19, 1774, in the debate on the repeal of tea-duty.

480.8–9   Mathew Brady's photographs] American photographer Mathew Brady (1823–1896) and his associates documented camp life, war preparations, and the grisly aftermath of battle during the Civil War.

480.21   Who's this? . . . only son!] *Henry VI*, Part III, II.v.61, 82–83.

480.26   "all the mad old joy"] Walt Whitman, "The Artilleryman's Vision," originally entitled "The Veteran's Vision," *Drum-Taps* (1865).

480.31–481.8   Curious I halt . . . he lies.] Walt Whitman, "A Sight in Camp in the Daybreak Gray and Dim," *Drum-Taps* (1865).

483.1–3   "not virtue . . . such organism . . ."] Edmund Wilson, *Patriotic Gore* (1962), Introduction.

483.18   "The Battle Hymn of the Republic."] Pro-Union, antislavery lyric written by American social reformer Julia Ward Howe (1819–1910) for the song "John Brown's Body," published in *The Atlantic* (1862).

484.1–6   Donald Worster has said . . . slave power . . ."] Comments made in the Chautauqua discussion at the Prairie Festival, The Land Institute, Salina, Kansas (1999).

485.7–9   imagination . . . definitions by Coleridge and Blake] See Coleridge's *Biographia Literaria* (1817), Chapter XIII, where he differentiates between fancy (the lower, associative faculty) and imagination (the higher, creative faculty, which shapes and unifies). For Blake, imagination is more than an artistic faculty: "Imagination is the real and eternal world of which this vegetable universe is but a faint shadow."

485.16–21   as if the earth . . . eat filth . . .] William Carlos Williams, "To Elsie," *Spring and All* (1923).

485.22–25   "like a chicken . . . to death . . ."] William Carlos Williams, *In the American Grain* (1925).

486.33–37   Robert Ulanowicz says . . . everywhere!"] Ulanowicz's comments were made in a public lecture, "Naturalism and/or Immanent Divine Action?," in Santa Barbara, California (January 26, 2006).

489.25–26   James Still's *River of the Earth*, written half a century later] Still's novel, set in the 1930s coalfields of eastern Kentucky, appeared in 1940.

489.29–30   Ernest J. Gaines inherited the harder . . . racial history] Gaines (b. 1933), an African American writer, grew up on the River Lake Plantation in Pointe Coupee Parish, Louisiana.

492.10–13   Jane Kenyon's poems . . . Donald Hall's . . . out of the same story] American poet Jane Kenyon (1947–1995), who grew up in the Midwest, moved with her husband, the American poet and essayist Donald Hall (1928–2018), her former teacher at the University of Michigan, to Eagle Pond Farm in Wilmot, New Hampshire, Hall's grandparents' former home. For Berry's account of his own return home to Henry County, Kentucky, see "The Long-Legged House" and "A Native Hill."

496.23   Bert Hornback] Bert G. Hornback (b. 1935), a member of the English faculty at the University of Michigan from 1964 to 1992.

498.26–28   "In art . . . saying nothing."] See the selection of Ludwig Wittgenstein's notebooks entitled *Culture and Value* [Vermischte Bemerkungen], published in German in 1977 and in an English translation by Peter Winch in 1980. Berry quotes from the Winch translation.

499.33   like Ruth of old] Moabite woman of the Old Testament and Hebrew Bible's Book of Ruth who after her husband's death moved with her mother-in-law, Naomi, to Judah, where she remarried and began a new life.

500.7–8   a disparagement . . . in the *Hudson Review*] Robert Phillips, "Poems, Mostly Personal, Some Historical, Many Unnecessary," *The Hudson Review* (Winter 1997).

503.5   "things fall apart"] W. B. Yeats, "The Second Coming," published in *The Dial* (November 1920) and first collected in *Michael Robartes and the Dancer* (1921).

504.7–10   Come live . . . mountains yields.] Christopher Marlowe, "The Passionate Shepherd to His Love" (1599).

504.12–13   The flowers . . . reckoning yields . . .] Sir Walter Raleigh, "The Nymph's Reply to the Shepherd" (1600).

504.15–16   "of being versed in country things"] Robert Frost, "The Need of Being Versed in Country Things," published in *Harper's Magazine* (December 1920) and first collected in *New Hampshire* (1923).

*505.1–18   O, when degree . . . eat up himself.] *Troilus and Cressida*, I.iii.101–108, 116–124.

507.3   Meliboe in *The Faerie Queene*] An old shepherd. A pastoral interlude in Book VI of *The Faerie Queene* brings Sir Calidore to the house of old Meliboe and his foster daughter, Pastorella.

507.33   "pathetic fallacy,"] A term coined by English art critic John Ruskin (1819–1900), which he defined as the "extraordinary, or false appearances" assumed by objects "when under the influence of emotion, or contemplative fancy."

509.8–9   Spenser's Colin Clout or Hardy's Hodge] Colin Clout, a pastoral persona making appearances in several of Edmund Spenser's works, including *The Shepheardes Calendar* (1579), *Colin Clouts Come Home Againe* (1595), and *The Faerie Queene* (1590–96); Hodge, the subject of Thomas Hardy's poem "Drummer Hodge or the Dead Drummer" (1899), first collected in *Poems of Past and Present* (1901).

509.12   "A trewe swinkere and a good . . ."] Chaucer, *The Canterbury Tales*, Prologue, line 531.

511.25   "The least of these my brethren"] Matthew 25:40.

525.35   Christ's agony in Gethsemane] Described in Matthew 26:36–46.

525.38   story of the Prodigal Son] Recounted in Luke 15:11–32.

528.40   "satisfy the society's impulse to renewal."] J. A. Bryant, Jr., *Shakespeare and the Uses of Comedy* (1986).

533.23–31   A thousand tymes . . . it preve.] See the opening lines of the Prologue to Chaucer's *The Legend of Good Women*.

535.9–11   But will God . . . I have builded?] 1 Kings 8:27.

535.16–19   For we . . . wait for it.] Romans 8:24–25.

538.18–21   The *Iliad* . . . What we are.] Hayden Carruth, "The Impossible Indispensability of the Ars Poetica," published in *Toward the Distant Islands: New and Selected Poems* (2006).

From IT ALL TURNS ON AFFECTION

545.1   *It All Turns on Affection*] Delivered as the 41st Annual Jefferson Lecture of the National Endowment for the Humanities at the John F. Kennedy Center for the Performing Arts, Washington, D.C. (April 23, 2012).

545.18   The depression of the 1890s] One of the worst economic crises in American history, with unemployment exceeding 10 percent for five or six consecutive years. A declining economy combined with surplus production caused numerous foreclosures on mortgaged farms.

545.35 American Tobacco Company] Incorporated in 1890 after the takeover and merger of the nation's leading tobacco manufacturers, J. B. Duke's American Tobacco Company controlled four-fifths of the U.S. tobacco industry (excluding cigars) by 1906.

546.6–7 my teacher, Wallace Stegner] Berry was a Stegner Fellow in the Stanford Creative Writing Program in the school year 1958–59.

546.8–12 "boomers" and . . . made it in."] Wallace Stegner, *Where the Bluebird Sings to the Lemonade Springs: Living and Writing in the West* (1992), Introduction.

550.15–16 Land, as Wes Jackson has said . . . as exhaustible as oil or coal.] A common theme of Jackson in his and Berry's conversations.

551.17 "the land-community"] Aldo Leopold, *A Sand County Almanac and Sketches Here and There* (1949), Part 3.

556.1 "carried to the heart"] Allen Tate, "Ode to the Confederate Dead," published in *Mr. Pope and Other Poems* (1928) and reprinted in *Collected Poems, 1919–1976*.

556.39–557.2 "a statement . . . limitations of mankind."] John Lukacs, *Last Rites* (2009).

557.24–25 "the equall mother . . . unto brother."] *The Faerie Queene*, VII, canto vii, stanza 14.

557.27–29 "In short . . . citizen of it."] Leopold, *A Sand County Almanac*, Part 3.

558.14–16 "Imagination implied . . . a detail"] One of Wallace Stevens's aphorisms from his "Adagia" (1934–40), published in *Opus Posthumous* (1957).

563.17–18 a speech . . . "Revitalizing Rural Communities,"] Delivered by Frederick Kirschenmann at the forty-third annual meeting of the Northern Plains Conference of the United Church of Christ, Bismarck, North Dakota (June 10, 2006), reprinted in Kirschenmann's *Cultivating an Ecological Conscience: Essays from a Farmer Philosopher* (2010).

563.26–28 "The human mind . . . loves them . . ."] Keith Critchlow, *The Hidden Geometry of Flowers: Living Rhythms, Form, and Number* (2011).

566.2 Matt Rothschild] Senior editor at *The Progressive* from 1994 to 2014.

566.22 I committed . . . an act of civil disobedience] On the weekend of February 12, 2011, Berry and nineteen other protestors occupied Kentucky governor Steve Beshear's outer office to protest surface mining for coal in the Kentucky mountains.

568.5–7 Ken Kesey once said . . . *you* dependent on evil.] Kesey in conversation with Berry.

568.8   Edward Abbey once said . . . good hobby] According to Berry, a response by Abbey to an invitation to protest. Abbey makes a similar remark in *Postcards from Ed* (2006): "Saving the world is only a hobby. Most of the time I do nothing."

From OUR ONLY WORLD

574.1–2   "It is not good . . . be alone."] Genesis 2:18.

575.32–33   "God guard me . . . mind alone. . . ."] W. B. Yeats, "A Prayer for Old Age," published in *Parnell's Funeral and Other Poems* (1935).

575.34–35   "We only believe . . . whole body"] See Yeats's Introduction to *Certain Noble Plays of Japan* (1916), edited by Ernest Fenollosa and Ezra Pound.

576.4–12   "He saw . . . younger generations."] Philip Sherrard, *The Wound of Greece: Studies in Neo-Hellenism* (1978).

576.16–26   "You may dissociate . . . missing something."] John Crowe Ransom, "Poets Without Laurels," published in his collection of essays *The World's Body* (1938).

581.2   the day of the bombing in Boston] April 15, 2003. Two bombs were detonated on Boylston Street near the finish line of the Boston Marathon, killing three people and injuring 264 others.

582.34–35   Thomas L. Friedman announced in the *Times*] In an op-ed piece by Friedman entitled "Bring on the Next Marathon," April 17, 2003. All quotes attributed to Friedman come from this piece.

584.29–30   Liberty Hyde Bailey, J. Russell Smith, Hugh Hammond Bennett, Albert Howard, Aldo Leopold] Liberty Hyde Bailey, see note 323.12–13; Joseph Russell Smith, see note 323.12–13; Hugh Hammond Bennett (1881–1960), American soil conservationist and founder of the Soil Conservation Service (now the Natural Resources Conservation Service); Albert Howard, see note 112.17; Aldo Leopold (1887–1948), American conservationist, ecologist, forester, and writer best known for his book *A Sand County Almanac* (1949).

589.22   *Suffolk* horses] Heavy chestnut-colored breed of English draft horses developed specifically for the purpose of farmwork.

603.37–39   The Land Institute, Tilth, the Quivira Coalition, the Southern Agriculture Working Group, the Land Stewardship Project] The Land Institute, see note 217.19–20; the Tilth Association (now the Tilth Alliance), incorporated in 1977, a sustainable agricultural movement and organization in the Pacific Northwest inspired by Berry's speech at the "Agriculture for a Small Planet" symposium in Spokane, Washington, July 1, 1974; the Quivira Coalition, an organization dedicated to restoring ecological and social health to the working landscapes of the American West and finding a "middle way" between ranchers and environmentalists; the Southern Agriculture Working Group, operating across thirteen southern states, was formed in 1991 "to foster a movement towards a more sustainable farming and food system"; the Land Stewardship Project, founded in 1982, a Minnesota-based organization dedicated to fostering "an ethic of stewardship for farmland, to promote sustainable agriculture and to develop sustainable communities."

605.1–2   *Local Economies to Save the Land and the People*] Originally delivered as a speech for Kentuckians for the Commonwealth, Carrollton, Kentucky (August 16, 2013).

605.16   Harry Caudill] See note 53.21–23.

607.29–30   the Appalachian Group to Save the Land and the People.] An early grassroots effort to put an end to strip-mining in eastern Kentucky, organized in Hindman, Kentucky, in 1965, before Dan Gibson's standoff with strip miners at Clear Creek.

612.2–4   to borrow language from John Todd, must be "elegant solutions . . . place."] John Todd, "Tomorrow Is Our Permanent Address," published in *The Book of the New Alchemists*, edited by Nancy Jack Todd (1977).

617.10–15   Communism is not . . . *become* political.] Thomas Merton, "The Pasternak Affair," published in his collection of essays *Disputed Questions* (1960).

620.6–7   "How can we . . . than victory"] William E. Hull, *Beyond the Barriers* (1981).

625.22   24th and 104th Psalms as scriptural norms] Psalms 24:1: "The earth is the Lord's, and the fulness thereof; the world, and they that dwell therein." Psalms 104:31: "The glory of the Lord shall endure for ever: the Lord shall rejoice in his works."

626.1–3   Ken Kesey once saw . . . me one?"] According to Berry, Kesey "told me that in conversation just before or just after I told him of a warning I had seen on a condom machine in a men's room: 'Don't buy this gum. It tastes like rubber.'"

626.24–27   For mercy, pity, peace . . . and care.] William Blake, "The Divine Image," published in *Songs of Innocence* (1789).

628.12–20   Conjugal love . . . a quiet spirit.] See Volume II of *Either/Or* (1843), section on "Aesthetic Validity of Marriage," by Danish philosopher Søren Kierkegaard (1813–1855). Berry quotes from the 1958 revised English translation of *Enten-Eller* by Walter Lowrie.

629.15–18   "Woman, where . . . no more."] John 8:7, John 8:10–11.

632.1–13   Many voices . . . so afflicted . . . ?] Harry M. Caudill, *The Watches of the Night* (1979).

633.28–39   the spirit of independence . . . and poverty.] From *The History of the Highland Clearances* (1883) by Alexander Mackenzie (1838–1898), Scottish ploughman, editor, and historian.

634.12–14   "them that join . . . the earth!"] Isaiah 5:8.

634.20   England's Enclosure Acts] A series of acts passed by Parliament— many of them coming from 1750 to 1860—that empowered the "enclosure" of open fields and common land in England and Wales, stripping the rights of local people to land they had worked for generations. Evicted rural laborers and poor farmers often migrated to cities to find work.

634.20–21   The same dispossession of country people . . . Stalin's Russia] Under collectivization, carried out by the Soviet government from 1928 to 1940, peasant farmers were forced to give up their land and join larger collective farms.

634.21–22   taking place now in China.] Collectivization of agriculture in China began in 1952 and was complete by 1956; after Mao Zedong's death in 1976, farming communes began to be dismantled, as part of the Four Modernizations campaign. Berry refers to China's more recent efforts to convert agricultural land to industrial sites and the resulting displacement of farmers.

634.26   semi-official agricultural policy] Beginning in the 1950s, the Committee for Economic Development championed the idea that there were too many farmers. Eisenhower's secretary of agriculture, Ezra Taft Benson (1899– 1994), admonished farmers "to get big or get out," which would be echoed in the 1970s by Secretary of Agriculture Earl Butz (1909–2008) under Nixon and Ford.

637.21–23   Wes Jackson's . . . "eyes-to-acres ratio"] A formulation by Jackson and often quoted by Berry to indicate the proper scale of farm management.

638.10–21   [T]he essential eye . . . saturated lea!] Robert B. Weeden, letter to Wendell Berry (January 23, 2013). Weeden (b. 1933) is an ecologist and writer who worked in Alaska.

640.31–32   farming fitted to the land, as J. Russell Smith said it should be.] J. Russell Smith, *Tree Crops: A Permanent Agriculture* (1929), Chapter XXIII. See also note 323.12–13.

643.21   Oliver Goldsmith's poem] Oliver Goldsmith (1728–1774), Irish poet, dramatist, and novelist who emigrated to England. "The Deserted Village" appeared in 1770 as the English countryside was being transformed by enclosure.

643.25   John Maynard Keynes] British economist (1883–1946) whose refutation of laissez-faire economics proved influential in the practices of Western democracies.

647.3–4   "that take . . . to use."] From an unpublished paper by Robert B. Weeden entitled "Small Forays into Big Spaces."

656.18 "spill" of 4-methylcyclohexane methanol] On January 9, 2014, 10,000 gallons of crude MCHM, a foaming agent used to wash coal, was released into West Virginia's Elk River.

656.19–23 a *New York Times* editorial noticed . . . "meaningful reform."] "Contaminated Water in West Virginia," *New York Times* (January 17, 2014).

658.21 "creative destruction,"] A term coined by Austrian political economist Joseph Schumpeter (1883–1950) to refer to the way in which "industrial mutation" in capitalist economies creates new ways of doing things even as it destroys the old ways. According to Berry, "It is often quoted still to justify any ecological or human cost of industrial development such as fracking."

659.23 "Luddite"] See note 55.25.

661.28–30 Albert Howard, Aldo Leopold . . . Wes Jackson] Albert Howard, see note 112.17; Aldo Leopold, see note 584.29–30; Wes Jackson, see note 15.21–23.

662.1–2 F. H. King's *Farmers of Forty Centuries*] Franklin Hiram King (1848–1911), American agricultural scientist and author of *Farmers of Forty Centuries: Or Permanent Agriculture in China, Korea and Japan* (1911), drawn from observations made during his extensive travels to Asia.

662.2–3 Vandana Shiva] Indian physicist, ecofeminist, and food sovereignty advocate (b. 1959). Her many books include *Biopiracy: The Plunder of Nature and Knowledge* (1997), *Stolen Harvest: The Hijacking of the Global Food Supply* (1999), and *Earth Democracy: Justice, Sustainability, and Peace* (2005).

662.7–8 "The small landholders . . . a state."] Thomas Jefferson, letter to James Madison, October 28, 1785.

663.10 the 50-Year Farm Bill] Proposed by Wes Jackson and The Land Institute, fruitlessly advocated by Jackson and Berry in a *New York Times* op-ed piece of January 2, 2009. Later that year, accompanied by Fred Kirschenmann of Iowa State University's Leopold Center for Sustainable Agriculture, the two men traveled to Washington, D.C., to lobby for the proposed bill.

665.19–20 "Take . . . no thought for the morrow . . ."] Matthew 6:34.

670.26–27 "The kingdom of heaven is at hand"] Matthew 4:17.

From THE ART OF LOADING BRUSH

691.34–35 Agricultural Adjustment Act] The nation's first omnibus farm bill, signed into law on February 16, 1938.

707.13–14 an essay of 1945, "The Outlook for Farm Wildlife,"] Contributed by Aldo Leopold for the 10th North American Wildlife Conference, scheduled to be held in New York, February 26–28, 1945, but cancelled that year due to government bans on all conventions.

709.8–10   "Of all the pantheon . . . hardest to kill."] C. S. Lewis, *Studies in Words* (1960), Chapter 2.

713.16–17   fear and trembling] Cf. Psalms 55:5 and Philippians 2:12.

714.16–17   "this noble goddesse Nature"] Chaucer, *The Parliament of Fowls*, line 303.

714.17   "great dame Nature"] Edmund Spenser, *The Faerie Queene*, VII, canto vii, stanza 5.

714.19   Alanus De Insulis, or Alan of Lille] French theologian and poet (1128–1202/3). All quotations of Alan of Lille are from his allegorical prose-and-verse work *De Planctu Naturae* (1168–70), translation and commentary by James J. Sheridan: *The Plaint of Nature* (1980).

716.24   self-love as in Matthew 22:39] "Thou shalt love thy neighbour as thyself."

716.36   Genesis 1:31] "And God saw every thing that he had made, and, behold, it was very good. And the evening and the morning were the sixth day."

718.3–4   "the decorations . . . note of delight."] C. S. Lewis, *The Allegory of Love* (1936), Chapter 2.

718.38–719.5   Sheridan says . . . increase and multiply . . ."] James J. Sheridan, in his commentary on Alan's *The Plaint of Nature* (1980).

719.31–34   "Does it never . . . they *knew* . . ."] Thomas Carlyle, *Past and Present* (1843), Book II, Chapter 2.

719.36   Maurice Telleen] See note 471.11–12.

723.17–18   once with acknowledgement of . . . "the Pleynt of Kynde."] Chaucer, *The Parliament of Fowls*, line 316.

724.14–15   "The byldere ok . . . piler elm,"] *The Parliament of Fowls*, lines 176–177.

724.20   Ronsard] Pierre de Ronsard, French poet of noble birth (1524–1585). Berry published a partial translation of Ronsard's Elegy XXIV (1584) in *The Hudson Review* (Spring 1979), under the title "Ronsard's Lament for the Cutting of the Forest of Gastine."

725.4–5   I wrote the introduction for a new printing of . . . Stanwell-Fletcher's *Driftwood Valley*] In his Introduction (1989), Berry says that as a teenage boy it was "the only book I read for a year or two, the end serving only to permit a new return to the beginning." Originally published in 1946, *Driftwood Valley* recounts the life and work of Theodora Stanwell-Fletcher (1906–2000), an American naturalist, in remote northern British Columbia before World War II.

725.8–10   "to Mrs. Stanwell-Fletcher's . . . feelings."] From Berry's Introduction to *Driftwood Valley*.

726.20   old Corycian farmer] See note 466.15.

726.21–22   "cottage economy" . . . advocated by William Cobbett] William Cobbett (1763–1835), an English journalist, farmer, and political reformer, championed traditional rural English practices versus many changes brought by industrialization. His book *Cottage Economy* (1821), first published as a series of pamphlets, is a practical guide to economic self-sufficiency, offering instructions on "the brewing of beer, making of bread, keeping of cows, pigs, bees, ewes, goats, poultry, and rabbits, and . . . [on] other matters deemed useful in the conducting of the affairs of a labourer's family."

727.4–5   "Kek kek! . . . quek quek!"] Chaucer, *The Parliament of Fowls*, line 499.

727.8–9   "There are more stars . . . by light."] *The Parliament of Fowls*, lines 595, 599.

727.12–13   "such array"] *The Parliament of Fowls*, line 318.

727.16   "than any creature"] *The Parliament of Fowls*, line 301.

727.22   "wrought" according to her design.] *The Parliament of Fowls*, line 305.

727.27–33   Come, my Corinna . . . neatly enterwove . . .] See "Corinna's Going A-Maying" by English poet and cleric Robert Herrick (1591–1674), published in his magnum opus, *Hesperides* (1648).

727.34   "noble emperesse . . . grace"] *The Parliament of Fowls*, line 319.

728.3–5   the vicaire . . . of accord . . .] *The Parliament of Fowls*, lines 379–381.

728.40–729.4   "vicar general," in charge . . . "counterfeited."] See the first two stanzas of Chaucer's "The Physician's Tale."

729.6–7   "was and is . . . up to Nature . . ."] *Hamlet*, III.ii.21–22.

729.33–34   "the highest him . . . of Nature."] Spenser, *The Faerie Queene*, VII, canto vi, stanza 35.

729.37–38   "well disposed" . . . "Natures Sergeant."] *The Faerie Queene*, VII, canto vii, stanza 4.

730.4–34   Then forth issewed . . . all beheld . . ."] *The Faerie Queene*, VII, canto vii, stanzas 5, 6, 7, and 13.

731.16   "through a glass darkly."] 1 Corinthians 13:12.

731.23   Francis Bacon] Early English empiricist philosopher (1561–1626).

731.29–30   old under the sun] Cf. Ecclesiastes 1:9.

731.34–39   Who Right . . . vnto brother.] *The Faerie Queene*, VII, canto vii, stanza 14.

732.31–39   I well consider . . . states maintaine.] *The Faerie Queene*, VII, canto vii, stanza 3.

733.1–2   C. S. Lewis speaks . . . those lines.] See Lewis's *The Allegory of Love*, Chapter 7, in which he discusses Spenser's poem.

733.4–11   For, all that . . . strange disguise . . .] *The Faerie Queene*, VII, canto vii, stanza 18.

735.1–2   Langland . . . read *Plaint of Nature*. C.S. Lewis assumed he had.] William Langland, presumed author of *Piers Plowman* (c. 1332–c. 1386). Lewis makes this observation in *The Allegory of Love*, in a footnote in his chapter on Chaucer.

735.13–14   Of the flowers . . . the causes] Langland, *The Vision of Piers the Plowman: A Critical Edition of the B-Text*, edited by A.V.C. Schmidt (1978), passus XI, lines 320–366.

735.15   "putteth it in hir ere,"] Langland, *The Vision of Piers the Plowman: A Critical Edition of the B-Text*, edited by A.V.C. Schmidt (1978), passus XII, lines 218–228.

735.40–736.2   "a stately palace . . . all dainties."] John Milton, *Comus*, stage direction after line 658.

736.8–9   "The grave's . . . there embrace."] Andrew Marvell, "To His Coy Mistress," lines 31–32, written from 1650 to 1652 but not published until 1681, three years after the poet's death.

736.13   "the sage . . . Virginity."] Milton, *Comus*, lines 786–787.

736.17–27   Wherefore did Nature . . . yet despised . . .] *Comus*, lines 710–14, 720–724.

736.38–737.1   "something holy lodges in that breast."] *Comus*, line 246.

737.4–12   Means her provision . . . even proportion . . .] *Comus*, lines 765–773.

738.19–21   "huge heaps . . . a breeze."] Alexander Pope, "Epistle to Burlington" (1731), lines 109, 108. See also note 323.17.

738.21–22   "Good Sense . . . Heaven."] Pope, "Epistle to Burlington," line 43.

738.26–27   "In all, let Nature . . . Place in all."] "Epistle to Burlington," lines 50 and 57.

738.36   "lavish cost, and little skill,"] "Epistle to Burlington," line 167.

739.10–11   'Tis Use alone . . . from Sense.] "Epistle to Burlington," lines 179–180.

739.37–740.2 "the din . . . the world."] William Wordsworth, "Lines Composed a Few Miles Above Tintern Abbey," lines 25–26, 39–40, 52–53, the concluding poem in Wordsworth and Coleridge's *Lyrical Ballads* (1798).

740.13–15 with an eye . . . of things.] Wordsworth, "Tintern Abbey," lines 47–49.

740.18–21 A presence . . . setting suns . . .] "Tintern Abbey," lines 94–97.

740.28–31 In nature . . . moral being] "Tintern Abbey," lines 108–111.

741.3–6 Nature never . . . joy to joy . . .] "Tintern Abbey," lines 122–125.

741.8 "The dreary intercourse of daily life."] "Tintern Abbey," line 131.

741.12–13 "maketh his sun . . . the unjust."] Matthew 5:45.

741.15–16 "worshipper of Nature."] "Tintern Abbey," line 152.

742.5–7 neither gay . . . eager industry.] Wordsworth, "Michael," lines 120–122, first published in the 1800 edition of *Lyrical Ballads*.

742.13–15 "was keen . . . all winds."] "Michael," lines 44–48.

742.18–21 "Something between . . . age of ten.] "Michael," lines 189, 194–198.

742.25–26 "in surety . . . brother's son,"] "Michael," line 211.

742.30–31 "A prosperous man . . . in trade."] "Michael," lines 249–250.

742.34–37 He quickly . . . be gained?] "Michael," lines 252–255.

743.2 "loving letters"] "Michael," line 433.

743.4–9 Luke began . . . the seas.] "Michael," lines 442–447.

743.15 "into a stranger's hand."] "Michael," line 475.

745.10–13 departed, caring little . . . impoverished time.] Hayden Carruth, "Marshall Washer," section 2, published in *Brothers, I Loved You All* (1978) and reprinted in *Collected Shorter Poems, 1946–1991* (1992).

745.20–23 farming is . . . knowledge useless.] Hayden Carruth, "Marshall Washer," section 6.

746.31 a Carter family song, "The Homestead on the Farm"] Written by Herbert S. Lambert and Harry J. Lincoln, "The Homestead on the Farm" was recorded by the Carter Family on November 22, 1929, and released as a single by Victor Records in February 1930.

747.17–30 With Usury . . . Nature's increase.] Ezra Pound, *The Cantos*, Canto LI.

748.1–3 Dress 'em . . . the homestead.] Pound, Canto XCIX.

748.4–5   "Chao-Kong . . . his labour"] Pound, Canto LIII.

748.8–10   Pull down . . . true artistry . . .] Pound, Canto LXXXI.

748.36   Dogen] Dōgen Zenji (1200–1253), Japanese Buddhist monk who founded Soto Zen in Japan, based on the practice of sitting meditation.

749.3–13   Clear running stream . . . in me.] Gary Snyder, "Running Water Music II," published in *The Last Wildlands* (1978) and reprinted in *No Nature: New and Selected Poems* (1992). Snyder (b. 1930) is an American poet of the Pacific Rim, essayist, and environmental activist, whose many works include the 1974 Pulitzer Prize–winning poetry collection *Turtle Island*. *Distant Neighbors* (2015), edited by Chad Wriglesworth, is a selection of Berry and Snyder's correspondence.

751.25–27   "Their achievement . . . on purpose."] Jacques Maritain, *Art and Scholasticism with Other Essays* (1947), translated by J. F. Scanlan.

753.1   Gene Logsdon] See note 471.11–12.

753.5–8   "Four-fifths . . . into clothes."] G. C. Coulton, *Medieval Panorama: The English Scene from the Conquest to the Reformation* (1938), Chapter IX.

753.16–20   "Froissart shows . . . in arms."] G. C. Coulton, *Medieval Panorama*, Chapter XX.

754.2   "creative destruction."] See note 658.21.

757.39–40   an article, by a "semi-retired" rancher] Walt Davis, "Consider the Cow, and Her Purpose on the Ranch," *Beef Producer* (October 2015).

758.25–26   Henry Besuden of Vinewood Farm] See Berry's essay "A Talent for Necessity," where he tells Besuden's story at greater length. Several of Besuden's quoted remarks come from that essay.

760.9   F. H. King] See note 662.1–2.

760.21–22   "cargoes of . . . mineral fertilizers."] See the Introduction to F. H. King's *Farmers of Forty Centuries* (1911).

760.27–30   "the people . . . mountain land."] King, *Farmers of Forty Centuries*, Introduction.

760.33–35   "through our modern . . . their fields."] *Farmers of Forty Centuries*, Introduction.

761.14–17   "Nothing," King wrote . . . making purchases."] *Farmers of Forty Centuries*, Chapter X.

761.27–29   "Forest—field . . . plow agricultures . . ."] J. Russell Smith's *Tree Crops: A Permanent Agriculture* (1929), Chapter I.

761.31–34   The slope . . . gutted countryside.] *Tree Crops*, Chapter I.

761.37–762.3   "Man has . . . cereals are weaklings."] *Tree Crops*, Chapter II.

762.11   "two-story agriculture,"] *Tree Crops*, Chapter II.

762.20–22   "I see a million . . . the hills."] *Tree Crops*, Chapter XXIII.

762.39–763.1   "*farming should fit . . . point of view*."] *Tree Crops*, Chapter XXIII.

763.21–23   "What are . . . woods and forests."] Albert Howard, *An Agricultural Testament* (1940), Introduction.

763.28–36   Since the Industrial Revolution . . . of erosion.] Howard, *An Agricultural Testament*, Preface.

764.22–32   The main characteristic . . . against disease.] *An Agricultural Testament*, Introduction.

764.35–37   "work and experience . . . Great Britain."] *An Agricultural Testament*, Preface.

764.39–765.1   "of high reputation locally,"] See *Sir Albert Howard in India* (1953) by Louise E. Howard (1880–1969), a classics scholar and second wife of Albert Howard.

765.20–31   It needs . . . dead and decayed.] Albert Howard, *The Soil and Health* (1947), Chapter 2.

766.26–28   "The using up . . . pure and simple."] *The Soil and Health*, Chapter 15.

767.1–2   "instinctive awareness . . . natural principle"] L. E. Howard, *Sir Albert Howard in India*.

767.7–9   "one great subject . . . and man."] *The Soil and Health*, Introduction.

767.11–15   "revolt against . . . the product."] L. E. Howard, *Sir Albert Howard in India*.

767.22–24   Health is the capacity . . . tends otherwise.] Aldo Leopold, *A Sand County Almanac* (1949), Part IV.

769.21   The Land Institute] See note 217.19–20.

772.20–21   "Elegant solutions . . . uniqueness of place."] See note 612.2–4.

Index

842

INDEX

Nuclear energy, 251, 310–11, 566–67, 659

Nuclear waste, 152, 155

Nuclear weapons, 242, 310–11

Objectivity, 146, 149, 157–58, 191, 299, 379, 709, 716

Oceans, 361

Ogowe region, 293

Ohio, 680, 697

Ohio River, 475, 477, 566, 652, 680

Oligarchy, 282–83

Organic farming, 273–74, 603, 663

Orgel, Stephen, 524, 530

Originality, 171, 175–76, 178–79, 181, 228–29, 499–500

Outsourcing, 428, 444

Owen County, Ky., 96

Ozick, Cynthia, 187, 192

Ozone depletion, 421

Palmer, Samuel, 466

Pantheism, 741

Parks, 15, 642

Parliament, British, 55

Pasternak, Boris: *Doctor Zhivago*, 617

Patriotism, 14, 241, 243, 248–49, 266, 326, 446, 643

Patronage, 175, 184, 194, 216

Paul (apostle), 33, 35, 74, 535

Peace, 245–47, 253–55, 260, 262–65, 268, 310

Pearl Harbor attack, 240, 259

Pecore, Marshall, 107

Peiper, Joachim, 44

Pendleton County, Ky., 407

Pennsylvania, 586, 591

Pennsylvania State University, 586

Perry, Harry Erdman (grandfather), 465

Peru, 471

Pesticides/herbicides, 92

Petroleum, 244

Philadelphia, Pa., 607

Philanthropy, 548

Pioneer Forest, 448

Pledge of Allegiance, 149

Pluralism, 84–87

Poetry, 491–92, 498–99, 749, 752

Point Coupée Parish, La., 486

Poland, 4

Police force, 258

Pollution, 119, 256, 280, 722; and acceptable risk, 17, 303–4, 388–89; air, 17, 251, 313–14, 364, 421, 539, 566, 659; and climate change, 667, 710; and corporations, 223, 289, 738; from food production, 347; from industrialization, 738; land, 17, 251, 659; toxic, 244, 364, 447; water, 17, 251, 324, 335, 364, 421, 539, 655–56, 659, 664, 681, 744

Pope, Alexander, 323, 717, 735, 737, 751, 759, 765; "Epistle to Burlington," 738–39

Popular culture, 296

Population decline, 435–37, 635, 649, 678, 701–2, 753–54

Population growth, 5, 693

Pornography, 50, 76–78

Porter, Eduardo: "Moving On from Farm and Factory," 678–80

Port Royal, Ky., 5–7, 546, 648–49, 652

"Port William" narratives (Berry), 464–66, 474

Postal Service, U.S., 660

Postman, Neil: *Technopoly*, 121

Pound, Ezra, 387, 493; *ABC of Reading*, 179–81; Canto XLV, 432–33; Canto LI, 432, 747; Canto LIII, 748; Canto LXXXI, 748; Canto XCIX, 748; *Certain Noble Plays of Japan*, 575

Poverty, 256, 266, 271, 288, 444, 633, 738

Prague, Czech Republic, 182

Predictions, 141–42, 151

Preemptive attack, 240, 252

Privacy, 51, 80–84, 191, 258, 266, 559, 617

Productivity, 10–11, 108, 173–74, 271, 397–98, 401, 584, 592, 602, 653, 691, 701, 768

Professionalism, 139, 142, 145, 173–74, 184, 190, 196, 220–21, 577, 716

Progress, 8, 11, 145, 177–79, 193–95, 221, 225, 257, 263, 265–66, 269, 436, 439–40, 443, 452, 550–53, 579, 613, 615, 632–33, 659, 676, 678, 680, 695, 703, 706, 766

Prohibition, 615, 622

Promiscuity, 50, 63, 76

Propaganda, 48, 168, 262, 277, 282, 286

Stalin, Joseph, 617, 634
Stamp Act, 94
Stanwell-Fletcher, Theodora:
    Driftwood Valley, 725
Stegner, Wallace, 221, 466, 484, 546
Stein, Gertrude, 493
Stevens, Wallace, 301; "Adagia," 558;
    "Le Monocle de Mon Oncle," 59
Stewardship, 4, 9, 31, 100, 134, 222–23,
    249, 277, 285, 292, 336, 349, 368,
    403, 554, 592, 642, 650, 656, 660
"Stickers," 221, 227, 546–47, 550, 560
Stiglitz, Joseph E.: *Creating a
    Learning Society*, 674
Still, James, 466; *River of Earth*, 489
Stout, David, 357
Strip mining, 6–7, 364, 569
Strohman, Richard C., 189, 230
Stumbo, Grady, 700
Sturt, George: *The Wheelwright's Shop*,
    228–29
Supreme Court, U.S., 357, 618, 623
Surgeon General, U.S., 681
Surowiecki, James, 674–75
Surplus production, 285, 673, 692–94,
    696
Survival value, 207–8, 253
Susquehannock State Forest, 595
Sustainability, 11, 13, 234, 248, 364,
    425; of agriculture, 16, 91–92, 94,
    273, 397, 554, 673–74, 679, 686,
    689–90, 768; of forests, 16, 586, 588,
    590, 592–93, 603–4; and
    industrialization, 320; of natural
    resources, 16; and wilderness, 106
Sympathetic Mind, 298–317
Synthesis/integration, 153, 157, 161

Tate, Allen, 468; "Notes on Liberty
    and Property," 552–53; "Ode to the
    Confederate Dead," 556; "Remarks
    on the Southern Religion," 318, 323
Taxes, 368, 445–47
Technology, 104, 157, 168, 178, 182–83,
    192, 229, 256–57, 316, 348, 560;
    agricultural, 222, 251–52, 403, 635–
    36, 676, 679–80, 687, 693–94, 764;
    coal mining, 703; and community,
    112; and conservation, 353;
    determinism, 340; and ecology, 144;
    for efficient energy, 250–51;

electronic, 133; and exploitation, 48,
    100, 561, 564; failures, 152–53; global,
    289; industrial, 56–57, 100, 333–35,
    436, 479, 559, 561, 613, 631, 657–59,
    704; labor-saving machines, 104,
    114, 284–85, 400–402, 639, 660;
    limitlessness of, 706; and
    materialism, 539; medical, 395–96;
    and progress, 225, 263, 265, 706;
    revolution in, 577; salvation through,
    5, 8, 10–11, 19, 141, 551–52; and
    science, 710–11; of warfare, 137, 241–
    42
Television, 50–51, 77
Telleen, Maurice, 471, 719–20
Tenant farmers, 697–98
Ten Commandments, 33, 415, 525
Tennessee, 550
Tennyson, Alfred: "Ulysses," 169–70,
    185
Terminator genes, 222
Terrorism, 239, 241–44, 250, 254, 256,
    258–59, 264, 268, 295, 308–10, 313,
    315, 341, 357, 360
Third World, 57
Thomas, Clarence, 46, 52, 58
Thoreau, Henry David, 35, 466, 508;
    Walden, 207
Tilth, 603
Tobacco, 6, 96, 98, 101, 119, 352, 545,
    548, 613, 681–83, 693–99
Todd, John, 323, 612, 772
Topsoil, 280, 361, 364, 539
Tower of Babel, 311–12
Tractors, 103, 230, 399–400, 402–3,
    639, 646, 755–56
Trafalgar, battle of, 392–94
Trees, 96–97, 107–8, 585, 587, 597–98,
    651–52
Trump, Donald, 703
Tuscany, 471
Twain, Mark (Samuel Clemens), 466;
    Huckleberry Finn, 43, 210, 212, 473
"Two cultures," 171, 178, 197, 215, 219,
    221

Ulanowicz, Robert, 486
"Unbridled Energy" summit, 450–51
United States Steel Corporation, 553
University of California, Davis, 287
University of Kentucky, 467, 569

This book is set in 10 point ITC Galliard, a face designed
for digital composition by Matthew Carter and based
on the sixteenth-century face Granjon. The paper is acid-free
lightweight opaque that will not turn yellow or brittle with age.
The binding is sewn, which allows the book to open easily and lie flat.
The binding board is covered in Brillianta, a woven rayon cloth
made by Van Heek–Scholco Textielfabrieken, Holland.
Composition by Dedicated Book Services.
Printing by Sheridan Grand Rapids, Grand Rapids MI.
Binding by Dekker Bookbinding, Wyoming MI.
Designed by Bruce Campbell.

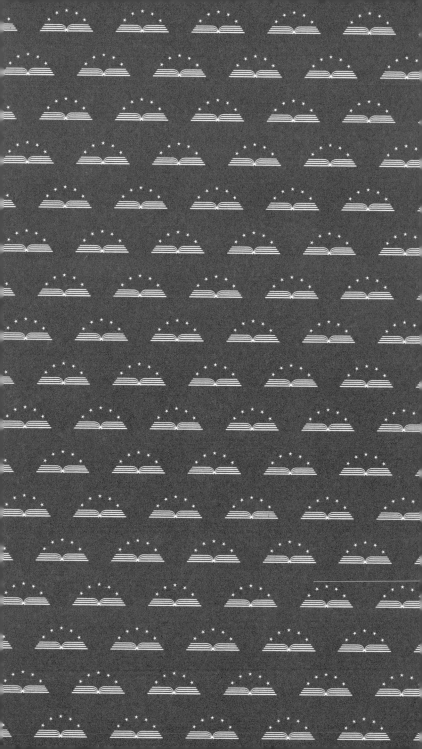